THIS BOOK BELONGS
TO
E. SHKOLNIK

HANDBOOK OF THERMAL ANALYSIS OF CONSTRUCTION MATERIALS

HANDBOOK OF THERMAL ANALYSIS OF CONSTRUCTION MATERIALS

by

V.S. Ramachandran, Ralph M. Paroli, James J. Beaudoin, and Ana H. Delgado

Institute for Research in Construction
National Research Council of Canada
Ottawa, Ontario, Canada

NOYES PUBLICATIONS

WILLIAM ANDREW PUBLISHING
Norwich, New York, U.S.A.

Library of Congress Catalog Card Number: 2002016536
ISBN: 0-8155-1487-5
Printed in the United States

Published in the United States of America by
Noyes Publications / William Andrew Publishing
13 Eaton Avenue
Norwich, NY 13815
1-800-932-7045
www.williamandrew.com
www.knovel.com

10 9 8 7 6 5 4 3 2 1

NOTICE

To the best of our knowledge the information in this publication is accurate; however the Publisher does not assume any responsibility or liability for the accuracy or completeness of, or consequences arising from, such information. This book is intended for informational purposes only. Mention of trade names or commercial products does not constitute endorsement or recommendation for use by the Publisher. Final determination of the suitability of any information or product for use contemplated by any user, and the manner of that use, is the sole responsibility of the user. We recommend that anyone intending to rely on any recommendation of materials or procedures mentioned in this publication should satisfy himself as to such suitability, and that he can meet all applicable safety and health standards.

Library of Congress Cataloging-in-Publication Data

Handbook of thermal analysis of construction materials / edited by V.S. Ramachandran ...[et al.].
p. cm. -- (Construction materials science and technology series)
Includes bibliographical references and index.
ISBN 0-8155-1487-5 (alk. paper)
1. Building materials--Thermal properties--Handbooks, manuals, etc. I. Ramachandran, V. S. (Vangipuram Seshachar) II. Series

TA418.52 .H36 2002
691'.028'7--dc21 2002016536

CONSTRUCTION MATERIALS SCIENCE AND TECHNOLOGY SERIES

Editor

V. S. Ramachandran, National Research Council Canada

CONCRETE ADMIXTURES HANDBOOK; Properties, Science and Technology, *Second Edition:* edited by V. S. Ramachandran

CONCRETE ALKALI-AGGREGATE REACTIONS: edited by P. E. Grattan-Bellew

CONCRETE MATERIALS; Properties, Specifications and Testing, *Second Edition:* by Sandor Popovics

CORROSION AND CHEMICAL RESISTANT MASONRY MATERIALS HANDBOOK: by W. L. Sheppard, Jr.

HANDBOOK OF ANALYTICAL TECHNIQUES IN CONCRETE SCIENCE AND TECHNOLOGY; Principles, Techniques, and Applications: edited by V. S. Ramachandran and James J. Beaudoin

HANDBOOK OF CONCRETE AGGREGATES; A Petrographic and Technological Evaluation: by Ludmila Dolar-Mantuani

HANDBOOK OF FIBER-REINFORCED CONCRETE; Principles, Properties, Developments, and Applications: by James J. Beaudoin

HANDBOOK OF POLYMER MODIFIED CONCRETE AND MORTARS; Properties and Process Technology: by Yoshihiko Ohama

HANDBOOK OF THERMAL ANALYSIS OF CONSTRUCTION MATERIALS: by V. S. Ramachandran, Ralph M. Paroli, James J. Beaudoin, and Ana H. Delgado

LIGHTWEIGHT AGGREGATE CONCRETE; Science, Technology, and Applications: by Satish Chandra and Leif Berntsson

WASTE MATERIALS USED IN CONCRETE MANUFACTURING: edited by Satish Chandra

Foreword

Material science and technology have played a preeminent role in the construction industry. Some examples include the development of novel materials, development of materials with improved properties, new techniques for evaluation purposes, techniques to predict the long term behavior, improvement of standard methods, and assessment of wastes and by-products for incorporation in the construction industry.

The material scientists at the National Research Council of Canada have established one of the best thermal analysis laboratories in the world. Various types of thermal analysis techniques have been applied successfully to the investigation of inorganic and organic construction materials. These studies have provided important information on the characterization of raw as well as finished materials, quality control, quantitative estimation, interrelationships between physical, chemical, mechanical, and durability characteristics.

Information on the application of thermal analysis to construction materials is dispersed in literature and hence the IRC scientists embarked on producing a handbook, the first of its kind, incorporating the latest knowledge available in this field of activity. Almost all important construction materials have been included.

The book should be a valuable addition to the literature in construction science and technology and should be useful to the material scientist, technologist, student, engineer, analytical chemist, architect, manufacturer, and user of construction materials.

Institute for Research in Construction
National Research Council Ottawa
Ottawa, Canada

Sherif Barakat
Director General

Preface

A substance subjected to thermal treatment may undergo physicochemical processes involving weight changes, crystalline transitions, mechanical properties, enthalpy, magnetic susceptibility, optical properties, acoustic properties, etc. Thermal techniques follow such changes, generally as a function of temperature, that could extend from subzero to very high temperatures. Several types of thermal techniques are in use and examples include thermogravimetry, differential thermal analysis, differential scanning calorimetry, thermomechanical analysis, derivative thermogravimetry, dynamic thermal analysis, dielectric analysis, and emanation thermal analysis. A related technique that is extensively applied to investigate inorganic construction materials is called *conduction calorimetry* which measures the rate of heat changes, as a function of time or temperature.

Thermal analysis techniques have been employed to study various types of inorganic and organic construction materials. They have been applied more extensively to the investigation of inorganic materials. Useful information generated by the use of these techniques includes: characterization, identification of compounds, estimation of materials, kinetics of reactions, mechanisms, synthesis of compounds, quality control of raw materials, rheological changes, glass transitions, and causes leading to the deterioration of materials. Thermal techniques are also used in combination with other techniques such as chemical analysis, x-ray diffraction, infrared analysis, and scanning electron microscopy.

There is no book at present that provides a comprehensive treatise on the application of thermal analysis techniques to various types of construction materials. This book comprises sixteen chapters and includes information on almost all important construction materials. Four chapters, Chs. 2, 4, 9 and 13, are devoted to the general introduction of these materials because of the complex nature and behavior of these materials.

The first chapter describes the more common thermoanalytical techniques that are adopted in the study of construction materials. The general principles and types of equipment used are given with typical examples. The described techniques include differential thermal analysis, differential calorimetry, thermogravimetry, thermomechanical analysis, dynamic mechanical analysis, dielectric analysis, and conduction calorimetry.

The physicochemical characteristics of concrete depend on the behavior of the individual components of portland cement as well as on the cement itself. The second chapter provides essential information on cement and cement components so that the information presented in subsequent chapters can easily be followed. In this chapter, the formation of cement, the hydration of individual cement compounds and cement itself, physicochemical processes during the formation of the pastes, the properties of the cement paste, and the durability aspects of concrete are discussed.

The information presented in Ch. 3 clearly demonstrates the extensive applicability of thermal techniques for investigations of raw materials for the manufacture of cement, clinker formation, hydration of cement compounds and cement, the oxide systems of relevance to cement chemistry, and durability processes. Some examples of the usefulness of associated techniques for these investigations are also given

Incorporation of chemical and mineral admixtures in concrete results in many beneficial effects such as enhanced physical and mechanical properties and durability. Many types of admixtures are currently in the market and their effect on concrete is determined by complex factors. Hence, Ch. 4 has been included to describe types of admixtures and their roles in concrete technology. This chapter should serve as an introduction to the subsequent chapters devoted to the application of thermal analysis techniques for the investigation of the role of admixtures in concrete.

The versatility of the thermal analysis techniques such as TG, DTG, DTA, DSC, and conduction calorimetry for evaluating the role of admixtures in concrete is demonstrated in Chs. 5 through 8. The actions of accelerators, retarding/water-reducing admixtures, superplasticizers and supplementary cementing, and other admixtures are described in Chs. 5, 6,

7, and 8, respectively. Various types of valuable information may be derived by applying these techniques. Examples include: heats of hydration, mechanisms of reactions, composition of the products, cement-admixture interactions, compatibility of admixtures with cement, prediction of some properties, abnormal behavior of concrete, material characterization, development of new admixtures and techniques, and quick assessment of some properties. In many instances, the results obtained by thermal techniques can be related to strength development, microstructure, permeability, and durability aspects in cement paste and concrete. Thermal analysis techniques are shown to be eminently suited to characterize supplementary cementing materials and for determining the potential cementing properties of wastes and by-products. The relative activities of supplementary materials such as silica fume, slag, pozzolans, etc., from different sources may be quickly assessed by thermal methods.

Portland cement-based concretes are extensively used in the construction industry. Non-portland cement based systems, although not produced to the same extent as portland cement, have found applications especially for repair of concrete structures. Chapter 9, an introduction to non-portland cements, provides a description of the hydration and engineering behaviors of cements such as oxychloride/oxysulfate cements, calcium aluminate cement, portland-calcium aluminate blended cement, phosphate cement, regulated set cement, and gypsum. Chapter 10 provides information on the application of thermal techniques such as DTA, DSC, DTG, TG, and conduction calorimetry to selected groups of rapid setting cements. Studies on the degree of hydration at different temperatures, identification and estimation of products, and heats of hydration are discussed in this chapter.

Gypsum is an essential ingredient in portland cement. Calcined gypsum finds many uses in the construction industry. It is also used as an insulating material. Thermal methods are shown to be applicable to the rapid evaluation of these systems. Chapter 11 deals with the studies on gypsum and α and β forms of $CaSO_4 \cdot \frac{1}{2}H_2O$. The effect of environmental conditions on the determination of various forms of calcium sulfate is also given along with the development of recent techniques. A subchapter on the industrial products such as portland cement stucco, gypsum-based cement, sedimentary rocks, plasters, and expanding cement is also included.

One of the first applications of thermal techniques was related to the characterization of clay minerals. Extensive work has been carried out on thermal analysis of clay products. Identification and characterization of clay raw materials and accessory minerals, reactions that occur during the firing process, and durability aspects of clay products can be examined conveniently by DTA, TG, TMA, and dilatometry and these aspects are discussed in Ch. 12.

There is a great potential for the application of thermal analysis techniques to study the behavior of organic construction materials such as adhesives, sealants, paints, coatings, asphalts, and roofing materials. Different types of polymers constitute these materials. Chapter 13 is an introduction to the organic construction materials and provides essential information on aspects such as the sources, structure, classification, general characteristics, applications, and durability. Next, Chs. 14, 15, and 16, discuss the application of thermal analysis techniques for studies pertaining to sealants/adhesives, roofing materials, and paints/coatings, respectively.

Many physical and chemical processes are involved in the degradation of sealants and adhesives. Thermal analysis techniques have been used to characterize polymeric adhesives and sealant formulations and also to study the processes of degradation when they are exposed to natural elements. The application of techniques such as TG, DSC, DTG, Dynamic Mechanical Analysis, Dynamic Mechanical Thermal Analysis, Thermomechanical Analysis, and Dynamic Load Thermomechanical Analysis for such materials has been discussed in Ch. 14.

Although bituminous and modified bituminous roofing materials are well known in the construction industry, several types of synthetic polymers such as PVC, EPDM, KEE, TPO, and polyurethane are also adopted in various applications. Many types of thermal techniques have been applied to investigate glass transition temperatures, vulcanization reactions, oxidation stability, weight, and dimensional, rheological and phase modifications in the roofing material systems. These techniques have also provided useful information on the degradation processes. Chapter 15 provides several examples of the applicability of thermal analysis techniques for investigating the traditional as well as new types of roofing materials.

Thermal analysis techniques also find applications in the study of paints and coatings. Chapter 16 describes the utilization of these techniques for investigations related to characterization, drying phenomenon, decomposition and cross linking, thermal stability, mechanism of decomposition, degree of curing, kinetics of reactions, influence of impurities, differences in crystallinity during pigment formation, heats of reaction or mixing, effects of environmental conditions, and waste utilization.

This comprehensive book containing essential information on the applicability of thermal analysis techniques to evaluate inorganic and organic materials in construction technology should serve as a useful reference material for the scientist, engineer, construction technologist, architect, manufacturer, and user of construction materials, standard-writing bodies, and analytical chemists.

February 5, 2002
Ottawa, Ontario

V.S. Ramachandran
Ralph M. Paroli
James J. Beaudoin
Ana H. Delgado

Table of Contents

1

Thermoanalytical Techniques

1.0 INTRODUCTION

Thermal analysis has been defined by the International Confederation of Thermal Analysis (ICTA) as a general term which covers a variety of techniques that record the physical and chemical changes occurring in a substance as a function of temperature.[1][2] This term, therefore, encompasses many classical techniques such as thermogravimetry (TG), evolved gas analysis (EGA), differential thermal analysis (DTA), and differential scanning calorimetry (DSC), and the modern techniques, such as thermomechanical analysis (TMA) as well as dynamic mechanical analysis (DMA), and dilatometry, just to name a few. The application of thermal analysis to the study of construction materials stems from the fact that they undergo physicochemical changes on heating.

2.0 CLASSICAL TECHNIQUES

Ever since the invention of DSC, there has been much confusion over the difference between DTA and DSC. The exact ICTA definition of *DTA* is a method that monitors the temperature difference existing between a sample and a reference material as a function of time and/or temperature assuming that both sample and reference are subjected to the same environment at a selected heating or cooling rate.[1][2] The plot of ΔT as a function of temperature is termed a DTA curve and endothermic transitions are plotted downward on the y-axis, while temperature (or time) is plotted on the x-axis. *DSC,* on the other hand, has been defined as a technique that records the energy (in the form of heat) required to yield a zero temperature difference between a substance and a reference, as a function of either temperature or time at a predetermined heating and/or cooling rate, once again assuming that both the sample and the reference material are in the same environment.[1][2] The plot obtained is known as a DSC curve and shows the amount of heat applied as a function of temperature or time. As can be seen from the above definitions, the two techniques are similar, but not the same. The two yield the same thermodynamic data such as enthalpy, entropy, Gibbs' free energy, and specific heat, as well as kinetic data. It is only the method by which the information is obtained that differentiates the two techniques. A brief history on the development and a comparison of the two techniques are given.*

2.1 Differential Thermal Analysis and Differential Scanning Calorimetry

A little over a hundred years ago, two papers were published by Le Châtelier dealing with the measurement of temperature in clays; the first entitled *On the Action of Heat on Clays* and the second *On the Constitution of Clays*.[20][21] The experiment described in these papers was not a truly differential one since the difference in temperature between the clay and reference material was not measured. The apparatus consisted of a Pt-Pt/10%-Rh thermocouple embedded in a clay sample, which in turn was packed into a 5 mm diameter Pt crucible. The crucible was then placed in

*For a more detailed history, comparison, and theoretical description, consult the references listed in Refs. 3–19.

a larger crucible, surrounded with magnesium oxide and inserted into an oven. Le Châtelier used a heating rate of 120 K min^{-1} and recorded the electromotive force of the thermocouple on a photographic plate at regular time intervals. As long as no phase change occurred in the clay, the temperature rose evenly and the lines on the plate were evenly spaced. If, however, an exothermic transformation took place, then the temperature rose more rapidly, and, therefore, the lines were unevenly spaced and closer together. An endothermic transition, on the other hand, caused the measured temperature to rise more slowly, and the spacing between the lines was much larger. To ensure that the measured temperatures were correct, he calibrated his instrument with the aid of boiling points of known materials such as water, sulfur, and selenium, as well as the melting point of gold. Since Le Châtelier's experiment does not fit the ICTA definition of DTA, his main contribution to the development of DTA was the automatic recording of the heating curve on a photographic plate. True differential thermal analysis was actually developed twelve years later (in 1899) by Roberts-Austen.[22]

Roberts-Austen connected two Pt-Pt/10%-Ir thermocouples in parallel which, in turn, were connected to a galvanometer. One thermocouple was inserted into a reference sample consisting of a Cu-Al alloy or of an aluminum silicate clay (fireclay). The other thermocouple was embedded into a steel sample of the same shape and dimensions as the reference. Both the sample and reference were placed in an evacuated furnace. A second galvanometer monitored the temperature of the reference. The purpose of the experiments was to construct a phase diagram of carbon steels and, by extension, railway lines. Since his method was a true differential technique, it was much more sensitive than Le Châtelier's. The DTA design used today is only a slight modification of Roberts-Austen's, and the only major improvements are in the electronics of temperature control and in the data processing, which is now handled by computers (see Fig. 1).

It took about fifty years for the DTA technique to be considered not only qualitative, but also as a quantitative means of analyzing and characterizing materials. Moreover, it was only then that the Roberts-Austen setup was modified by Boersma.[23] The modification was in the placement of the thermocouples. Rather than placing the thermocouples into either the sample or the reference, Boersma suggested that they be fused onto cups and that sample and reference be placed into these cups. This modification eliminated the necessity of diluting the sample with reference materials and reduced the importance of sample size. The vast majority of today's DTA instruments are based on the Boersma principle in that only the crucibles are in contact with the thermocouples.

Boersma's DTA configuration, Fig. 1b, can be considered as the missing link between differential thermal analysis and differential scanning calorimetry. Some even feel that this configuration is, in fact, a DSC instrument. This is the major reason behind the confusion as to the differences between DTA and DSC.

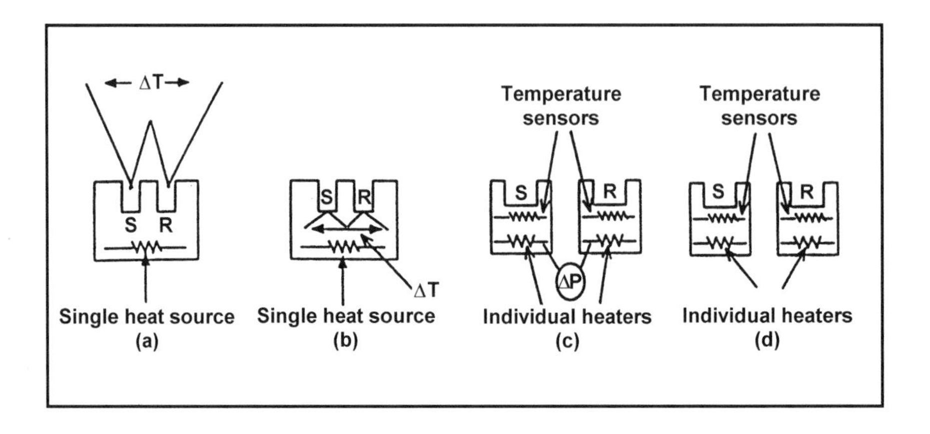

Figure 1. Schematic diagrams of different instruments used in thermal analysis to detect energy changes occurring in a sample: *(a)* conventional DTA, *(b)* Boersma[23] DTA, *(c)* power-compensation DSC, and *(d)* heat-flux DSC.

The two most crucial differences between the two techniques are: (a) in DSC, the sample and reference have their own heaters and temperature sensors as compared to DTA where there is one common heater for both; (b) DTA measures ΔT versus temperature, and, therefore, must be calibrated to convert ΔT into transition energies, while DSC obtains the transition energy directly from the heat measurement. The confusion is also partly due to the fact that there are at least three different types of DSC instruments: a DTA *calorimeter*, a heat-flux type (Fig. 2c), and a power compensation (Fig. 1d) one. This, in turn, arises from the fact that some define calorimetry as quantitative-DTA. As opposed to conventional DTA, the thermocouples in a DSC instrument do not come into contact with either the sample or reference. Instead, they either surround the sample (thermopiles) or are simply outside the sample (thermocouples). Furthermore, the sample and reference weights are usually under 10 mg.

2.2 DSC

The DTA calorimeter, sometimes called DSC, was developed by David in 1964.[24][25] The term DTA calorimeter is more appropriate since this system actually measures ΔT directly from the experiment. Unlike conventional DTA however, the experiment is performed at quasi-equilibrium conditions, i.e., sample mass is less than 10 mg, slow cooling/heating rate, and only one calibration coefficient needs to be measured for the entire temperature range. This, therefore, yields quantitative data but by definition remains a DTA instrument. The other two categories of DSC apparatus are true calorimetric instruments in that the calorimetric information is obtained directly from the measurement, i.e., no conversion factor is required to convert ΔT into readily used energy units as the thermometric data is obtained directly. A constant is still required to convert the energy term into more suitable units. The main goal of any enthalpic experiment, which is to determine the enthalpy of a sample as a function of temperature, is attained by measuring the energy obtained from a sample heated at a constant rate with a linear temperature or time programming. These two DSC instruments are based on the method developed by Sykes in the mid-1930s.[26][27] Sykes' apparatus was designed so that the temperature of the metal block, which contained the sample, was slightly lower than the temperature of the sample itself. To maintain the sample at the same temperature as the block, power was supplied to the sample. The main disadvantage of this apparatus was that a correction factor had to be applied to account for the heat transfer between the surrounding medium and the block. Both the heat flux and power-compensation DSC instruments overcome this drawback because, as the name suggests, they are *differential* instruments. The heat-flux instruments measure the flux across a thermal resistance, whereas the power compensating differential scanning calorimeters measure the energy applied to the sample (or the reference) by an electrical heater in order to maintain a zero-temperature differential.

The first commercial DSC instrument was introduced by Watson and his co-workers at Perkin-Elmer (Model DSC-1) in 1964.[28] Watson, et al., also appear to be the first to have used the nomenclature *differential scanning calorimetry*. Their instrument, a power-compensating DSC, maintained a zero temperature difference between the sample and the reference by supplying electrical energy (hence, the term *power-compensation*) either to the sample or to the reference, as the case may be, depending on whether the sample was heated or cooled at a linear rate. The amount of heat required to maintain the sample temperature and that of the reference material isothermal to each other is then recorded as a function of temperature. Moreover, in power-compensation DSC, an endothermic transition, which corresponds to an increase in enthalpy, is indicated as a peak in the upward direction (since power is supplied to the sample), while an exothermic transformation, a decrease in enthalpy, is shown as a negative peak. This, therefore, differs from the DTA curve since the peaks are in opposite direction and the information obtained is heat flow, rather than ΔT, as a function of temperature (see Fig. 2). Also, as will be shown later, the integration of a DSC curve is directly proportional to the enthalpy change.

The heat-flux DSC instrument is very often based on the Tian-Calvet calorimeter. The original calorimeter, built in the early 1920s by Tian,[29] consisted of a single compensation vessel and the measurement was via a thermopile. Calvet modified this setup about twenty-five years later by making it a twin calorimeter, i.e., applying the differential technique.[29] The energy measuring device is a thermopile consisting of approximately 500 Pt-Pt/10%-Rh thermocouples which are equally spaced and connected in series. This arrangement enables the electromotive force (emf) to be directly proportional to the amount of heat lost by the sample and reference holders. Essentially, this type of calorimeter measures the difference in temperature between the sample and reference as a function of time, and since the temperature varies linearly with time, as a function of temperature as well. The heat-flux is actually derived from a combination of the $\Delta T(t)$ curve and the $d\Delta T(t)/dt$, both of these are transparent to the user since the electronics used yield a direct heat flux value from these terms. If temperature compensation is required, then it is done by Joule heating (for an endothermic process) or by Peltier effect (for an exothermic process). As in the DTA case, an endothermic signal is in the negative direction, while an exothermic signal is the upward direction (see Fig. 2).

Both the heat-flux calorimeters and power-compensation calorimeters have their advantages and disadvantages, but, the end result is the same, the two will yield the same information. The advantage of the heat-flux type is that it can accommodate larger sample volumes, has a very high sensitivity, and can go above 1100 K. The disadvantage is that it cannot be scanned at rates faster than 10 K min^{-1} at high temperatures and not faster than 3 K min^{-1} at sub-ambient temperatures. The main advantage of the power-compensation calorimeter is that it does not require a calibration in that the heat is obtained directly from the electrical energy supplied to the sample or reference compartment (a calibration is still necessary, however, to convert this energy into meaningful units) and that very fast scanning rates can be obtained. The disadvantage of this system is that the electronic system must be of extremely high sensitivity and large fluctuations in the environment must be absent so as to avoid compensating effects which are not due to the sample. Also, the complexity of the electronics prevents the system from being used above ~1100 K.

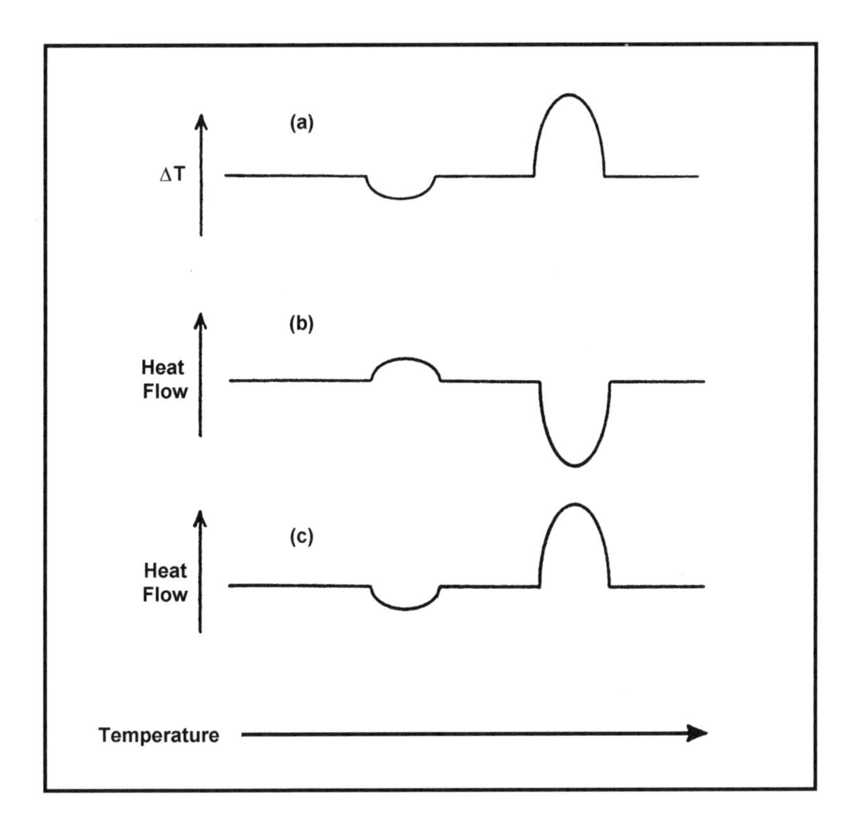

Figure 2. Comparison of curves obtained on heating by *(a)* DTA, *(b)* power-compensating DSC, and *(c)* heat-flux DSC.

2.3 Calibration of DTA and DSC

The calibration of a DSC or DTA instrument is crucial for various reasons. Firstly, for the determination of the temperature, and secondly, to convert the dissipated power into useful energy units, e.g., joules or calories. The temperature calibration is of vital importance since, in most cases, a calibrated thermometer cannot be used for the temperature measurement. As for the calibration of energy, it too is important as, in many cases, the amplitude of the signal of dissipated power is affected by the heating and cooling rates. Based on these facts, it is obvious that the accuracy of the measurement is generally lower than the degree of reproducibility.

There are quite a few different methods for the calibration of DSC instruments, of which the most popular are: (a) calibration by Joule-effect and (b) calibration by heats of fusion.[12][15][30] The Joule-effect calibration is relatively simple and straight-forward in that it consists of an electrical heater inserted into the sample and reference compartments. A pulse of predetermined duration and intensity is sent to the sample, and the dissipated power is then measured. The disadvantage of this method is that some heat flux can be dissipated in the heater wires, and, therefore, not truly measured. Furthermore, the electrical heater is not necessarily composed of the same material as the sample and reference holders. Still, the accuracy of this calibration technique is better that 0.2%.

The heats of fusion calibration method affords two simultaneous calibrations. Pure substances, which undergo phase transformations at very well-characterized temperatures, are used. Since the enthalpy of fusion and temperature of fusion of the calibrant are well known, both a temperature and enthalpic calibration can be performed with the same substance. Ideally, more than one compound and more than one scanning rate should be utilized (or if only one scanning rate is employed, then the scanning rate should correspond to that which will be used for the experiment) since the sensitivity of the measurement is not only temperature dependent, but also scan rate dependent. Since the thermal conductivity might play an important role in the measured response, the mass of the calibrant should be as close as possible to the sample mass. The following criteria should be used when choosing a calibrant:

a) The substance must be available in high purity.

b) The transition temperature and enthalpy of transition should be known with a high degree of accuracy.

c) The substance should not show any tendency to superheating.[4][5][12]

The major drawback of this method is that since transitions are very temperature specific, one substance might be suitable for only one temperature range, hence the need to use more than one calibrant (or one must assume that the calibration will hold for the entire range being studied). Another calibration method is with the use of radioactive materials since they generate constant heat (i.e., power), which is independent of temperature. Some of these materials, however, are not suitable at high temperatures as they might diffuse through the sample holder. The most often used radioactive material as a calibrant appears to be plutonium.[31]

The integration of a DSC (and a DTA) curve is directly proportional to the enthalpy change,[32]

Eq. (1) $$\text{Area} = Km\Delta H$$

where K is the calibration coefficient, m the sample mass, and ΔH the heat of transition. Unlike DTA, however, in DSC, K is temperature independent. As is the case for DTA,* the term dH/dt for DSC is given by three measured quantities,[32]

Eq. (2) $$dH/dt = -(dq/dt) + (C_s - C_r)dT_p/dt + RC_s d^2q/dt^2$$

where dq/dt is the area, $(C_s - C_r)dT_p/dt$ is the baseline contribution, and $RC_s d^2q/dt^2$ is the peak slope. The differences between the two techniques are quite apparent; firstly, the area under the curve is $\Delta q = -\Delta H$, i.e., the enthalpy and secondly, the thermal resistance, R, only shows up in the third term of the equation. Although a calibration coefficient is still required, it is only needed as a means of converting the area (heat flow) into an acceptable energy unit, such as joules or calories, and it is not a thermal constant.[26]

Phases, which are thermodynamically stable, have a finite number of degrees of freedom. Each phase is separated by a boundary where the phase change occurs. As one crosses the boundary, a new phase appears to the detriment of the other, and, since the overall free energy of the process is zero, the thermodynamic parameters such as ΔS, and ΔH must change in a quantitative manner at the border. Since different types of phase boundaries are encountered, different types of enthalpies are obtained, for example, ΔH_f, entropy of fusion; enthalpy of transition, ΔH_t; etc. The previous discussion shows that a great deal information can be obtained from a DSC curve, and that the interpretation of such a curve can yield valuable insight into the nature of the material being investigated. It is important to be able to identify what type of phase transition is occurring in the substance by looking at the curve itself, and, therefore, what follows is a brief explanation on phase transformations in general, and how they can be identified from a DSC (or DTA) curve.

---

*In DTA[32]: $R(dH/dT) = (T_s - T_r) + R(C_s - C_r)dT_r/dT + RC_s d(T_s - T_r)/dt$

Oscillating DSC (ODSC). The characterizing ability of DSC can be greatly enhanced using dynamic DSC measurements known as modulated (MDSC™),[33]–[35] oscillating (ODSC),[36] or dynamic (DDSC).[37] The dynamic DSC measurement is a fairly new technique, which was first developed jointly by TA Instrument and ICI Paints in the early 1990s and followed by Seiko Instruments (ODSC), and Perkin-Elmer (DDSC). It combines aspects of both DSC and AC Calorimetry. In this technique, the temperature program (linear heating or isothermal or cooling) is modulated by some form of perturbation. This approach provides new information on reversing (C_p) and non-reversing (kinetic) characteristics of thermal events. This information helps to interpret thermal events and provides unique insights into the structure and behavior of materials.[38]

As mentioned earlier in DSC, thermocouples are utilized to measure the quantity or heat flow difference (ΔQ) between the sample and the reference. For example, when a sample melts during heating, energy is absorbed by the sample. If a material crystallizes or cures, its temperature becomes greater than that of the reference, and heat evolves.[36] This heat flow can be mathematically expressed by the equation:[39]

Eq. (3) $$dQ/dt = -C_p \cdot (dT/dt) + f(T,t)$$

where dQ/dt = heat flow out of the sample
C_p = thermodynamic heat capacity
dT/dt = heating rate
T = temperature
t = time
$f(T,t)$ = the function governing the kinetic response of any physical or chemical transformation

With the dynamic DSC technique, the same DSC furnace assembly and cell is utilized, but a different heating or cooling profile is applied to the sample and reference by the same furnace assembly. An oscillating time/temperature sinusoidal signal is superimposed onto the conventional linear heating ramp. This yields a heating profile where the sample temperature profile, on the average, is still increasing in a constant manner with respect to time. However, on a short-term examination, the increase is not linear, but sinusoidal in nature.[36]

The temperature increases at a rate which is sometimes faster than the average, underlying heating ramp. The time-temperature profile obtained via dynamic DSC is continuously accelerating and decelerating during the course of a heating experiment. There are three parameters that control the variations of the heating profile: frequency of the time-temperature oscillation; the amplitude of the oscillation; and the average, underlying heating rate. Therefore, the application of the oscillating time-temperature wave to the heating ramp will have a great impact on the resulting heat flow signal.

In dynamic DSC the temperature program is represented by:[39]

Eq. (4) $$T = bt + B \cdot \sin(wt)$$

where
w = frequency
b = heating rate
B = amplitude of temperature program

Assuming a small temperature excursion and a linear response of the rate of the kinetic process to temperature, Eq. (4) can be expressed as:[39]

Eq. (5) $$dQ/dt = C_p\,[b + Bw \cdot \cos(wt)] + f''(T,t) + C \cdot \sin(wt)$$

where
$f''(T,t)$ = is the underlying kinetic function after subtraction of the sine wave modulation
C = amplitude of kinetic response to the sine wave modulation
$[b + Bw \cdot \cos(wt)]$ = measured dT/dt

Thus, as can be seen in Eq. (5), the heat flow signal will contain a cyclic component which is dependent on amplitude of kinetic response (C), amplitude of temperature (B), and frequency (w).

The periodic integration of the original dynamic DSC will result in a deconvoluted DSC heat flow signal, which is equivalent to the heat flow data obtained by traditional DSC. The subtraction of the C_p component from the deconvoluted signal yields the kinetic component data. The C_p component gives information on *reversible* thermal events, such as T_g while the kinetic component provides data on the *irreversible* aspects of thermal transitions such as evaporation, decomposition, crystallization, relaxation, or curing.

The new and unique capabilities of the dynamic technique include:[36]

- Improved resolution of closely occurring and overlapping transitions.
- Increased sensitivity for low energy or subtle transitions.
- Heat capacity measurements (under low heating/cooling rate conditions).

2.4 Thermogravimetry

Thermogravimetry (TG) measures the change in mass of a material as a function of time at a determined temperature (i.e., isothermal mode), or over a temperature range using a predetermined heating rate. Essentially, a TG consists of a microbalance surrounded by a furnace. A computer records any mass gains or losses. Weight is plotted against a function of time for isothermal studies and as a function of temperature for experiments at constant heating rate. Thus, this technique is very useful in monitoring heat stability and loss of components (e.g., oils, plasticizers, or polymers).

Thermogravimetry is also widely used both in studies of degradation mechanisms and for methods for service lifetime prediction measurements.[40] TG lifetime prediction routines are available from instrument manufacturers. The routine calculates the activation energy by using a form of an Arrhenius equation (Eq. 6).

Eq. (6) $dx/dt = A \exp(-E_a/RT)(1-x)^n$

where
x = degree of conversion
t = time
dx/dt = reaction rate
n = reaction order
A = pre-exponential factor
E_a = activation energy
R = gas constant
T = temperature (K)

Taking the logarithm of the above equation, the following expression is obtained:

Eq. (7) $\ln[1/(1-x)^n] = (-E_a/R)(1/T) + \ln A - \ln(dx/dt)$

For a given degree of conversion, x_i, and temperature, T_i,

Eq. (8) $\ln[1/(1-x_i)^n] = (-E_a/R)(1/T_i) + \ln A - \ln(dx_i/dt)$

Since the reaction rate is constant,

$$\ln A - \ln(dx_i/dt) = \text{constant} = \beta$$

Therefore, a plot of the logarithm of the heating rate versus reciprocal temperature gives a straight line with a slope equal to $-E_a/R$ and an intercept equal to β, assuming a first order reaction.

Thermoanalytical methods, such as TG, where degradation of a material can be measured under conditions that accelerate its rate and the resulting parameters extrapolate to predict a service lifetime could have great commercial importance[41] in the construction industry. They could be used not only for planning economic replacement before catastrophic failure occurs or avoiding premature replacement, but also for developing specifications for quality assurance and control tests and formulations.

If the E_a/R value and the rate, at a given temperature, are known, rates at any other temperature may be obtained and failure predictions can be made. A typical computer routine calculates the activation energies using Eq. (8). Once a failure criterion is selected (e.g., 5% weight loss), the logarithms of the times to reach failure are calculated at various temperatures. These plots are used to predict times to failure at service temperatures that are outside the range of experimental temperature measurements. Such predictions depend on the reaction mechanism remaining unchanged over the entire range of extrapolation. However, these routines are frequently questionable.[40] Weight loss usually reaches a measurable rate only when temperatures are high enough for considerable molecular movement to occur. Therefore, extrapolation of kinetic equations parameters obtained at these temperatures, through temperature ranges where phase and large viscosity changes take place down to service temperature where material diffusion limits the kinetics, results in false predictions.[41]

According to Flynn,[40] differential scanning calorimetry and thermomechanical analysis techniques may give more reliable correlation between natural and accelerated aging than TG. Therefore, the accelerated aging experiments should take into account factors such as determining the property whose deterioration is responsible for failure; chemical groups or morphological characteristics susceptible to attack; attacking agents; and factors accelerating the deterioration through intensification, sensitivity of the technique, and reliability of the measurements as well as relevance of the extrapolation. Also, it is important to validate the procedure used by comparing the predictions from the proposed method with those from methods that measure another physical property, data from actual service, or from long-term aging experiments.

2.5 High Resolution TG

Reactions investigated by TG are, by nature, heterogeneous. Therefore, experimental results are affected by weight, geometry, and particle size of the specimen. Moreover, temperature calibration and thermal gradient in the material can also affect the results. Hence, low heating rates should be used to alleviate the problem and to obtain good resolution under non-isothermal conditions.

With complex systems such as polymers and fiber reinforced composites, good resolution is essential to obtaining reliable results and kinetic parameters that can be used to compare the stability of different systems and assess their lifetime. Since, low heating rates lengthen the experiment time, a novel TG mode, high resolution TG (Hi-Res^{TM} TGA)[42] was introduced by TA Instrument. This technique provides a means to increase the resolution while often decreasing the time required for experiments. The technique has two novel non-isothermal modes of operating: variable heating rate mode and constant reacting rate mode. In the variable heating rate mode, the heating rate is dynamically and continuously varied to maximize resolution whereas in the reacting rate mode, an attempt is made to keep the reaction at a specified constant value by changing the heating rate.

Using the Hi-Res^{TM} TGA technique, a simplified method has been developed by Salin, et al.,[42] to extract kinetic parameters from variable heating experiments by using a mathematical function which takes into account resolution, sensitivity, and initial heating rate. These parameters affect the overall heating rate and can be controlled by the operator.

As shown by Eq. (8), the kinetics governing a thermal decomposition event depend on time, temperature, and rate of decomposition. TG experiments performed at a constant heating rate allow temperature and time to be interchanged in the case of first order kinetics and one-step decompositions.[43] Hi-Res^{TM} TGA allows the determination of kinetic parameters such as activation energy and reaction order for each step in multiple component materials using four different TG approaches:[44] constant heating rate, constant reaction rate, dynamic heating rate, and stepwise isothermal.

As discussed previously, the constant heating rate approach is based on the Arrhenius Eq. (6) and requires different heating rates. Flynn and Wall[45] rearranged the equation to obtain Eq. (9)

Eq. (9) $$E_a = \left(\frac{-R}{b}\right)\frac{d(\ln Hr)}{d(1/T)}$$

where b = constant for $n = 1$
Hr = heating rate (°C/min)

Using a point of equivalent weight loss beyond any initial weight loss due to evolution of volatiles, a plot of $\ln(Hr)$ versus $1/T$ can be constructed to obtain E_a and the pre-exponential factor (A). The results from this approach plotted as estimated lifetime versus temperature can provide useful information.[46]

In the constant reaction rate approach, developed by Rouquero[47] and improved by Paulik, et al.,[48] the heating rate is adjusted as required by the instrument to maintain a constant rate of weight loss. This is a high resolution approach, which has proved to be very useful for samples which decomposed reversibly,[44] such as inorganic materials, which lose ligand molecules (e.g., water, CO_2). Assuming a first order reaction, the last two terms in Eq. (8) are constant, hence, the E_a can be obtained by plotting $\ln\ [1/(1 - x)^n]$ vs $1/T$. The advantages of this approach are the ability to evaluate multiple component materials and the need for only a single experiment.[44]

The dynamic heating rate approach consists in varying continuously both the heating rate and the rate of weight loss, but the heating rate is decreased as the rate of weight loss increases. This results in enhanced resolution and faster experiments.

According to Sauerbrunn, et al.,[44] kinetic parameters can be obtained from dynamic heating rate experiments using the equation developed by Saferis, et al.,[42]

Eq. (10) $$\ln\left(\frac{H'r}{T^2}\right) = -\frac{E_a}{RT} - \ln\left[\frac{AR}{E_a} n(1-x)^{n-1}\right]$$

where $H'r$ = heating rate at the peak (°C/min)
T = temperature at the peak (K)
A = pre-exponential factor
R = gas constant
n = reaction order
E_a = activation energy
x = degree of conversion

Assuming x is constant; $dHr/dt = 0$ at the peak maximum; $d(dx/dT)/dT = 0$ and reaction order (n) is equal to one, Eq. (10) can be written as:

Eq. (11) $$\ln\left(\frac{H'r}{T^2}\right) = -\frac{E_a}{RT} - \ln\left(\frac{AR}{E_a}\right)$$

where $$\ln\left(\frac{AR}{E_a}\right) = Z(\text{constant})$$

Eq. (12) $$\ln\left(\frac{H'r}{T^2}\right) = -\frac{E_a}{RT} - z$$

In a dynamic heating rate approach, the activation energy value, E_a, can be calculated from a plot of ln (Hr/T^2) versus $1/T$, where at least three experiments with different maximum heating rates have been used. The calculation of E_a is independent of reaction rate and mechanism.[44]

Sichina[49] reported that although the variable heating rate approach offers some advantages in improving resolution, care must be taken to ensure that the resulting data is displayed in a correct manner to avoid visual artifacts. To demonstrate the importance of time and temperature in the variable heating rate approach, he heated copper sulfate pentahydrate using a series of heating ramps coupled with isothermal holds. His purpose was to produce data compression and decompression regions.

Figure 3 displays the TG curve using this approach as a function of temperature. Between room temperature and 100°C, the TG curve in Fig. 3 appears to have four well-resolved weight losses presumably resulting from the water of hydration. Previous research[50] has reported that by using variable heating rates, five well-defined waters of hydration can be identified in $CuSO_4\,5H_2O$.

Sichina[49] reported that the four well-resolved weight losses observed in Fig. 3 are an artifact because a plot of the same data as a function of time (Fig. 4) only shows two weight losses. Furthermore, he plotted the first weight loss (%), which stoichiometrically corresponds to simultaneous evolution of two waters, as a function of time (Fig. 5). Hence, in this plot, all data are equally spaced. However, if the same weight loss is plotted as a function of temperature (Fig. 6), data compression and decompression occur. As a result, the TG curve appears to have two resolved events due to two waters of hydration. This was attributed to a visual artifact.

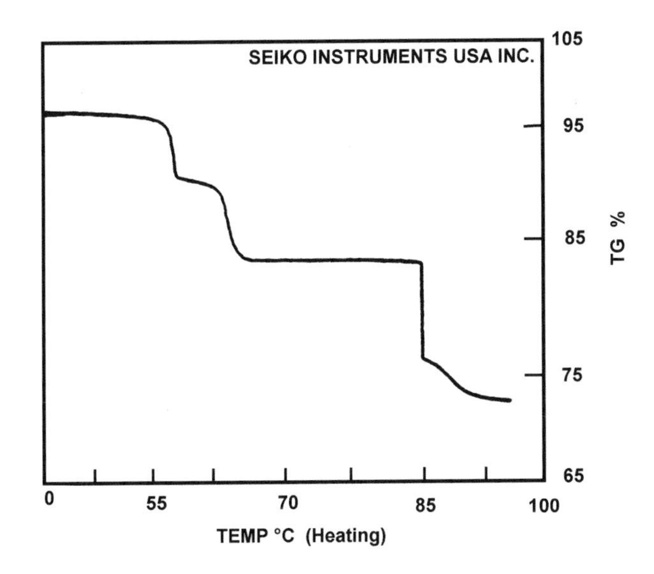

Figure 3. Specially programmed variable heating rate data for copper sulfate pentahydrate plotted as a function of temperature. TG data appears to show four resolved mass loss events. *(Reprinted with permission.)*[49]

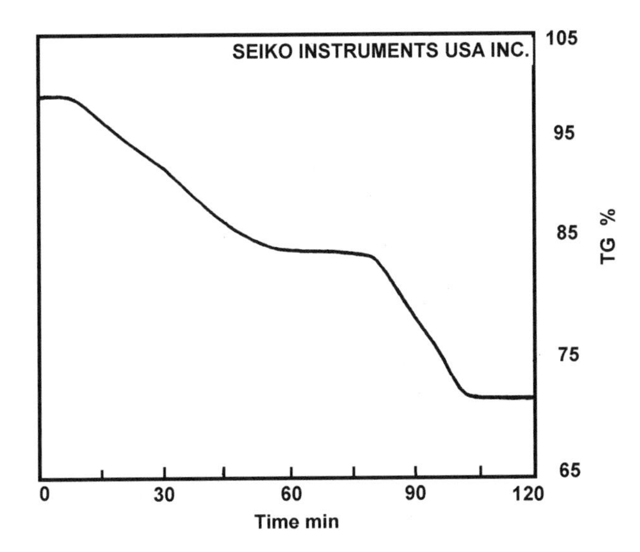

Figure 4. Same data shown in Fig. 5 displayed as a function of time. The TG trace shows only two-well resolved steps. *(Reprinted with permission.)*[49]

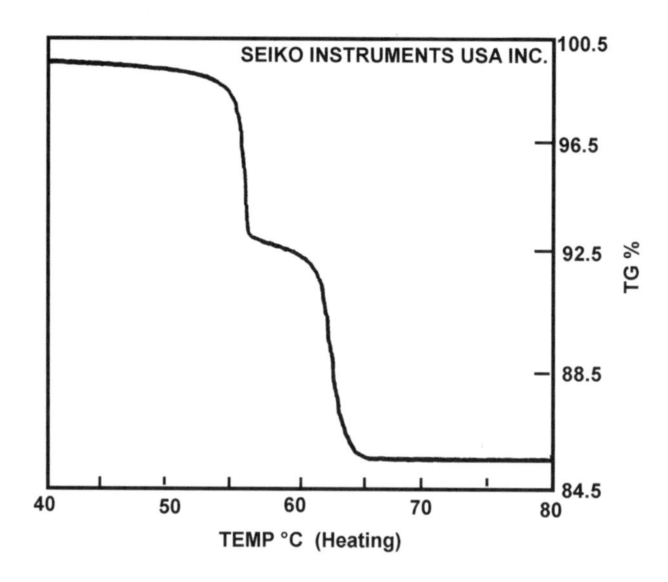

Figure 5. First weight loss event of $CuSO_4\ 5H_2O$ plotted as a function of time. *(Reprinted with permission.)*[49]

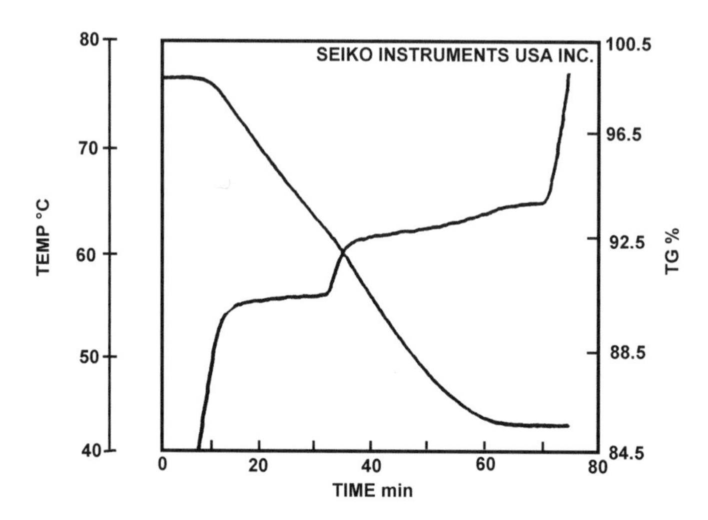

Figure 6. First weight loss event plotted as a function of temperature showing creation of an artificial step due to data compression. *(Reprinted with permission.)*[49]

After all these technical considerations in the variable heating rate approach, Sichina[49] highlighted the following:

- In the variable heating rate approach, the heating rate controlled at any given time by the instrument is dependent upon the rate of sample volatilization, but the decomposition is dependent upon experimental factors such as initial sample mass, geometry and physical nature of the sample, surrounding atmosphere, purge gas, flow rate, heating rate, etc. Therefore, this may affect the precision of the resulting data because the experimental variables associated with the variable heating rate approach may have a larger effect on the decomposition kinetics as compared to experiments performed at constant heating rates.
- Decomposition of a material is a kinetically controlled, time-based phenomenon. Hence, resolution of any analytical experiment should be properly defined on a time basis rather than a temperature basis because time is always the factor in any experiment. Changes in the heating rate during a decomposition event may result in artifacts in the TG data when plotted as a function of temperature.
- Separations of decomposition events plotted on a time basis are always real, but resolution of events plotted on a temperature basis may not necessarily be real.
- Since the time-based quantity is always equivalent to the rate of mass loss, the derivative of weight loss should be displayed on a time (dc/dt) rather than a temperature basis.

The stepwise isothermal approach was first introduced by Sorenson.[51] In this approach, a maximum heating rate and two weight loss per minute thresholds are defined by the operator. The instrument ramps at the maximum heating rate until the sample starts to lose weight and reaches the maximum specified threshold, stops and then goes to the next segment where the temperature is held isothermally until the rate of decomposition falls below the minimum threshold. The method is repeated until all the weight losses have been observed.

The approach has a kinetic treatment similar to Eq. (6), but the term (E_a/RT) is constant during an isothermal experiment. Hence, if n is equal to one, a plot of dx/dt versus $(1 - x)$ yields a straight line. For reaction orders different from one, Eq. (6) is written as:[44]

Eq. (13) $\ln (dx/dt) = n \ln (1 - x) + C$

where C = constant
n = reaction order
x = fraction of decomposition or conversion

Plotting dx/dt versus $(1-x)$ results in a line with a slope equal to the reaction order. The stepwise approach allows the calculation of the reaction order for each step of the multiple step decomposition from a single TG experiment.[42]

A variation of the stepwise TG method was also developed by Sichina.[43] The approach, called automated stepwise, consists of heating a sample at a constant heating rate until a significant weight loss occurs, as determined when the rate of decomposition exceeds a pre-selected "entrance" threshold level. Then, the instrument automatically holds the sample isothermally until the rate of reaction decreases below a pre-selected "exit" threshold level. The heating then is resumed at a constant rate until the next weight loss is encountered. This sequence is repeated for each weight loss during the experiment. The stepwise TG method has shown to be a valuable technique in resolving transitions, which are closely spaced with regards to temperature.

3.0 MODERN TECHNIQUES

3.1 Thermomechanical Analysis (TMA)

Thermomechanical analysis (TMA), as defined by ASTM E473-85, is a method for measuring the deformation of a material under a constant load as a function of temperature while the material is under a controlled temperature program. The measuring system consists of a linear voltage differential transformer (LVDT) connected to the appropriate probe (Fig. 7). Various probes are available and the measurements can be done in either compression, expansion, penetration, flexure, or in tension mode. It is this variety of probes which allows for the measurement on samples of different configurations. Any displacement of the probe generates a voltage that is then recorded. The dimensional change of a sample with an applied force is measured as a function of time or temperature. The plot of expansion (or contraction) versus temperature (or time) can then be used to obtain T_g, the coefficient of thermal expansion (CTE), softening temperature, and Young's modulus.

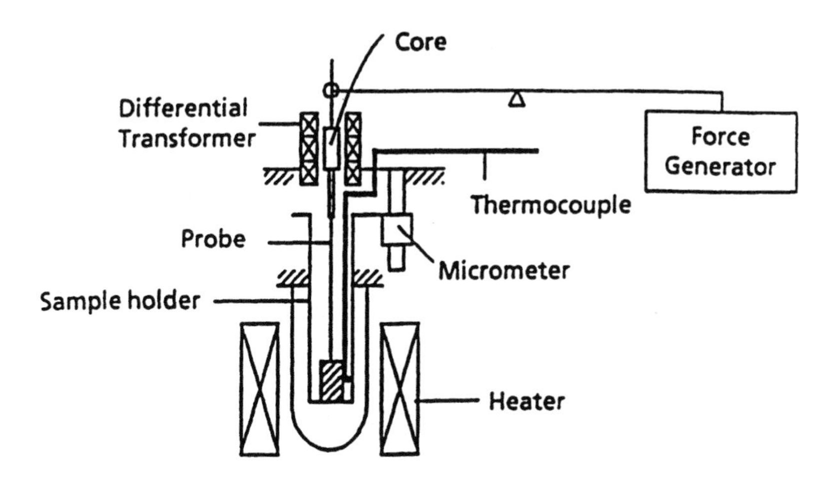

Figure 7. TMA measurement principle. *(Reprinted with permission from Seiko TMA Manual.)*

The change in linear dimension as a function of temperature can be described by the following:

Eq. (14) $$L_2 = L_1(1 + \int_{T_1}^{T_2} \alpha_l \, dT)$$

where α_l is the coefficient of linear expansion, and L_1 and L_2 are the lengths of the specimen at temperatures (or time) T_1 and T_2 respectively. If the difference between T_2 and T_1 is relatively small, then the equation can be represented by:

Eq. (15) $$L_2 - L_1 = L_1 \alpha_l (T_2 - T_1)$$

or it can be rewritten as:

Eq. (16) $$\alpha_l = \frac{1}{L_1} \frac{\Delta L}{\Delta T}$$

Therefore, the slope of the curve of length versus temperature yields $\alpha_l L_1$ and the coefficient of linear thermal expansion is obtained by dividing by L_1.

There are some drawbacks with thermomechanical analysis. Proper calibration is required to obtain reliable and reproducible data. Other sources of errors include slippage of the probe on the specimen and specimens undergoing creep in addition to length changes.

3.2 Dynamic Mechanical Analysis (DMA)[52]–[59]

The following equations describe the stress-strain relationship (Fig. 8) as measured by DMA:

Eq. (17) $\sigma = \sigma_o \sin(\omega t) \cos \delta + \sigma_o \cos(\omega t) \sin \delta$

Eq. (18) $\varepsilon = \varepsilon_o \sin(\omega t)$

where σ is the stress, ω the angular frequency, t is the time, δ is the phase angle, and ε is the strain.

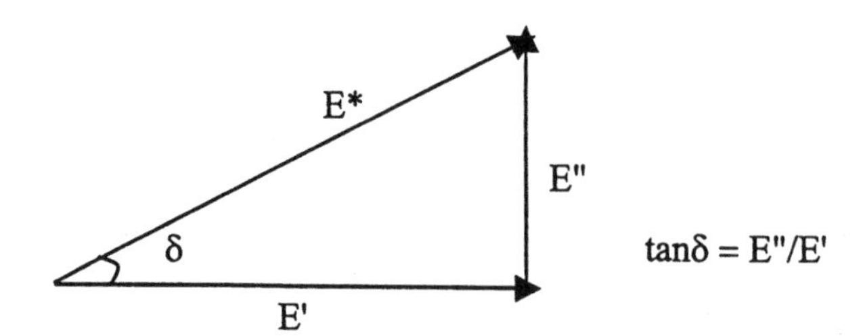

Figure 8. Stress-strain relationship measured by DMA.

The *real* component, $\sigma_o \cos \delta$, occurs when stress is in-phase with strain. The *imaginary* component is 90° out-of-phase with strain and corresponds to $\sigma_o \sin \delta$. The stress-strain components can be resolved and real and imaginary components of modulus are obtained:

Eq. (19) $\sigma = \varepsilon_o E' \sin(\omega t) + \varepsilon_o E'' \cos(\omega t)$

where $E' = (\sigma_o/\varepsilon_o) \cos \delta$ is a measure of recoverable strain energy in a deformed body and is known as the *storage* modulus. $E'' = (\sigma_o/\varepsilon_o) \sin \delta$, where E'' is the loss modulus and is associated with the loss of energy as heat due to the deformation of the material. The loss tangent or damping factor, tan δ, is defined as the ratio of E''/E.

In a DMA experiment, E' and/or E'' and/or tan δ are plotted as a function of time or temperature (see Fig. 8). T_g, an α-transition, is usually obtained from the most intense peak observed for either the E'' or tan δ (and significant inflection for E') curve. Amorphous polymers have a more intense α-peak than semi-crystalline polymers because the former are less rigid.

The glass transition temperature, determined by DMA, is dependent on the heating-rate and frequency. Therefore, T_g values obtained by this dynamic technique are generally different from that obtained by static techniques, such as differential scanning calorimetry (DSC). Moreover, the temperature of a polymer can also be increased by subjecting the material to high frequency and high amplitude oscillations. Thus, when studying dynamic mechanical properties, low frequencies and low strain amplitudes should be used. Low strain amplitude is associated with the linear region of a stress-strain curve, but if a large stress or strain amplitude is applied to a viscoelastic material, high internal heat due to molecular vibration is generated. This results in a nonlinear viscoelastic response that is quite complex to analyze. Also, in nonlinear viscoelastic regions, the material is permanently modified. For example, microscopic crack formation or failure due to fatigue can result.

Clamping will affect modulus results, and, therefore, absolute modulus values are obtained with great difficulty using DMA. If care is taken, results within a given laboratory will be reproducible, hence comparison amongst various materials is feasible. Although DMA is weak with respect to the accuracy of absolute modulus, the transition temperatures can routinely be determined with great accuracy. The method used to obtain T_g (i.e., E'' or tan δ peak temperature) affects the value, and, therefore, the parameter must be specified . As long as the same parameter is used throughout a study, the trend observed will be the same regardless of the parameter used.

3.3 Dielectric Analysis (DEA)

Dielectric analysis (DEA) or *dielectric thermal analysis* (DETA) is another important thermoanalytical technique that is rapidly evolving. This technique measures two fundamental electrical characteristics of a material—capacitance and conductance—as a function of time, temperature, and frequency. The *capacitive* nature of a material is the ability to store electric charge whereas the *conductive* nature is the ability to transfer electric charge. The parameters measured in dielectric analysis are permitivity (ε') and the loss factor (ε'').[60] The former is the alignment of the molecular dipoles in the material and the latter represents the energy required to align the dipoles or move trace ions.

DEA is used in the characterization of thermoplastics, thermosets, composites, adhesives, and coatings, and it is complementary to other thermoanalytical techniques such DSC, DMA, TG, and TMA. DEA is an important technique because it has high inherent sensitivity, wide frequency range, and the ability to easily detect rheological changes that occur during heating of uncured materials.[61]

The mobility of ions and dipoles is measured by applying a sinusoidal voltage to the sample and measuring the current (Fig. 9).[62] Process behavior, the physical and chemical structure of polymers, and other organic materials can be investigated through the measurement of their electric properties. The charged sites found in organic and inorganic polymers are typically ions and dipoles. Dipoles in the material will attempt to orient themselves with the applied electric field, while charged ions, usually present as impurities, will move towards the electrodes of opposite polarity. Changes in the degree of alignment of dipoles and in the ion mobility provide information about physical transitions in the material and about material properties such as viscosity, rigidity, reaction rate, and degree of cure.[62]

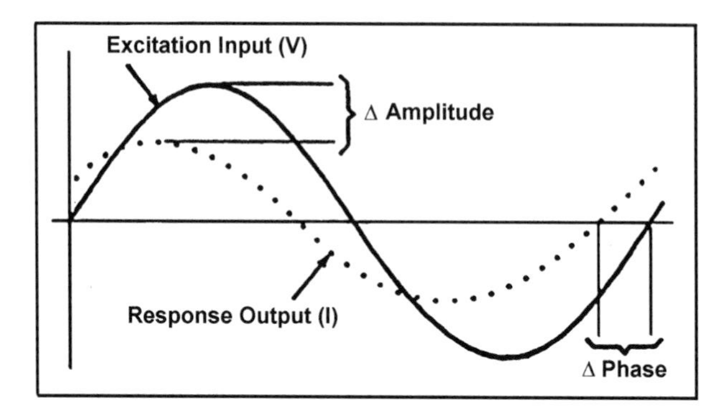

Figure 9. DEA excitation and response. The mobility of ions and dipoles is measured applying a sinusoidal voltage to the sample and measuring the current. *(Reprinted with permission.)*[62]

Dielectric analysis (DEA) may be used for characterizing high performance thermoplastics such as poly(ether ether ketone) (PEEK),[63] which is used as a matrix in advance composites. It may also be used for monitoring the plasticizing effects of moisture on polymer transitions,[60] electronics, pharmaceutical,[64] petrochemical, and food products.[65] Despite the wide range of applications, no work has been reported in the literature on the use of DEA in the characterization of construction materials. The technique may have potential in the new field of conductive concrete recently developed by Xie, et al.[66] DEA could also be used to monitor the curing of construction adhesives and sealants.

One of the advantages of DEA application over conventional thermoanalytical techniques is that it has the ability to handle highly brittle and very soft samples. For example, DMA measurements require mechanical deformation of the sample when measuring the viscoelastic properties of the material. If the material is too brittle, it will crack during deformation. On the other hand, characterization of very soft materials requires support, and, in some cases, cannot be characterized with some DMA instruments. However, DEA can overcome these difficulties since it requires neither clamping nor mechanical deformation. Hence, DEA can yield excellent material data on both brittle and soft samples, including tar and waxes. It is quite difficult to analyze these materials by DMA because they become very brittle below their glass transition temperature and very soft when heated above the melting point. The difficulty in using DEA though is that good contact with the sample is needed, which is not always easy with hard samples.

In addition, DEA is effective for rheological studies because by monitoring the movement of ions in a single test, it can identify key events affecting rheological changes such as time or temperature corresponding to the minimum viscosity, the onset of the flow, the onset of cure and its completion, the maximum rate of reaction and transition to a glassy phase.[67]The technique has a higher sensitivity than DSC for analyzing the stages of cure as well as good stability for liquid evaluation.

As can be seen, DEA has much potential. It helps scientists and technology in achieving a new understanding of the structure and behavior of materials.[64] It is anticipated that this technique will be widely used in construction within the next ten years.

3.4 Conduction Calorimetry

The reactions of various types of cements and their components with water is an exothermic process. The intensity of heat liberated with time depends on the type of chemical, surface area, reactivity, etc. Measurement of the total heat and rate of heat development provides information on the kinetics of hydration, degree of hydration, mechanism of hydration, the effect of additives, setting phenomenon, etc. Conduction calorimetry finds extensive applications in concrete technology.

Development. The first conduction calorimeter was developed in 1923.[68] Subsequently, Carson applied this technique to the investigation of cement hydration.[69] A systematic investigation of the effect of gypsum on heat evolved during the hydration of cement was carried out by Lerch in 1946.[70] Important conclusions were drawn by Stein who applied the conduction calorimetric technique to studying the effect of organic and inorganic additions on cement hydration.[71] A highly sensitive conduction calorimeter known as a Wexham calorimeter was developed by Forrester to investigate cement hydration.[72] Bensted[73] applied a Setaram heat flux calorimeter to a study of oil well cements. This equipment permitted in situ mixing of cement with water and enabled recording of the initial reactions that occur soon after water comes into contact with the cement. In this calorimetric technique, the mixing cell withstands internally generated pressures up to 1.03 MPa (150 psi). Experiments could be carried out at temperatures up to 180°C. A calorimeter containing six cells has also been developed. This equipment consists of a water tight block with six chambers, a constant temperature bath, a thermopile, and a recording system. Each chamber in the block contains a teflon coated sample holder into which a known amount of sample and water are mixed. The heat developed from the instant water comes into contact with the sample is registered by injecting water through a syringe. The heat given off by the reaction at different times is carried by thermopiles, and the signals are amplified and registered by a computer system.

Applications. Conduction calorimetry has been widely used for a study of the hydration reactions of various cementitious systems. Tricalcium silicate, being the dominant compound in portland cement, determines to a large extent the strength and other properties of concrete. Conduction calorimetric curves of tricalcium silicate and portland cement show five steps during the hydration process (Fig. 10).[74] In the first stage, as soon as the silicate or cement comes into contact with water, Ca and OH ions are released into the solution phase. This is followed by a rapid release of heat that ceases within 10–12 minutes. This is called the pre-induction period. In the second stage, the reaction is slow, and it is known as the dormant or induction period. This may be extended or shortened by a few hours by the addition of a small amount of chemicals, known as chemical admixtures. In the third stage, the reaction proceeds rapidly and accelerates with time, releasing a maximum amount of heat at the end of the acceleratory period. At this stage, a rapid crystallization of calcium hydroxide occurs. In the fourth stage, there is a slow deceleration. At the final stage, there is only a limited formation of products, and at this stage the reaction is diffusion controlled. Thus, conduction calorimetry permits determination of the rate and amount of hydration as a function of temperature, the water:cement (w/c) ratio, the type of admixture added, the particle size of the starting material, etc.

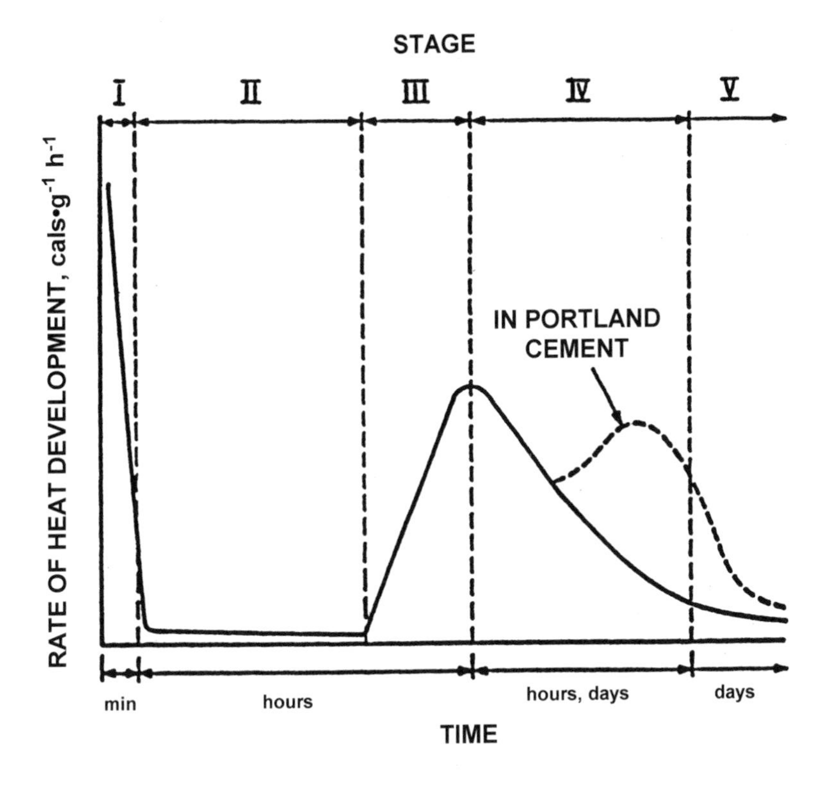

Figure 10. Conduction calorimetry curves of hydrating tricalcium silicate and cement.

Chemical admixtures in small amounts are added to concrete to enhance the physical, mechanical, chemical, and durability characteristics of concrete. *Superplasticizers* are admixtures that have the ability to increase the workability of concrete (for easy placement), but also produce high strength concretes. The relative rates of hydration of cement containing superplasticizers at different temperatures are conveniently followed by conduction calorimetry. In Fig. 11 both the rates of hydration and the cumulative amounts of heat developed in cement pastes hydrated at temperatures of 20, 40, and 55°C are plotted as a function of time.[75] In the figure, SMF refers to the superplasticizer based on sulfonated melamine formaldehyde. The addition of the superplasticizer retards the hydration of cement. The retardation increases with the dosage of the superplasticizer. Also, the retardation effect becomes less significant as the temperature of hydration is increased.

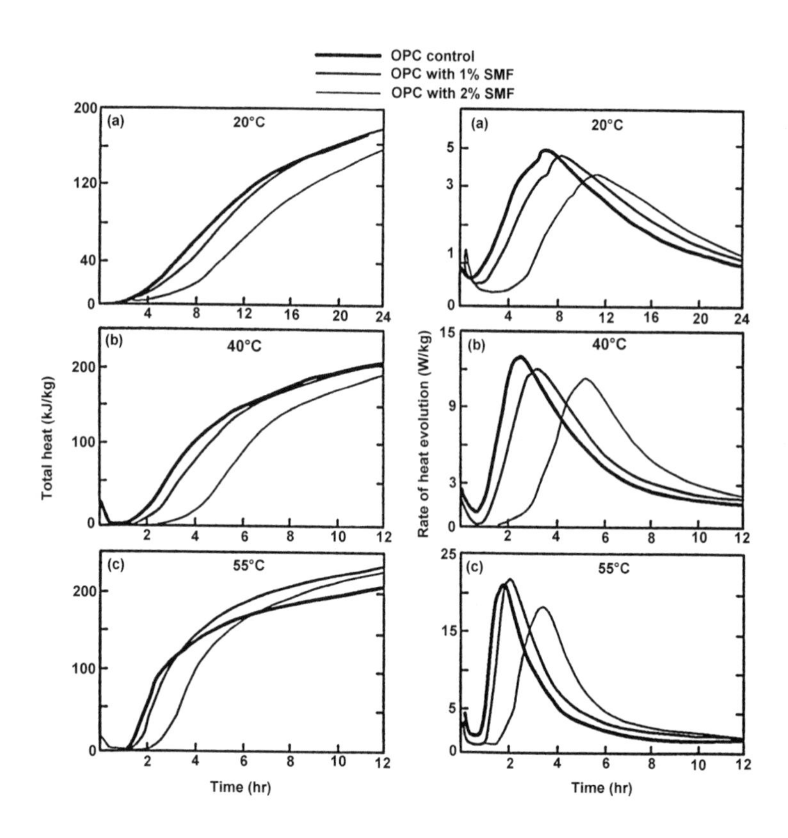

Figure 11. Conduction calorimetric curves of superplasticized cement at different temperatures.

Many industrial waste products have been used in concrete with success. Some examples include fly ash, slag, silica fume, rice husk ash, phosphogypsum, etc. Concrete, containing these minerals will have improved durability, workability, and other properties. Fly ash, a high surface area byproduct obtained as a glassy material, is separated from flue gases during the combustion of pulverized coal in thermal plants. Conduction calorimetry is used with success to follow the rate of hydration of cement containing various types of fly ashes. In one study, the rate of hydration of cement-fly ash (30%) was studied by using fly ash (FA) of surface areas of 200, 450, and 650 m^2/kg.[75] These mixtures are designated PFA-0, PFA-1, and PFA-2 respectively in Fig. 12. The fly ash cements exhibit sharper peaks at earlier times at 11–12 hrs compared to the reference cement. These peaks become sharper as the surface area increases. The peak effects are associated with the hydration of C_3A after the depletion of gypsum in the system. The initial exothermal effect due to the hydration of the calcium silicate phase in cement is retarded by the fly ashes. Several important applications of conduction calorimetry in the field of cements are to be found in the publications of Ramachandran and co-workers.[74][77][78]

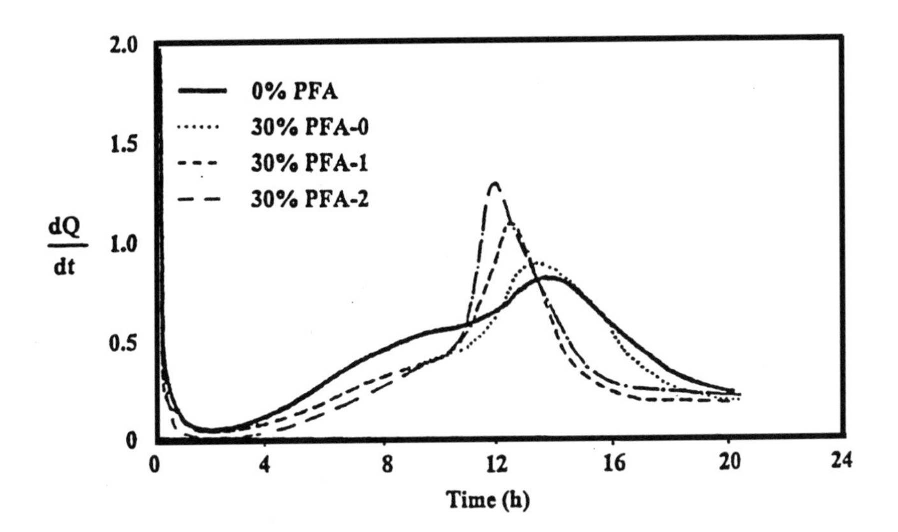

Figure 12. The influence of particle size of fly ash on the hydration of cement.

REFERENCES

1. Mackenzie, R. C., *Talanta,* 16:1227 (1969)
2. Mackenzie, R. C., Keattch, C. J., Dollimore, D., Forrester, J. A., Hodgson, A. A., and Redfern, J. P., *Talanta,* 19:1079 (1972)
3. Arntz, H., Schneider, G. M., *Farad. Discuss. Chem. Soc.,* 69:139 (1980)
4. Barrall, E. M., and Johnson, J. F., Differential Scanning Calorimetry Theory and Applications, in: *Techniques and Methods of Polymer Evaluation,* Vol. 2, Ch. 1 (1970)
5. Judd, M. D., and Pope, M. I., *J. Inorg. Nucl. Chem.,* 33:365 (1971)
6. Mackenzie, R. C., Basic Principles and Historical Development, in: *Differential Thermal Analysis,* (R. C. Mackenzie, ed.), Vol. 1, Ch. 1, Academic Press, New York (1970)
7. Mackenzie, R. C., and Laye, P. G., *Chemistry in Britain,* 22:1005–1006 (1986)
8. Mackenzie, R. C., and Mitchell, B. D., Technique, in: *Differential Thermal Analysis,* (R. C. Mackenzie, ed.), Vol. 1, Ch. 4, Academic Press, New York (1970)
9. Mackenzie, R. C., and Mitchell, B. D., *Analyst,* 87:420 (1962)
10. Meisel, T., *J. Thermal Anal.,* 29:1379 (1984)
11. Redfern, J. P., Low-Temperature Studies, *Differential Thermal Analysis,* (R. C. Mackenzie, ed.), Vol. 2, Ch. 30, Academic Press, New York (1972)
12. Rouquerol, J., and Boivinet, P., Calorimetric Measurements, in: *Differential Thermal Analysis,* (R. C. Mackenzie, ed.), Vol. 2, Ch. 27, Academic Press, New York (1972)
13. Pennington, S. N., *Rev. Anal. Chem.,* 1:113 (1972)
14. Sestak, J., Holba, P., and Lombardi, G., *Ann. Chim. (Rome),* 67:73 (1977)
15. Skinner, H. A., Theory, Scope, and Accuracy of Calorimetric Measurements, in: *Biochem. Microcalorimetry,* (H. D. Brown, ed.), Ch. 1, Academic Press, New York (1969)
16. Suzuki, H., and Wunderlich, B., *J. Thermal Anal.,* 29:1369 (1984)
17. Vold, M. J., *Anal. Chem.,* 21:683 (1949)
18. White, M. A., *Thermochimica Acta,* 74:55 (1984)
19. Wilhoit, R. C., Thermodynamic Properties of Biochemical Substances, in: *Biochem. Microcalorimetry,* (H. D. Brown, ed.), Ch. 2, Academic Press, New York (1969)
20. Le Châtelier, H., *Compt. Rend. Hebd. Séanc. Acad. Sci. Paris,* 104:1443 (1887)

21. Le Châtelier, H., *Bull. Soc. Fr. Mineral. Cristallogr.,* 10:204 (1887)
22. Roberts-Austen, W. C., *Proc. Inst. Mech. Eng.,* Parts 1–2, pp. 35–101 (1899)
23. Boersma, S. L., *J. Am. Ceram. Soc.,* 38:281 (1955)
24. David., D. J., *Anal. Chem.,* 36:2162 (1964)
25. Murrill, E., and Breed, L. W., *Thermochimica Acta,* 1:409 (1970); Pacor, P., *Anal. Chim. Acta,* 37:200 (1967)
26. Sidorova, E. E., and Berg, L. G., Determination of Thermal Constants, in: *Differential Thermal Analysis,* (R. C. Mackenzie, ed.), Vol. 2, Ch. 26, Academic Press, New York (1972)
27. O'Neill, M. J., *Anal. Chem.,* 36:238 (1964)
28. Watson, E. M., O'Neill, M. J., Justin, J., and Brenner, N., *Anal. Chem.,* 36:1233 (1964)
29. Calvet, E., *J. Chim. Phys.,* 59:319 (1962)
30. Callanan, J. E., and Sullivan, S. A., Private Communication.
31. Radenac, A., and Berthaut, C., Bull. Inform. Sci. Tech., Commis. Energ. At., (Fr.), 180:43 (1973)
32. Gray, A. P., A Simple Generalized Theory for the Analysis of Dynamic Thermal Measurement, in: *Analytical Calorimetry,* (R. S. Porter, J. F. Johnson, eds.), pp. 209–218, Plenum Press (1968)
33. Reading, M., Luget, A., and Wilson, R., *Thermochimica Acta,* 238:295–307 (1994)
34. Schawe, J. E. K., *Thermochimica Acta,* 260:1–16 (1995)
35. Schawe, J. E. K., *Thermochimica Acta,* 261:183–194 (1995)
36. Sichnia, W. J., and Nakamura, N., *The TA Journal,* pp. 1–6 (Winter, 1993)
37. Casell, B., DiVitio, M., and Goodkowsky, S., New Dynamic DSC Methodology Used to Obtain Thermodynamic Data on Polymers, *1995 Pittsburgh Conf. Proc., New Orleans, LA* (March 5–10, 1995)
38. Gill, P. S., Sauerbrunn, S. R., and Reading, M., *J. Therm. Anal.,* 40:931–939 (1993)
39. Reading, M., Elliott, D., and Hill, V. L., *J. Therm. Anal.,* 40:949–955 (1993)
40. Flynn, J. H., *Encyclopedia of Polymer Science and Engineering,* 2nd Ed., Supplement Volume, pp. 692–701, John Wiley & Sons, Inc., USA (1989)
41. Flynn, J. H., *Thermochimica Acta,* 131:115, *SPE Conf. Proc., ANTEC88, Atlanta, GA, (Apr. 18–21, 1988),* Society of Plastics Engineers, Brookfield, CT., pp. 93–932 (1988)
42. Salin, I., and Seferis, J. C., Kinetic Analysis of High Resolution TGA Variable Heating Rate Data, *J. Applied Polymer Sci.,* 47:847–856 (1993)

43. Sichina, W. J., Automated Stepwise Thermogravimetric Analysis, *Proc. 21st North Am. Thermal Anal. Soc. Conf., Atlanta, GA,* pp. 214–217 (1992)
44. Sauerbrunn, S., and Gill, P., Decomposition Kinetics Using High Resolution TGA™, *Proc. 21st North Am. Thermal Anal. Soc. Conf., Atlanta, GA,* pp. 13–16 (1992)
45. Flynn, J., and Wall, L. A., *Polymer Lett.*, 19:323 (1966)
46. Estimation of Polymer Lifetime by TGA Decomposition Kinetics, TA Instruments Application Brief TA-125 (1990)
47. Rouquerol, J., *Bull. Soc. Chim.,* p. 31 (1964)
48. Paulik, F., and Paulik, J., *Anal. Chim. Acta,* 56:328 (1971)
49. Sichina, W. J., Considerations in Variable Heating Rates Thermogravimetric Experiments, *Proc. 21st North Am. Thermal Anal. Soc. Conf., Atlanta, GA,* pp. 206–211 (1992)
50. Sauerbrunn, S. R., *Proc. 20th NATAS Conf., Minneapolis,* pp. 299–306 (1991)
51. Sorensen, S., *J. Therm. Anal.,* 13:429 (1978)
52. Murayama, T., *Dynamic Mechanical Analysis of Polymeric Materials,* Elsevier Scientific Publishing Company, New York (1978)
53. Wendlandt, W. W., Chemical Analysis 19, *Thermal Analysis,* 3rd Ed., John Wiley and Sons (1986)
54. Dutt, O., Paroli, R. M., Mailvaganam, N. P., and Turenne, R. G., Glass Transition in Polymeric Roofing Membranes—Determination by Dynamic Mechanical Analysis, *Proc. 1991 Int. Symp. on Roofing Technol.,* pp. 495–501 (1991)
55. Paroli, R. M., and Dutt, O., Dynamic Mechanical Analysis Studies of Reinforced Polyvinyl Chloride (PVC) Roofing Membranes, *Polymeric Mater. Sci. and Eng.,* 65:362–363 (1991)
56. Paroli, R. M., Dutt, O., Delgado, A. H., and Mech, M., The Characterization of EPDM Roofing Membranes by Thermogravimetry and Dynamic Mechanical Analysis, *Thermochimica Acta,* 182:303–317 (1991)
57. Delgado, A. H., Paroli, R. M., and Dutt, O., Thermal Analysis in the Roofing Area: Applications and Correlation with Mechanical Properties, CIB World Congress, Montreal (May 18–22, 1992)
58. Paroli, R. M., and Penn, J. J., Measuring the Glass-Transition Temperature of EPDM Roofing Materials: Comparison of DMA, TMA and DSC Techniques, *Symp. Assignment of the Glass Transition,* (Rickey J. Seyler, ed.), ASTM STP 1249, pp. 269–279 (1994)
59. Crompton, T. R., *Analysis of Polymers: An Introduction,* Pergamon Press (1989)

60. Keating, M., Evaluation of Moisture Effects on the Dielectric Properties of Polymer Films, Thermal Analysis Application Brief Number TA-124, TA Instruments
61. Sichina, W. J., Take, M., Nakamura, N., Teramoto, Y., and Okubo, N., New Instrumentation for the Measurement of Dielectric Properties. Communication, Seiko Instruments USA, Seiko Instruments Inc., Japan
62. Cassel, R. B., Twonbly, B., and Shepard, D. D., Simultaneous Dynamic Mechanical Analysis and Dielectric Analysis of Polymers, *Proc. 23rd North Am. Thermal Anal. Soc. Conf., Toronto,* pp. 470–475 (Sept. 25–28, 1994)
63. Characterization of PEEK Film Using Dielectric Analysis, Thermal Analysis Application Brief Number TA-109, TA Instruments
64. Smith, N. T., and Shepard, D. D., Theory and Applications of Dielectric Cure Analysis in Industry, *Thermal Trends,* 4:1, 6, 8, 12, and 14 (Winter, 1997)
65. Sichina, W., and Leckenby, J., Dielectric Analysis Aapplications from Coatings to Chocolate, *American Laboratory,* pp. 72, 74, 76–80, (Oct. 1989)
66. Xie, P., Gu, P., Fu, Y., Beaudoin, J. J., Conductive Concrete-Based Compositions, p. 10, US Patent 5,447,564 (1995)
67. Sichina, W., DEA Takes the Guesswork Out of Composite Manufacture, *Materials Engineering,* 106:49–53 (1989)
68. Tian, A., Use of Dynamic Chemical Calorimetric Method, Employment of a Compensation Calorimeter, *Bull. de la Soc. Chimique de France,* 33:427–428 (1923)
69. Carlson, R. W., The Vane Calorimeter, *Proc. Amer. Soc. Testing Materials,* 34:322–328 (1934)
70. Lerch, W., The Influence of Gypsum on the Hydration and Properties of Portland Cement Pastes, *Proc. Amer. Soc. Testing Materials,* 46:1252–1292 (1946)
71. Stein, H. N., The Influence of Some Additives on the Hydration Reactions of Portland Cement, *J. App. Chem.,* 11:474–492 (1961)
72. Forrester, A., A Conduction Calorimeter for the Study of Cement Hydration, *Cement Tech.,* 1:95–99 (1970)
73. Bensted, J., and Aukett, P. N., Applications of Higher Temperature Conduction Calorimetry of Oil Cements, *World Cement,* 21:308–320 (1990)
74. Ramachandran, V. S., *Concrete Admixtures Handbook,* p. 1152, Noyes Publications, NJ, USA (1995)
75. Singh, N. B., and Prabha Singh, S., Superplasticizers in Cement and Concrete, *J. Sci. Industr. Res.,* 52:661–675 (1993)

76. Bouzoubaa, N., Zhang, M. H., Malhotra, V. M., and Golden, D. M., Blended Fly Ash Cements—A Review, *6th CANMET/ACI Int. Conf. on Fly Ash, Silica Fume, Slag and Natural Pozzolans in Concrete,* Supplementary Papers, pp. 717–749 (1998)

77. Ramachandran, V. S., and Beaudoin, J. J., *Handbook of Analytical Techniques in Concrete Science and Technology,* p. 964, Noyes Publ., William Andrew Publishing, LLC, New York (2001)

78. Ramachandran, V. S., Malhotra, V. M., Jolicoeur, C., and Spiratos, N., *Superplasticizers: Properties and Applications in Concrete*, p. 404, Minister of Public Works and Government Services, Canada (1998)

2

Introduction to Portland Cement Concrete

Concrete, made from cement, aggregates, chemical admixtures, mineral admixtures, and water, comprises in quantity the largest of all synthesized materials. The active constituent of concrete is cement paste and the performance of concrete is largely determined by the nature of the cement paste. Admixtures are chemicals that are added to concrete for obtaining some beneficial effects such as better workability, strength, durability, acceleration, retardation, air entrainment, water reduction, plasticity, etc. Mineral admixtures, such as blast furnace slag, fly ash, silica fume, and others, are also incorporated into concrete to improve its quality.

The performance of concrete depends on the quality of the ingredients, their proportions, placement, and exposure conditions. For example, the quality of the raw materials used for the manufacture of clinker, the calcining conditions, the fineness and particle size of the cement, the relative proportions of the cement phases, and the amount of mixing water influence the physico-chemical behavior of the hardened cement paste in concrete. In addition, the cement type, nature of fine and coarse aggregates, water, temperature of mixing, admixture, and the environment will determine the physical, chemical, and durability aspects of concrete. Thermal analysis techniques are widely applied to investigate the physico-chemical behaviors of cement compounds, cement, and concrete subjected to various conditions.

Although added in small amounts, admixtures may influence many of the properties of concrete, from the time water comes into contact with the dry ingredients of concrete, to its long term behavior. Generally, concrete contains one or more admixtures and their role has been studied extensively by thermal techniques and, hence, a separate chapter is devoted to describe their application.

1.0 PRODUCTION OF PORTLAND CEMENT

According to ASTM C-150 portland cement is a hydraulic cement produced by pulverizing clinker consisting essentially of hydraulic calcium silicates, usually containing one or more types of calcium sulfate as an interground addition.

The raw materials for the manufacture of portland cement contain, in suitable proportions, silica, aluminum oxide, calcium oxide, and ferric oxide. Since the compounds in cement science are complex, a simplified representation is often used in cement nomenclature: C = CaO; S = SiO_2; A = Al_2O_3; F = Fe_2O_3; H = H_2O; $\bar{S}$ = SO_3; K = K_2O; and N = Na_2O. In addition, "W" or "w" represents water, "C" or "c," cement, and "S" or "s," solid. Thus, "W/C" or "w/c" is water:cement ratio and "W/S" or "w/s" is water:solids ratio. A source of lime is provided by calcareous ingredients such as limestone or chalk and the source of silica and aluminum oxide being shales, clays, or slates. The iron bearing materials are iron and pyrites. Ferric oxide not only serves as a flux but also forms compounds with lime and alumina. The raw materials also contain small amounts of other compounds such as magnesia, alkalis, phosphates, fluorine compounds, zinc oxide, and sulfides. The cement clinker is produced by feeding the crushed, ground, and screened raw mix into a rotary kiln and heated at a temperature of about 1300–1450°C. Approximately 1100–1400 kcal/g of energy is consumed in the formation of clinker. The sequence of reactions is as follows.

- At a temperature of about 100°C (drying zone), free water is expelled.
- In the pre-heating zone (100–750°C), firmly bound water from the clay is lost.
- In the calcining zone (750–1000°C), calcium carbonate is dissociated.

- In the burning zone (1000–1450°C), partial fusion of the mix occurs, with the formation of C_3S, C_2S, and clinker.
- In the cooling zone (1450–1300°C), crystallization of melt occurs with the formation of calcium aluminate and calcium aluminoferrite.

After firing the raw materials for the required period, the resultant clinker is cooled and ground with about 4–5% gypsum to a specified degree of fineness. Grinding aids, generally polar compounds are added to facilitate grinding.

2.0 COMPOSITION

The major phases of portland cement are:

- Tricalcium silicate ($3CaO \cdot SiO_2$)
- Dicalcium silicate ($2CaO \cdot SiO_2$)
- Tricalcium aluminate ($3CaO \cdot Al_2O_3$)
- Ferrite phase of average composition ($4CaO \cdot Al_2O_3 \cdot Fe_2O_3$)

In a commercial clinker these phases do not exist in a pure form. The $3CaO \cdot SiO_2$ phase is a solid solution containing Mg and Al and is called *alite.* In the clinker, it consists of monoclinic or trigonal form, whereas synthesized $3CaO \cdot SiO_2$ is triclinic. Alite is the most important constituent of normal portland cement, constituting 50–60% and promoting strength development. The $2CaO \cdot SiO_2$ phase occurs in the β (belite) forms and contains, in addition to Al and Mg, some K_2O. Four forms, α, α', β, and γ, of C_2S are known although in clinker only the β form with a monoclinic unit cell exists. It reacts slowly with water and contributes little to strength development in the first 28 days. At one year, pure alite and belite yield the same strengths. The aluminate phase C_3A constitutes 4–12% in most portland cements and is substantially modified by ionic substitution. In some clinkers small amounts of calcium aluminate of formula NC_8A_3 may also form. The ferrite phase, designated C_4AF, is a solid solution of variable composition from C_2F to C_6A_2F. Potential components of this compound are C_2F, C_6AF_2, C_4AF, and C_6A_2F. The MgO content in cement is usually limited to 4–5% because in the form of crystalline periclase it may cause slow expansion. Free lime behaves similarly. Excessive SO_3 can also lead to expansion. Alkalis such as K_2O and Na_2O in excess of 0.6% equivalent

Na_2O are not permitted as they promote expansion with certain types of aggregates.

ASTM C-150 describes five major types of portland cement. They are:

- Normal—Type I—is used for most purposes and when special properties specified for any other type are not required
- Moderate Sulfate Resistant or Moderate Heat of Hydration—Type II
- High Early Strength—Type III
- Low Heat—Type IV
- Sulfate Resisting—Type V

The general composition, fineness, and compressive strength characteristics of these cements are shown in Table 1.[1][31]

Portland cement may be blended with other ingredients to form blended hydraulic cements. ASTM C-595 covers five kinds of blended hydraulic cements. The portland blast furnace slag cement consists of an intimately ground mixture of portland cement clinker and granulated blast furnace slag or an intimate and uniform blend of portland cement and fine granulated blast furnace slag in which the slag constituent is within specified limits. The portland-pozzolan cement consists of an intimate and uniform blend of portland cement or portland blast furnace slag cement and fine pozzolan. The slag cement consists mostly of granulated blast furnace slag and hydrated lime. The others include pozzolan-modified portland cement (pozzolan $<$ 15%) and slag-modified portland cement (slag $<$ 25%).

3.0 INDIVIDUAL CEMENT COMPOUNDS

3.1 Tricalcium Silicate

Knowledge of the hydration behavior of individual cement compounds and their mixtures forms a basis for interpreting the complex reactions that occur when portland cement is hydrated under various conditions. For a given particle size distribution and water:solid ratio, tricalcium silicate and alite harden in a manner similar to that of a typical portland cement.

Table 1. Compound Composition, Fineness, and Compressive Strength Characteristics of Some Commercial U.S. Cements[1]

ASTM Type	ASTM Designation	Composition				Fineness cm^2/g	Compressive Strength % of Type I Cement		
		C_3S	C_2S	C_3A	C_4AF		1 day	2 days	28 days
I	General purpose	50	24	11	8	1800	100	100	100
II	Moderate sulfate resistant-moderate heat of hydration	42	33	5	13	1800	75	85	90
III	High early strength	60	13	9	8	2600	190	120	110
IV	Low heat	26	50	5	12	1900	55	55	75
V	Sulfate resisting	40	40	4	9	1900	65	75	85

(With permission, Noyes Publications, Concrete Admixtures Handbook, *2nd Edition, 1995.)*

Tricalcium silicate and dicalcium silicate together make up 75–80% of portland cement (Table 1). In the presence of a limited amount of water the reaction of C_3S with water may be represented as follows:

$$3CaO{\bullet}SiO_2 + xH_2O \rightarrow yCaO{\bullet}SiO_2{\bullet}(x+y\text{-}3)H_2O + (3\text{-}y)Ca(OH)_2$$

or typically,

$$2[3CaO{\bullet}SiO_2] + 7H_2O \rightarrow 3CaO{\bullet}2SiO_2{\bullet}4H_2O + 3Ca(OH)_2$$

The above chemical equation is somewhat approximate because it is not easy to estimate the composition of C-S-H (the C/S and S/H ratio) and there are also problems associated with the determination of $Ca(OH)_2$. In a fully hydrated cement or C_3S paste about 60–70% of the solid comprises C-S-H. The C-S-H phase is poorly crystallized containing particles of colloidal size and gives only two very weak peaks in XRD at 0.27–31 nm and 0.182 nm.

The direct methods of determining C/S ratios are based on electron optical methods such as electron microprobe, other attachments, or by electron spectroscopy (ESCA). Although several values are reported the usual value for C/S ratio after a few hrs of hydration of C_3S is given as 1.4–1.6.[2] The C/S ratio of the C-S-H phase may be influenced by admixtures.

The C/S ratio of C-S-H in a fully reacted C_3S may be calculated if CH and CO_2 contents are known. Quantitative XRD is used to determine unreacted C_3S. Methods such as TG, DTG, semi-isothermal DTG, thermal evolved analysis, DTA, DSC, XRD, IR spectroscopy, image analysis of back scattered electron images, and extraction methods have been applied to estimate lime. Variations in the estimated values are possible depending on the technique. In one study in which C_3S paste was prepared at a W/S ratio of 0.45 and stored wet for 25 years at 25°C, the TGA results indicated that 1.15 moles of CH were formed for one mole of C_3S. The calculation showed the C/S ratio of C-S-H to be 1.81.[3]

There are problems associated with the determination of H_2O chemically associated with C-S-H. It is difficult to differentiate this water from that present in pores. It has been proposed that drying to 11% RH is a good base for studying the stoichiometry of calcium silicate hydrate. At this condition, the estimate of adsorbed water can be made with some confidence. This does not mean that higher hydrates do not exist above

11% RH. Feldman and Ramachandran[4] have estimated that the bottled hydrated C-S-H equilibrated to 11% RH (approached from 100% RH) has a composition of $3.28CaO{:}2SiO_2{:}3.92H_2O$. In another procedure, known as d-drying, the sample is equilibrated with dry ice at -79°C. In this method, some chemically bound water is lost from the interlayer. It may, however, be used as an empirical method to measure the degree of hydration. The water content of fully hydrated C_3S is given as 20.4–22.0% and this corresponds to a C-S-H composition $1.7\ CaO \bullet SiO_2 \bullet 1.3–1.5H_2O$.

The rate of hydration of C_3S has been determined by following the consumption of C_3S, non-evaporable water or $Ca\ (OH)_2$ with time. In Fig. 1, the general form of curve relating the fraction of C_3S consumed as a function of time in a paste hydrated at a W/S ratio of 0.5 is given.[5] Various stages of reaction could be related to the conduction calorimetric curves. At 120 days, more than 80% hydration has taken place. At one year presumably all C_3S has been consumed.

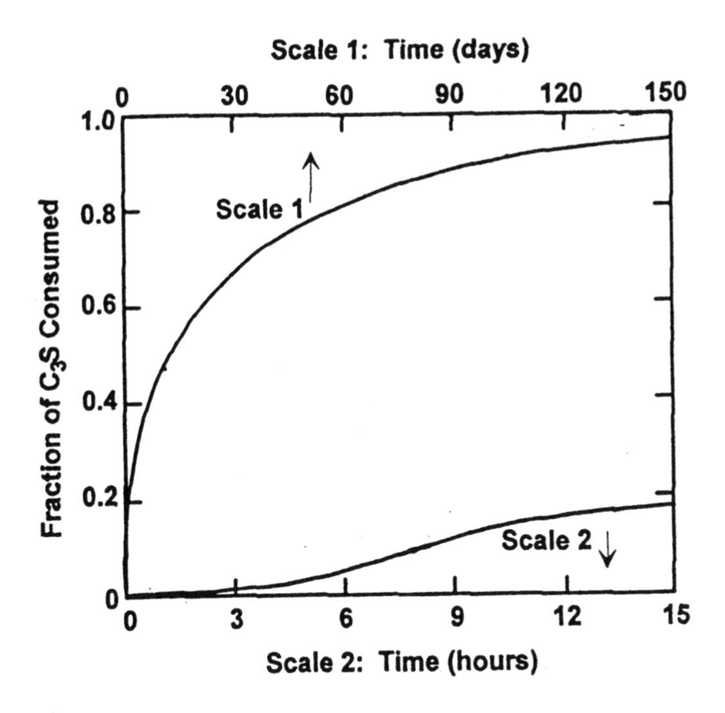

Figure 1. Fraction of tricalcium silicate consumed as a function of time in a paste.

In spite of a large amount of work, even the mechanism of hydration of C_3S, the major phase of cement, is not clear. Any mechanism proposed to explain the hydrating behavior of C_3S should take into account the following steps through which the hydration proceeds. Five stages can be discerned from the conduction calorimetric studies (Fig. 2).[31] In the first stage, as soon as C_3S comes into contact with water there is a rapid evolution of heat and this ceases within 15–20 minutes. This stage is called the *pre-induction period.* In the second stage the reaction rate is very slow. It is known as the *dormant* or *induction period.* This may extend for a few hours. At this stage the cement remains plastic and is workable. In the third stage, the reaction occurs actively and accelerates with time reaching a maximum rate at the end of this accelerating period. Initial set occurs at about the time when the rate of reaction becomes vigorous. The final set occurs before the end of the third stage. In the fourth stage there is slow deceleration. At the fifth stage, the reaction is slow and is diffusion controlled. An understanding of the first two stages of the reaction has a very important bearing on the subsequent hydration behavior of the sample. Admixtures can influence these steps. Retarders such as sucrose, phosphonic acid, calcium gluconate and sodium heptonate extend the induction period and also decrease the amplitude of the acceleration peak whereas accelerators such as calcium chloride, formate, and nitrite increase the amplitude of the peak.

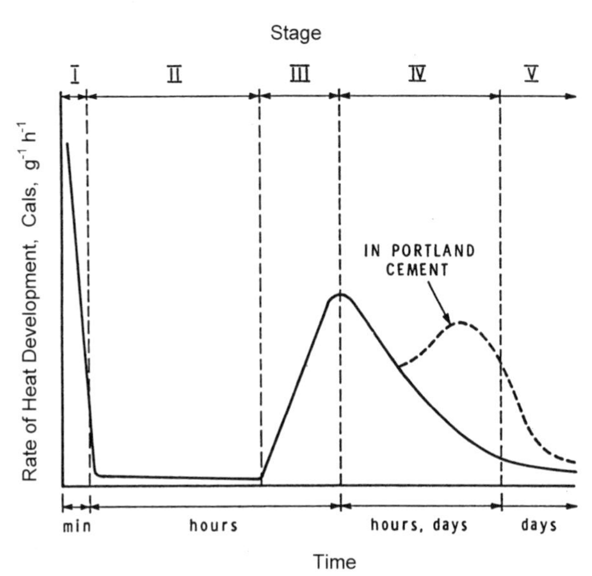

Figure 2. Rate of heat development during the hydration of tricalcium silicate and portland cement. *(With permission, Noyes Publications,* Concrete Admixtures Handbook, *V. S. Ramachandran, ed., 2nd Ed. 1995)*

It is generally thought that initially a reaction product forms on the C_3S surface that slows down the reaction. The renewed reaction is caused by the disruption of the surface layer. According to Stein and Stevels,[6] the first hydrate has a high C/S ratio of about 3, and it transforms into a product of lower C/S ratio of about 0.8–1.5 through loss of calcium ions into solution. The second product has the property of allowing ionic species to pass through it, thus enabling a rapid reaction. The conversion of the first to the second hydrate is thought to be a nucleation and growth process. Although this theory is consistent with many observations, there are others which do not conform to this theory.

The end of the induction period has been explained by the delayed nucleation of CH. It does not explain the accelerated formation of C-S-H. Tadros, et al.,[7] found the zeta potential of the hydrating C_3S to be positive, indicating the possibility of the chemisorption of Ca ions on the surface resulting in a layer that could serve as a barrier between C_3S and water. There are other mechanisms based on the delayed nucleation of C-S-H to explain the end of this induction period. Other theories have been proposed to fit most of the observations. Although they appear to be separate theories they have many common features and have been reviewed by Pratt and Jennings.[8] A detailed discussion of the mechanisms of hydration of cement and C_3S has been presented by Gartner and Gaidis.[9]

3.2 Dicalcium Silicate

The hydration of β-C_2S proceeds in a similar way to that of C_3S but is much slower. Typically, 30% is reacted in 28 days and 90% in one year. As the amount of heat liberated by C_2S is very low compared to that of C_3S, the conduction calorimetric curve will not show well defined peaks as indicated in Fig. 1. Accelerators will enhance the reaction rate of C_2S.

Just as in the hydration process of C_3S, there are uncertainties involved in determining the stoichiometry of the C-S-H phase found in the hydration of C_2S. The hydration of dicalcium silicate phase can be represented by the equation.

$$2\,[2CaO{\bullet}SiO_2] + 5H_2O \rightarrow 3CaO{\bullet}2SiO_2{\bullet}4H_2O + Ca(OH)_2$$

The amount of $Ca(OH)_2$ formed in this reaction is less than that produced in the hydration of C_3S. The dicalcium silicate phase hydrates much more slowly than the tricalcium silicate phase.

Figure 3 compares the rates of hydration of C_3S and C_2S.[31] The absolute rates differ from one sample to the other. It is general knowledge C_3S is much more reactive than β-C_2S.

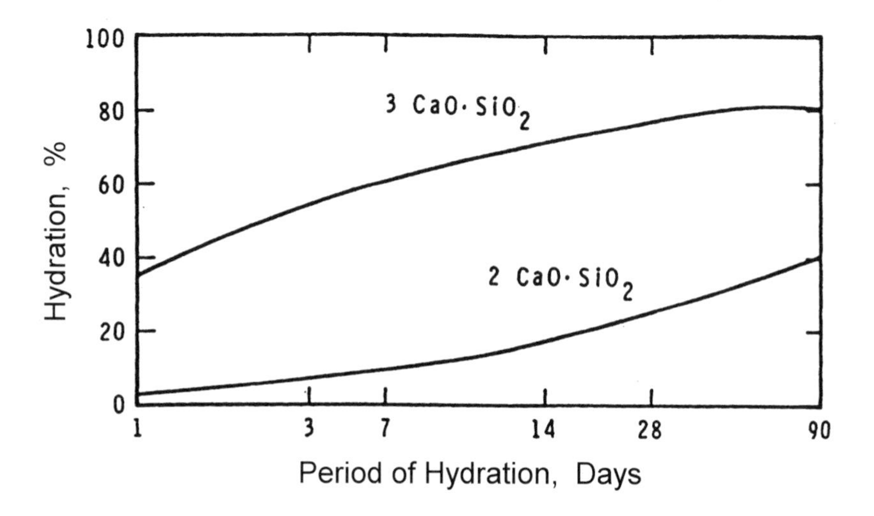

Figure 3. The relative rates of hydration of 3 CaO•SiO$_2$ and 2 CaO•SiO$_2$. *(With permission, Noyes Publications,* Concrete Admixtures Handbook, *V. S. Ramachandran, ed., 2nd Ed., 1995.)*

3.3 Tricalcium Aluminate

In portland cement, although the average C_3A content is about 4–11%, it influences significantly the early reactions. The phenomenon of *flash set*, the formation of various calcium aluminate hydrates, and calcium carbo-and sulfoaluminates involve the reactions of C_3A. Higher amounts of C_3A in portland cement may pose durability problems. For example, a cement which is exposed to sulfate solutions should not contain more than 5% C_3A.

Tricalcium aluminate reacts with water to form C_2AH_8 and C_4AH_{13} (hexagonal phases). These products are thermodynamically unstable so that without stabilizers or admixtures they convert to the C_3AH_6 phase (cubic phase). In a paste, hydration is slightly retarded in the presence of CH. In dilute suspensions the first hydrate formed is C_4AH_{19}.

In portland cement, the hydration of the C_3A phase is controlled by the addition of gypsum. The flash set is thus avoided. The C_3A phase reacts with gypsum in a few minutes to form ettringite as follows:

$$C_3A + 3C\bar{S}H_2 + 26H \rightarrow C_3A{\bullet}3C\bar{S}H_{32}$$

After all gypsum is converted to ettringite, the excess C_3A will react with ettringite to form the low sulfoaluminate hydrate. If calcium hydroxide is present the compound C_4AH_{13} will also form.

$$C_3A{\bullet}3C\bar{S}H_{32} + 2C_3A + 4H \rightarrow 3[C_3A{\bullet}C\bar{S}H_{12}]$$

The effect of KOH or C_3S on the formation of ettringite at different temperatures up to 80°C has been reported.[3] Elevated temperatures do not prevent the formation of ettringite but the amounts depend on the concentration of other additives. Potassium hydroxide retards ettringite formation. At concentrations above 1 mol/l syngenite is also formed. In the presence of C_3S ettringite is formed at all temperatures. Only small amounts of ettringite are formed in the presence of 0.5 mol/l of KOH and C_3S.

3.4 The Ferrite Phase

The ferrite phase constitutes about 8–13% of an average portland cement. In portland cement the ferrite phase may have a variable composition that can be expressed as $C_2(A_nF_{1-n})$ where $O < n < 0.7$.

Of the cement minerals, the ferrite phase has received much less attention than others with regard to its hydration and physico-mechanical characteristics. This may partly be ascribed to the assumption that the ferrite phase and the C_3A phase behave in a similar manner. There is evidence, however, that significant differences exist.

The C_4AF phase is known to yield the same sequence of products as C_3A. The reactions are slower, however. In the presence of water C_4AF reacts as follows:

$$C_4AF + 16H \rightarrow 2C_2(A,F)H_8$$

$$C_4AF + 16H \rightarrow C_4(A,F)H_{13} + (A,F)H_3$$

Amorphous hydroxides of Fe and Al form in the reaction of C_4AF. The thermodynamically stable product is $C_3(A,F)H_6$ and this is the conversion product of the hexagonal hydrates. Seldom does the formation of these hydrates cause flash set in cements.

In cements, C_4AF reacts much slower than C_3A in the presence of gypsum. In other words gypsum retards the hydration of C_4AF more efficiently than it does C_3A. The rate of hydration depends on the composition of the ferrite phase, that containing higher amounts of Fe exhibits lower rates of hydration. The reaction of C_4AF with gypsum proceeds as follows:

$$3C_4AF + 12C\bar{S}H_2 + 110H \rightarrow 4[C_6(A,F)\bar{S}H_{32}] + 2(A,F)H_3$$

The low sulfoaluminate phase can form by the reaction of excess C_4AF with the high sulfoaluminate phase.

$$3C_4AF + 2[C_6(A,F)\bar{S}H_{32}] \rightarrow 6[C_4(A,F)\bar{S}H_{12}] + 2(A,F)H_3$$

The above equations involve formation of hydroxides of Al and Fe because of insufficient lime in C_4AF. In these products F can substitute for A. The ratio of A to F need not be the same as in the starting material. Although cements high in C_3A are prone to sulfate attack, those with high C_4AF are not. In high C_4AF cements, ettringite may not form from the low sulfoaluminate possibly because of the substitution of iron in the monosulfate. It is also possible that amorphous $(A,F)_3$ prevents such a reaction. Another possibility is that the sulfoaluminate phase that forms is produced in such a way that it does not create crystalline growth pressures. The ferrite phase in cement may behave differently from pure C_4AF because of the differences in the composition of the ferrite phases and the influence of other compounds present in cement.

4.0 RELATIVE BEHAVIORS OF INDIVIDUAL CEMENT MINERALS

The variation in the rate of strength development in individual cement compounds was determined by Bogue and Lerch in 1934.[10] The comparison of reactivities and strength development of these compounds was not based on adequate control of certain parameters such as particle size distribution, water:solid ratio, specimen geometry, method of estimation of the degree of hydration, etc. Beaudoin and Ramachandran[11] have re-assessed the strength development in cement mineral pastes, both in terms of time and degree of hydration. Figure 4 compares the results of Bogue and Lerch with those of Beaudoin and Ramachandran.[11] Significant differences in the relative valuesof strengths developed by various phases were found. At 10 days of hydration the strength values were ranked as follows by Beaudoin and Ramachandran: $C_4AF > C_3S > C_2S > C_3A$. At 14 days, the relative values were in the order: $C_3S > C_4AF > C_2S > C_3A$. The Bogue-Lerch strength values both at 10 and 14 days were: $C_3S > C_2S > C_3A > C_4AF$. At one year, the corresponding values were: $C_3S > C_2S > C_4AF > C_3A$ (Beaudoin-Ramachandran) and $C_3S > C_2S > C_3A > C_4AF$ (Bogue-Lerch). Comparison of strengths as a function of the degree of hydration revealed that at a hydration degree of 70–100%, the strength was in the decreasing order: $C_3S > C_4AF > C_3A$.

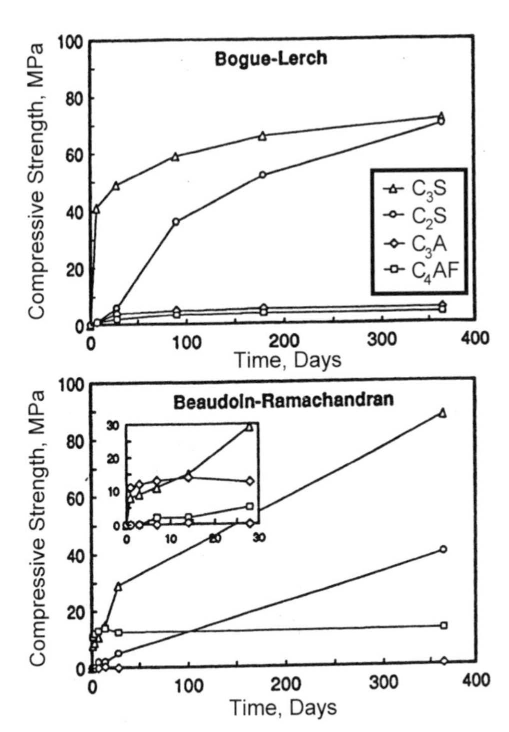

Figure 4. Compressive strength of hydrated cement compounds. *(Reprinted from Beaudoin, J. J., and Ramachandran, V. S.,* Cement and Concrete Res., *22:689–694, 1992, with kind permission from Elsevier Science Ltd, The Boulevard, Langford Lane, Kiddlington OX51GB.)*

5.0 HYDRATION OF PORTLAND CEMENT

Although hydration studies of the pure cement compounds are very useful in following the hydration processes of portland cement itself, they cannot be directly applied to cements because of complex interactions. In portland cement, the compounds do not exist in a pure form but are solid solutions containing Al, Mg, Na, etc. The rate of hydration of alites containing different amounts of Al, Mg, or Fe has shown that at the same degree of hydration Fe-alite shows the greatest strength. There is evidence the C-S-H formed in different alites is not the same in composition.[12] The hydration process of C_3A, C_4AF, and C_2S in cement is affected because of changes in the amounts of Ca^{2+} and OH^- in the hydrating solution. The reactivity of C_4AF can be influenced by the amount of SO_4^{2-} ions consumed by C_3A. Some SO_4^{2-} ions may be depleted by being absorbed by the C-S-H phase. Gypsum is also known to affect the rate of hydration of calcium silicates. Significant amounts of Al and Fe are incorporated into the C-S-H structure. The presence of alkalis in portland cement also has an influence on the hydration of the individual phases.

It is generally believed that the rate of hydration in the first few days of cement compounds in cements proceeds in the order of $C_3A > C_3S > C_4AF > C_2S$. The rate of hydration of the compounds depends on the crystal size, imperfections, particle size, particle size distribution, the rate of cooling, surface area, the presence of admixtures, the temperature, etc. After ninety days, little or no alite or aluminate phase is detectable.

Quantitative x-ray diffraction has been used to determine the degree of reaction of individual cement compounds present in cement. Some errors in these estimations are recognized. Figure 5 shows the fractional amounts of alite, belite, aluminate, and ferrite phases that hydrate in cement when hydrated for different times.[3] These rates are not the same when the individual compounds are hydrated.

In a mature hydrated portland cement, the products formed are C-S-H gel, $Ca(OH)_2$, ettringite (AFt phase), monosulfate (AFm phase), hydrogarnet phases, and possible amorphous phases high in Al^+ and SO_4 ions. A small amount of cryptocrystalline CH may be intimately mixed with C-S-H phase.

The C-S-H phase in cement paste is amorphous or semicrystalline calcium silicate hydrate, the hyphens denoting that the gel does not necessarily consist of 1:1 molar $CaO:SiO_2$. The C-S-H phase of cement pastes gives powder patterns very similar to that of C_3S pastes. The composition of C-S-H (in terms of C/S ratio) is variable depending on the time of hydration. At 1 day, the C/S ratio is about 2.0 and 1.4–1.6 after several years. The C-S-H can take up substantial amounts of Al^{3+}, Fe^{3+}, and SO_4^{2-} ions.

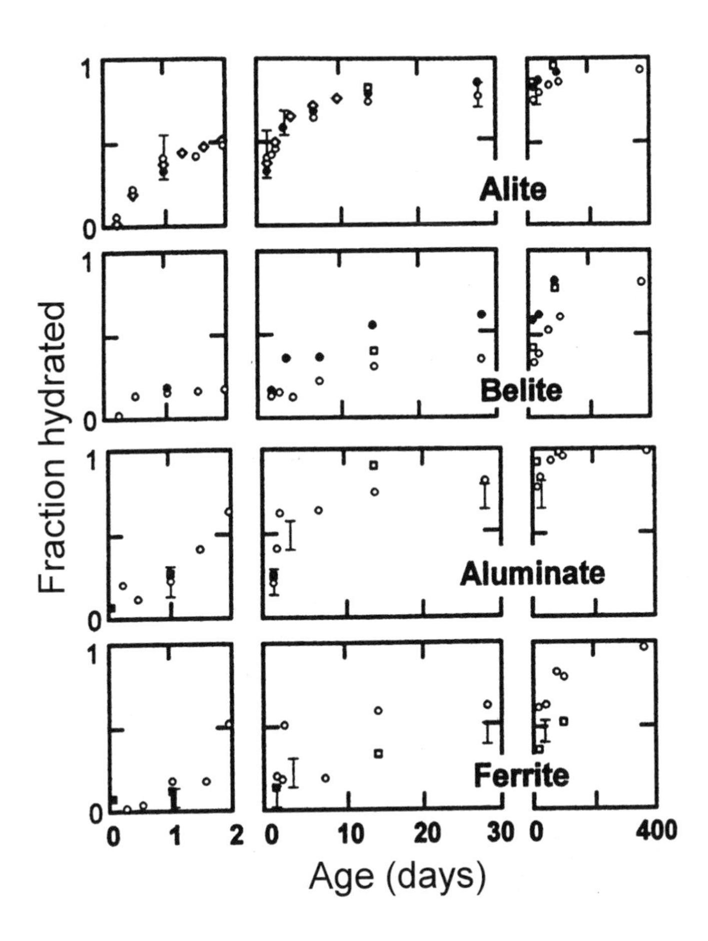

Figure 5. Fraction of cement phases hydrated in cement pastes at different times.

Recent investigations have shown that in both C_3S and portland cement pastes, the monomer that is present in the C_3S and C_2S compounds (SiO_4^{4-} tetrahedra) polymerizes to form dimers, and larger silicate ions as hydration progresses. The gas liquid chromatographic analysis of the trimethyl silylation derivatives has shown that anions with 3 or 4 Si atoms are absent. The polymer content with five or more Si atoms increases as the hydration proceeds and the amount of dimer decreases. In C_3S pastes the disappearance of monomer results in the formation of polymers. In cement pastes even after the disappearance of all C_3S and C_2S, some monomer is detected possibly because of the modification of the anion structure of C-S-H through replacement of some Si atoms by Al, Fe, or S. Admixtures can influence the rate at which the polymerization occurs in portland cement and C_3S pastes.

The minimum water:cement ratio for attaining complete hydration of cement has been variously given from 0.35 to 0.40, although complete hydration has been reported to have been achieved at a water:cement ratio of 0.22.[13] In a fully hydrated portland cement $Ca(OH)_2$ constitutes about 20–25% of the solid content.

The ettringite group, also called AFt phase in cement paste, stands for Al-Fe-tri (tri = three moles of CS) of the formula $C_3A{\bullet}3C\bar{S}{\bullet}H_{32}$ in which Al can be replaced by Fe to some extent. The AFt phase forms in the first few hours (from C_3A and C_4AF) and plays a role in setting. After a few days of hydration only a little amount of it may remain in cement pastes. It appears as stumpy rods in SEM and the length does not normally exceed a few micrometers. The principle substitutions that exist in the AFt phase are Fe^{3+} and Si^{4+} for Al^{3+} and various anions such as OH^-, CO_3^{2-}, and silicates for SO_4^{2-}.

The monosulfate group, also known as the AFm phase, is represented by the formula $C_4A\bar{S}H_{12}$ or $C_3A{\bullet}C\bar{S}{\bullet}H_{12}$. AFm stands for Al-Fe-mono, in which one mole of $C\bar{S}$ is present. In portland cement this phase forms after the AFt phase disappears. This phase may constitute about 10% of the solid phase in a mature cement paste. In SEM, this phase has a hexagonal morphology resembling that of $Ca(OH)_2$ and the crystals are of sub-micrometer thickness. The principle ionic substitutions in the AFm phase are Fe^{3+} for Al^{3+}, and OH^-, CO_3^{2-}, Cl^-, etc. for SO_4^{2-}. The density of this phase is 2.02 g/ml. The amount of crystalline hydrogarnet present in cement paste is less than 3%.[14] It is of the type $Ca_3Al_2(OH)_{12}$ in which part of Al^{3+} is replaced by Fe^{3+} and $4OH^-$ by SiO_4^{4-} [e.g., $C_3(A_{0.5}F_{0.5})SH_4$]. It may be present in small amounts in mature cement pastes and is also formed at higher temperatures. The crystal structure of this phase is related to $C_3A\bar{S}_3$ (garnet). Hydrogarnet is decomposed by CO_2 with the formation of $CaCO_3$ as a product.

It is the opinion of some workers that the lowest sulfate form of calcium sulfohydroxyaluminate hydrate, a crystalline solid solution phase in the system CaO-Al_2O_3-$CaSO_4$-H_2O, is also formed in cement pastes.

The mechanisms that have already been described for pure cement compounds form a basis for a study of the hydration mechanism of portland cement. The conduction calorimetric curves of C_3S and portland cement are similar except portland cement may yield a third peak for the formation of monosulfate hydrate (Fig. 1). The detailed influence of C_3A and C_4AF on the hydration of C_3S and C_2S in cement is yet to be worked out. The delayed nucleation models and the protective layer models, taking into account the possible interactions, have been reviewed.[8] Although the initial process is not clear for C_3S (in cements), it appears that C_3A hydration products form through solution and topochemical processes.

6.0 PROPERTIES OF CEMENT PASTE

6.1 Setting

A mixture of cement and water, mixed in certain proportions, causes setting and hardening, and it is called a *cement paste.*

The stiffening times of cement paste or mortar fraction are determined by setting times. The setting characteristics are assessed by *initial set* and *final set.* When the concrete attains the stage of initial set it can no longer be properly handled and placed. The final set corresponds to the stage at which hardening begins. At the time of the initial set, the concrete will have exhibited a measurable loss of slump. Admixtures may influence the setting times. The retarders increase the setting times and accelerators decrease them.

At the time of the initial set of cement paste, the hydration of C_3S will have just started. According to some investigators, the crystallization of ettringite is the major contributing factor to the initial set. The final set generally occurs before the paste shows the maximum rate of heat development, i.e., before the end of the third stage in conduction calorimetry.

False or flash set is known to occur in concrete. When stiffening occurs due to the presence of partially dehydrated gypsum, false set is noticed. Workability is restored by remixing. False set may also be caused by excessive formation of ettringite especially in the presence of some retarders and an admixture such as triethanolamine. The formation of syngenite ($KC\bar{S}_2H$) is reported to cause false set in some instances. The setting times of cement can be determined by Gillmore (ASTM C 266) or the Vicat apparatus (ASTM 191).

6.2 Microstructure

Many of the properties of the cement paste are determined by its chemical nature and microstructure. Microstructure constitutes the nature of the solid body and that of the non-solid portion, viz., porous structure. Microstructural features depend on many factors, such as the physical and chemical nature of the cement, type and the amount of admixture added to it, temperature and period of hydration, and the initial w/c ratio. The solid phase study includes examination of the morphology (shape and size), bonding of the surfaces, surface area, and density. Porosity, pore shape, and pore size distribution analysis is necessary for investigating the non-solid phase. Many of the properties are interdependent, and no one property can adequately explain the physico-mechanical characteristics of cement paste.

A study of the morphology of the cement paste involves observation of the form and size of the individual particles, particularly through high resolution electron-microscopes. The most powerful techniques that have been used for this purpose are: Transmission Electron Microscopy (TEM), Scanning Electron Microscopy (SEM), the High Voltage Transmission Electron Microscope using environmental cells, the Scanning Transmission Electron Microscope (STEM) of ion-beam thinned sections, and the High Resolution SEM using STEM instruments in reflection mode.

The morphology of C-S-H gel particles has been divided into four types and described by Diamond.[15]

- Type I C-S-H, forming elongated or fibrous particles, occurs at early ages. It is also described as spines, acicular, aciculae, prismatic, rod-shaped, rolled sheet, or by other descriptions. These are a few micrometers long.
- Type II C-S-H is described as a reticular or honey-combed structure and forms in conjunction with Type I. It does not normally occur in a C_3S or C_2S paste, unless it is formed in the presence of admixtures.
- In hardened cement pastes, the microstructure can be nondescript and consist of equant or flattened particles (under 1000 Å in largest dimension); such a morphological feature is described as Type III.
- Type IV, a late hydration product, is compact and has a dimpled appearance, and is believed to form in spaces originally occupied by cement grains. This feature is also found in C_3S pastes.

The above list is not exclusive because other forms have also been reported.

6.3 Bond Formation

Cementitious materials, such as gypsum, portland cement, magnesium oxychloride, and alumina cement form porous bodies, and explanations of their mechanical properties should take into account the nature of the void spaces and the solid portion. If the solid part determines strength, then several factors should be considered including the rate of dissolution and solubility of the cement, the role of nuclei and their growth, chemical and physical nature of the products, and energetics of the surface and interfacial bonds.

The C-S-H phase is the main binding agent in portland cement pastes. The exact structure of C-S-H is not easily determined. Considering the several possibilities by which the atoms and ions are bonded to each other in this phase, a model may be constructed. Figure 6 shows a number of possible ways in which siloxane groups, water molecules, and calcium ions may contribute to bonds across surfaces or in the interlayer position of poorly crystallized C-S-H material.[16] In this structure, vacant corners of silica tetrahedra will be associated with cations, such as Ca^{++}.

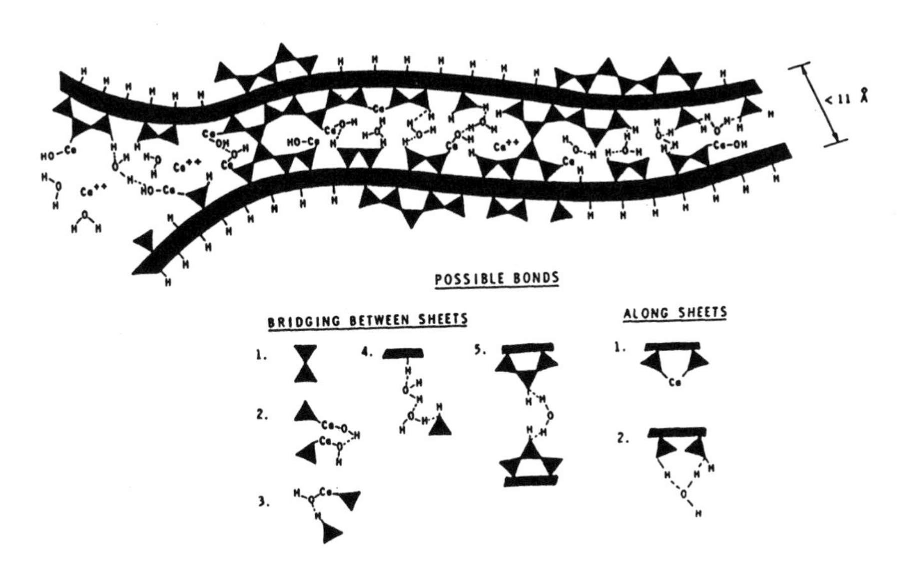

Figure 6. Suggested C-S-H structure.

6.4 Density

The density value quoted in the literature for a given material is accepted without question because it depends simply on mass and volume at a given temperature. An accurate assessment of density, however, is one of the most important factors in determining porosity, assessing durability and strength, and estimating lattice constants for the C-S-H phase in hydrated portland cement.

Traditionally, density of hydrated portland cement was measured in the d-dried state by pycnometric methods, using a saturated solution of calcium hydroxide as a fluid. Since the d-dried hydrated portland cement rehydrates on exposure to water, this method is of questionable value. More realistic values can be obtained by proper conditioning of the sample and using fluids that do not affect the structure of the paste.

Drying to 11% RH and measuring with a saturated solution of $Ca(OH)_2$ gives an uncorrected value of 2.38 g/cc, as compared to a corrected value of 2.35 g/cc and 2.34 g/cc using helium. At the d-dried state, a higher value obtained by the $Ca(OH)_2$ solution technique is due to the penetration of water into the interlayer positions of the layered structure of the crystallite.

6.5 Pore Structure

Pore structure influences the strength development and durability of concrete. Pore structure is modified in the presence of admixtures. Porosity and pore size distribution are usually determined using mercury porosimetry and nitrogen or water adsorption isotherms. Total porosity is obtained by using organic fluids or water as a medium. But water cannot be used as it may interact with the body. Mercury porosimetry involves forcing mercury into the vacated pores of a body by the application of pressure. The technique measures a range of pore diameters down to about 3 nm. Beaudoin[17] measured total porosity by Hg porosimetry using pressures up to 408 MPa and concluded that the porosimetry and He pycnometry methods could be used interchangeably to determine porosity of cement paste formed at a w/c ratio equal to or greater than 0.40.

6.6 Surface Area and Hydraulic Radius

Surface area is the area available to gases or liquids by way of pores and the external area. Hydrated portland cement is very complex and there is controversy over the significance of H_2O as an adsorbate in determining surface area. With water as an adsorbate, the surface area is estimated at 200 m^2/g and remains constant for different w/c ratio pastes. The surface area varies with w/c ratio when using nitrogen, methanol, isopropanol, and cyclohexane as adsorbates.[18] With nitrogen, the values vary from 3 to 147 m^2/g.

The average characteristic of a pore structure can be represented by the *hydraulic radius* which is obtained by dividing the total pore volume by the total surface area. The pore volume of d-dried paste determined by nitrogen, helium, or methanol, is due to capillary porosity, and the hydraulic radius is known to vary from 30 to 107 Å for w/c ratios from 0.4 to 0.8.

6.7 Mechanical Properties

Hydrated portland cement contains several types of solid phases, and the theoretical treatment of such a material is complex.

Many observations have led to the conclusion that the strength development of hydrated portland cement depends on the total porosity, P. Most data can be fitted to an exponential dependence term, e^{-bP}, with b values associated with different types of pores. Porosity and grain size effects on strength become clearly separable as pores approach or become smaller than the grain size. Uniform distributions of different types of pores will have similar exponential strength-porosity trends, but the b values will change. They will depend on the pore location, size, and shape. The latter two are important only when the pore causing failure is large in comparison with the grain size or with the specimen size. For small pores, its location is important. Pores at grain boundaries are more critical than pores within grains.

Correlation of porosity with mechanical property values has led to several types of semi-empirical equations, the most common being that due to Ryshkewitch:[19]

$$M = M_0 \exp(-bP)$$

where M is the mechanical strength property at porosity P, M_0 is the value at zero porosity, and b is a constant. As stated previously, b is related to the pore shape and orientation. This equation shows good agreement with experimental values at lower porosities. Another equation, due to Schiller, is as follows:[20]

$$M = D \ln \frac{P_{CR}}{P}$$

where D is a constant and P_{CR} is the porosity at zero strength. It shows good agreement at high porosities.

Feldman and Beaudoin[16] correlated strength and modulus of elasticity for several systems over a wide range of porosities. The systems included pastes hydrated at room temperature, autoclaved cement paste with and without additions of fly ash, and those obtained by other workers. Porosity was obtained by measurement of solid volume by a helium pycnometric technique and apparent volume, through the application of Archimedes' principle.

Several attempts have been made to relate the strength of cement paste to the clinker composition. A series of equations was proposed by Blaine, et al., in 1968[21] to predict strength against a number of clinker compositions, ignition loss, insoluble fraction, and air and alkali contents. Other investigators have also proposed equations expressing the relationship between the clinker composition and the 28 day strength.[22]

The data on the effect of clinker composition on strength are rather conflicting, although it is recognized the multiple regression equations reflect reasonably well the relationship for narrow ranges of cement composition. It is recognized that other effects such as the texture, presence of minor components, particle size distribution, and amount of gypsum will have a significant influence on the potential strength of cement.

7.0 PERMEABILITY OF CEMENT PASTE

The rate of movement of water through concrete under a pressure gradient, termed *permeability,* has an important bearing upon the durability of concrete. The measure of the rate of fluid flow is sometimes regarded as a direct measure of durability.

It is known that the permeability of hardened cement paste is mainly dependent on the pore volume. However, pore volume, at different water/cement ratios and degrees of hydration, does not uniquely define the pore system and, thus, is not uniquely related to the permeability.

8.0 DIMENSIONAL CHANGES

Aging, within the context of surface chemical considerations, refers to a decrease in surface area with time. For hydrated portland cement, this definition can be extended to include changes in solid volume, apparent volume, porosity, and some chemical changes (excluding hydration) which occur over extended periods of time.

The volume of cement paste varies with its water content, shrinking when dried and swelling when re-wetted. It has been found that the first drying shrinkage (starting from 100% RH) for a paste is unique in that a large portion of it is irreversible. By drying to intermediate relative humidities (47% RH), it has been observed that the irreversible component is strongly dependent on the porosity of the paste, being less at lower porosities and w/c ratios.[23]

Concrete exhibits the phenomenon of creep, involving deformation at a constant stress that increases with time. Creep of concrete (basic creep) may be measured in compression using the ASTM C512 method. There are two types of creep: *basic creep* in which the specimen is under constant humidity conditions and *drying creep* when the specimen is dried during the period under load.

Creep of a cement paste increases at a gradually decreasing rate, approaching a value several times larger than the elastic deformation. Creep is, in part, irrecoverable, as is drying shrinkage. On unloading, deformation decreases immediately due to elastic recovery. This instantaneous recovery is followed by a more gradual decrease in deformation due to creep recovery. The remaining residual deformation, under equilibrium conditions, is called the *irreversible creep.* Creep increases with w/c ratio and is very sensitive to relative humidity and water content. It may also be affected by admixtures.

Many theories have been proposed over the years to account for creep mechanisms in cement paste, and each is capable of accounting for some of the observed facts. The descriptions and mechanisms are based on seepage, change of solid structure, and interlayer space.

9.0 MODELS OF HYDRATED CEMENT

In order to predict the performance of concrete, it is important to have a model of cement paste that incorporates its important properties and explains its behavior. There are two main models. In the Powers-Brunauer model, the cement paste is considered a poorly crystalline gel and layered. The gel has a specific surface area of 180 m^2/g with a minimum porosity of 28%. The gel pores are assumed to be accessible only to water molecules because the entrance to these pores is less than 0.4 nm in diameter. Any space not filled with cement gel is called *capillary* space. The mechanical properties of the gel are described using this model. The particles are held together mainly by van der Waal's forces.[24] Swelling on exposure to water is explained by the individual particles separating, due to layers of water molecules existing between them. Creep is the result of water being squeezed out from between the particles during the application of stress This model recognizes the existence of some chemical bonds between the particles and the existence of layers.

In the Feldman-Sereda model,[16][24] the gel is considered as a poorly crystallized layered silicate and the role of water is much more complex (Fig. 7) than is recognized by the Powers-Brunauer model. Water does not re-enter the interlayer after d-drying. Water, in contact with the d-dried gel, acts in several ways:

a) It interacts with the free surface, forming hydrogen bonds.

b) It is also physically adsorbed on the surface.

c) It also enters the collapsed layered structure of the material even at humidities below 10% RH.

d) It fills large pores by capillary condensation at higher humidities.

According to this model creep is a manifestation of aging, i.e., the material moves towards a lower total energy by aggregation of sheets due to the formation of more layers. Surface area is reduced by this process. Aggregation is accelerated by stress and facilitated by the presence of interlayer water.

10.0 MATHEMATICAL MODELS

Recently, there has been a significant interest in the development of computer-based models for the microstructure, hydration, and structural development in cement-based materials. Factors such as composition and shape of cement particles, w/c ratio, and curing conditions have been considered for obtaining mass and volume fraction of phases in cement paste, non-evaporable water, heat evolution, strength, porosity, permeability, etc. Garboczi and Bentz[25] describe the computer-based model of microstructure and properties as *"...a theoretical construct, which is made using valid scientific principles expressed in mathematical language, that can be used to make quantitative prediction about a material's structure and/or properties."* The computer-based model is, thus, used to numerically represent the amount and spatial distribution of different phases of the material being studied and, thus, predict from the numerical representation of microstructure, properties that can be derived from actual experiments. Simulation of the interfacial zone model has also been carried out. Details of the application of the models have been reviewed recently.[26][27] These models also have to consider that the properties of concrete depend on the fine structure of C-S-H as well as that of coarse aggregate. It is also important to determine the microstructural characteristics of the material as it deforms due to rheology, creep, shrinkage, and fracture.

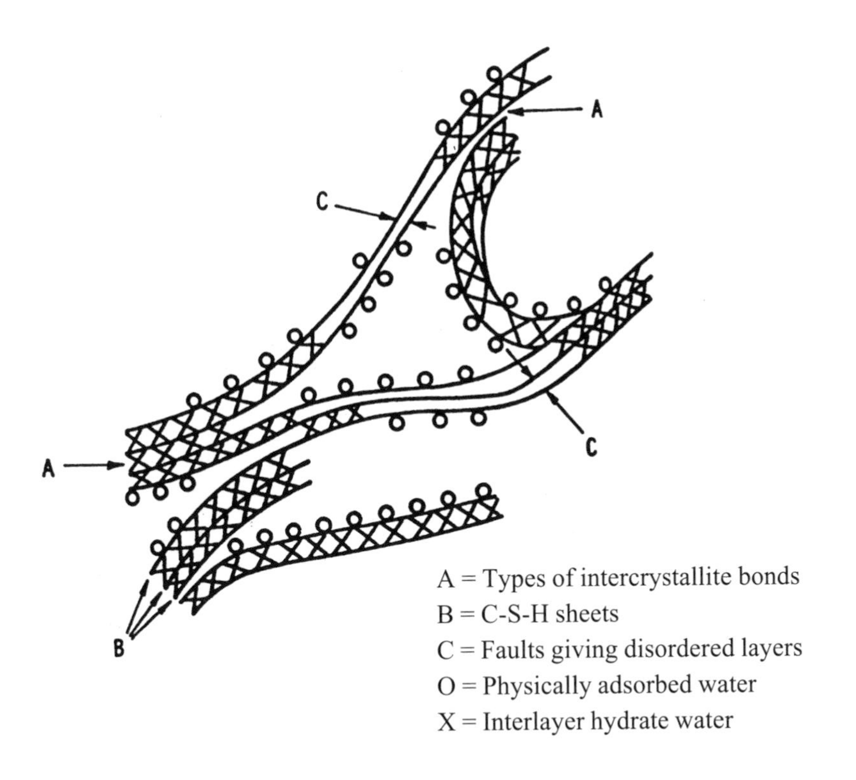

Figure 7. Structure of C-S-H gel according to the Feldman-Sereda model.

11.0 CONCRETE PROPERTIES

The role of pore structure and cement paste has already been described. Aggregates, occupying 60–80% of the volume of concrete influence its unit weight, elastic modulus, dimensional stability, and durability. Generally, aggregates are stronger than the matrix. Coarse aggregates are larger than 4.75 mm and fine aggregates are smaller than 4.75 mm. Typically, fine aggregates comprise particles of size in the range of 75 mm to 4.75 mm, whereas coarse aggregates are of sizes from 4.75 mm to 50 mm. Mass concrete, the size of coarse aggregates reaching 150 mm, is generally composed of sand, gravel, and crushed rock. The synthetic aggregates, i.e., thermally processed materials such as expanded clay/shale, slag, and fly ash, are also used. Natural silica is used extensively as a fine aggregate. ASTM-294 provides a descriptive nomenclature of the commonly occurring minerals in rocks.

Some minor constituents of fine or coarse aggregates such as clay lumps, friable particles, coal, lignite, chert, etc. may adversely affect the workability, setting, handling, and durability characteristics of concrete. A list of harmful substances and permissible limits is given in ASTM C-33.

11.1 Workability

The quality of fresh concrete is determined by the ease and homogeneity with which it can be mixed, transported, compacted, and finished. It has also been defined as the amount of internal work necessary to produce full compaction.[28] The rheological behavior of concrete is related to the rheological terms such as plasticity and visco-elasticity of cement paste. As the workability depends on the conditions of placement, the intended use will determine whether the concrete has the required workability. A good workable concrete should not exhibit excessive bleeding or segregation. Thus, workability includes properties such as flowability, moldability, cohesiveness, and compactibility. One of the main factors affecting workability is the water content in the concrete mixture. Workability may also be improved by the addition of plasticizers and air-entraining admixtures. The factors that affect workability include quantities of paste and aggregates, plasticity of the cement paste, maximum size and grading of the aggregates, and shape and surface characteristics of the aggregate.

11.2 Setting

The setting of concrete is determined by using the mortar contained in it. A penetrometer is used for determining the initial and final setting times of mortar. A needle of appropriate size has to be used. The force required to penetrate one inch in depth is noted. The force divided by the area of the bearing surface of the needle yields the penetration resistance. The initial setting time is the elapsed time after the initial contact of cement and water required for the mortar sieved from the concrete to reach a penetration resistance of 500 psi (3.5 MPa). The corresponding resistance for the final setting time is 4000 psi (27.6 MPa).

11.3 Bleeding and Segregation

In a freshly placed concrete which is still plastic, settlement of solids is followed by the formation of a layer of water on the surface. This is known as bleeding. In lean mixtures, localized channels develop and the seepage of water transports some particles to the surface. Bleeding may, thus, give rise to *laitance,* a layer of weak, nondurable material containing diluted cement paste and fines from the aggregate. If bleeding occurs by uniform seepage of water, no undesirable effects result and such a bleeding is known as *normal bleeding.* Bleeding is not necessarily harmful. If undisturbed, the water evaporates so that the effective water:cement ratio is lowered with a resultant increase in strength.

The amount of bleeding can be reduced by using proper amounts of fines, increasing cement content and admixtures such as pozzolans, or air-entraining admixtures. Bleeding characteristics are measured by bleeding rate or bleeding capacity applying the ASTM C232 standard.

During the handling of concrete mix, there may be some separation of coarse aggregates from the mixture resulting in a non-uniform concrete mass. This is known as *segregation.* Segregation may lead to flaws in the final product, and honeycombing may occur in some instances. By proper grading of the constituents and handling, this problem can be controlled.

11.4 Mechanical Properties

The hardened concrete has to conform to certain requirements for mechanical properties. They include compressive strength, splitting-tensile strength, flexural strength, static modulus of elasticity, Poisson's ratio, mechanical properties under triaxial loads, creep under compression, abrasion resistance, bond development with steel, penetration resistance, pull out strength, etc.

The mechanical behavior of concrete should be viewed from the point of view of a composite material. A composite material is a three dimensional combination of at least two chemically and mechanically distinct materials with a definite interface separating the components. This multiphase material will have different properties from the original components. Concrete qualifies as such a multiphase material. Concrete is composed of hydrated cement paste (C-S-H, CH, aluminate, and ferrite-based compounds) and unhydrated cement, containing a network of a mixture of different materials. In dealing with cement paste behavior, basically it is considered that the paste consists of C-S-H and CH with a capillary system. The model of concrete is simplified by treating it as a matrix containing aggregate embedded in a matrix of cement paste. This model provides information on the mechanical properties of concrete.

The factors that influence the mechanical behavior of concrete are: shape of particles, size and distribution of particles, concentration, their orientation, topology, composition of the disperse and continuous phases, and that between the continuous and disperse phase and the pore structure. The important role played by cement paste is already described.

12.0 DURABILITY OF CONCRETE

One of the most important requirements of concrete is that it should be durable under certain conditions of exposure. Deterioration can occur in various forms such as alkali-aggregate expansion reaction, freeze-thaw expansion, salt scaling by de-icing salts, shrinkage and enhanced attack on the reinforcement of steel due to carbonation, sulfate attack on exposure to ground waters containing sulfate ions, sea water attack, and corrosion caused by salts. The addition of admixtures may control these deleterious effects. Air entrainment results in increased protection against freeze-thaw action; corrosion inhibiting admixtures increase the resistance to corrosion; inclusion of silica fume in concrete decreases the permeability; consequently, the rate of ingress of salts and the addition of slags in concrete increases the resistance to sulfate attack.

13.0 ALKALI-AGGREGATE EXPANSION

Although all aggregates can be considered reactive, only those that actually cause damage to concrete are cause for concern. Experience has shown that the presence of an excessive amount of alkalis enhances the attack on concrete by an expansion reaction. Use of marginal quality aggregate and the production of high strength concrete may also produce this effect.

The alkali-aggregate reaction in concrete may manifest itself as map cracking on the exposed surface, although other reactions may also produce such failures. The alkali-aggregate reaction, known as alkali-silica type, may promote exudation of a water gel, which dries to a white deposit. These effects may appear after only a few months or even years.

Three types of alkali-aggregate reactions are mentioned in the literature:

- *Alkali-silica reaction.* Alkali-silica reactions are caused by the presence of opal, vitreous volcanic rocks, and those containing more than 90% silica.
- *Alkali-carbonate reaction.* The alkali-carbonate reaction is different from the alkali-silica reaction in forming different products. Expansive dolomite contains more calcium carbonate than the ideal 50 % (mol) proportion and frequently also contains illite and chlorite clay minerals.
- *Alkali-silicate reaction.* The alkali-silicate reaction has not received general recognition as a separate entity. The alkali-silicate reaction was proposed by Gillott.[29] The rocks that produced this reaction were greywackes, argillites, and phyllites containing vermiculites.

The preventive methods to counteract alkali-aggregate expansion include replacement of cement with pozzolans or blast-furnace slag and use of low alkali cement.

14.0 FROST ACTION

Frost action is defined as the freezing and thawing of the moisture in materials and the resultant effects on these materials. Essentially, three kinds of defects are recognized:

- *Spalling* occurs as a definite depression caused by the separation of surface concrete.
- *Scaling* occurs to a depth of 25 mm from the surface, resulting in local peeling or flaking.
- *Cracking* occurs as d- or map-cracking and is sometimes related to the aggregate performance.

Good resistance to frost expansion can be obtained by proper mix proportioning and choice of materials. In addition to w/c ratio, quality of aggregate and proper air entrainment, the frost resistance depends on the exposure conditions. Dry concrete will withstand freezing-thawing, whereas highly saturated concrete may be severely damaged by a few cycles of freezing and thawing.

According to many workers, frost damage is not, necessarily, connected with the expansion during freezing of water, although it can contribute to damage. When a water-saturated porous material freezes, macroscopic ice crystals form in the coarser pores and water, which is unfrozen in the finer pores, and migrates to the coarser pores or the surfaces.[30] The large ice crystals can feed on the small ice crystals, even when the larger ones are under constraint and cause expansion.

Deterioration of plain concrete due to de-icing salts may generally be termed *salt scaling;* it is similar in appearance to frost action but more severe. Any theory on salt scaling should account for this increased damage. The most widely used test for assessing the resistance of concrete to freezing and thawing is the ASTM test on *Resistance of Concrete to Rapid Freezing and Thawing* (ASTM C666). The general approach to preventing frost attack in concrete is to use an air-entraining admixture. Tiny bubbles of air are entrapped in concrete due to the foaming action developed by the admixture during mixing.

15.0 SEA WATER ATTACK

The deterioration of concrete due to sea water attack is the result of several simultaneous reactions. However, sea water is less severe on concrete than can be predicted from the possible reactions associated with the salts contained in it. Sea water contains 3.5% salts by weight. They include NaCl, $MgCl_2$, $MgSO_4$, $CaSO_4$, and possibly $KHCO_3$.

The deterioration of concrete depends on the exposure conditions. Concrete not immersed but exposed to marine atmosphere will be subjected to corrosion of reinforcement and frost action. Concrete in the tidal zone, however, will be exposed to the additional problems of chemical decomposition of hydrated products, mechanical erosion, and wetting and drying. Parts of the structure permanently immersed are less vulnerable to frost action and corrosion of the reinforcing steel.

16.0 CORROSION OF REINFORCEMENT

Corrosion of steel in concrete is probably the most serious durability problem of reinforced concrete in modern times, and therefore, a clear understanding of the phenomenon is of crucial importance. The phenomenon itself is an electro-chemical reaction. In its simplest form, corrosion may be described as a current flow from anodic to cathodic sites in the presence of oxygen and water. This is represented by the following equations:

At anode: $Fe \rightarrow Fe^{2+} + 2e^-$

At cathode: $\frac{1}{2}O_2 + H_2O + 2e^- \rightarrow 2(OH)^-$

These reactions would result in the formation of oxide at cathodic sites. The high alkalinity of cement paste, however, provides protection for the steel reinforcement. Corrosion is increased in the presence of chloride salts. It contributes, together with CO_2 ingress, to the depression of the pH of the pore fluid and increases the electrical conductance of the concrete, allowing the corrosion current to increase.

Several methods of corrosion prevention have been tried over the years. These include protective coatings, placement of impermeable concrete overlays, cathodic protection or the use of corrosion resistant steels, and galvanized or epoxy coated bars. Recent work has shown that galvanized steel may be of benefit if used in low chloride bearing concrete.

17.0 CARBONATION OF CONCRETE

The corrosion of depassivated steel in reinforced concrete has focused attention on the reactions of acidic gases such as carbon dioxide with hydrated cement and concrete. As a result of the reaction of carbon dioxide, the alkalinity of concrete can be progressively reduced, resulting in a pH value below 10.

The process of carbonation of concrete may be considered to take place in stages. Initially, CO_2 diffusion into the pores takes place, followed by dissolution in the pore solution. Reaction with the very soluble alkali metal hydroxide probably takes place first reducing the pH and allowing more $Ca(OH)_2$ into the solution. The reaction of $Ca(OH)_2$ with CO_2 takes place by first forming $Ca(HCO_3)_2$ and finally $CaCO_3$. The product precipitates on the walls and in crevices of the pores. This reduction in pH also leads to the eventual breakdown of the other hydration products, such as the aluminates, C-S-H gel, and sulfoaluminates. Generally, it is found that low w/c ratio, good compaction, and proper curing cause significant improvements in concrete permeability and resistance to carbonation.

Several workers have concluded that carbonation depth is proportional to the square root of time. The proportionality constant is a coefficient related to the permeability of the concrete. Factors, such as w/c ratio, cement content in concrete, CO_2 concentration in the atmosphere, and the relative humidity, in addition to normal factors such as concrete density, affect the value of this coefficient.

18.0 DELAYED/SECONDARY ETTRINGITE FORMATION

The potential for concrete deterioration as a consequence of the delayed ettringite formation in the precast industry has recently been recognized. One of the important factors required for this type of reaction is high temperature curing of concrete such as that occurring in the precast industry. The delayed formation of ettringite is attributed to the transformation of monosulfoaluminate to ettringite when steam curing is followed by normal curing at later ages. In recent work, it was indicated that sulfate may be bound by the C-S-H gel that is released at later ages. Increased temperature is expected to accelerate the absorption of sulfate by the silicate hydrate. It has also been confirmed that the ettringite crystals are usually present in cracks, voids, and the transition zone at the aggregate-binder interface, causing expansion and cracking. Further, ASTM Type III cement is more vulnerable to deterioration due to the delayed ettringite formation than ASTM Type I or Type V cement. Thermal drying after high temperature curing intensifies the deterioration. In the secondary ettringite formation, calcium sulfate formed from the decomposition of AFt or AFm phase, as a consequence of severe drying, may dissolve upon rewetting and migrate into cracks to react with the local Al-bearing materials to cause expansion.

REFERENCES

1. Bresler, B., *Reinforced Concrete Engineering,* Wiley-Interscience, New York (1974)
2. Taylor, H. F. W., Portland Cement: Hydration Products, *J. Edn. Mod. Mater. Sci. Eng.,* 3:429–449 (1981)
3. Taylor, H. F. W., *Cement Chemistry,* 2nd Ed., Telford, London (1997)
4. Feldman, R. F., and Ramachandran, V. S., Differentiation of Interlayer and Absorbed Water in Hydrated Portland Cement of Thermal Analysis, *Cem. Concr. Res.,* 1:607–620 (1971)
5. Taylor, H. F. W., et al., The Hydration of Tricalcium Silicate, *Mater. Const.,* 17:457–468 (1987)
6. Stein, H. N., and Stevels, J., Influence of Silica on Hydration of $3CaO{\bullet}SiO_2$, *J. Appl. Chem.,* 14:338–346 (1964)
7. Tadros, M. E., Skalny, J., and Kalyoncu, R., Early Hydration of C_3S, *J. Am. Ceram. Soc.,* 59:344–347 (1976)
8. Pratt, P. L., and Jennings, H. M., The Microchemistry and Microstructure of Portland Cement, *Ann. Rev. Mater. Sci.,* 11:123–149 (1981)
9. Gartner, E. M., and Gaidis, W. R., *Hydration Mechanisms in Materials Science of Concrete: I* (J. Skalny, ed.), pp. 95–125, American Ceramics Society (1989)
10. Bogue, R. H., and Lerch, W., Hydration of Portland Cement Compounds, *Ind. Eng. Chem.,* 26:837–847 (1934)
11. Beaudoin, J. J., and Ramachandran, V. S., A New Perspective on the Hydration Characteristics of Cement Pastes, *Cem. Concr. Res.,* 22:689–694 (1992)
12. Mascolo, G., and Ramachandran, V. S., Hydration and Strength Characteristics of Synthetic Al-, Mg-, and Fe-Alites, *Mater. Constr.,* 8:373–376 (1975)

13. Beaudoin, J. J., and Ramachandran, V. S., Physico-Chemical Characteristics of Low Porosity Cement Systems, Ch. 8, *Materials Science of Concrete,* (J. Skalny, ed.), 3(8):362, American Ceramic Society (1992)
14. Raffle, J. F., The Physics and Chemistry of Cements and Concretes, *Sci. Prog.,* Oxf., 64:593–616 (1977)
15. Diamond, S., *Cement Paste Microstructure - An Overview at Several Levels in Hydraulic Cement Pastes-Their Structure and Properties,* Conference, University of Sheffield, p. 334 (1976)
16. Ramachandran, V. S., Feldman, R. F., and Beaudoin, J. J., *Concrete Science, A Treatise on Current Research,* p. 334, Heyden & Son Ltd., UK (1981)
17. Beaudoin, J. J., Porosity Measurements of Some Hydrated Cementituous Systems by High Pressure Mercury Intrusion - Microstructural Limitations, *Cem. Concr. Res.,* 9:771–781 (1979)
18. Mikhail, R. S., and Selun, S. A., Adsorption of Organic Vapors in Relation to the Pore Structure of Hardened Portland Cement Pastes, *Symposium on Structure of Portland Cement Paste and Concrete,* Special Report 90, HRB:123–134 (1966)
19. Ryshkewitch, E., Compression Strength of Porous Sintered Alumina and Zirconia, *J. Amer. Ceram. Soc.,* 36:65–68 (1953)
20. Schiller, K. K., Strength of Porous Materials, *Cem. Concr. Res.,* 1:419–422 (1971)
21. Blaine, R. L., Arni, H. T., and Defore, M. R., *Interaction Between Cement and Concrete Properties, Building Science Series 8,* Natl. Bur. Stand., Washington, DC (1968)
22. Odler, I., Strength of Cement, *Materials & Structures,* 24:143–157 (1991)
23. Verbeck, G., and Helmuth, R. A., Structures and Physical Properties of Cement Pastes, *Proc. 5th. Int. Symp. Chem. of Cement, Tokyo,* III:1–31 (1968)
24. Feldman, R. F., and Sereda, P. J., The New Model for Hydrated Portland Cement and Its Practical Implications, *Eng. J.,* 53:53–57 (1970)
25. Garboczi, E. J., and Bentz, D. P., Fundamental Computer-Based Models of Cement-Based Materias, *Materials Science of Concrete,* (J. Skalny, and S. Mindess, eds.), American Ceramics Society, Westerville, OH (1991)
26. Garboczi, E. J., and Bentz, D. P., Computer-Based Models of the Microstructure and Properties of Cement-Based Materials, *Proc. 9th Int. Congress on Cement Chemistry, New Delhi,* VI:3–15 (1992)
27. Coverdale, R. T., and Jennings, H. M., Computer Modelling of Microstruture of Cement Based Materials, *Proc. 9th Int. Congress on Chemistry of Cement, New Delhi, India*, pp. 16–21 (1992)

28. Neville, A. M., *Properties of Concrete,* Pitman Publishing Co., London (1995)
29. Gillott, J. E., Practical Implications of the Mechanisms of Alkali-Aggregate Reactions, *Symp. Alkali-Aggregate Reaction, Reykjavik,* pp. 213–230 (1975)
30. Everett, D. H., The Thermodynamics of Frost Damage to Porous Solids, *Trans. Faraday Soc.,* 57:1541–1551 (1961)
31. Ramachandran, V. S., *Concrete Admixtures Handbook,* p. 1153, Noyes Publications, NJ (1995)

3

Formation and Hydration of Cement and Cement Compounds

1.0 INTRODUCTION

An understanding of the complex physico-chemical phenomena associated with the formation and behavior of cementitious compounds is facilitated through the application of many different types of investigative methods. Techniques such as NMR, XRD, neutron activation analysis, atomic absorption spectroscopy, IR/UV spectroscopy, electron microscopy, surface area techniques, pore characterization, zeta potential, viscometry, thermal analysis, etc., have been used with some success. Of the thermal analysis techniques the Differential Thermal Analysis (DTA), Thermogravimetric Analysis (TG), Differential Scanning Calorimetry (DSC), and Conduction Calorimetric methods are more popularly used than others. They are more adaptable, easier to use, and yield important results in a short span of time. In this chapter the application of these techniques will be highlighted and some of the work reported utilizing other related methods will also be mentioned with typical examples.

A substance subjected to thermal treatment may undergo physical or chemical changes as in dimension, magnetic susceptibility, weight, crystalline transition, mechanical property, acoustic property, and heat effects, etc. In thermal analysis these changes are followed as a function of temperature. It has been suggested that thermal analysis should also be extended to allow for rapid heating of the sample to some temperature followed by a measurement of the property with time under an isothermal condition. In the quasi-static thermal analysis method, a substance is heated at known intervals of temperature for a few hours and a particular property is measured. In the dynamic method, such as DTA and TG the property of a material is followed by continuous heating at a uniform rate.

The detailed description of DTA, TG, DSC, and Conduction Calorimetric Techniques has been presented in Ch. 1. A very brief account of these techniques is given below.

In DTA the difference in temperature ΔT between the sample and a reference material such as α-Al_2O_3 is recorded while both are subjected to the same heating program. Generally, the temperature (x-axis) is plotted against the ΔT on the y-axis. The exothermal effects are shown upward and the endothermal effects downward with respect to the base line. In the DTA of cementitious materials, thermal effects are reported in terms of the characteristic temperature, peak temperature, temperature range of the peak, peak width, peak amplitude or height, and peak area. By determining the nature of the peak (endothermic or exothermic), the temperature of the characteristic peak, and other general characteristics, it is possible to utilize DTA for both qualitative and quantitative purposes. By heating the binary or ternary mixtures in the DTA apparatus, the sequence of reactions during heating may also be followed. Thermal methods are used for identifying new compounds, estimating the products, formulating the mechanism of reactions, synthesis of compounds, durability assessment, trouble shooting, and quality control, etc. These methods are also used in conjunction with other physical, chemical, and mechanical techniques. Many factors such as the type and size of sample holder, furnace, thermocouple, rate of heating, sensitivity of the recording system, degree of dryness of the sample, the amount of sample, particle size and crystallinity, packing density, thermal conductivity, and shrinkage or swelling of the sample, will affect the results. The usefulness of DTA is further enhanced with the development of multipurpose types of equipment which incorporate one or more types of adjunct techniques to DTA. Examples are: DTA-Effluent Gas Analyzer, DTA-Mass Spectrometer, DTA-DTG-TG, DTA-TG-Radioactive Emission, DTA-TG-Dilatometer, DTA-XRD (X-Ray Diffraction), etc.

DSC (Differential Scanning Calorimetry) has also been used in cement science investigations to some extent. It is based on a power compensated system. In this technique the reference and the sample under investigation are maintained at a constant temperature throughout the heating schedule. The heat energy required to maintain the isothermal condition is recorded as a function of time or temperature. There are some similarities between DTA and DSC including the appearance of thermal curves. DSC can be used to measure the heat capacities of materials. DSC measures directly the heat effects involved in a reaction.

In thermogravimetric analysis (TG), the weight changes are determined as the sample is heated at a uniform rate. It differs from the semi-static or static method in which the sample is held at a constant temperature for a required period of time. In concrete investigations, TG is commonly used with DTA to follow the hydration reactions. The first derivative of change of mass (DTG) can also be used for identification purposes as it yields sharper peaks. TG cannot detect crystalline transitions as they do not involve weight losses.

Conduction calorimetry is another technique that is extensively used for following the hydration reactions of cement and cement compounds. In this method, heat evolved during the hydration reactions is followed as a function of time from the moment water comes into contact with the cement. The curves are obtained under isothermal conditions. This technique can also be used to study the rate of hydration at different temperatures. Conduction calorimetry has been used to determine kinetics of hydration and for studying the role of admixtures, relative setting times of cement, and for identification purposes.

A few other thermal techniques such as thermomechanical analysis, dilatometry, emanation analysis, etc., are only used to a limited extent in concrete investigations. Several publications have appeared which exclusively deal with the application of thermal analysis to the investigation of cementitious systems.[1]–[9]

2.0 RAW MATERIALS

The raw mix for the production of cement clinker comprises calcareous and siliceous materials. They are characterized by techniques such as DTA, XRD, chemical analysis, volatility test, burnability test, TG, etc.

The siliceous clay raw material is a source of both aluminum and silica. It may contain one or more types of clay minerals. Other sources of silica are quartz, chalcedony, opal, feldspar, etc. Table 1 shows the types of clay minerals and their suitability as raw mixes for cement manufacture.[5]

Table 1. Clay Minerals, Mineralogy, and Their Suitability for Cement Manufacture

Group	Minerals	Suitability
Kaolinite	Allophane Kaolinite Halloysite Dickite	Suitable if they do not contain alkalis and chloride.
Montmorillonite	Montmorillonite Beidellite Nontronite Glauconite	Usually not suitable because of swelling characteristics and presence of Na or Mg. Glauconite can be used if it has a low K content.
Illite	Illite	Suitable if it does not contain excessive amounts of alkalis.
Chlorites	Ferrogenous Chlorite Low Iron Chlorite	Suitable if free of alkalis.
Mica	Muscovite Biotite	Both not used because they have high alkali contents.
Amphiboles	Sepiolite Attapulgite	Both not advocated because of higher Mg contents.

DTA can be applied to identify and estimate various types of clays and calcareous minerals (Fig. 1).[6] These minerals can be differentiated by determining the temperature at which the endothermic and exothermic peaks occur. For example, kaolinites exhibit an endothermic peak at

550–600°C due to the dehydration effect, followed by an exothermic peak at about 980°C associated with the formation of α-Al_2O_3 or nucleation of mullite. Illites exhibit endothermal effects at 100–200°C, 500–650°C, and at about 900°C, and an exothermic peak immediately following the third endothermal peak. The first endothermal peak corresponds to the loss of interlayer, the second and the third to expulsion of water from the lattice and destruction of illite lattice, respectively. The exothermic peak is probably associated with the formation of a spinel. Illite differs from kaolinite in showing additional endothermal peaks at 100–200° and 900°C.

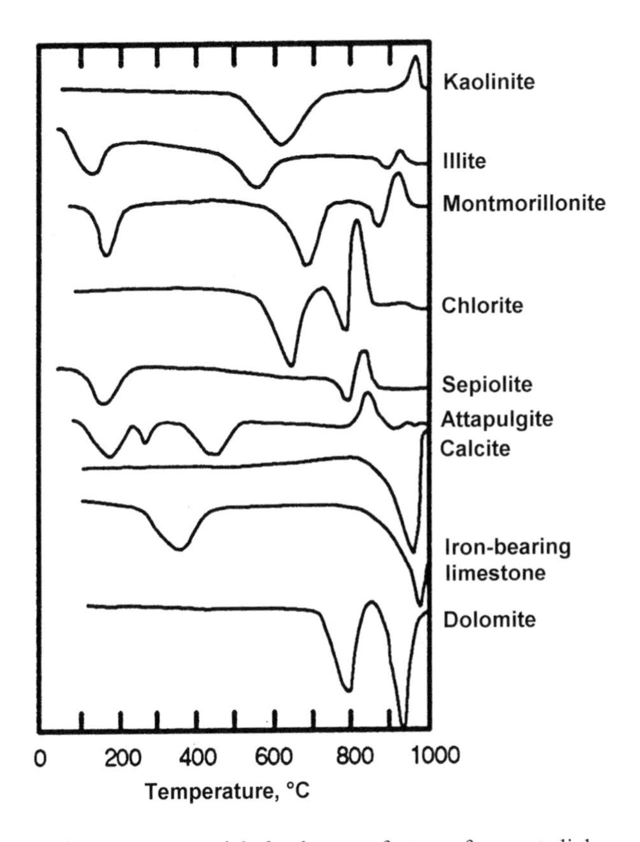

Figure 1. DTA of some raw materials for the manufacture of cement clinker.

Limestone is commonly used as a calcareous source. Others are chalk, marl, caliche (W. USA), naki (Middle East), kankar (India), reef sand (Australia and Bahamas), sea shells, calcareous sea sands, marbles, and others. They may contain impurities such as chalcedony, opal, pyrite, siderite, goethite, dolomite, magnesite, gypsum, etc. Thermal methods have been applied to characterize such minerals. Limestone gives an endothermal peak at about 800°C for de-carbonation and iron-bearing limestone gives an additional peak at about 350°C. Dolomite exhibits two endothermal effects for the decomposition of magnesium carbonate and calcium carbonate (Fig. 1). DTA technique is ideally suited to estimate even very small amounts of dolomite (0.2%) in limestone (Fig. 2).[10] Discernible endothermal effect for magnesium carbonate decomposition is evident even at a dolomite content of 0.3%. The activation energy of limestone decomposition has a strong bearing on the clinker formation. The de-carbonation of limestone varies, depending on the type, particle size, and impurities. DTA has been used to test the effect of fineness on the disassociation of limestones from different sources. It has been concluded that the finer the grain sizes and more the impurities, the lower is the activation energy.[5]

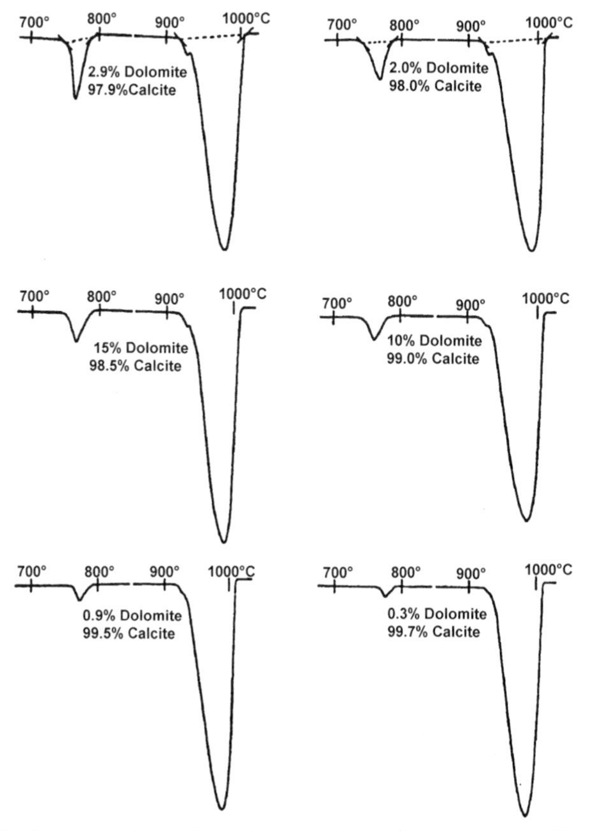

Figure 2. Determination of small quantities of dolomite by differential thermal analysis (DTA).

Ferrogenous minerals have been used as fluxes to form the melt and to facilitate the clinkerization at lower temperatures. The types of minerals used include hematite, magnetite, goethite, limonite, siderite, and ankerite. Thermal methods may be used to identify them.

Mineralizers are added to the raw feed to accelerate the kinetics of reactions by modifying the solid and liquid state sintering. The temperature of decomposition of calcium carbonate is lowered in the presence of mineralizers. In the synthesis of C_3S and C_2S, DTA has shown that some phosphates, carbonates, sulfates, and chlorides decrease the decarbonation temperature and that of the formation of the silicates.[20] Calcium fluoride acts both as a flux and a mineralizer in promoting the formation the tricalcium silicate phase. DTA thermograms have been applied to study the effect of mineralizers on the melting behaviors and crystallization temperatures of clinker. Early formation of liquid can be established by thermal techniques.[5]

3.0 CLINKERIZATION

The chemical processes that are involved in the formation of clinker are very complex. Thermal Analysis, XRD, and IR techniques have been applied for studying the reactions during the calcination of cement raw materials. The rates of reactions are influenced by many factors resulting in variations in temperatures at which the processes occur. Only approximate temperatures are assigned for the reactions. At a temperature of about 100°C evaporation of water takes place, at > 500°C, combined water from calcium hydroxide is expelled, and at > 800°C, calcium carbonate and magnesium carbonate decompose with the formation of CA and C_2F and C_2S begins to form. Between 800 and 900°C, $C_{12}A_7$ is formed, and at 900–1000°C, C_3A and C_4AF begin to appear and calcium carbonate is completely decomposed. The production of both C_3A and C_4AF is completed at about 1000–1200°C and the formation of C_2S reaches the maximum. At 1200–1280°C, liquid appears, with the occurrence of a substantial amount of C_3S between 1200 and 1450°C.[11] Co-existence of phases is also well documented. It has been found that the reactivity of clay minerals with calcium carbonate increases in the order: muscovite > montmorillonite > chlorite > illite > kaolin, and the reactivity of silica with CaO increases in the order: quartz > chalcedony > opal > cristobalite > tridymite-silica (from feldspar) > silica (from mica) > silica (from glassy slag). A comprehensive review of the application of thermal analysis to the clinkerization process has been published by Courtault.[12]

DTA-TG techniques have been applied to study calcination kinetics of raw materials, quantification of raw materials, determination of total heat for clinker formation, and prediction of material temperature profile in a kiln.[14]–[15]

High temperature DTA is a useful tool for studying the clinkering reactions. Figure 3 shows a typical DTA curve of an industrial raw meal.[16]

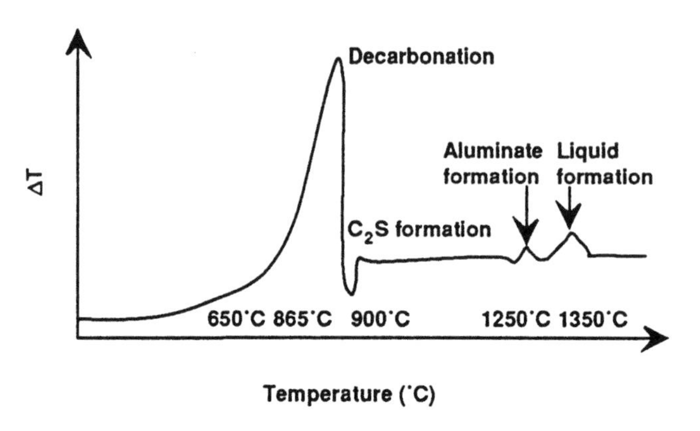

Figure 3. Typical DTA curve of an industrial portland cement raw mix.

Four thermal peaks are evident in the curve. The first endothermal effect below 900°C can be attributed to the de-carbonation of calcium carbonate. A small exothermal peak following this endothermal effect is ascribed to the formation of C_2S. The formation of aluminate and calcium aluminate ferrite phases is indicated by small heat effects and the liquid formation, by a high temperature endothermal effect. The position and the intensity of the peaks are good indicators of the burnability of the raw materials. In addition, by applying the thermodilatometry, the temperature of liquid formation and its quantity can be determined.[16] The liquid formation leads to shrinkage which can be determined by this method. In Fig. 4 dilatometric curves of two industrial raw mixes are given.[16] The temperature at which the liquid forms and its quantity can be used to determine the kinetics of clinkering reactions. DTA of white cement clinker has also been reported.[17] The curves show an exothermic peak at 1290°C for the belite formation and a main endothermic peak at 1360°C for melting and a small endothermic peak at 1380°C for $\alpha'_H \rightarrow \alpha$ transition of belite.

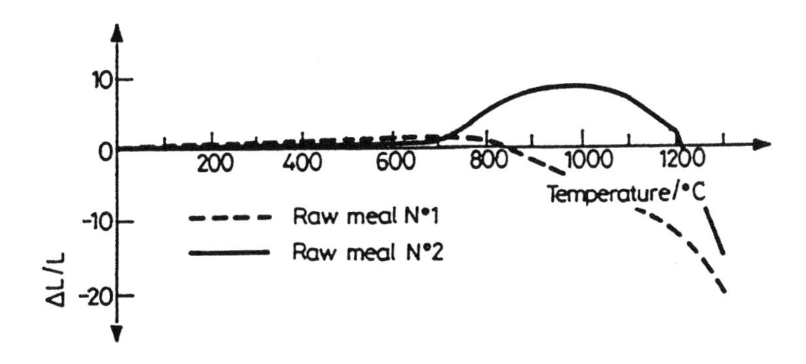

Figure 4. Dilatometric curves for two cement raw mixes.

Application of DTA for the determination of the melt content during clinkerization poses problems because the overlapping of exothermic crystallization process and endothermic peaks for melting leads to a sharp baseline shift. Reliable data are obtained by examining thermal curves on reheating or cooling. The five typical stages of crystallization of C_3A, C_4AF, and C_2S obtained by DTA are shown in Table 2.[5] These results vary since alkalis and minor compounds influence these processes.

Table 2. DTA Data of Devitrification of Glass of Clinker Melt Composition

DTA Peak	Reaction	Temperature
Endothermic	Nucleation of C_3A	650–770°C
Exothermic	Crystallization of C_3A	700–890°C
Endothermic	Nucleation of aluminoferrite	950°C
Exothermic	Crystallization of aluminoferrite	980–1000°C
Exothermic	C_2S crystallization	1130°C

An assessment of the burnability of raw mix is important because it denotes the amount of mass transfer of its constituents to the clinker phase. It is measured by determining the amount of uncombined lime after a certain time and a particular temperature. The high temperature thermal methods are useful tools to compare the burnability of different raw mixes. Burnability is influenced by the chemical, mineralogical, and granulometric compositions of the raw materials. DSC has been utilized to evaluate the burnability of raw meals. Figure 5 shows a good correlation between BI (burnability index) and A_1+A_2 (areas of peaks corresponding to the formation of belite).[13]

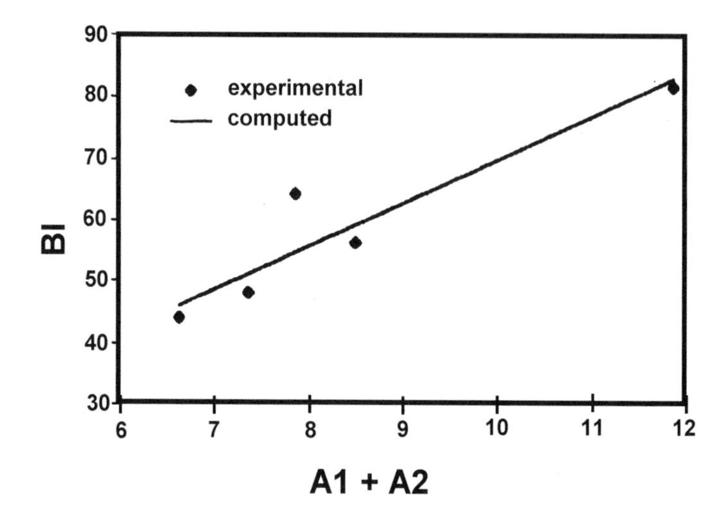

Figure 5. Correlation between burnability and DSC peak areas.

Although thermal curves for even the same raw materials vary, individual cement clinker minerals always form in sequence and over a certain narrow temperature range. Dilatometry is particularly useful in conjunction with DTA to identify transitional phases. For example, cement clinkers made by heating a mixture of class F fly ash, CaO, CaF_2, and gypsum have been analyzed by dilatometric and XRD techniques.[21] Pure fly ash shows shrinkage from 700–1194°C. In fly ash mixtures no shrinkage is observed between 950 and 1120°C that corresponds to the clinkering reaction. Only above 1120°C is shrinkage registered.

Specific heat is an important thermodynamic parameter needed for determining the heat balance of a reaction. The specific heat of crude cement and clinker has been measured by DSC.[18] The results demonstrate that the specific heat of crude powder is higher than that of the clinker. In the temperature range 50–300°C the specific heat value is about 0.025 cal/g/°C.

Another application of DTA/TG is to assess the quality of the unhydrated cement by identifying and estimating the amounts of CH, H_2O, and carbonates. Only minimum amounts of these (formed by aeration) should be present in the raw cement for it to be of good quality. The quality of oil cements has been evaluated applying this approach.[19]

In the production of sulfoaluminate cement from raw materials such as limestone, bauxite, and anhydrite, DTA has been used to monitor the reactions that take place when these mixes are heated to a temperature of 1025°C. By applying DTA and XRD it was found that the dehydration and decomposition of bauxite occurs at 530°C. At about 900°C calcium carbonate is decomposed to CaO and it reacts with α-SiO_2 to form C_2AS. At a higher temperature, C_2AS reacts with $CaSO_4$ to form calcium sulfoaluminate and α-C_2S.[22] Thermal analysis has also been applied to investigate the reactions occurring in the formation of a clinker from calcium carbonate mixed with CMS_2, C_3AS_3, and NAS_6.[23]

There has been a continued interest in producing cement clinker at lower temperatures than what has been the normal practice. A study of binary and ternary systems containing limestone, alkaline basalt and fluorite has been conducted to study the possibility of obtaining belite and alite at lower temperatures.[24] The reaction was followed by DTA, DTG, and TG. Several endothermal and exothermal peaks were obtained. Decarbonation occurred at about 650°C, and a peak at 1145°C could be linked to the formation of $C_{12}A_7$ and gehlenite. The peak at 1170°C was attributed to the formation of belite, and that at 1235°C to the formation of alite.

Extensive work has been carried out on the utilization of waste and by-products in concrete industry. The utilization of cement kiln dust is an example of a waste material that is also a large source of pollution at the cement plant. In an investigation, a mixture of kiln dust and kaolin was fired up to 1250°C and the resulting phases were studied by DTA and XRD.[25] Several peaks (two exothermal and five endothermal) were obtained in DTA. The DTA examination revealed that the dust consists of dolomitic limestones with some alkalis and quartz. At 1000°C, β-C_2S formed as the main phase with some $C_{12}A_7$ and C_4AF and at 1100°C spurrite decomposition occurred. At 1250°C gehlenite was formed. This study was useful in assessing the temperature to which the dust should be calcined so that the resultant product could be utilized for making porcelain.

Although most cement clinkers are manufactured utilizing the rotary kilns, in some countries vertical kilns have been used. In addition to the mechanism of clinkering, the heating and cooling schedules in these two types of kilns are different. A study was undertaken to determine the difference in reactivities of the clinkers produced by these two methods by applying the DTA technique.[26] By conducting DTA of clinkers hydrated for various periods, it was concluded that the rate of hydration, as determined from the peak intensities of $Ca(OH)_2$, was lower for clinkers made in the rotary kiln.

4.0 SYNTHESIS OF CEMENT PHASES

DTA of binary, ternary, and quarternary systems of relevance to clinkers has been studied. Mixtures containing various proportions of oxides such as calcium oxide, aluminum oxide, ferric oxide, and silica are heated to high temperatures in the DTA apparatus and the endothermal and exothermal effects that develop are analyzed.

In the heating of a kaolin-lime mixtures, gehlenite (C_2AS) and a small amount of β-C_2S have been detected at 900°C. At 1050°C, formation of $C_{12}A_7$ occurs. At the temperatures of 1100°C and 1400°C, C_3A and C_3S are formed respectively.[1] In the thermograms of $Ca(OH)_2$-SiO_2 mixtures, several endothermal and exothermal effects result that can be attributed to the formation of β-C_2S, γ-C_2S, $Ca(OH)_2$, etc.

DTA has been applied to investigate the efficiency of silicate materials to form C_3S when they are heated with CaO. In Fig. 6 thermograms of mixtures of CaO and siliceous materials such as silica gel, quartz, silica glass, α-CS or β-CS (synthetic glass), C_3S_2 and γ-C_2S are given.[27] The weak endotherms at about 1175K and 1225K may be attributed to C_3S transition. The endothermal inflection at about 1700K corresponds to the α' to α-C_2S transition. By determining the peak intensities of the endothermal effects due to C_3S transitions, it is possible to assess the relative amounts of C_3S formed in various mixes. The results show that the largest amount of C_3S is formed in the mixture γ-C_2S+CaO and the lowest, in the silica gel + CaO mixture.

It has been observed that the nature of the reactions in the system CaO-SiO_2 may not reflect that occurring in the kiln because of the prevalence of CO_2 in the kiln. Hence, some work has been carried out on the system containing $CaCO_3$ and SiO_2. This mixture when heated shows endothermal effects at 900 and 976°C that are attributed to triclinic-monoclinic-trigonal transformations in C_3S.[1] According to De Keyser's work, in the system $CaCO_3$-SiO_2, the formation of C_2S occurs at 808°C, transformation of γ-C_2S to α' at 912°C, and the transformation of α'-C_2S to β-C_2S at 965°C.[1]

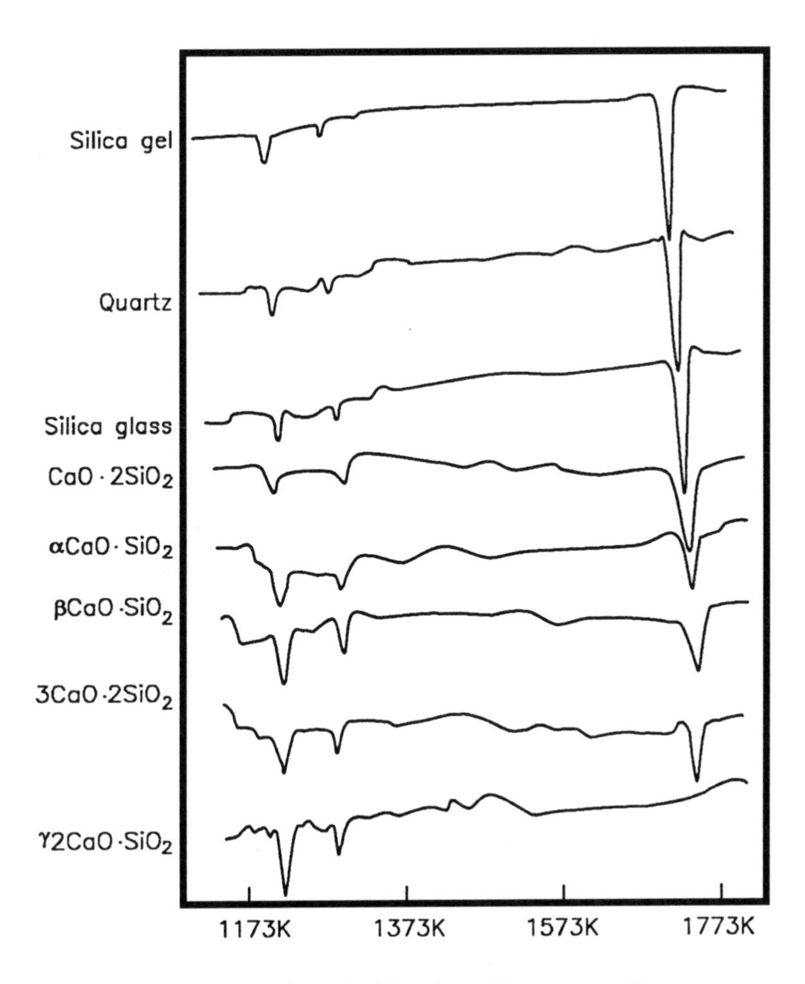

Figure 6. DTA curves of CaO heated with various siliceous materials.

The fluorine containing compounds are added to the raw cement mix to accelerate the formation of cement minerals. The effect of addition of fluorides such as LiF, NaF and KF on the reactions in the system $CaCO_3$-SiO_2 has been followed by DTA.[1] The results are shown in Fig. 7. In the absence of the additions, the curve (solid line) shows a slope at about 920°C while those treated with the fluorides exhibit a slope at a temperature of about 700°C, indicating early commencement of reactions. The effectiveness of other mineralizers have also been examined by DTA. The influence of mineralizers such as ZnO, CuO, and MnO_2 on the formation of alite from a mixture of calcium carbonate and a silica-bearing material such as the rice husk ash has also been studied by thermal analysis.[28] Lithium carbonate is used to lower the decomposition temperature of calcium carbonate in the reaction of $CaCO_3$ with quartz. DTA-DTG of mixtures of $CaCO_3$ and quartz containing different amounts of lithium carbonate (equivalent to 0.1–5% Li_2O) was examined up to a temperature of 1450°C.[29] The decomposition temperature of calcium carbonate was lowered as the amount of lithium carbonate was increased. The CaO formed from the decomposition of the carbonate, combined with SiO_2 to form C_2S. At 1% Li_2O, β-C_2S was detected even at a temperature of 750°C, but the reaction was completed only at 1350°C. At an additional level of 5%, the final reaction temperature decreased to 1250°C. In terms of the decomposition of $CaCO_3$, at Li_2CO_3 contents of 0.5, 1, and 5%, approximately 2, 6, and 80% $CaCO_3$ decomposed at about 600°C.

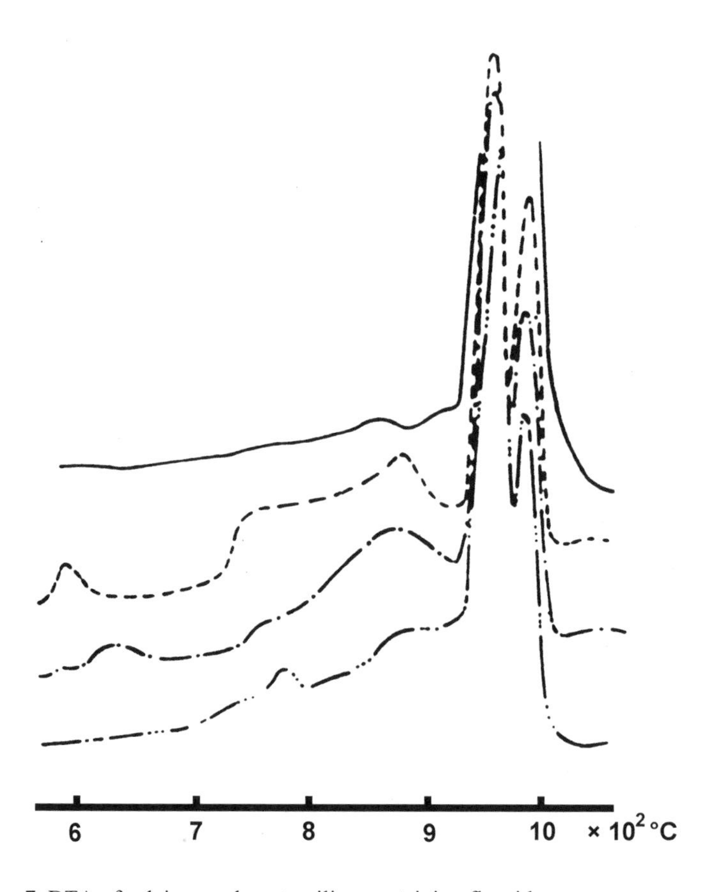

Figure 7. DTA of calcium carbonate-silica containing fluorides.

In the system $CaCO_3$-Al_2O_3, exothermal effects due to the formation of CA and $C_{12}A_7$ have been detected (Fig. 8).[1] Differential thermograms show an exothermal effect in the range 950–1000°C at all concentrations. This can be ascribed to a simultaneous formation of CA and $C_{12}A_7$. The endothermal effect at 1170°C is assigned to the change of γ-Al_2O_3 to the α form and the other at 1200°C denotes the onset of formation of C_3A. Melting effect in the system is indicated by an endothermic effect at about 1300°C.

In the system containing CaO and Fe_2O_3, independent of the proportion of CaO and Fe_2O_3, CF forms at about 950°C. Melting occurs at 1150°C, which is reflected as an endothermal effect. An endothermal effect at 1300°C signifies incongruent melting of the mix (Fig. 9).[1]

Some work has been carried out by Barta on the ternary system containing calcium oxide-aluminum oxide-ferric oxide mixtures (Fig. 10).[1] Formation of C_4AF with an exothermal peak around 980–1000°C and of C_2F with an endothermal peak at 1160°C is evident. At higher temperatures, $C_{12}A_7$ is formed. Many other ternary and quaternary systems have been studied by thermal analysis and are described in a publication.[1]

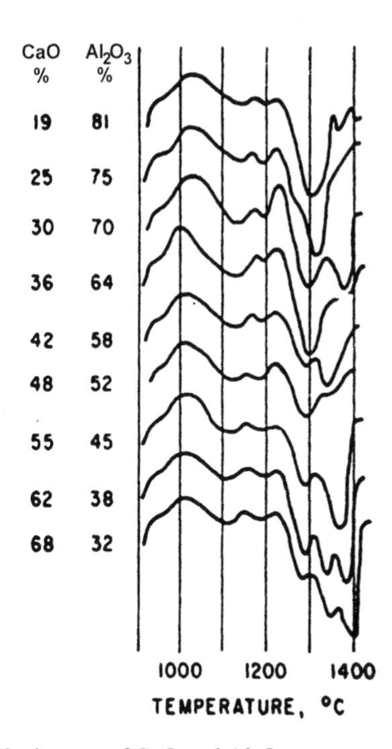

Figure 8. Thermograms of mixtures of CaO and Al_2O_3.

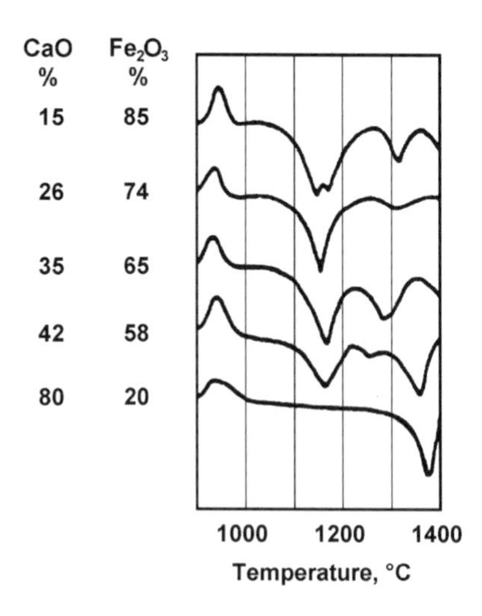

Figure 9. Thermograms of mixtures of CaO and Fe_2O_3.

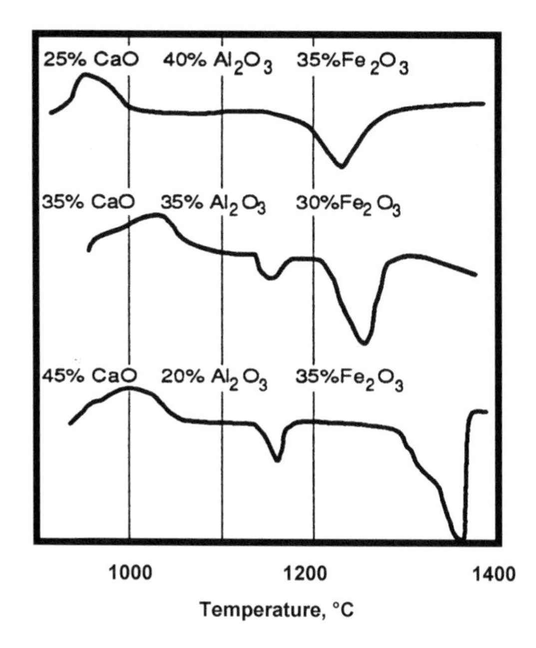

Figure 10. Thermograms of the ternary mixture containing CaO, Al_2O_3, and Fe_2O_3.

5.0 POLYMORPHISM IN SILICATES

Tricalcium silicate exists in several polymorphic forms. It can be triclinic, monoclinic, or trigonal. The stability of polymorphs and their transitions are temperature dependent and not easy to determine because many forms have small transition enthalpies. The triclinic form (T_I) can be stabilized with chromium oxide, the triclinic form (T_{II}) by ferric oxide, the monoclinic form (M_I) by MgO and monoclinic (M_{II}) form by zinc oxide.[5] The monoclinic form is present in commercial clinkers and only rarely has the rhombohedral form been detected.

Jeffery[1][30] was one of the first to carry out the thermograms of tricalcium silicate preparation and alites. Figure 11 compares the DTA curves of alite and a synthetically prepared tricalcium silicate.[1][30] The C_3S preparation exhibits as many as six endotherms at 464, 622, 750, 923, 980, and 1465°C. The peak at 464°C is caused by the calcium hydroxide formed from the hydration of free lime present in the preparation. Dicalcium silicate, present as an impurity exhibits three peaks at 622, 750, and 1465°C. The peaks at 923°C and 980°C are attributed to triclinic-to-monoclinic, and monoclinic-to-trigonal or triclinic-to-trigonal and trigonal-to-trigonal transition plus rotation of ions, respectively. The thermogram of alite differs from that of the synthetic silicate by having only two endotherms at 825 and 1427°C. The effect at 825°C is ascribed to monoclinic to trigonal transition and corresponds to a similar effect in synthesized tricalcium silicate. This effect occurs at a lower temperature owing to the solid solution effect. In the cooling cycle, the endothermal effects show up as exothermal effects, indicating reversibility. Regourd[31] reported DTA peaks corresponding to tricalcium silicate transitions at 600, 920, 980, 990, and 1050°C. The transition at 920°C was undetectable by XRD. All the transitions involve very slight displacement of atoms without disruption of any of the first co-ordination bonds. More recently other investigations have been carried out on calcium silicates by applying different techniques and the temperatures of transitions may be at some variance from other published data. The variation in such effects is expected because the behavior of the silicate depends on the type of stabilizer used. Foreign ions may substitute Si or Ca ions or may occupy the interstitial sites. Depending on the nature of these ions certain modifications may be stabilized.

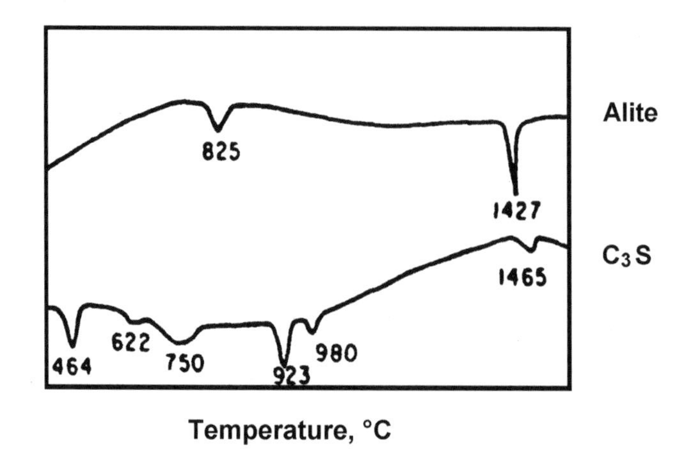

Figure 11. Thermograms of alite and synthesized tricalcium silicate.

The polymorphism of C_2S is much more complicated than that of C_3S. Several polymorphic forms of dicalcium silicate are reported. Except γ-C_2S all other forms are stable at high temperature. Basically the forms are α, α', β, and γ, and typically compounds such as calcium phosphate, strontium oxide, boron oxide, and alkalis are capable of stabilizing them. Thermal behavior of dicalcium silicate has been studied both during heating and cooling cycles[1][32] (Fig. 12). Curve A shows the thermogram of γ-C_2S. The endothermal hump between 780 and 830°C indicates a sluggish transition of γ to α' C_2S. At 1447°C, the sharp endothermal effect can be correlated with the conversion of α' to α-C_2S. The thermal effects are reversible, as indicated by the cooling curve B. The exothermic peak at 1425°C indicates conversion of α to α' form and that at 670°C is due to α' to a transition. In addition, there is an irregular exothermal dent starting at 525°C for the β to γ inversion and this effect is accompanied by a sudden increase in volume of the mass. Curve C shows the behavior of β-C_2S stabilized by CaO. The endotherms at 705° and 1447°C are respectively due to β to α' and α' to α transitions. The transition data of different forms of C_2S reported in the literature show some variation. The tricalcium aluminate does not show any polymorphic transformation.

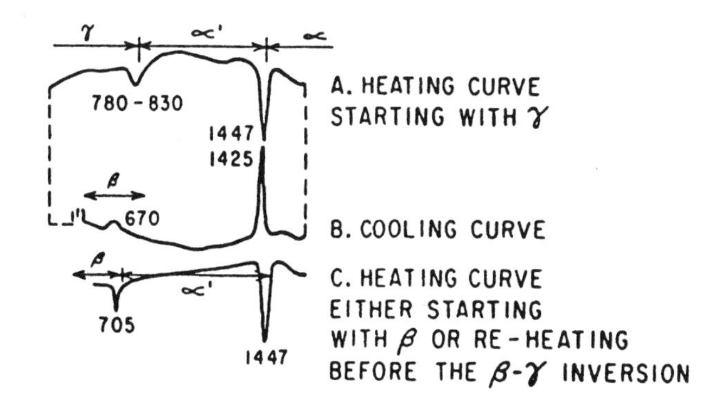

Figure 12. Inversions in dicalcium silicate.

6.0 HYDRATION

6.1 Calcium Silicates

A knowledge of the hydration of individual cement compounds and their mixtures forms a basis of interpreting the complex reactions that occur when portland cement is hydrated under various conditions. Tricalcium silicate and dicalcium silicate together make up 75–80% of portland cement. In the presence of water, the reaction products are calcium silicate hydrate (endothermal effects below 200°C) and calcium hydroxide, with an endothermal effect in the range 450–550°C. Some calcium carbonate may also be detected in the range 750–900°C by an endotherm. Under normal conditions of hydration it is difficult to prevent some carbonation of lime that is formed. At higher temperatures some peaks may occur due to crystalline transformations. The calcium silicate hydrate is poorly crystallized and gives only weak diffusion lines in XRD. During the course of hydration, the degree of hydration may be estimated by determining the amount of lime formed or nonevaporable water content or by the amount of tricalcium silicate that has reacted. Several types of methods have been adopted to determine the degree of hydration but each has its limitations. Thermal analysis techniques such as DTA, DSC, TG, and conduction calorimetry are found to be convenient, fast and accurate, and yield results that are not easily obtainable by some other methods. Thermal techniques have been used to study kinetics of hydration, mechanism of hydration, the influence of admixtures, identification of new compounds formed, estimation of products, etc.

Conduction calorimetry of C_3S and cement shows five steps during the hydration process (Fig. 13). In the first stage, as soon as C_3S or cement comes into contact with water Ca and OH ions are released into the solution phase. This is followed by a rapid release of heat that ceases within 10–12 minutes. This is called a *pre-induction period.* In the second stage, the reaction rate is slow and it is known as the *dormant* or *induction period.* This may be extended or shortened by a few hours by the use of admixtures. In the third stage, the reaction proceeds rapidly and accelerates with time, releasing a maximum amount of heat at the end of the *acceleratory period.* At this stage, a rapid crystallization of CH and C-S-H occurs. In the fourth stage, there is a slow deceleration. At the final stage, there is only limited formation of products and at this stage the reaction is diffusion-controlled. Conduction calorimetry permits determination of the rate of hydration as a function temperature, water:cement ratio, type of admixture, particle size, pH, etc.

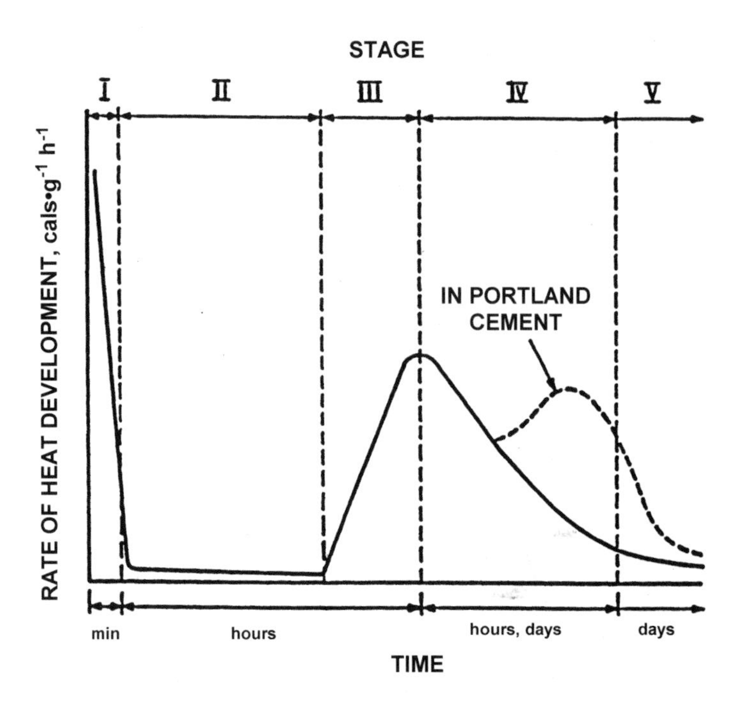

Figure 13. Conduction calorimetric curves of hydrating tricalcium silicate and cement.

Conduction Calorimetry may also be used to follow the hydration of C_3S at different temperatures. In Fig. 14, the calorimetric curves for C_3S hydrated at 30 and 80°C are given. The intense peak for the accelerated hydration of C_3S occurs with a peak at about 2 hrs at 80°C, but the hydration proceeds at a slower rate at 30°C as evident from a hump that occurs at about 7 hrs.[18]

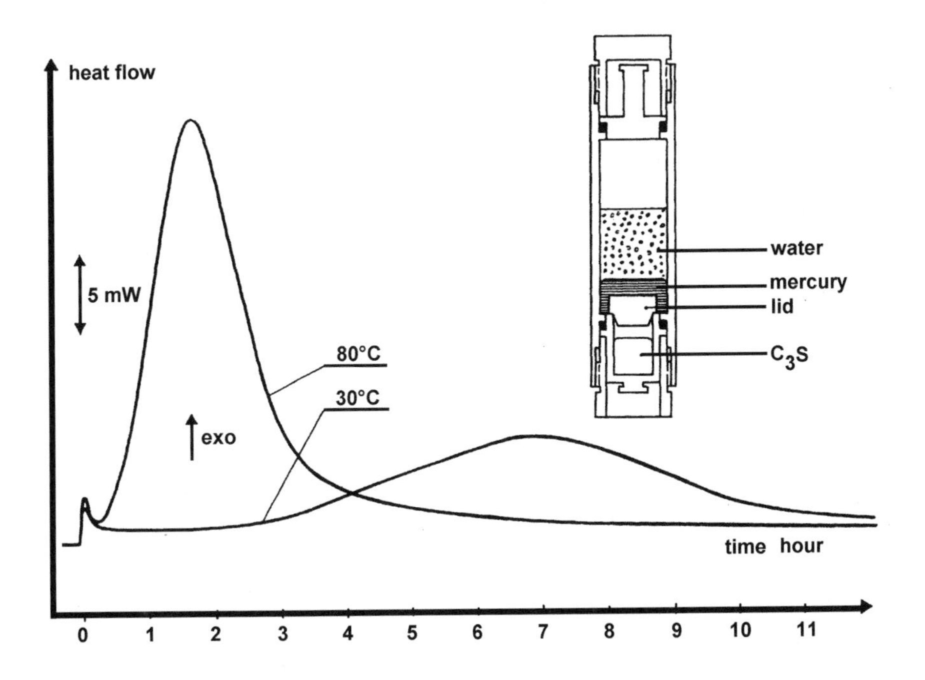

Figure 14. Conduction calorimetric curves for C_3S hydrated at two temperatures.

One of the methods of following the degree of hydration as a function of time is by the determination of the calcium hydroxide content. Figure 15 compares the relative amounts of calcium hydroxide formed at different times of hydration, using DTA and chemical methods. Although the general trend of the curves is similar, the values derived from the chemical analysis are somewhat higher than those from the thermal method. This may be due to attack of the C-S-H phase by the solvents used in the extraction of lime that is adopted in the chemical method.[33]

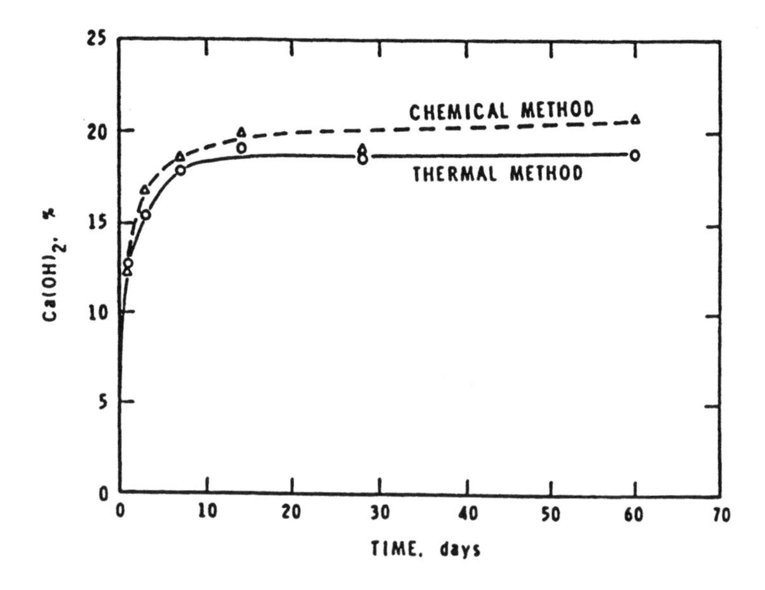

Figure 15. Amounts of $Ca(OH)_2$ formed in the hydration of C_3S.

DTA is a convenient method to follow the hydration of C_3S as a function of time. In Fig. 16, the onset of hydration is evident from the small endothermal effect below 200°C.[34] This effect is caused by the removal of loosely bound water as well as firmly held water from the C-S-H gel. The increase in the intensity of this effect with time is indicative of increased formation of the C-S-H product with time. A very small endothermal effect at about 480°C appears within a few minutes, becomes more evident at one hour, and is attributed to the dehydration of $Ca(OH)_2$. In the first eight hours, the amount of $Ca(OH)_2$ produced is about 25% of that formed in thirty days.

A direct method of determining the degree of hydration of C_3S is to estimate its amount during the progress of hydration. Ramachandran[35] has provided a method to estimate tricalcium silicate by adopting the DTA method. Tricalcium silicate exhibits several peaks when heated to a temperature of 1000°C. The transition of triclinic to monoclinic (or triclinic II to triclinic III) results in an intense peak at about 915°C. The amplitude of this peak can be used to estimate C_3S content in the hydrating C_3S. In the heating mode interference may occur due to the decomposition of calcium carbonate and recrystallization of the C-S-H phase. Figure 17 shows the peaks that occur during the heating and cooling cycles. In the cooling mode, two peaks appear at higher temperature. On reheating, a single endothermal peak appears that may be used to estimate unhydrated C_3S. The percentage of C_3S hydrated at different times determined by DTA was plotted against those obtained by XRD method. The correlation between the two methods is very good (Fig. 18).

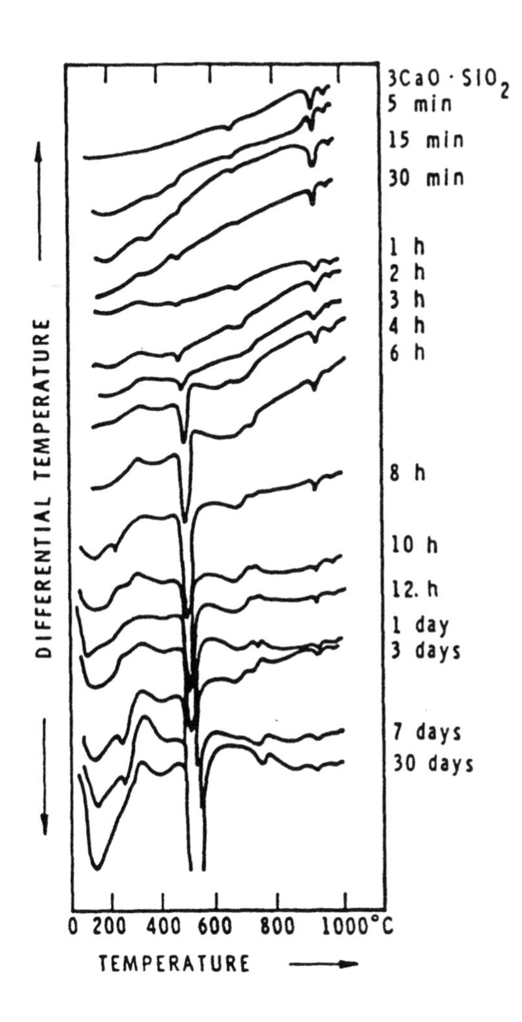

Figure 16. DTA Curves of $3CaO{\bullet}SiO_2$ hydrated in water.

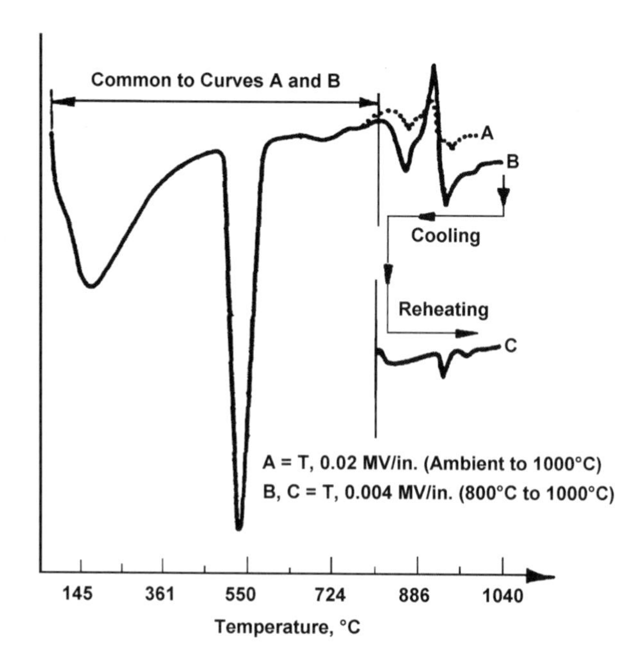

Figure 17. Elimination of interfering effects in hydrated C_3S.

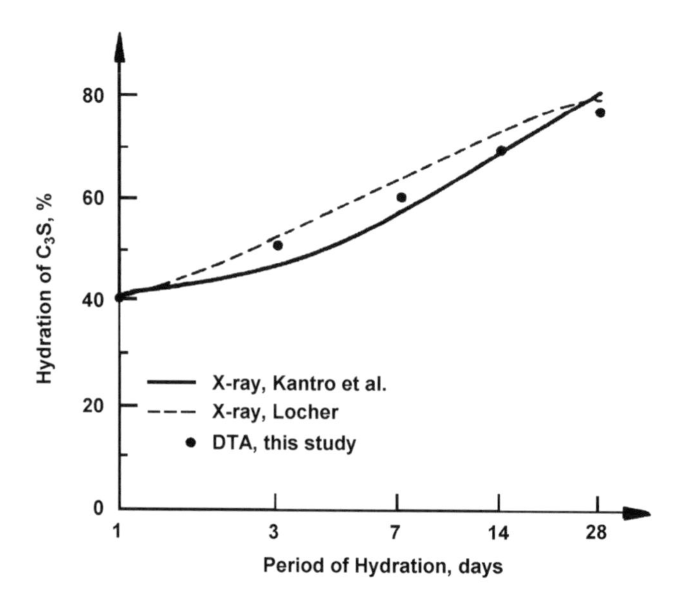

Figure 18. Comparison of the rate of hydration by DTA and XRD methods.

In portland cement, tricalcium silicate is a solid solution containing minor amounts of oxide and is known as an alite phase. Alites may contain Al, Mg, and Fe. Several alites were prepared containing Al, Mg, and Fe and were hydrated for different periods. The degree of hydration was determined according to the DTA method suggested above. In Fig. 19, the compressive strengths of these samples are plotted at different degrees of hydration.[36] Compared to pure C_3S, Fe-alite gives higher strength at any degree of hydration. This is attributable to the differences in the intrinsic nature of the hydration products.

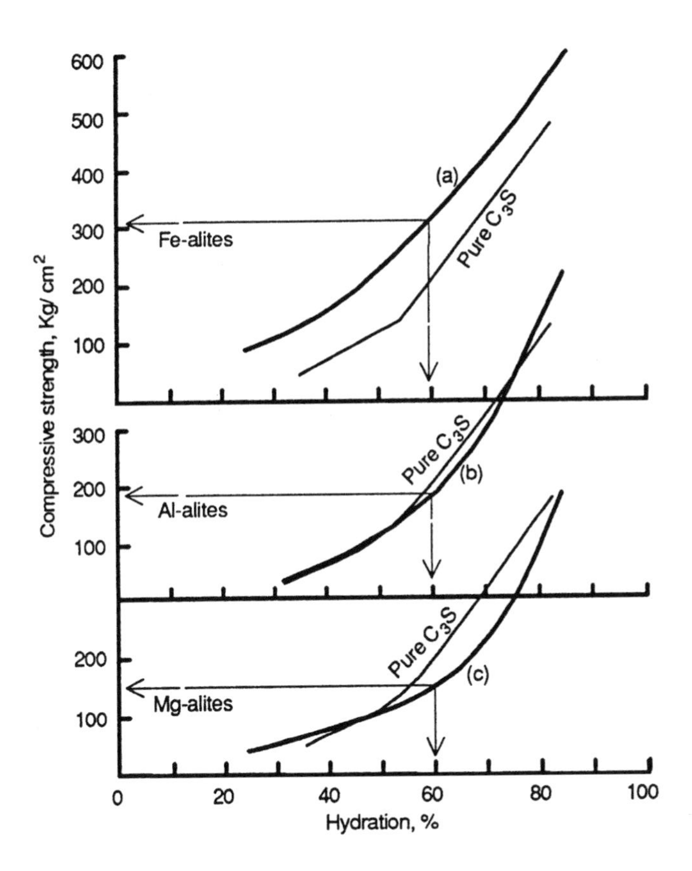

Figure 19. Kinetics of hydration of alites.

In the determination of the constitutional water associated with the C-S-H phase, it is important to differentiate the unreacted water from that which is bound by the gel. Feldman and Ramachandran[37] carried out TG/DTA of a bottle hydrated C_3S that was conditioned at several humidities for lengthy periods, starting from the d-dry, 100% RH and 11% RH. It was determined that the stoichiometry of the bottle hydrated C-S-H gel at 11% RH (approached from 100% RH) is $3.28CaO{\bullet}2SiO_2{\bullet}3.92H_2O$. It was also found that the thermograms could delineate free and interlayer water and the relative intensities of the peak effects of samples exposed to different conditions showed a good relationship with adsorption isotherms. A method called Dynamic Differential Calorimetry was adopted by Kurczyk and Schweite[1] to determine CaO/SiO_2 ratio of C_3S hydrated to different periods. The ratios at 1, 2, and 3 months of hydration were 1.88, 1.87, and 1.92 respectively. The DTG method can also be applied with success to estimate calcium hydroxide in the hydrated silicate system as the dehydration of calcium hydroxide is attended by a loss of water.

A mature hydrated silicate paste contains both the C-S-H and CH phases. Any characterization study of the C-S-H phase is not easy because of the possible interference due to the presence of calcium hydroxide. A procedure has been developed by Ramachandran and Polomark for extracting calcium hydroxide from the C_3S paste without affecting the C-S-H phase.[38] It involves exposing a thin disc of the paste to an aqueous solution of calcium hydroxide in such a way that the concentration of the solution remains in the range 9–12 millimole CaO/L throughout the extraction period. Calcium hydroxide remaining in the solid is monitored by DSC and further extraction is stopped when the amount of free lime remaining in the solid is almost nil. In Fig. 20, the amount of lime remaining in the C_3S pastes at different times of extraction is shown.

It is an established fact that C_3S is much more reactive than C_2S. The relative reactivities of Ca in various compounds of interest in cement science were studied by the extent to which they reacted with silver nitrate.[39] DTA was used in the cooling mode to estimate the unreacted silver nitrate by an exothermal transition at about 190°C. (See Table 3.) The reaction between CaO and $AgNO_3$ is almost stoichiometric. Calcium hydroxide also reacts stoichiometrically with silver nitrate. Only 0.81 mol out of 3 mols in C_3S has reacted with silver nitrate. This suggests that about 27% of tricalcium silicate is more reactive than the rest. Isothermal conduction calorimetric curves show that about the same amount of C_3S is relatively more reactive. Thus, it seems that all Ca ions are not the same in the silicate. Dicalcium silicate phase has reacted about 6 % only and this corresponds to 0.12 mol Ca out of the 2 mol present in the sample. All Ca in the C-S-H phase is equally reactive.

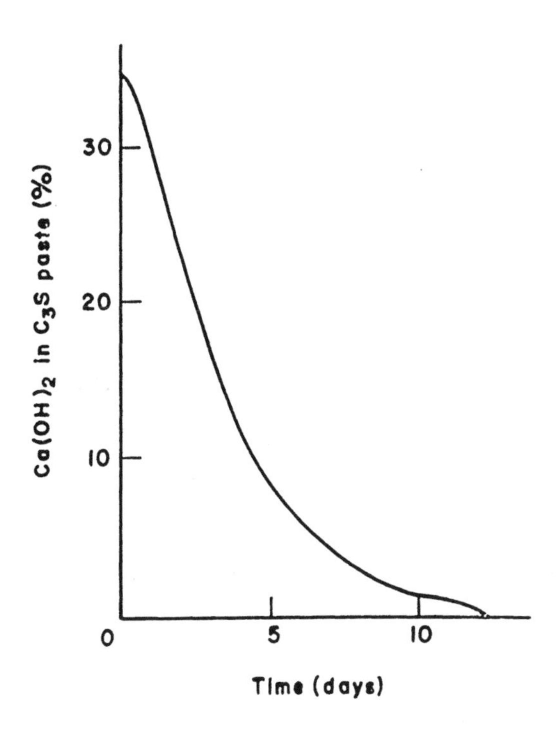

Figure 20. Amount of lime remaining in C_3S paste at different times of extraction.

Table 3. Extent of Reaction of $AgNO_3$ with Different Reactants *(with permission:* Nature*)*

Sample No.	Reactants	Extent of Reaction Reactant:$AgNO_3$ (Wt)	Amount of Ca in the Reactant (Mol)	Amount of Ca Reacted (Mol)
1	CaO	1:5.8	1	0.96
2	$Ca(OH)_2$	1:4.6	1	1.00
3	$3CaO{\bullet}SiO_2$	1:1.21	3	0.81
4	β-$2CaO{\bullet}SiO_2$	1:4.2	2	0.12
5	Hydrated $3CaO{\bullet}SiO_2$	1:3.74	6	5.99

Thermal analysis data on the hydration of dicalcium silicate are sparse because it is time consuming to follow the reaction of this phase which is very slow. The characteristic products obtained during its hydration are not much different from those formed in C_3S hydration. Also, the major strength development that occurs in cement in the first 28 days (a period of practical significance) is mainly due to the tricalcium silicate phase. TG, DTG, and DTA investigations of C_2S were carried out by Tamas.[40] The sensitivity of the instrument had to be increased substantially to detect the peaks due to the decomposition of calcium hydroxide and calcium carbonate, especially at earlier times. In Fig. 21, the DTA, DTG, and TG curves of C_2S hydrated for 21 days and 200 days are given. A comparison of these peaks with those obtained from C_3S pastes shows substantial differences in the intensity value of the peaks. The 200 day C_2S sample shows a weight loss of 4%, whereas C_3S hydrated for 21 days indicates a loss of 13%.

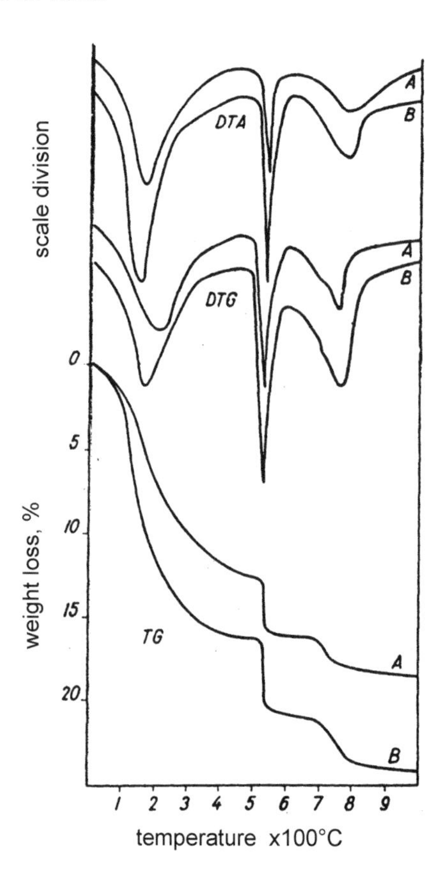

Figure 21. Thermograms of C_2S hydrated for 21 days and 200 days.

The reactivity of C_2S also depends on its polymorphic form. A thermal investigation of two C_2S samples heated to 1400 and 1100°C has indicated that the hydration rate is higher in the sample heated at higher temperatures.[41]

6.2 Calcium Aluminates

Tricalcium Aluminate. The aluminate phases, although present in small amounts, exert a significant effect on the setting and early strength development in cement pastes. In the hydration of tricalcium aluminate, the initial formation of hexagonal phases is identifiable by the endothermal effects at 150–200°C and 200–280°C. These are converted to a cubic phase of formula C_3AH_6. The cubic phase shows characteristic endothermal effects at 300–350°C and 450–500°C (Fig. 22).[7] The first peak exhibited by the cubic phase is ascribed to the decomposition to a mixture of $C_{12}A_7H$ and CH[17] for a bulk composition of $C_3AH_{1.43}$. At 450–500°C CH decomposes to CaO and on prolonged heating to 810°C, C_3A forms.[17] Thermogravimetric curves show major peaks at 250–310°C and at 450–530°C. DTA is so sensitive in the identification of the cubic phase that it is some times preferred to XRD procedures.

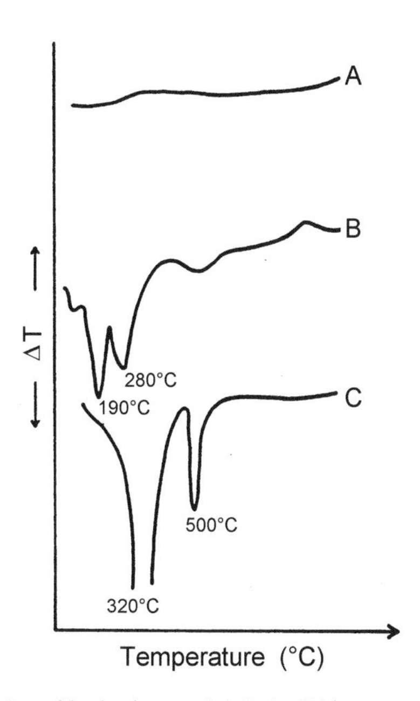

Figure 22. DTA of C_3A and its hydrates: *(A)* C_3A, *(B)* hexagonal phase, *(C)* cubic phase.

Variations in the peak temperatures of thermograms reported in the literature may be due to the differences in the purity of the sample, temperature of hydration, water:solid ratio, the type of equipment, rate of heating, etc.

Hydration characteristics of C_3A exposed to 20 or 80°C (w/s ratios of 0.12 and 1.0) have been studied applying DTA, TG, scanning electron microscopy, and microhardness techniques.[42] The DTA results are shown in Fig. 23. Whereas it takes 6 hrs before C_3AH_6 forms in considerable amounts at 20°C, it is evident even at 15 secs at 80°C. A similar trend was also obvious at higher water:solid ratios. The results indicate that at 80°C the cubic phase forms almost immediately on contact with water; hexagonal phases could not be detected at this temperature. At 20°C, the rate of hydration is much slower, the hexagonal phases forming initially and gradually converting to the cubic phase. Electron micrographs show that at 20°C the hydrated product consists of irregular as well as spherically shaped particles in the form of disconnected chunks. The product at 80°C has spherical particles connected or welded into a continuous network. This was caused by the direct bonding of C_3AH_6 products formed mainly on the original sites of C_3A. Microhardness values of C_3A hydrated at 80°C are more than fourfold those of the samples hydrated at 20°C. Feldman and Ramachandran.[43] applied DTA technique to follow the hydration of C_3A cured at 2, 12, 23, 52, and 80°C from a few seconds to 10 days. The rate of conversion of the hexagonal phases to the cubic phase increased with temperature. It was concluded that the reaction occurred at the surface of the aluminate phase, and the passage of water through the hexagonal aluminates controls the overall rate of reaction. In the system containing calcium hydroxide:cubic aluminate hydrate (5:1) the stability of the hexagonal hydrate formed is maintained even up to 28 days.[1]

The alkali in clinker is combined as a solid solution with the C_3A phase. The crystalline structure changes from cubic to orthorhombic or monoclininc structure, depending on the content of Na in the C_3A phase. Shin and Han[50] studied the effect of different forms of tricalcium aluminate on the hydration of tricalcium silicate by applying DTA, TG, and conduction calorimetry. It was concluded that the hydration of tricalcium silicate is accelerated when orthorhombic, monoclinic, or melt C_3A was present in the mixture. The cubic form of tricalcium aluminate was least effective for accelerating the hydration of the silicate phase.

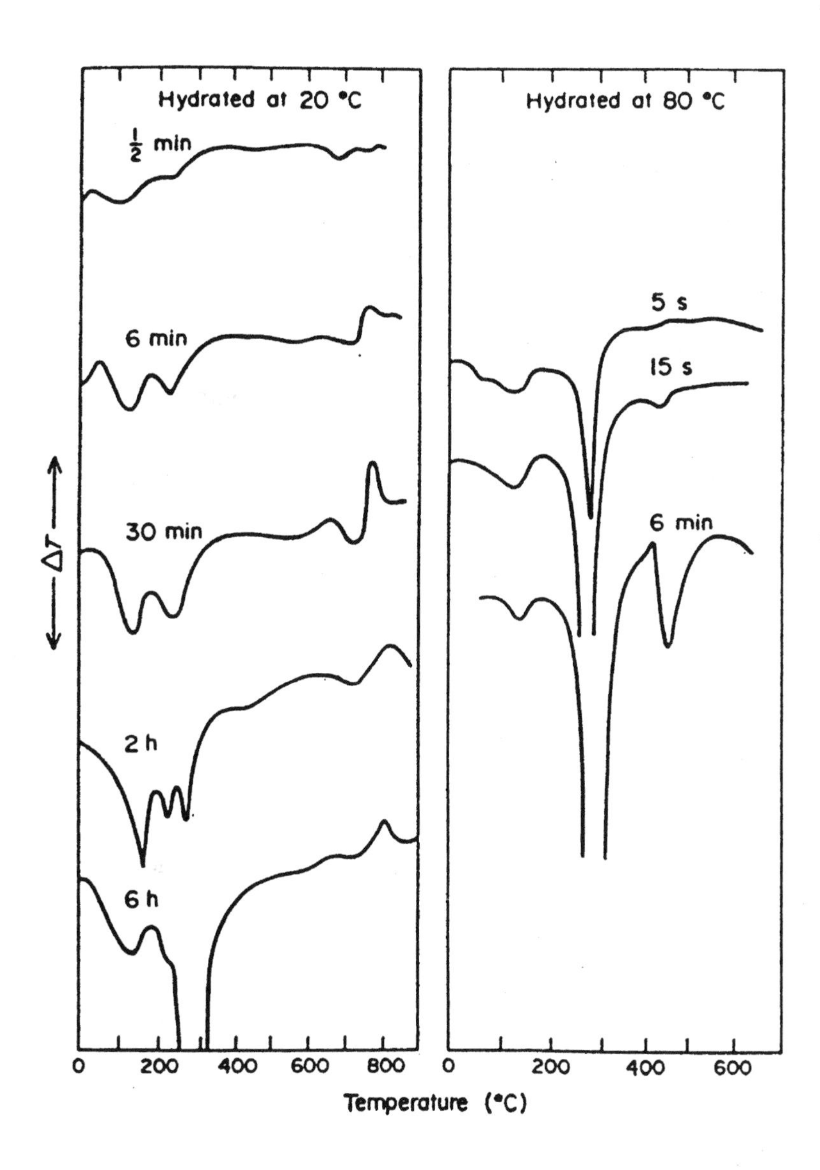

Figure 23. Thermograms of C_3A hydrated at 20 or 80°C (water:solid ratio: 0.12).

Tetracalcium Aluminoferrite. In the hydration of C_4AF, similar peaks to those of C_3A are indicated but the rate of hydration is slower. Hexagonal and cubic phases formed from the hydration of the ferrite phase contain Fe. Not much work has been carried out on the thermal analysis of C_4AF. The thermograms of C_4AF hydrated for various periods up to 180 days reported by Kalousek and Adams[1] are shown in Fig. 24. The sample hydrated for 7 days shows the presence of the cubic phase and the hexagonal phase through peaks at 360 and 220°C respectively. The peak effect for the cubic phase is maximum at 60 days after which it decreases and completely disappears at 180 days. Only the hexagonal phase is evident at this period. It is, thus, thought that the stable phase in the hydration of C_4AF is the hexagonal phase or related phase containing some Fe_2O_3, the cubic phase being metastable. Mossabauer spectroscopic work has indicated the presence of the hydrous iron hydroxide amongst the products of the hydration of C_4AF at 72°C.[17]

Ramachandran and Beaudoin[49] investigated the effect of temperature and w/s ratios on the hydration of C_4AF by applying various measurements and techniques such as DSC, TG, conduction calorimetry, XRD, length changes, specific surface area, porosity, microstructure, and microhardness. The samples were hydrated at w/s ratios of 0.08, 0.13, 0.3, 0.4, 0.5, and 1.0 and temperatures of 23, 80, and 216°C. The TG technique was used to estimate the amount of hexagonal and cubic phases formed. The weight loss up to about 250°C was attributed to the hexagonal phases and the loss beyond, up to 500°C was assigned the presence of the cubic phase. Generally, the TG curves exhibited losses corresponding to the endothermal effects in the DSC. In Table 4, the estimates of hexagonal and cubic phases, determined by TG, are given. The samples prepared at a w/c ratio of 0.13, or 0.3–1.0 was hydrated for 2 days, whereas that prepared at a w/s ratio of 0.08 was hydrated for 45 days. Of all the samples studied, those hydrated at a w/s ratio of 0.13 at 80°C indicated the highest value for the cubic/hexagonal phase ratio and the lowest value was obtained for the sample formed at a w/s ratio of 0.08 and hydrated at 23°C. The results showed that the formation of the cubic phase should not necessarily be detrimental to strength development. The cubic phase exhibits high strengths provided it is formed at a low w/s ratio. Maximum strengths were obtained in autoclave-treated samples. Direct formation of the cubic phase at the sites of the unhydrated particles may occur at higher temperatures and lower w/s ratios. In the micrographs, C_4AF has a closely welded structure similar to a vitrified body.

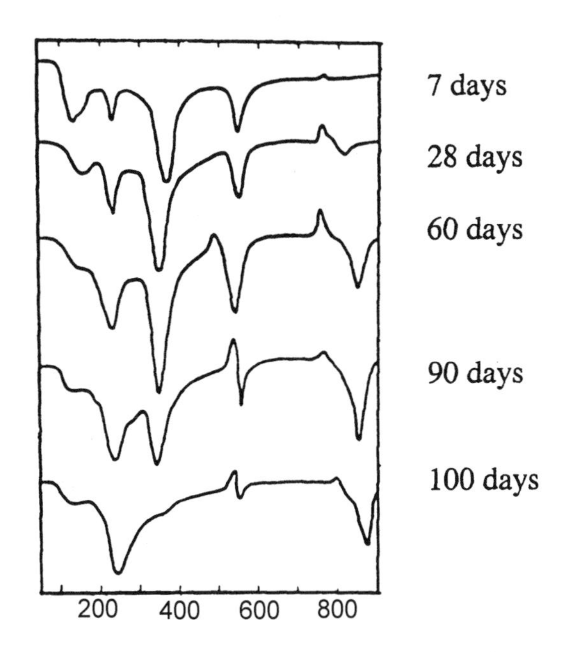

Figure 24. Thermal analysis curves of $4CaO{\bullet}Al_2O_3{\bullet}Fe_2O_3$ hydrated to different periods.

Table 4. Estimation of Hexagonal and Cubic Phases by TG Technique

W/S	Sample Preparation Temperature (°C)	Amounts of Phases (%) Hexagonal	Cubic
0.3	23	14.5	53.3
0.4	23	15.2	61.2
0.5	23	9.2	71.9
1.0	23	13.4	79.7
0.4	80	9.8	74.7
1.0	80	8.2	71.5
0.13	23	13.2	38.5
0.13	80	2.8	52.8
0.08	23	8.7	17.2
0.08	80	4.0	28.5

6.3 Calcium Aluminates Plus Gypsum

Tricalcium Aluminate Plus Gypsum. Several types of compounds are formed in the hydration of aluminate phases with gypsum. The peak effects are affected depending on the w/s ratio, particle size of the aluminate, temperature, relative proportions of gypsum, etc. Compounds detected in this system include, C_2AH_8, C_4AH_{13}, C_3AH_6, low sulfoaluminate hydrate, high sulfoaluminate hydrate, and their solid solutions. Feldman and Ramachandran[44] examined the sequence of products formed by hydrating C_3A mixed with 0, 0.25, 2.5, 10, and 20% gypsum and hydrated at 2, 12, 23, and 52°C for various periods. DTA, XRD, and length change measurement techniques were also adopted. The endothermic peak appearing at about 170–200°C was assigned to the high sulfoaluminate hydrate, the endotherms in the range 140–180°C and 240–285°C were assigned to the hexagonal C_2AH_8 and C_4AH_{13} compounds, respectively, and two endothermal effects at 290–300°C and 460–500°C, to the cubic C_3AH_6 phase. The compound C_4AH_{13} may be associated with the low sulfate aluminate solid solution. Thermograms revealed that the hexagonal hydroaluminates are formed immediately on exposure of the C_3A + gypsum to water. The formation of hexagonal hydroaluminates is retarded by gypsum. The conversion of the hexagonal to the cubic form is also delayed. In general the rate of hydration of C_3A and its reaction with gypsum are increased with temperature. As the temperature is increased higher amounts of gypsum are required to suppress the formation and prevention of conversion of the hexagonal phases.

As a part of an investigation of the effect of calcium carbonate on the C_3A-gypsum system, Ramachandran and Zhang Chun-Mei[45] studied the hydration of the C_3A-gypsum system. In Fig. 25, DSC curves for the C_3A + 12.5% gypsum mixture hydrated at 20°C are shown. Gypsum is almost consumed in 30 minutes, as is evident from the intensity of the peak at 130°C. Ettringite is characterized by an endothermic effect at 165–180°C. The intensity of the peak for ettringite increases after 30 minutes. The low sulfate hydrate, identifiable by an endothermal effect at 280–300°C, is present at 1 hour and later. With the addition of 25% gypsum, ettringite is detected at 5 minutes and its amount increases with time until 1–2 hours. Gypsum is mostly consumed within an hour, but at 12.5% gypsum, it is consumed in 30 minutes (Fig. 26).[45]

Thermogravimetric technique has also been applied to follow the formation of sulfoaluminate products hydrated in a mixture containing tricalcium aluminate and gypsum.[17]

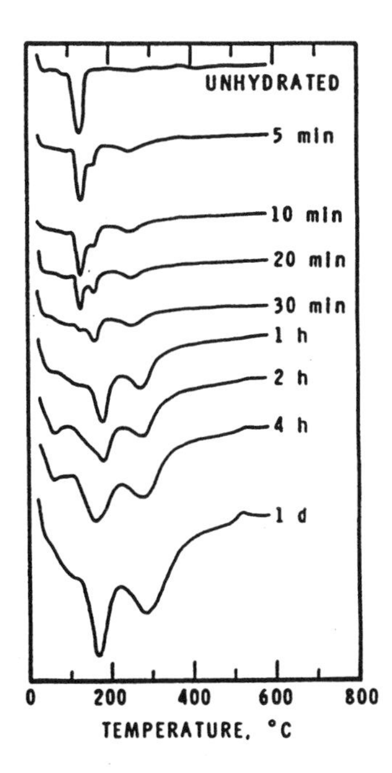

Figure 25. DSC curves of C_3A hydrated for different periods in the presence of 12.5% gypsum.

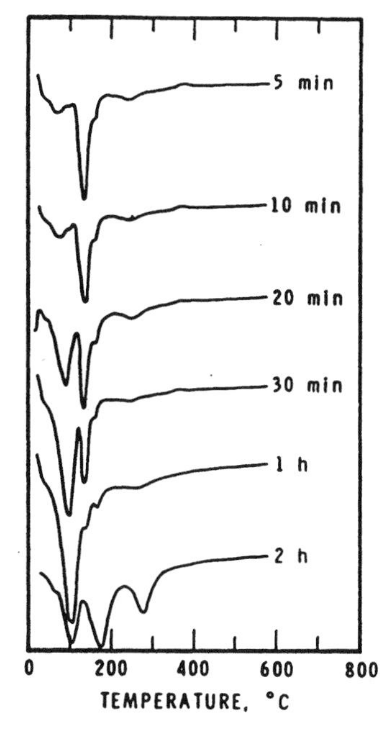

Figure 26. DSC curves of C_3A hydrated for different periods in the presence of 25% gypsum.

In the conduction calorimetric curves of the aluminate-gypsum system,the first stage is characterized by a peak at 30 minutes for the formation of ettringite. In the second stage, the peak occurs at 24–48 hours when the monosulfate phase forms by the reaction between ettringite and aluminate.

Kalousek, et al.,[1] in an extensive work on the thermal analysis of cements concluded that both low and high aluminate sulfate forms ultimately disappear and are incorporated into the C-S-H gel.

Collepardi, et al.,[46] studied the effect of gypsum and calcium hydroxide on the hydration of C_3A by applying DTA and isothermal calorimetry. It was found that the retardation of C_3A hydration occurred only when ettringite was formed. Almost no retardation resulted when sodium sulfate was substituted for gypsum. With sodium sulfate addition, no ettringite was formed. Largest retardation was observed when both gypsum and lime were present. This effect was attributed the formation of a colloidal sized ettringite crystals on the aluminate surface. Odler and Abdul-Maula [47] compared the DTA and XRD methods for estimating the low and high sulfoaluminates in hydrating systems.

A calcium fluoroaluminate of formula $11CaO{\bullet}7Al_2O_3{\bullet}CaX_2$, where X is a halogen, preferably fluorine, is used in combination with portland cement to produce a regulated set cement. This cement is useful where quick setting and early strength would be needed. DTA has been applied to study the hydration behavior of the halogenated compound hydrated for different times and temperatures.[54] The peak obtained at 90°C was attributed to the loss of adsorbed water and water from alumina gel. A peak at 300°C denoted the presence of C_3AH_6 and one at 260–270°C was caused by the conversion of AH_3 (gibbsite) to AH. Transformation of C_2AH_8 to C_3AH_6 and AH_3 was characterized by a peak at 250°C. At 15°C of hydration, initially CAH_{10} formed and after 2 hours C_2AH_8 and C_3AH_6 appeared. At 3 days, hydration was completed. At 25°C, the main hydration products were C_2AH_8, C_3AH_6, and alumina gel. The C_2AH_8 phase disappeared after 7 days and C_3AH_6 was well developed. At 30, 40, and 60°C, the only hydrates that existed were C_3AH_6 and alumina gel. In the presence of cement, ettringite and monosulfate hydrate were identified. The rate of strength development was rapid during the first 6 hours. Further strength was attained at later ages. The strength in the earlier period was due mainly to ettringite. Later, the C-S-H phase contributed to strength. It was concluded that in the presence of fluroaluminate, hydration of alite in portland cement is accelerated.

Attempts have been made to apply an *evolved gas analysis* technique (EGA) to investigate cement systems. In the EGA technique, a substance is heated at a uniform rate and the evolved gas is quantitatively measured, thus permitting the estimation of the components. For example, characteristic peaks have been obtained when sulfoaluminate hydrates, gypsum, $Ca(OH)_2$, etc., are subjected to EGA.[48] In this method, water produced by disassociation reactions is passed through P_2O_5 layer and adsorbed water is electrolyzed and measured coulometrically. The EGA traces of calcium monosulfate hydrate, ettringite, and gypsum are shown in Fig. 27.[48] The disassociation pattern of ettringite is similar to that obtained in DTA. The first sharp peak and the second broad peak are related to the loss of 24 and 6 mols of water respectively from ettringite. Four peaks are registered by monosulfate hydrate. Gypsum exhibits typical dual peaks for a stepwise dehydration.

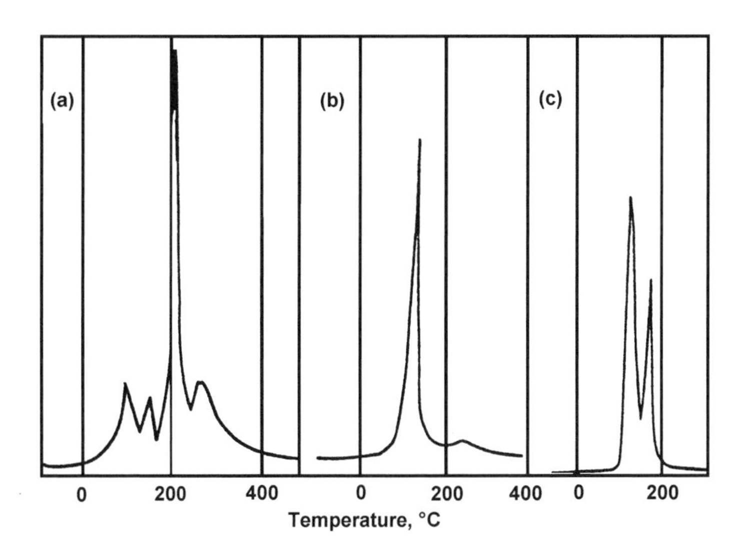

Figure 27. Typical EGA traces: *(a)* calcium aluminum monosulfate hydrate, *(b)* calcium aluminum trisulfate hydrate, *(c)* gypsum.

Tetracalcium Aluminoferrite Plus Gypsum. In the hydration of C_4AF + gypsum mixtures, depending on the initial properties, temperature and time of hydration, several types of compounds are detected. The compounds may consist of unhydrated C_4AF, $CaSO_4 \cdot 2H_2O$ and the hydration products, $C_4(AF)H_{13}$ (hexagonal phase), $C_3(AF)H_6$ (cubic phase), $C_3(AF) \cdot 3CaSO_4 \cdot H_{32}$ (ettringite), $C_3A(AF) \cdot CaSO_4 \cdot H_{12}$ (low sulfoaluminate), and a solid solution of sulfoaluminate with $C_4(AF)H_{13}$.

Ramachandran and Beaudoin[51] followed the hydration of C_4AF containing 5–30% gypsum, prepared at w/s ratios of 0.08, 0.13, 0.5, and 1.0 and exposed to a temperature of 25°C or 80°C. Techniques such as DTA, XRD, and conduction calorimetry were used. Compounds were identified by their characteristic thermal effects as follows. Gypsum indicated two peaks in the range 150–200°C representing the stepwise removal of water. At low concentrations, gypsum gave a single peak at about 150–160°C. Free water gave an endothermal effect at 100°C. The hexagonal phase exhibited an endothermal peak at about 160–175°C, the cubic form at 300–325°C. Ettringite was identified by an endothermal peak effect at 110–125°C and the low sulfoaluminate by an endothermal peak at about 200–210°C.

The relative amounts of unreacted gypsum and high sulfate aluminate hydrate contained in samples cured for different times in a sample hydrated at 25°C and a w/s ratio of 0.13 are given in Fig. 28.[53] Calculations suggest that within 1 hour relatively more gypsum has reacted to form ettringite in mixes containing larger amounts of gypsum. Conduction calorimetric investigation also indicates that in the first 30 minutes a larger amount heat of heat is developed in samples containing greater amounts of gypsum. Although almost all gypsum has reacted at 7 hours in samples containing 5–20% gypsum, substantial amounts are still present in the mixture prepared with 30% gypsum. All samples show a general decrease in the amount of ettringite after 3–7 hours of hydration and an increase in the amount of low sulfoaluminate phase. It is generally believed that ettringite begins to convert to the low sulfate form after all gypsum has been consumed. This may be valid at low gypsum contents. At a 30% gypsum level, although there is a decrease in the amount of ettringite and an increase in the amount of low sulfate form, there is still a large percentage of unreacted gypsum at 7 hrs. It appears that after this length of time the reaction between ettringite and ferrite phase to form low sulfoaluminate progresses at a faster rate than the reaction between gypsum and C_4AF to form ettringite. Length measurements as a function of time in admixture containing 30% gypsum show a higher expansion of 10.8% at 7 hrs compared to only 2.9% for that containing 20% gypsum. A higher expansion may result in a higher porosity and better availability of the C_4AF surface for the reaction. It was also found that at very low w/s ratios and a higher temperature the low monosulfate phase need not result from the conversion reaction involving ettringite.

In Fig. 29, the thermal curves of $C_4AF\text{-}CaSO_4\cdot 2H_2O$ mixes hydrated at a w/s ratio of 0.5 or 1.0 are given. Most samples hydrated for 2 days at 25°C contain mainly the low sulfoaluminate.[51] The interconversions seem to occur at a faster rate at higher w/s ratios. At 80°C and a low gypsum content, C_4AF hydrates to form the cubic phase. The conversion of ettringite to the monosulfate also occurs at a faster rate. A combination of higher w/s ratios and high temperatures is conducive to faster hydration and inter-conversion.

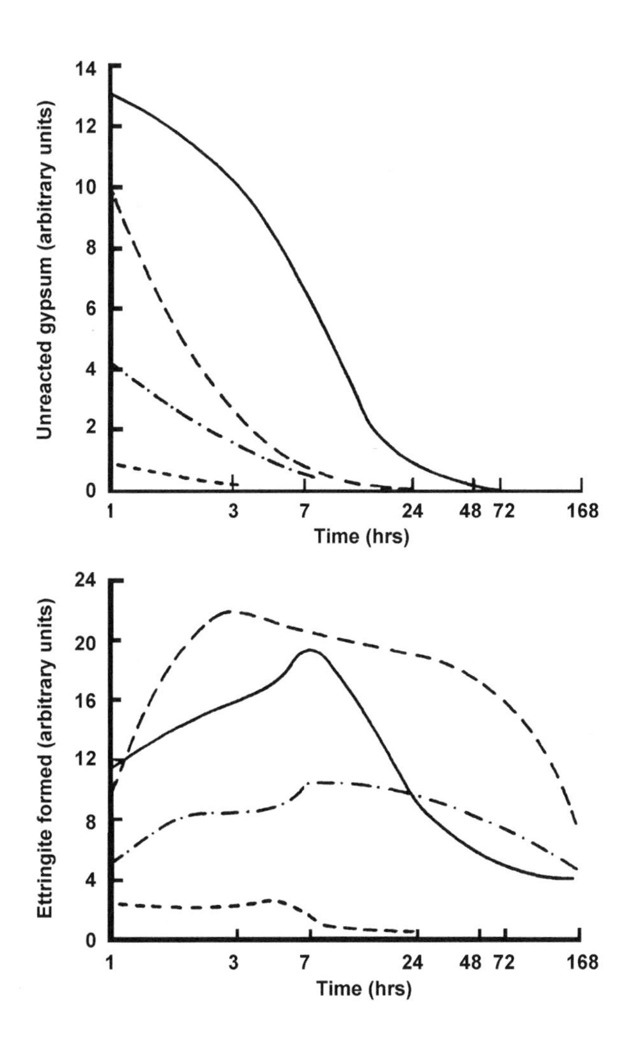

Figure 28. Relative amounts of gypsum and ettringite present in the C_4AF-$CaSO_4{\bullet}2H_2O$-water system.

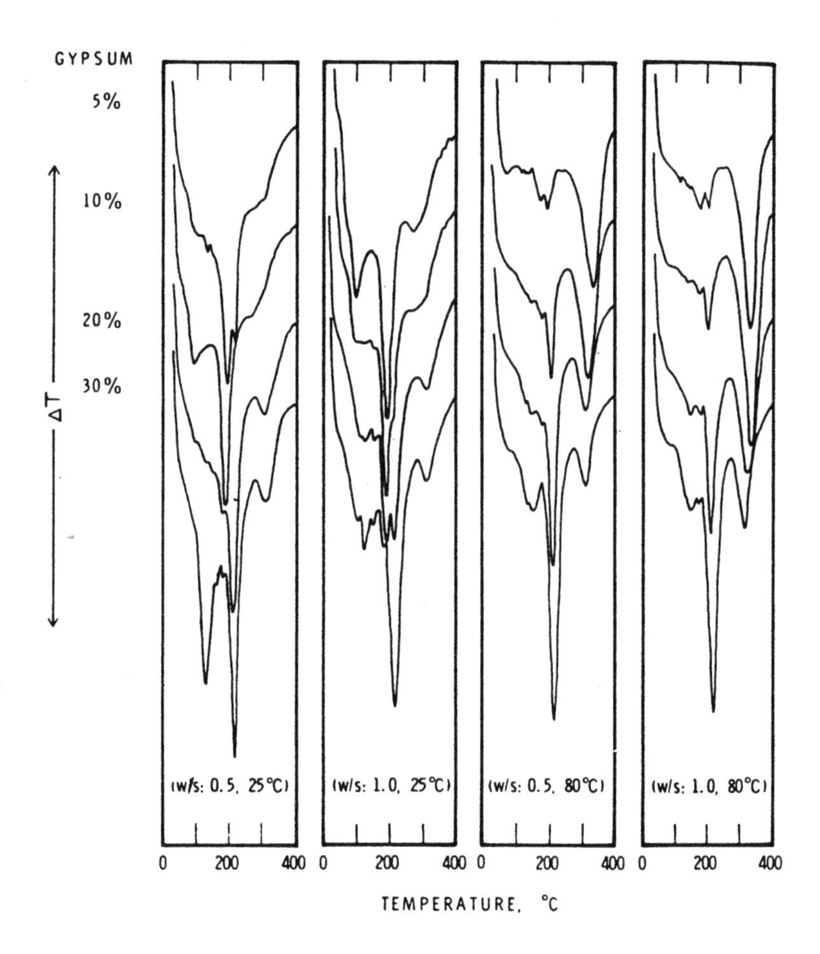

Figure 29. Differential thermal behavior of C_4AF-gypsum (w/s ratio 0.5 or 1.0) hydrated for 2 days at 25 or 80°C.

Chen and Shi[52] compared by DTG the hydration reaction of the ferrite phases C_4AF and C_6AF_2 in the presence of gypsum. In DTG curves they detected tricalcium aluminate sulfate, gypsum, low sulfoaluminate, and $Fe(OH)_2$ gel. The ettringite transformed to the low sulfate form after all gypsum was exhausted. The process of C_6AF_2 hydration was similar to that of C_4AF. The gel phase was found in greater amounts in C_6AF_2 and some $Ca(OH)_2$ was also generated. It was concluded that the rate of ettringite formation from C_6AF_2 was slower than that from C_4AF. When calcium hydroxide was also used with gypsum, the formation of ettringite was retarded.

7.0 PORTLAND CEMENT

In the hydration of portland cement, several products can be detected by applying DTA and TG techniques. Unreacted gypsum may be identified by endothermal peaks in the temperature range of 140–170°C, the C-S-H gel at temperatures in the range of 115–125°C, ettringite at temperatures of 120–130°C, AFm phase at 180–200°C, calcium hydroxide by an endotherm in the range of 450–550°C, and calcium carbonate at 750–850°C. Unhydrated cement may also exhibit a small endothermal peak at about 485°C for CH formed during storage. It has to be recognized that there are interference effects at low temperatures, depending on the drying procedures and the state of the material. In some instances, a small additional endotherm appears before the onset of the calcium hydroxide endotherm and this is attributed to the dehydration effect of the chemisorbed water on lime or to the finely divided form of lime. The lime formed at different times may be estimated by determining the endothermal area of the lime peak or weight loss. The amount of lime is nearly proportional to the degree of hydration of cement.

The sequence of hydration products formed in the hydration of cement at different periods has been followed by DTA (Fig. 30).[1][55] The endothermal effect at 140 and 170°C characterize gypsum and the endothermal peaks below 500°C and 800°C are due respectively to the decomposition of $Ca(OH)_2$ and $CaCO_3$. Within five minutes, an endothermic effect appearing at 130°C is caused by the dehydration of ettringite. Gypsum is partly consumed at this period as evident from the reduced intensity of the peaks at 140°C and 170°C. At one hour, there is an increase in the intensity of the ettringite peak at the expense of the gypsum peak. The onset of an endotherm at about 500°C, after four hours of hydration, is due to the $Ca(OH)_2$ formed from the hydration of the C_3S phase. A small endotherm registered below 500°C may be due to the chemisorbed water on free lime or to the finely divided $Ca(OH)_2$. The endothermal effect at 800°C is caused by the decomposition of calcium carbonate.

The DTA of low alkali and high alkali cements show interesting features. Bensted[68] carried out DTA investigations of low and high alkali cements hydrated from a few minutes to one year. The hydration sequence (with respect to gypsum component) proceeded as follows: hemihydrate→ dihydrate of gypsum→ ettringite→ monosulfate/C_4AH_{13}. These reactions did not proceed to completion at any one stage. Syngenite, appearing at very early stages, became depleted through reaction with sulfate to form ettringite.

It reappeared when alkalis were leached out into the aqueous phase. More than one type of C-S-H appears to have been detected. Small amounts of MgO were present at three days of hydration and beyond. At one year, phases identified included C-S-H, monosulfate/C_4AH_{13}, syngenite (only in the low alkali cement), brucite, and $Ca(OH)_2$.

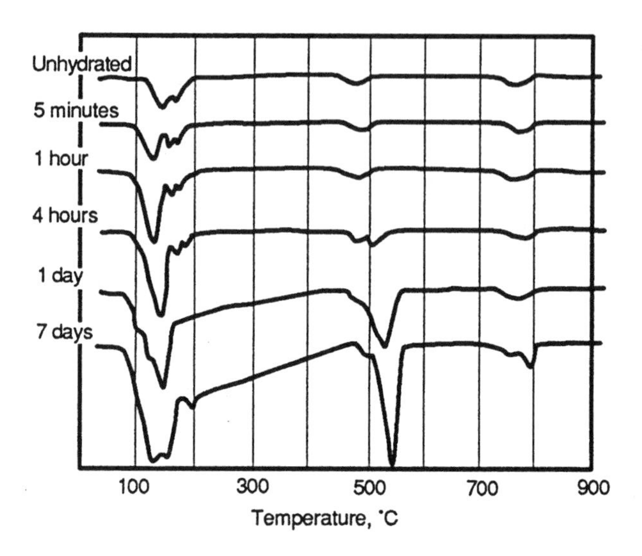

Figure 30. DTA of cement hydrated at different periods.

Odler and Abdul-Maula[47] concluded from their investigation of cement pastes that for the estimation of AFt phase, DTA was a better method than XRD. In Fig. 31, DTA curves of pure ettringite, monosulfate, and portland cement hydrated for different periods are given.[47] Neither gypsum, nor the low monosulfate, interferes with the ettringite determination. As the hydration progresses, C-S-H formation results in a broad endothermal effect. The interference due to this effect could be circumvented by placing a pre-hydrated gypsum free cement, rather than α-Al_2O_3 as a reference material.

The rate of hydration of cements can be determined through the heat development characteristics using a conduction calorimeter. In Fig. 13, the heat effect in the first few minutes is attributable to the heat of wetting and ettringite formation. Within a few hours, another strong exotherm appears due to the hydration of C_3S. In some cases, depending on the composition of the cement, an additional peak is observed after the C_3S peak. This is related to the reaction of C_3A to form the low sulfoaluminate hydrate.

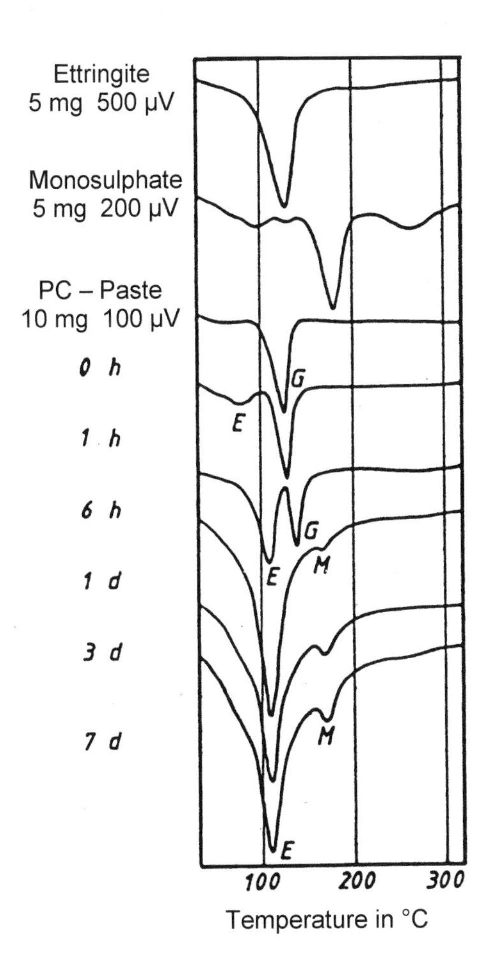

Figure 31. Thermograms of calcium aluminate sulfates and cement hydrated for different periods.

The isothermal conduction calorimetry offers a method to follow the rate of hydration of cement at different temperatures of curing. Calorimetric curves of cement hydrated at 25, 30, 40, 50, 60, and 80°C have been analyzed.[56][57] It was found that as the temperature increased the C_3S hydration peak appeared at earlier times. The shape of the curves also underwent changes. The apparent activation energy of hydration could be calculated.

In cement pastes, the calcium silicate hydrate phase contains adsorbed, as well as interlayer, water. In order to differentiate and estimate these two types of water, Feldman and Ramachandran[58] carried out DTA/TG of cement pastes under controlled humidity conditions. Two endothermal peaks emerged in samples exposed to different humidities, one due to the adsorbed water at 90–110°C and the other at 120–150°C caused by the interlayer water.

The mechanism of setting of cement pastes is of considerable interest. It is generally thought that the ettringite formation is a prerequisite for the setting of the paste. Some studies have suggested that some hydration of C_3S should take place before setting can occur. The evidence is based on the existence of an endothermal peak due to the dehydration of calcium hydroxide present at the time of setting. This was observed for several cement pastes containing various types of admixtures.[59]

The TG method has also been found to be suitable for following the hydration of portland cement.[17] The TG of a mature paste indicates a loss at 425–550°C primarily due to the CH decomposition, and the AFm phase also shows a step in the same temperature range. The loss at 550°C is partly caused by CO_2 and partly due to the final stage of dehydration of C-S-H and hydrated aluminate phase. For a typical portland cement cured for 3–12 months, the CH amount determined by thermal or XRD method is in the range of 15–25% on the ignited basis. Thermogravimetric method (static) was adopted by Feldman and Ramachandran[60] to construct adsorption-desorption isotherms in cement pastes. Water content was measured at different humidities by heating the material in a vacuum for 3 hrs at 100°C in a TG apparatus. This was equivalent to exposure to d-dry condition. In the d-drying method the sample is evacuated, using a trap with solid carbon dioxide. In Fig. 32, the isotherms of cement paste exposed to different humidity conditions are presented. Preparation A was a paste exposed to 11% RH for 6 months. Preparation B was d-dried and exposed to 11% RH. Preparation C was obtained by heating A at 70°C for 2 hrs. The sample contained 3% more water than preparation B. The samples were exposed to an equilibrium RH values of 11, 32, 44, 58, 66, 76, 84, and 100% and reconditioned from 100% to 11% RH. Scanning curves were also constructed at different humidities. In comparing the two isotherms for B and C it can be observed that the sorption boundary for curve B (d-dried) has a greater slope. A large hysteresis loop was found for preparation B. Since preparation C was not d-dried it did not allow complete regain of water when exposed at low or intermediate humidities. Interlayer water re-enters in the 0–33% range. The increased step in the BET region is due to the increased interlayer re-entry in this region. All the water, including hydrate water, is removed when the sample is d-dried or partially dried (C) from 11% RH, after re-wetting.

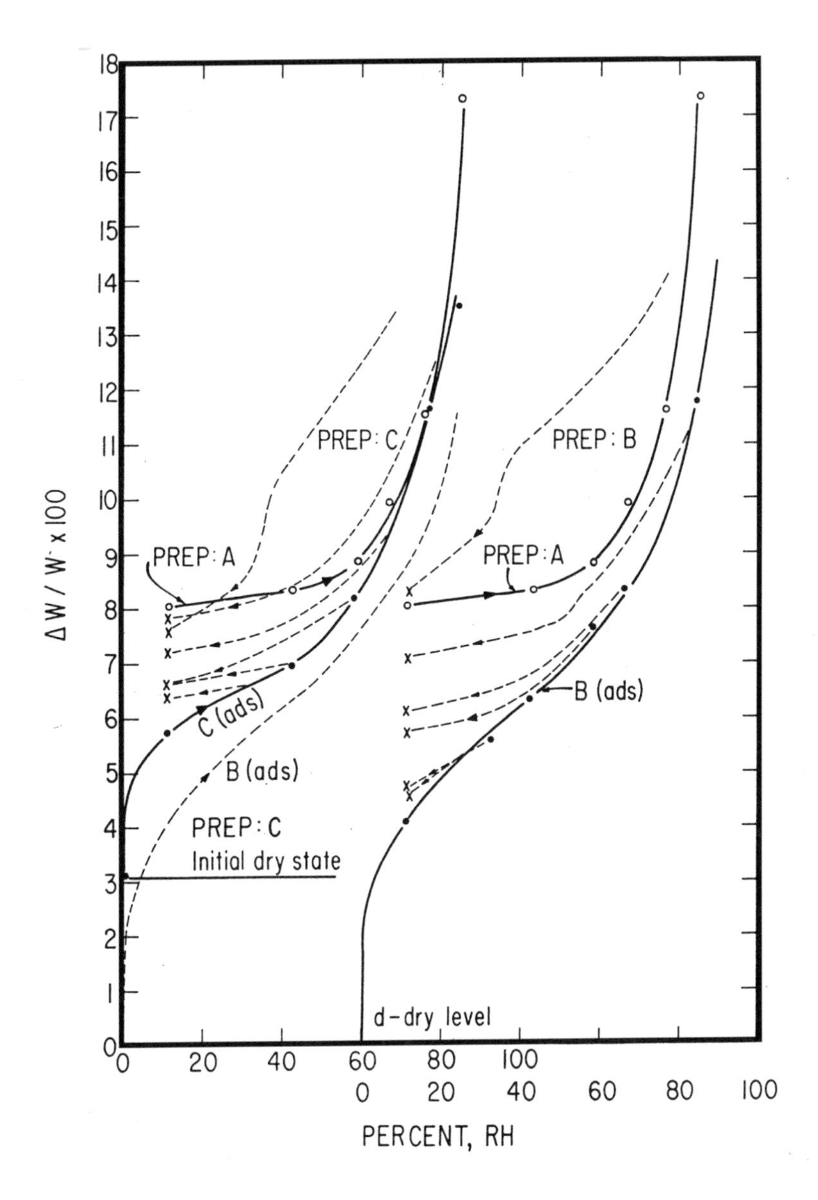

Figure 32. Isotherms of hydrated cement from static thermal balance.

A semi-isothermal method of derivative thermogravimetry has been suggested for better resolution of thermal effects and for quantitative analysis of calcium hydroxide in cement pastes.[61] The formation of several compounds during the hydration of cement results in interfering effects. Butler and Morgan[62] dehydrated cement at 200°C in N_2, followed by carbonation and subsequent thermal decomposition of $CaCO_3$. Only calcium hydroxide was carbonated. Although it was not possible to determine C-S-H accurately, it was possible to determine by DSC other products such as sulfoaluminate, hexagonal aluminate hydrates, cubic aluminate hydrate, and calcium hydroxide.

A cement exposed to moisture and carbon dioxide during storage may partially hydrate and carbonate. The kinetics of hydration and strength development of cement may be affected. DTA/TG techniques have been applied to identify the hydrated products in such a cement. Thiesen and Johansen[74] investigated the effect of prehydration of cement on compressive strength. The degree of adsorbed water was determined by TG and the compounds (mainly calcium aluminate-based compounds) were identified by DTA. DTA showed the formation of ettringite when the cement was exposed to a low temperature and a humid atmosphere. In the TG curve, weight loss occurred in four stages. From these losses the amount of combined water with clinker minerals was calculated. In commercial clinkers the combined water varied from 0.15 to 0.30%. Each cement has a critical value above which strength may deteriorate. Strengths, in certain severe cases were reduced by 30%.

It is important to acquire information on the hydration processes that occur when cement is subjected to low temperatures. Even at temperatures below zero cement hydrates slowly, as water does not freeze due to the presence of soluble materials in the pores. The products that form at low temperatures are calcium aluminate hydrates at earlier periods and after several months, the products of hydration of tricalcium silicate. A weak endothermal peak at about 330°C could be due to the presence of silica gel and this is preceded by the appearance of calcium hydroxide and C-S-H gel.[5]

Supersulfated cement has a lower heat of hydration and shows better resistance to sulfate attack than normal portland cement. It has lower CH contents and most of aluminum is bound as ettringite. This cement may contain 80–85% slag, 10–15% anhydrite, and 5% activator. The main hydration products are C-S-H and ettringite. A comparison of the conduction calorimetric curves of the supersulfated cement with that of normal cement made at different water:cement ratios indicates that the first intense inflection occurs much earlier in the supersulfated cement.[63] The high strength of supersulfated cement is due to the dense structure of ettringite produced at early times of hydration.

Oil well cements are employed in the exploration of oil and gas. They are used for sealing the annular space between the bore hole walls and steel casing. Individual well conditions would dictate the type of cement and admixture to be used. The Type G oil well cement is basically a sulfate-resisting cement coarsely ground to about 200–260 m^2/kg. Lota, et al.,[64] studied, by TG and DTA, the hydration of class G oil well cement at 5 and 20°C. In Fig. 33, thermograms of pastes hydrated from three hours to one year are presented. At 20°C and 3 hrs, an endotherm at 115°C indicates C-S-H which increases with time. At 135°C an endotherm is attributed to AFt phase which decreases after six hours. The peak at 435°C is caused by the decomposition of CH. Thermograms at 5°C also show similar peaks but of different intensities. A greater amount of AFt phase is indicated by a sharper peak. The CH peak is smaller in thermograms obtained at 5°C.

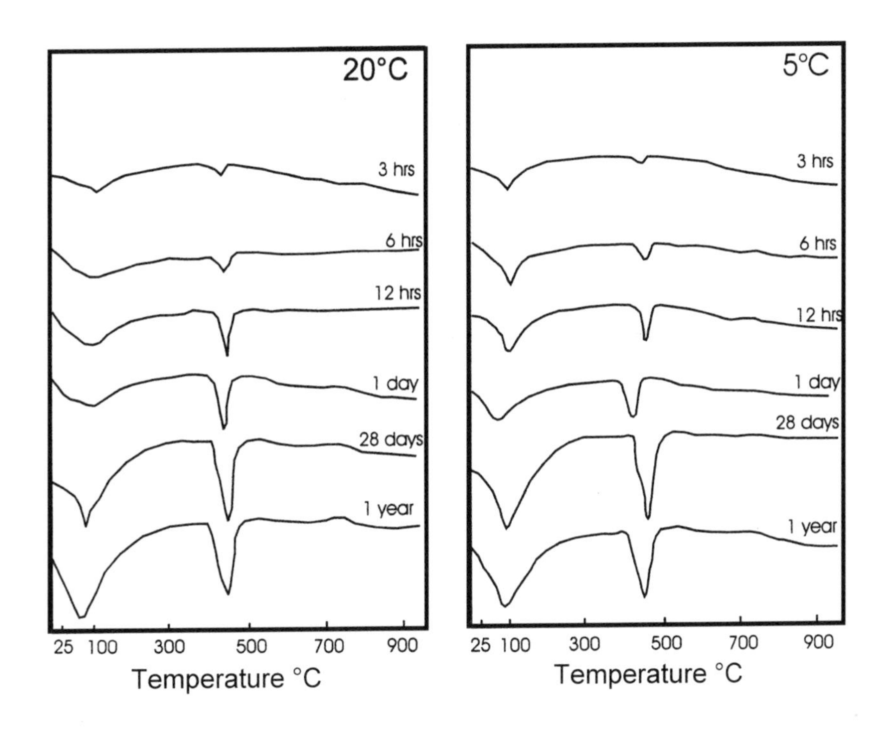

Figure 33. DTA curves of an oil well cement hydrated at 5 and 20°C.

Cements have been suggested as materials that could be used to immobilize heavy metals produced by various industries. Cadmium and its compounds are highly toxic and can effectively be retained in concrete provided the pH does not fall below 7. The main mechanism of Cd stabilization is related to its precipitation as cadmium hydroxide and physical entrapment. The possibility of Cd^{2+} substituting Ca^{2+} by solid diffusion or dissolution mechanism, forming a precipitate of $Ca{\bullet}Cd(OH)_4$ has been proposed by Goni, et al.,[69] based on TG/DTG studies.

The potentiality of the emanation thermal analysis for the investigation of the hydration of cement and cement compounds has been discussed by Balek.[67] This technique is based on the measurement of radioactive gases released from the hydrating phase. The amount of gas released depends on the physico-chemical processes taking place in the solid. A quantitative estimation of the rate of hydration of cement and C_3S at early stages has been obtained.

8.0 CaO-SiO_2-Al_2O_3-H_2O AND RELATED SYSTEMS

The binary, ternary, and quaternary systems containing oxides of relevance to cement science have been studied at ambient temperatures or under autoclaving conditions. Various hydrated products, crystalline as well as poorly crystallized, are formed.

The thermal analysis techniques were found to be of great value by Kalousek and co-workers for investigating these systems.[1] In the CaO-SiO_2-SiO_2 system studied at normal temperatures, the formation of C-S-H products with different C/S ratios gave peaks at different temperatures. In Fig. 34, reported by Kalousek,[1] the C-S-H compounds with C/S ratios of 1.0 and 1.33 show distinct exothermal peaks beyond 800°C. The peak denotes a crystallization effect. The peak is shifted as the C/S ratio increases. Even after subjecting the samples to extraction by acetoacetic ester the curves (dotted) are not modified. This suggests that lime in these compounds is strongly bound. Generally, in the system containing lime-silica-water, with age the peak due to lime disappears and an exothermal peak caused by the crystalline effect at higher temperatures increases in intensity.

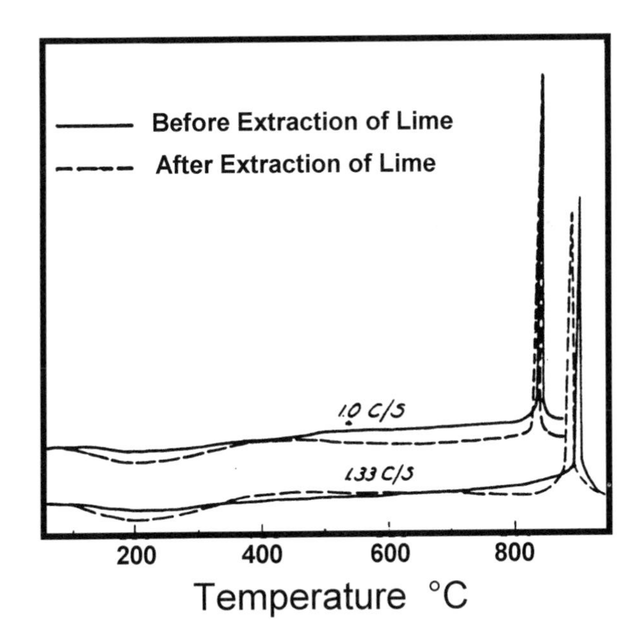

Figure 34. DTA curves of 1.0 and 1.33 C/S solids before and after extraction of free lime with acetoacetic ester.

Autoclaving is employed to produce articles of high early strength and reduced shrinkage. The strength attained in 24 hrs may be equivalent to that obtained at 28 days of normal curing. Many types of materials can be mixed with lime or cement and processed in high-pressure steam to produce concrete products of good strength. The materials that have been used are silica flour, pumice, scoria, tuff, cinder, calcined clay, fly ash, and blast furnace slag.

Several investigators have investigated the effect of autoclave curing on the products formed with mixtures of cement/lime , silica, and water. In an investigation, the effect of quartz (of different surface areas) added in different amounts to cements was investigated by DTA/TG techniques.[65] The main phases identified included CH (endothermal peak at 453°C), α-C$_2$SH (endothermal peak at 523°C), and C-S-H (a broad endothermal effect in the range of 100–700°C). An exothermal effect at 843–846°C indicated the crystallization of β-wollastonite, β-CS due to aluminous tobermorite or C-S-H of C/S ratio 0.8 to 1.0. The intensity of this peak increased with an increase in fineness of quartz (Table 5). A larger exothermal peak implied that higher amounts of C-S-H (I) or aluminous tobermorite was formed. For quantitative measurements peak area rather than peak height was found to be more accurate.

Table 5. The Effect of Surface Area of Quartz (cm^2/g) on the DTA Exothermal Area

Blaine Surface Area	Peak Height	Peak Area (Relative)
2200	0.69	1.26
2600	0.85	1.47
3600	0.95	1.47
5200	0.99	1.46
6750	0.92	1.52
(With permission from Elsevier Science)		

It has also been reported that the second derivative DTA is more useful to examine the products formed during the autoclaving of cement-quartz-metakaolin mixtures. Klimesch and Ray[66] subjected a mixture of quartz (38.5%) and cement (61.5%) containing different amounts of metakaolin and autoclaved them for 8 hrs at 180°C. It was found that the second derivative differential thermal curve provided a more detailed information, particularly in temperatures of 800–1000°C. In Fig. 35, DTA and second derivative curves for cement-quartz-metakaolin pastes are compared. The exotherms occur at 840, 903, and 960°C due to the formation of wollastonite from C-S-H, aluminum-substituted tobermorite, and anorthite from the hydrogarnet residue respectively. The small endotherm at 828°C preceding the first exotherm is probably caused by well crystallized calcite.

In the manufacture of autoclave products, cement, lime, or both, and silica-bearing materials form the raw mix. Various hydration products are identified when cement or calcium silicate is autoclaved. The compounds include: C-S-H gel, α-C_2SH, tricalcium silicate hydrate, CH, C_3AH_6, and a modified form of γ-C_2S hydrate. In the cement-aggregate system, products that may form are: C-S-H(I), C-S-H(II), 11 Å tobermorite,

α-C_2SH, xonotlite, gyrolite, afwillite, hillerbrandite, and 14 Å tobermorite. Various reactions that occur under autoclaving conditions have been reviewed by Ramachandran.[1] DTA of some of the hydrated products formed by autoclaving of siliceous systems with lime is shown in Fig. 36.

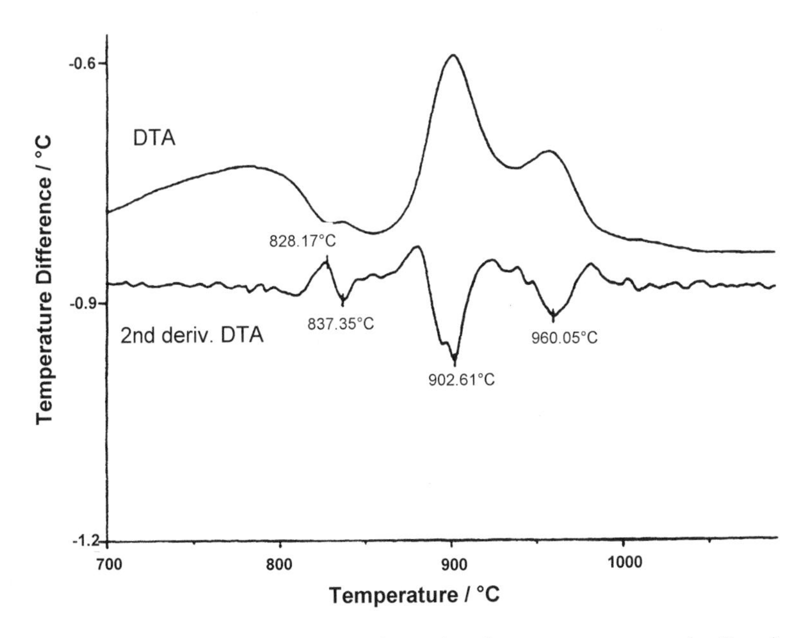

Figure 35. DTA and second derivative of autoclaved cement-quartz-metakaolin mixes.

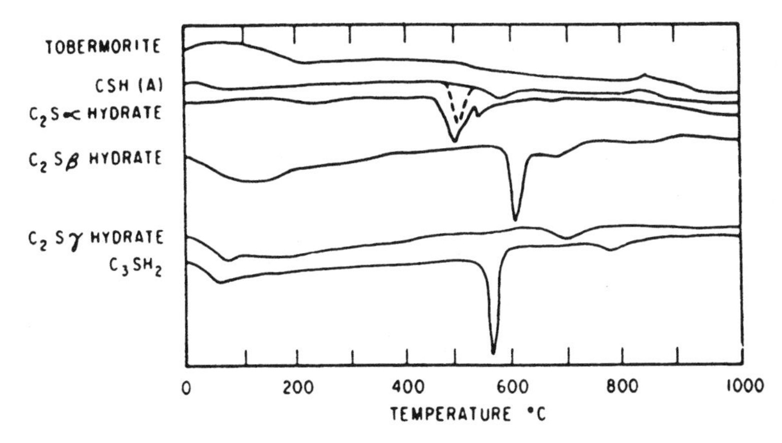

Figure 36. Differential thermograms of calcium silicate hydrates formed under autoclaving conditions.

The application of DTA to various binary, ternary, and quaternary systems of relevance to cement chemistry has been discussed in a book by Ramachandran.[1]

9.0 DURABILITY ASPECTS

Concrete may deteriorate if adequate precautions are not exercised to protect it from adverse effects that could result from exposure to natural or artificial conditions. Several physical, chemical, and electrochemical processes are known to induce cracking of concrete. Concrete can have durability problems as a consequence of its exposure to sea water, sulfates, chlorides, freeze-thaw action, carbon dioxide, etc., or when it is attacked by artificially induced processes such as exposure to acids and salts in chemical plants or to fire. In recent years, a new type of durability problem was encountered that involved use of steam cured concrete products. The distress was caused by the formation of delayed ettringite. If the raw materials in concrete are not carefully controlled, there may be an eventual failure of concrete elements, e.g., the presence of excess alkali in concrete that promotes alkali-aggregate expansion reaction, harmful impurities in the aggregates, or the presence of excess amounts of dead-burnt MgO. Thermal techniques in combination with others have been employed with success to examine the raw materials as well as the failed concrete. The knowledge gained from such work has been applied to produce more durable concrete.

9.1 Aggregates

Some organic and inorganic compounds present in small amounts in aggregates may affect concrete strength and durability. In Fig. 37, the thermograms of some of the harmful impurities that may be present in concrete are given.[1] Pyrite exhibits two exothermic peaks between 400 and 500°C caused by oxidation. Gypsum shows two characteristic effects at 180 and 220°C for dehydration effects. Montmorillonitic clay mineral is characterized by three endothermal effects at 160, 660, and 900°C due respectively to the release of interlayer, dehydroxylation and destruction of lattice. An exothermal effect at about 950°C is attributed to the formation of a spinel. Humus gives strong exothermal peaks over a wide range of temperatures (200–600°C) and lignite exhibits three exothermal peaks caused by oxidation. Opal shows an endothermic peak at 80°C for the loss of water and another at 170°C due to a structural transformation. Marcasite manifests three exothermal peaks at 450, 470, and 520°C.

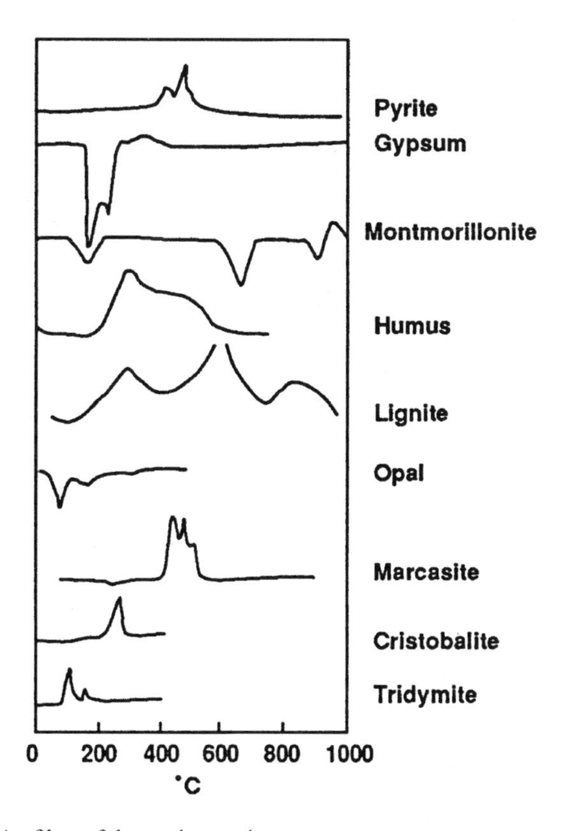

Figure 37. DTA of harmful constituents in aggregates.

The causes leading to the failure of concrete near the construction joints have been evaluated by TG.[70] The concrete sample exhibited peaks at 705–745°C and 905–940°C caused by the decomposition of calcite and dolomite, respectively. The slopes of the weight loss plot prior to the dolomite and calcite decompositions correlated with the field performance. In general it was found that in durable dolomite aggregate, carbon dioxide loss occurs at 570°C and continues up to 705°C for the decomposition of magnesium carbonate. Carbon dioxide from calcium carbonate is driven off at temperatures greater than 905°C. Poor durability dolomite showed very little weight loss until 700°C.

DTA, TG, IR, and X-ray microanalysis techniques were applied to identify the materials that formed around the rim of sandstone or silt stone aggregate in a thirty-year old concrete.[17] An alkali-substituted okenite ($C_5S_9H_9$), a precursor phase characterized by a 1.22 nm XRD spacing, was identified.

9.2 Magnesium Oxide

Magnesium oxide in cement exists mainly in a free state and its content does not exceed 6%. At the clinkering temperature of 1400–1500°C, free MgO is in a dead-burnt state in the form of periclase. Under normal conditions of exposure it may take years for periclase to hydrate. The conversion of MgO to $Mg(OH)_2$ is attended by an expansion in volume, hence there is a possibility of concrete to crack. Most specifications place a limit on the amount of MgO in cements and also a limit on the volume expansion of cements exposed to an autoclave treatment. Ramachandran[6][71] applied thermal techniques to explain the effect of curing conditions, viz, 50°C, boiling in water and autoclaving, on expansion and degree of hydration in cements containing different amounts of dead-burnt MgO. It was found that at MgO contents $< 2\%$ the expansion was lower and then there was a steep increase at higher concentrations (Fig. 38).[6] The amount of expansion depended on the strength of the matrix. At any particular concentration of MgO, cement B showed higher expansion than cement A. Cement A contained higher amounts of C_3S, had better strength and, hence, was able to resist expansive forces.[71]

The effect of various treatments on the conversion of MgO to $Mg(OH)_2$ was also examined by Ramachandran.[71] The computed results from DTA are shown in Table 6.[6][71] Curing at 50°C was found to be ineffective for hydrating MgO unless curing is extended to several days. Expansion is lower in pastes cured under non-autoclaved conditions. At 5% MgO addition, autoclaving (3 hours), boiling (2 days), steam curing (1 day) and curing at 50°C (2 days) produced length changes of 1.3, 0.3, 0.09, and 0.3, respectively in cement A. The corresponding values for cement B were 0.52, 0.3, -0.13, and -1.10%, respectively. These results demonstrate how the curing conditions and the type of cement influence the length changes.

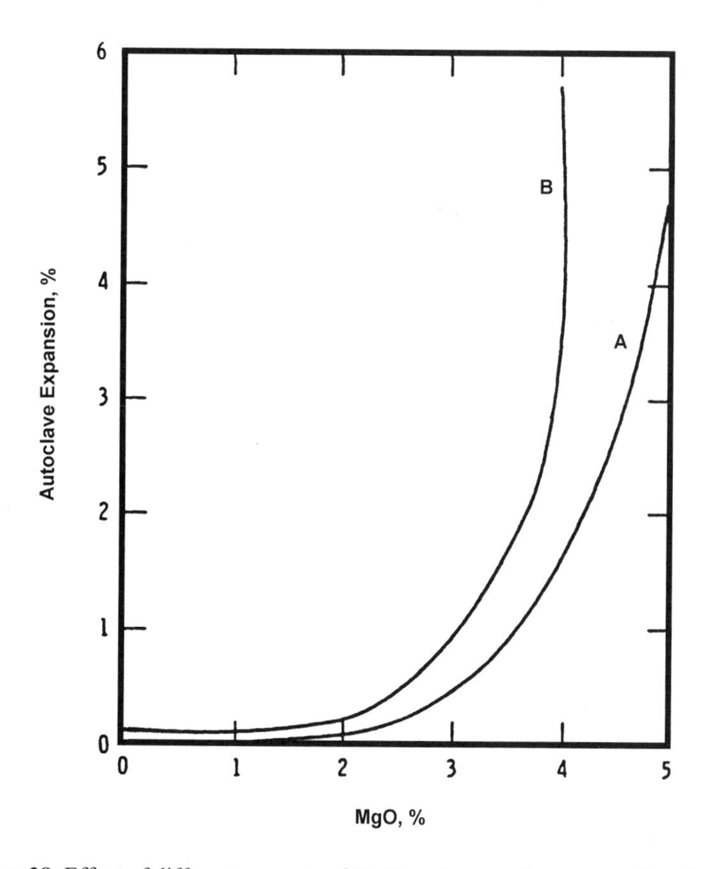

Figure 38. Effect of different amounts of MgO on the autoclave expansion of cements.

Under exceptional situations the aggregates may be contaminated with MgO. A contaminated coarse aggregate containing dead-burnt CaO and MgO was inadvertently added to concrete. After a few years the concrete exhibited large pop outs. DTA could be used to estimate the amount of $Mg(OH)_2$ present in the samples.[72] In another example, DTA[73] was applied to estimate the amount of unhydrated MgO present in a fifteen year old concrete. The amount of $Mg(OH)_2$ was determined by the endothermic peak at 400°C. By estimating the amount of $Mg(OH)_2$ before and after autoclave treatment, the amount of unhydrated MgO present in the sample could be computed.

Table 6. Effect of Curing Procedures on $Mg(OH)_2$ Produced

Curing Procedure	Cement A	Cement B
Autoclaved 2 MPa, 3 hrs	4.9	4.3
Boiling, 8 hrs	2.0	1.0
Boiling, 2 hrs	4.6	3.4
Steam Curing, 1 day	3.1	2.4
50°C, 2–3 days	0.0	0.0

9.3 High Temperature Effects

Concrete, having a relatively low thermal conductivity and high specific heat, provides protection to the steel against fire. At low temperatures concrete expands and by 300°C contraction due to the water loss occurs. Aggregates continue to expand and create stresses in the concrete. Quartz expands sharply at 573°C due to phase transition and decomposition of calcite leads to contraction. During the cooling period, calcium oxide begins to hydrate and causes expansion. Accidental fire causes damage to structural concrete elements. Assessment of the condition after a fire is important for recommendations for rehabilitation of concrete. DTA/TG techniques are useful to assess the temperature ranges to which the elements could have been exposed.[75] Damaged concrete (at different depths) has been analyzed by DTA/TG. The damaged concrete did not exhibit dehydroxylation peak of $Ca(OH)_2$ indicating that such concrete was exposed to temperatures above 500°C. The undamaged concrete contained $Ca(OH)_2$, as evidenced by the endothermal peak. A new thermal technique has been devised to monitor thermal expansion of cementitious materials as a function of temperature.[76] Higher temperatures also chemically alter the concrete performance.

Fiber reinforcement is an established means of improving the mechanical properties of a variety of matrices. Sarvaranta, et al.,[77] studied by TG/DSC the thermal behavior of polypropylene and two types of polyacrylonitrile fibers. A DSC examination revealed that the polypropylene fiber exhibits an endothermic effect at 160–170°C caused by melting. This is followed by an exothermal effect around 230°C due to oxidation and an endothermal peak at 300°C for the decomposition. The polyacrylonitrile did not melt but showed an exothermic effect around 300°C caused by the nitrile group polymerization and other side reactions. The non-melting of this fiber together with its rapid exothermic degradation effect may increase the risk of dangerous spalling in mortar exposed to rapid thermal exposure.

9.4 Freezing-Thawing Processes

Concrete is vulnerable to cracking when subjected to increasing number of freezing-thawing cycles. The freezing processes occurring in the various types of pores in the cement paste are complex and low temperature thermal analysis methods involving DTA and DSC have been applied with some success in explaining the mechanism of expansion and contraction.

The increased damage that occurs when concrete is exposed to the freeze-thaw cycles and salts has been investigated by many workers.[79] Experiments involving measurements of length changes and heat changes that occur when the cement paste is exposed to temperatures between 15 and -70°C have yielded information on the mechanism of salt scaling.[79][80] Figures 39 and 40 present results of length changes and DTA of cement pastes saturated with NaCl (brine) of different concentrations (0, 5, 9, 13, 18, and 26%). During cooling cycles (Fig. 39), especially at lower concentrations, there is an expansion of 0.5% at 5% concentration compared to 0.08% in water and 0.01% at 26% concentration. On warming a significant contraction occurs with highly concentrated salts at about -21°C, the melting point of the eutectic. In the cooling cycle (Fig. 40), the first exothermic peak is attributed to the commencement of formation of ice on the external surface. According to Litvan,[79][80] this represents a freezing of the disordered liquid from the pores of the cement paste. The second exothermal effect at -22°C is ascribed to the freezing of the solution of eutectic composition. On warming, two processes occur, one at about -21°C and another at a higher temperature, increasing in magnitude with increasing salt concentration. The peak at the lower temperature represents melting of eutectic mixture and that at the higher temperature, the melting of pure ice. In addition to the physical mechanism of increased volume at low temperatures due to the presence of salts, chemical reactions may also occur, especially at higher temperatures. For example, at a calcium chloride concentration of above 22%, the strength of concrete is reduced and salt scaling is increased.[81] The DTA method has been used to determine the reduction in $Ca(OH)_2$ content and increase in the content of complex $C_3A{\bullet}CaCl_2{\bullet}10H_2O$, depending on the exposure conditions and severity of attack. The existence of a complex salt, $CaCl_2{\bullet}Ca(OH)_2{\bullet}10H_2O$, with a characteristic peak at 580°C was also confirmed.

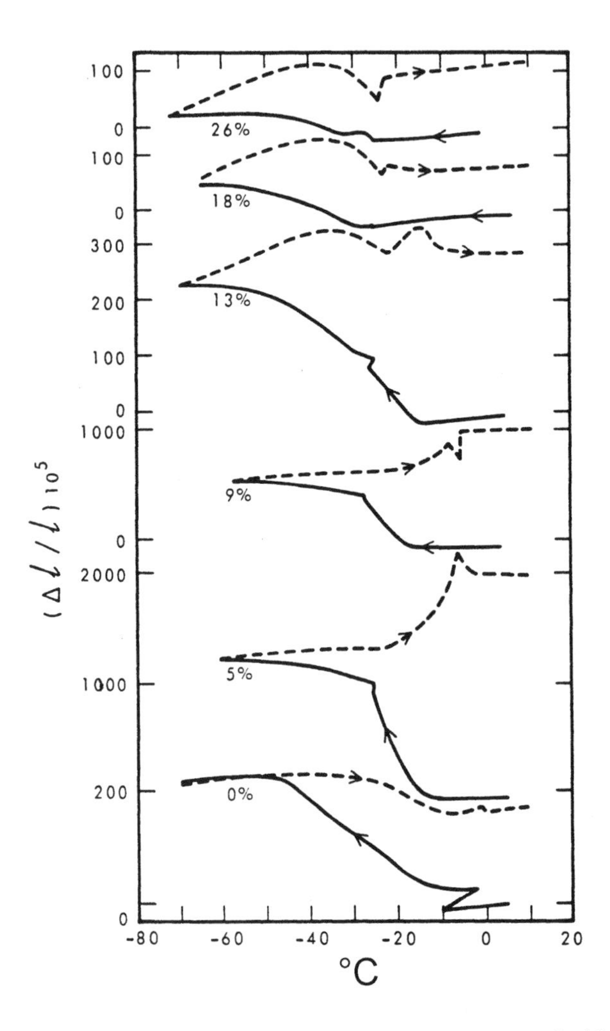

Figure: 39. Length changes of air-entrained cement paste saturated with brine of various concentrations.

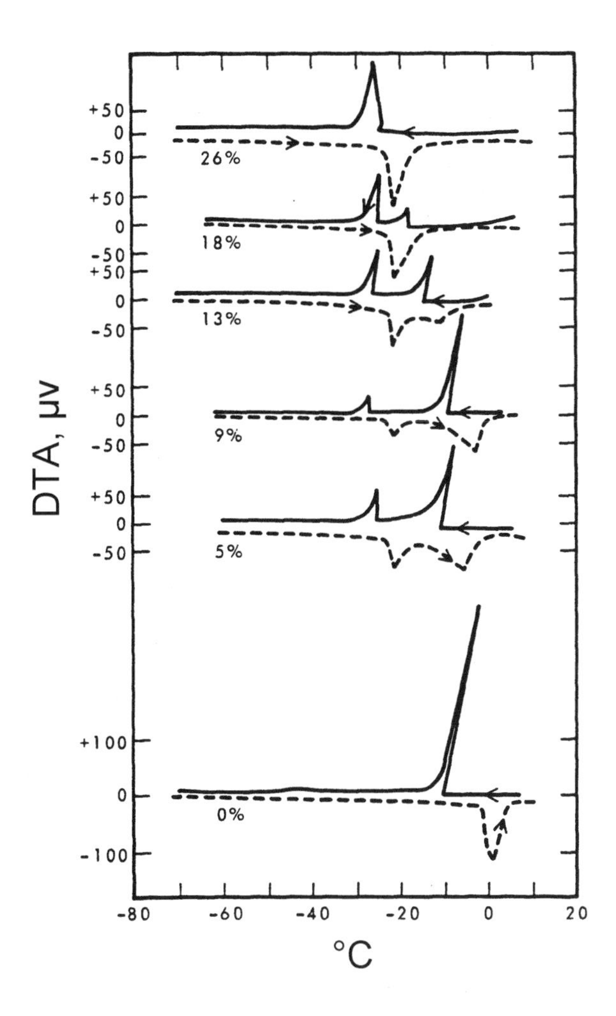

Figure 40. DTA of air-entrained 0.5 W/C cement paste saturated with brine of various concentrations.

Beddoe and Setzer[82] carried out extensive investigations on the freezing phenomenon in cement pastes, applying the DSC technique. Themograms were obtained in the temperature range of 20 to -175°C. The pores in the paste were classified into three groups, viz., macro, meso, and micro of radii > 50, 2–50, and < 2 nm, respectively. Figure 41 shows DSC curves of a cement paste containing different amounts of NaCl. Three endothermal effects occur in some samples. A general displacement of peaks towards lower temperatures is observed as the chloride concentration is increased. The sharp peak at -16°C is mainly attributed to the hetergeneous nucleation of supercooled bulk water in macropores. The lower transition at -38°C is due to freezing of aqueous solution in gel pores of radius of about 4 nm. The endothermal effects in the "macroscopic freezing area" in the presence of chloride is due to freezing chloride ions diffusing into smaller pores containing unfrozen solution. The locally enhanced concentration results in further supercooling and nucleation processes. The gel pore solution in the paste subjected to chloride solution exhibits the behavior of aqueous solution in the small mesopores. At concentrations > 1.3 mol Cl/l supercooled bulk water in the pores freezes at more than one temperature. Thermal methods have also been applied by Sudoh, Stockhause, and others to investigate the phase transitions in cement pastes exposed to lower temperatures.[5]

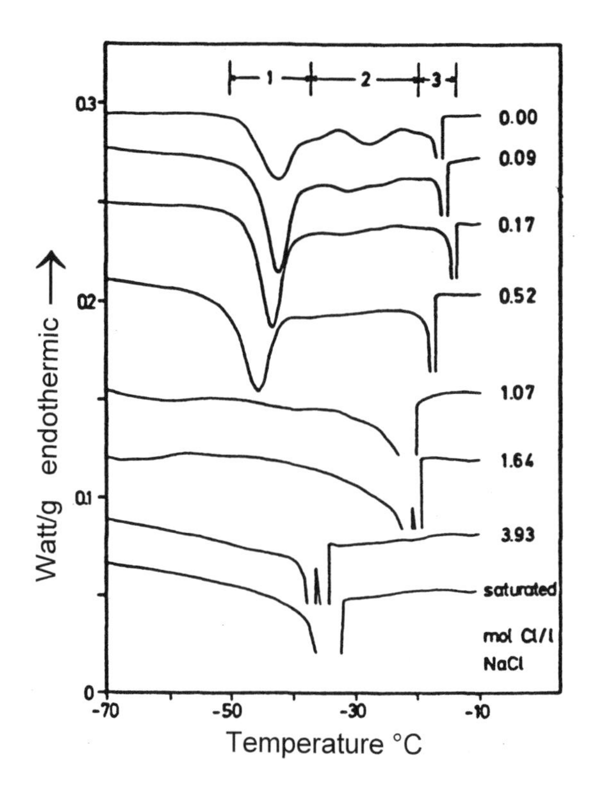

Figure 41. DSC curves of cement paste stored in NaCl solutions.

In addition to the physical effects that cause frost damage, the possibility of chemical changes playing a role has been considered by Ludwig and Stark.[78] The influence of low temperatures (freeze-thaw cycles) on the rate of formation of AFm and AFt phases in cements containing low C_3A (1.7%) or high C_3A (10.7%) was examined by DTG. At 150 freeze-thaw cycles, cement with a low C_3A content was found to contain 53% AFm phase and 17% AFt phase and the corresponding figures for the C_3A rich cement were 34% and 38%, respectively. It is not clear, however, the extent to which the rate and amount of formation of these sulfoaluminates influence the damage due to freezing and thawing process.

9.5 Carbonation

Carbonation involves the reaction of CO_2 with the hydrated cement components. This will result in shrinkage. In addition, carbonation also decreases the pH of the system making the reinforcing bars more prone to corrosion. It is not easy to assess the extent to which the C-S-H phase undergoes carbonation and shrinkage on exposure to CO_2 because it is not easy to remove CH from the paste without affecting the C-S-H phase. In a method involving continuous leaching and monitoring of the lime remaining in a hydrated tricalcium silicate paste by DTA, the time to terminate extraction was determined. Ramachandran[38] was, thus, able to prepare a lime-free C-S-H. The resultant C-S-H exhibited a much higher shrinkage than calcium hydroxide (Fig. 42).[84]

The carbonation effect on synthesized tobermorite has been examined by XRD / DTA.[83] An endothermal effect at about 180°C indicated the dehydration of tobermorite and an exothermal effect at about 800°C was caused by the formation of wollastonite. The carbonated samples exhibited a broad endothermal effect in the range of 450–600°C and a sharp peak at about 600–650°C. The former indicated the presence of vaterite and the latter, calcite. The carbonation decreased the peak due to wollastonite. Thermal methods can also be used to estimate the amount of carbonation that has occurred at different depths in concrete. These techniques estimate the amount of calcium carbonate formed and also the amount of calcium hydroxide remaining in an uncarbonated form.

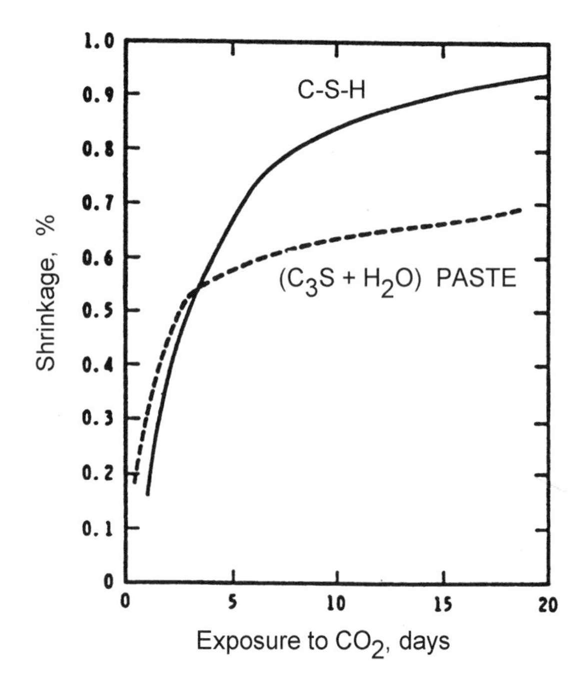

Figure 42. Carbonation shrinkage of C-S-H and hydrated C_3S paste.

In addition to calcium carbonate, the formation of scawtite ($Ca_7Si_6O_{21}H_6 \cdot CO_3$) and its decomposition with the evolution of CO_2 at lower temperatures was reported through TG investigations.[5] It has also been reported that by applying DTA/TG techniques the calcium carbonate formed by exposure of cement paste to CO_2 migrates to the surface of concrete in the form of layers.[5] The rate of carbonation on low and high alkali cements has not been found to be different and that the C-S-H and ettringite phases are also carbonated.[85]

The depth of carbonation in portland cement concrete may be assessed by a phenolphthalein indicator. The color changes from colorless to purple red in the pH range 8.3 to 10. The pH of non-carbonated concrete is about 12.6 and the initiation of corrosion of steel may occur at pH below 11.0. DTA/DTG techniques were used for a quantitative measurement of CH and $CaCO_3$ at different depths of carbonation.[86] In the neutralized depth both calcium carbonate and calcium hydroxide were identified. Thus, thermal techniques can be applied to indicate the depth at which all lime has been carbonated.

Several techniques, including TG, chemical analysis, and pore size distribution, were utilized to determine the extent of carbonation of granulated blast furnace slag cement concrete which was exposed for twenty years.[87] The weight loss at 450–550°C was attributed to the loss of water from $Ca(OH)_2$, and that at 780°C, to $CaCO_3$ decomposition. The amount of calcium carbonate formed at different depths from the interior of the building and from the exposed surface is compared in Fig. 43. The extent of carbonation is maximum at the surface which decreases to a low level from a distance of 40 mm from the surface.

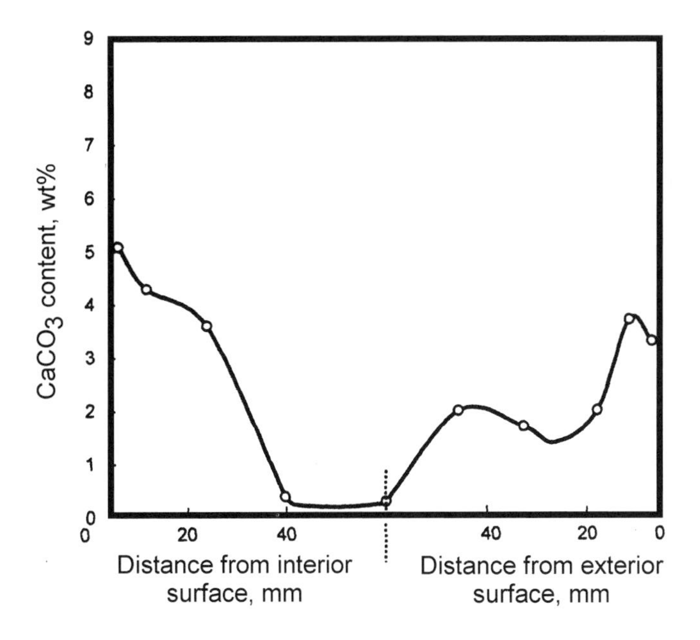

Figure 43. The $CaCO_3$ amount as a function of distance from the surface.

The potential for corrosion of steel is increased in a chloride environment that is subjected to CO_2. In the presence of $CaCl_2$, Friedel's salt of formula $3CaO{\bullet}Al_2O_3{\bullet}CaCl_2{\bullet}10\ H_2O$ and its ferrite analogue are formed. If these chlorides disassociate during the service life of concrete, the release of chloride and reduced pH may pose even a more increased risk for steel corrosion. In an examination of concrete exposed to CO_2, DTA was applied to identify and estimate the products of reaction.[88] Friedel's salt was identified by an endothermal peak in the range of 300–350°C. The presence of C_4AH_{13} was indicated by a series of inflections below 300°C. Calcium carbonate exhibited an endotherm at about 800°C. Ettringite, $Ca(OH)_2$, and α-quartz gave endothermal effects at 140, 500, and 573°C, respectively. Ettringite was absent in a severely carbonated concrete. Solubility of Friedel's salt increased with the degree of carbonation. Thermal peak values of calcium hydroxide and calcium carbonate were found to be ideally suited for estimating the degree of carbonation in concrete elements.

9.6 Chemical Attack

Exposure of concrete to natural elements or to industrial chemicals may result in its deterioration. In many chemical industries the chemicals produced may react with concrete that is used as a construction material. For example, the influence of formic acid on concrete was studied by applying thermal techniques.[60] DTA curves of samples showed that as the exposure of concrete to the acid is increased, the peaks due to lime and calcium carbonate are decreased. This gave an indication of the severity of attack by acids. Another example pertains to a fertilizer plant. Ammonium nitrate fertilizer contained 75% NH_4NO_3 and 25% $CaCO_3$ as a filler. Concrete used in this plant was attacked by nitrate. The lime was found converted to calcium nitrate. Also there was a reaction between calcium aluminate hydrate and ammonium nitrate resulting in the formation of calcium nitrate and calcium aluminate nitrate complex. These compounds could be identified by endothermal effects appearing in DTA.[89]

Another example is related to two damaged chimneys at a power station. Wesolowski,[5] applying thermal techniques, found that in addition to the sulfate attack there was evidence of incomplete hydration of cement due to the heating effects at the early curing periods. The chemical corrosion potential cement mortars subjected to organic chemicals can be studied by thermal methods. The interaction between phenols and lime in cement were established through the application of DTA/TG by Kovacs.[5]

In many countries, de-icing salts consisting of calcium chloride and sodium chloride are applied on roads in winter. Such concrete roads are vulnerable to attack by these chemicals. A series of mortars containing mineral admixtures was exposed to 30% calcium chloride solution at 5–40°C.[90] Mortars containing mineral admixtures were found to be effective in reducing the chloride penetration. XRD and DTA indicated that the deterioration was primarily due both to the dissolution of lime and formation of Friedel's salt. Lime and Friedel's salt were detected by peaks at 460–570° and 280°C, respectively.

Several complex reactions occur when concrete is exposed to sea water. Compounds such as aragonite, calcium bicarbonate, calcium monosulfate hydrate, ettringite, Ca-Mg silicate hydrates, magnesium silicate, thaumasite, etc., have been identified by the application of thermal analysis in conjunction with other techniques. Thus, the factors leading to the deterioration of concrete can be established.

Several cases of distress of concrete railway ties have been reported. Examination of cracks in such concretes has revealed that they were filled with secondary products. Ettringite was a prominent constituent, accompanied by CH, $CaCO_3$, and alkali-silica gel. DTA data indicated the presence of a substantial amount of secondary ettringite in many samples but petrographic and XRD did not reveal any ettringite.[91] Thermograms were used to quantitatively estimate the amounts of ettringite formed in failed samples.

Thaumasite, is a mineral of relevance in concrete technology. In a pure form it has the composition $Ca_6[Si(OH)_6]_2(CO_3)_2(SO_4)\bullet 24H_2O$. This compound may form through a combination of sulfate attack and carbonation. It can cause damage to concrete by decomposing the C-S-H phase. Its formation can be rapid in the presence of finely divided $CaCO_3$. DTA and TG techniques have been adopted to identify thaumasite in concrete.[92] In DTA the decomposition starts at 110°C, with a peak at 150°C. A small exothermic peak at 710°C has been attributed to a disorder-order type transition, akin to devitrification. TGA shows decomposition of mineral thaumasite starting at 110°C. A small loss of CO_2 occurs simultaneously with water up to 550°C. Most CO_2 is lost between 950 and 980°C.

9.7 Aged Concrete

Many studies have been carried out on old concretes to determine the reactions that could be responsible for deterioration. Sarkar, et al.,[93] examined a seventy-five year old stone building containing mortar that had shown signs of distress. The presence of gypsum (endothermic effect at 133°C), quartz (endothermal peak 573°C), calcium carbonate (endothermal effect at 900°C), and tharndite (endotherm at 880°C) could be identified. It was concluded that one of the main causes of deterioration was the interaction of SO_2 from the atmosphere with mortar and sandstone. In another study,[94] a fifty year old concrete was subjected to examination by DTA, TG, SEM (EDX), porosimetry, and chemical analysis. TG showed a loss in weight at 60°C for the loss of adsorbed water, and other losses occurred due to the dissociation of C-S-H, $Mg(OH)_2$, $Ca(OH)_2$, and $CaCO_3$. The w/c ratio that was used to make concrete did not influence the total amount of combined water. Higher amounts of carbonate were found in concretes made at lower w/c ratios. It was also concluded that the transition zone contained C-S-H and $Ca(OH)_2$, the relative ratios depending on the type of cement and cement:aggregate: water ratio.

REFERENCES

1. Ramachandran, V. S., *Applications of Differential Thermal Analysis in Cement Chemistry,* p. 308, Chemical Publication Co., New York (1969)
2. Ramachandran, V. S., and Garg, S. P., *Differential Thermal Analysis as Applied to Building Science* (with an annotated bibliography), p. 182, Central Building Research Institute, India (1959)
3. Longuet, P., Application of Thermogravimetry to the Chemistry of Cement, *Rev. Materiaux de Construction,* 537:538–540 (1960)
4. Mackenzie, R. C., (ed.), *Differential Thermal Analysis,* 2:607, Academic Press, London (1972)
5. Ghosh, S. N., (ed.), *Advances in Cement Technology,* p. 804, Pergamon Press, London (1983)
6. Ramachandran, V. S., and Beaudoin, J. J., (eds.), *Handbook of Analytical Techniques in Cement Science and Technology,* p. 964, Noyes Publications, NJ (2001)
7. Ramachandran, V. S., Elucidation of the Role of Chemical Admixtures in Hydrating Cements by DTA Technique, *Thermochimica Acta,* 3:343–366 (1972)
8. Bhatty, J. I., Review of Application of Thermal Analysis to Cement-Admixture Systems, *Thermochimica Acta,* 189:313–350 (1991)
9. Ramachandran, V. S., Feldman, R. F., and Sereda, P. J., Applications of Thermal Analysis in Cement Research, *Highway Res. Rec.,* 62:40–61 (1964)
10. Rowland, R. A., and Beck, C. W., Determination of Small Quantities of Dolomite by Differential Thermal Analysis, *Am. Mineralogist,* 37:76–82 (1952)
11. Ghosh, S. N., and Mathur, V. K., *Testing and Quality Control in Cement Industry,* p. 497, Akademia Books Int., New Delhi (1997)

12. Courtault, B., Etude des Reactions a l'etat Solide Jusqu'a de L'analyse Thermique Differentielle, Applications a la Chimie des Ciments, *Rev. Mater. Constr. Trav. Publ.,* 569:37–47; 570:67–78; 571:110–124; 572:143–156; 573:190–203 (1963)

13. Kakali, G., Chaniotakis, E., Tsivilis, S., and Danassis, E., Differential Scanning Calorimetry: A Useful Tool for Prediction of the Reactivity of Cement Raw Meal, *J. Thermal Anal. and Calorimetry,* 52:871–879 (1998)

14. Handa, S. K., and Raina, S. J., Cement Raw Mix Characterization by Differential Thermal Analysis, *Thermochimica Acta,* 93:609–612 (1985)

15. Bhattacharyya, A., Simulating Minerals Pyroprocessing by Thermal Analysis, *J. Therm. Analysis,* 40:141–149 (1993)

16. Sorrentino, F., and Castanel, R., Applications of Thermal Analysis to the Cement Industry, *J. Thermal Anal.,* 38:2137–2146 (1992)

17. Taylor, H. F. W., *Cement Chemistry,* p. 459, Thomas Telford, Oxford (1997)

18. SETARAM, File 2, Cements and Plasters, Sheet 1, p. 188, Caluire Cedex, France (Not Dated)

19. Lota, J. S., Bensted, J., and Pratt, P. L., Characterization of an Unhydrated Class G Oil Well Cement, *Ind. Ital. Cemento,* 68:172–183 (1998)

20. Lagzdina, S., and Sedmalis, S., Kinetics of $2CaO•SiO_2$ and $3CaO•SiO_2$ Formation in the Presence of Sodium and Potassium Compounds, *Proc. 10th Int. Congr. Cement Chem.,* Vol. I, Goethenberg, Sweden (1997)

21. Fang, Y., Ray, D. M., Chan, Y., and Silsbee, M. R., Cement Clinker Formation from Fly Ash by Microwave Processing, *Proc. 10th Int. Congr. Cement Chem.,* Vol. I, Goethenberg, Sweden (1997)

22. Su, M., Deng, J., Wu, Z., and Liu, X., Research on the Chemical Composition and Microstructures of Sulphoaluminate Cement Clinker, *Proc. 9th Int. Congr. Chem. Cements,* II:94–100, New Delhi, India (1992)

23. Akhmetov, I. S., and Miryuk, O. A., Phase Transformations When Synthesizing Clinkers Produced of Technologeous Raw Materials, *Proc. 9th Int. Congr. Chem. Cement,* II:74–80, New Delhi, India (1992)

24. Radnaassediin, S., Formation of Low Temperature Portland Cement Clinker by Using Non-Traditional Aluminosilicate Rock and Mineralizer, *Proc. 18th Int. Congr. Chem. Cements,* II:315–321, New Delhi, India (1992)

25. Abdel-Fattah, W. I., and El-Didamony, H., Thermal Investigation on Electrostatic Precipitator Kiln Dust, *Thermochimica Acta,* 51:297–306 (1981)

26. Raina, K., and Bhargava, R., Effect of Thermal Treatment on the Reactivity of Clinker Minerals Produced from Vertical Kiln, *Proc. 9th Int. Congr. Chem. Cement,* II:132–138, New Delhi, India (1992)
27. Kurdowski, W., and Deja, J., Formation of C_3S in the Mixtures Containing Different Silicate Raw Materials, *Proc. 9th Int. Congr. Chem. Cements,* II:255–260, New Delhi, India (1992)
28. Misra, K. C., and Borthakur, P. C., Role of Mineralizers in Alite Formation in Rice Husk Ash-$CaCO_3$ Raw Mix, *Proc 9th Int. Congr. Chem. Cements,* II:301–307, New Delhi, India (1992)
29. Mathur, V. K., Gupta, R. S., and Ahluwalia, S. L., Lithium as Intensifier in the Formation of Dicalcium Silicate Phase, *Proc. 9th Int. Congr. Chem. Cements,* II:406–412, New Delhi, India (1992)
30. Jeffery, J. W., The Tricalcium Silicate Phase, *3rd Int. Symp. Chem. Cements,* pp. 30–38, London (1952)
31. Regourd, M., Mineralogy of Portland Cement, *2nd Nordic Cement Colloquim,* pp. 11–39, Copenhagen (1972)
32. Nurse, R. W., The Dicalcium Silicate Phase, *3rd Int. Symp. Chem. Cements,* pp. 56–77, London (1952)
33. Ramachandran, V. S., Differential Thermal Method of Estimating Calcium Hydroxide in Calcium Silicate and Cement Phases, *Cem. Concr. Res.,* 9:677–684 (1979)
34. Ramachandran, V. S., Kinetics of Hydration of Tricalcium Silicate in the Presence of Calcium Chloride by Thermal Methods, *Thermochimica Acta,* 2:41–55 (1971)
35. Ramachandran, V. S., Estimation of Tricalcium Silicate Through Polymorphic Transition, *J. Therm. Anal.,* 3:181–190 (1971)
36. Mascolo, G., and Ramachandran, V. S., Hydration and Strength Characteristics of Synthetic Al, Mg and Fe Alites, *Mater. & Struct.,* 8:373–376 (1975)
37. Feldman, R. F., and Ramachandran, V. S., A Study of the State of Water and Stoichiometry of Bottle Hydrated Ca_3SiO_5, *Cem. Concr. Res.,* 4:155–166 (1974)
38. Ramachandran, V. S., and Polomark, G. M., Extraction of $Ca(OH)_2$ from Portland Cement and $3CaO \bullet SiO_2$ Pastes, *J. Chem. Biotech.,* 32:946–952 (1982)
39. Ramachandran, V. S., and Sereda, P. J., Applications of Hedvall Effect in Cement Chemistry, *Nature,* 233:134–135 (1971)
40. Tamas, F., The Hydration of Portland Cement and Clinker Minerals Investigated by Thermal Methods, *Proc. 6th Conf. Silicate Industry,* pp. 425–436, Budapest (1961)

41. Firens, P., and Tirloq, J., Effect of Synthesis Temperature and Cooling Conditions of β-C_2S and its Hydration Rate, *Cem. Concr. Res.,* 13:41–48 (1983)

42. Ramachandran, V. S., and Feldman, R. F., Significance of Low Water/Solid Ratio and Temperature on the Physico-Chemical-Mechanical Characteristics of Hydrates of Tricalcium Aluminate, *J. Appl. Chem. Biotech.,* 23:625–633 (1973)

43. Feldman, R. F., and Ramachandran, V. S., Character of Hydration of $3CaO{\bullet}Al_2O_3$, *J. Am. Ceram. Soc.,* 49:268–273 (1966)

44. Feldman, R. F., and Ramachandran, V. S., The Influence of $CaSO_4{\bullet}2H_2O$ Upon the Hydration Character of $3CaO{\bullet}Al_2O_3$, *Mag. Concr. Res.,* 18:185–196 (1966)

45. Ramachandran, V. S., and Chun-Mei, Z., Thermal Analysis of the $3CaO{\bullet}Al_2O_3$-$CaSO_4{\bullet}2H_2O$-$CaCO_3$-H_2O System, *Thermochimica Acta,* 106:273–282 (1986)

46. Collepardi, M., Baldini, G., Pauri, M., and Corradi, M., Tricalcium Aluminate Hydration in the Presence of Lime, Gypsum or Sodium Sulfate, *Cem. Concr. Res.,* 8:571–580 (1978)

47. Odler, I., and Abdul-Maula, S., Possibilities of Quantitative Determination of the AFt (Ettringite) and AFm (Monosulphate) Phases in Hydrated Cement Pastes, *Cem. Concr. Res.,* 14:133–141 (1984)

48. Forrester, J. A., Some Applications of Evolved Gas Analysis to the Study of Portland Cement, *Chem. and Industry,* pp. 1244–1246 (Sept. 1969)

49. Ramachandran, V. S., and Beaudoin, J. J., Significance of Water/Solid Ratio and Temperature on the Physico-Chemical and Mechanical Characteristics of Hydrating $4CaO{\bullet}Al_2O_3{\bullet}Fe_2O_3$, *J. Mater. Sci.,* 11:1893–1910 (1976)

50. Shin, G. Y., and Han, K. S., The Effect of C_3A Polymorphs on the Hydration of C_3S, *9th Int. Congr. Chem. Cements,* IV:90–96, New Delhi, India (1992)

51. Ramachandran, V. S., and Beaudoin, J. J., Hydration of C_4AF+Gypsum: Study of Various Factors, *7th Int. Congr. Chem. Cements,* II:25–30, Paris (1980)

52. Chen, Y., and Shi, L., Rate of Ettringite Formation from Calcium Aluminoferrite Hydration, *Proc. 10th Int. Congr. Chem. Cements,* p. 8, Gothenburg (1997)

53. Ramachandran, V. S., and Beaudoin, J. J., Physico-Mechanical Characteristics of Hydrating Tricalcium Aluminoferrite System at Low Water:Solid Ratios, *J. Chem. Tech. Biotech.,* 34A:154–160 (1984)

54. Yu, Q., and Feng, X., Hydration Characteristics of $11CaO{\bullet}7Al_2O_3{\bullet}CaF_2$, *10th Int. Cong. Chem. Cements,* 2:8, Gothenberg (1997)

55. Greene, K. T., Early Hydration Reactions of Portland Cement, *4th Int. Symp. Chem. Cements,* p. 359, Washington, DC (1960)

56. Scrivener, K. L., and Wieker, W., Advances in Hydration at Low, Ambient and Elevated Temperatures, *9th Int. Congr. Chem. Cem.,* I:449–482, New Delhi, India (1992)

57. Zacedatelev, I. B., Thermochemical Characterisitics of Cement and Acceleration of Concrete Setting, *7th Int. Congr. Chem. Cem.,* I:71, Paris (1980)

58. Feldman, R. F., and Ramachandran, V. S., Modified Thermal Analysis Equipment and Technique for Study Under Controlled Humidity Conditions, *Thermochimica Acta,* 2:393–403 (1971)

59. Ramachandran, V. S., Evaluation of Concrete Admixtures Using Differential Thermal Technique, *ACI/RILEM Symposium 1985 on Technology of Concrete When Pozzolans, Slags and Chemical Admixtures are Used,* pp. 35–52, Monterry, Mexico (1985)

60. Feldman, R. F., and Ramachandran, V. S., Differentiation of Interlayer and Adsorbed Water in Hydrated in Portland Cement by Thermal Analysis, *Cem. Concr. Res.,* 1:607–620 (1971)

61. El-Jazairi, B., and Illston, J. M., The Hydration of Cement Paste Using the Semi-Isothermal Method of Derivative Thermogravimetry, *Cem. Concr. Res.,* 10:361–366 (1980)

62. Butler, F. G., and Morgan, S. R., A Thermoanalytical Method for Determination of the Amount of $Ca(OH)_2$ Contained in Hydrated Portland Cement, *Proc. 7th Int. Congr. Chem. Cem.,* II:43–46, Paris (1980)

63. Inn, C. D., and Lee, D. Y., The Engineering Characteristics and Hydration of High Calcium Sulfate Cement Concrete, *Int. Symp. on Advances in Concrete,* II:327–343, Am. Concr. Inst. (1995)

64. Lota, S. S., Bensted, J., Munn, J., and Pratt, P. L., Hydration of Class G Oil Well Cement at 20 and 5°C., *Industr. Ital. Cemento,* 725:776–798 (1997)

65. Klimesch, D. S., and Ray, A., The Use of DTA/TGA to Study the Effects of Ground Quartz with Different Surface Area in Autoclaved Cement:Quartz Mixtures, *Thermochimica Acta,* 289:41–54 (1996)

66. Klimesch, D. S., and Ray, A., Use of Second Derivative Differential Thermal Curve in the Evaluation of Cement–Quartz Paste with Metakaolin Addition, Autoclaved at 180°C, *Thermochimica Acta,* 307:167–176 (1997)

67. Balek, V., The Hydration of Cement Investigated by Emanation Thermal Analysis, *Thermochimica Acta,* 72:147–158 (1984)

68. Bensted, J., Some Instrumental Investigation of Portland Cement Hydration-Part II Differential Thermal Analysis (DTA), *Il Cemento,* 76:117–126 (1979)

69. Goni, S., Macias, A., Madrid, J., and Diez, J. M., Characterization of a New Ca-Cd Hydroxide Hydrothermally Synthesized and Its Implications for Cement Isolation of Cd., *J. Mater. Res.,* 13:16–21 (1998)
70. Dubborke, W., and Marks, V. J., Thermogravimetric Analysis of Carbonate Aggregates, *Trans. Res. Rec.,* 1362:38–43 (1982)
71. Ramachandran, V. S., A Test for Unsoundness of Cement Containing MgO, *3rd Int. Conf. Durability of Bldg. Mater. and Comp.,* 3:46–54, Espoo, Finland (1984)
72. Scanlon, J. M., and Connolly, J. D., Laboratory Studies and Evaluation of Concrete Containing Dead-Burnt Dolomite, *Durability of Concrete,* Am. Concr. Inst. SP-145, p. 1224 (1994)
73. Ramachandran, V. S., Estimation of $Ca(OH)_2$ and $Mg(OH)_2$ -Implications in Cement Chemistry, Israel, *J. Chem.,* 22:240–246 (1982)
74. Thiesen, K., and Johansen, V., Prehydration and Strength Development, *Am. Ceram. Soc. Bull.,* 54:787–791 (1975)
75. Handoo, S. K., Agarwal, S., and Maiti, S. C., Application of DTA/TGA for the Assessment of Fire Damaged Concrete Structures, *Proc. Symp. Thermal. Anal.,* pp. 19–21, Bhubaneswar, India (1991)
76. Malek, R. I., Equipment to Measure Thermal Expansion of Cementitious Materials as a Function of Temperature, *9th Int. Congr. Chem. Cements,* V:427–430, New Delhi, India (1992)
77. Sarvaranta, L., Elomaa, M., and Jarvela, E., A Study of the Spalling Behavior of PAN Fibre-Reinforced Concrete by Thermal Analysis, *Fire and Materials,* 17:225–230 (1993)
78. Ludwig, H. M., and Stark, J., Effects of Low Temperature and Freeze Thaw Cycles on the Stability of Hydrated Products, *9th Int. Congr. Chem Cements,* IV:3–9, New Delhi, India (1992)
79. Litvan, G. G., Phase Transitions of Adsorbates: Effect of Deicing Agents on The Freezing of Cement Paste, *J. Am. Ceram. Soc.,* 58:26–30 (1975)
80. Ramachandran, V. S., *Calcium Chloride in Concrete,* p. 216, Appl. Sci. Publ., London (1976)
81. Yang, Q. B., Wu, X. L., and Huang, S. Y., Mechanisms of De-Icer Scaling of Concrete, *9th Int. Congr. Chem. Cements,* V:282–288, New Delhi, India (1992)
82. Beddoe, R. E., and Setzer, M. J., A Low Temperature DSC Investigation of Hardened Cement Paste Subjected to Chloride Action, *Cement. Concr. Res.,* 18:249–256 (1988)

83. Goto, S., and Ikeda, S., Effects of Carbonation on the Thermal Properties of Tobermorite, *9th Int. Congr. Chem. Cements,* IV:304–309, New Delhi, India (1992)
84. Ramachandran, V. S., Feldman, R. F., and Beaudoin, J. J, *Concrete Science,* p. 427, Appl. Sci. Publ., London (1981)
85. Gaztanaga, T., Goni, S., Guerrero, A. M., and Hernandez, M. S., Carbonation of Hydrated Aluminate Cement and Ordinary Portland Cements Varying in Alkali Contents, *Proc.10th Int. Congr. Chem. Cements, Gothenburg, Sweden,* p. 8 (1997)
86. Fukushima, T., Yoshizaki, Y., Tomosawa, F., and Takahashi, K., Relationship Between Neutralization Depth and Concentration Distribution of $CaCO_3$-$Ca(OH)_2$ in Carbonated Concrete, *4th Int. CANMET/ACI/JCI Conf. Advances in Concr. Techn.,* SP-179, p. 1109, Tukushima, Jpn (1998)
87. Litvan, G. G., Carbonation in Slag Cements, *Int. Workshop on Granulated Blast Furnace Slag in Concrete,* pp. 301–327, Toronto (1987)
88. Suryavanshi, A. K., and Narayana Swamy, R., Stability of Friedel's Salt in Carbonated Concrete Structure Elements, *Cement & Concrete,* 26:729–741 (1996)
89. Bajza, A., Corrosion of Hardened Cement Pastes by Formic Acid Solutions, *9th Int. Congr. Chem. Cements,* V:402–408, New Delhi, India (1992)
90. Toril, K., Sasatani, T., and Kawamura, M., Effects of Fly Ash, Blast Furnace Slag and Silica Fume on Resistance of Mortar to calcium Chloride Salts, *5th CANMETY/ACI Int. Conf. on Fly Ash, Silica Fume, Slag and Natural Pozzolans in Concrete,* II(SP-153):931–949, Milwaukee (1995)
91. Mielenz, R. C., Marusin, S. L., Hime, W. G., and Jugovic, Z. T., Investigation of Prestressed Concrete Railway Tie Distress, *Concr. Int.,* 17:62–68 (1995)
92. Bensted, J., and Varma, P., Studies of Thaumasite, *Silicates Ind.,* 39:11–19 (1974)
93. Sarkar, S. L., Bhadra, A. K., and Mandal, P. K., Investigation of Mortar and Stone Deterioration in the Victoria Memorial, Calcutta, *Mater. & Struct.,* 27:548–556 (1994)
94. Concotto, M. A., and John, V. M., Some Characteristics of a 50 Year Old Concrete, *9th Int. Congr. Chem. Cements,* V:452–458, New Delhi, India (1992)

4

Introduction to Concrete Admixtures

1.0 INTRODUCTION

Concrete admixtures are defined as materials other than hydraulic cement, water, or aggregates that are added immediately before or during mixing. Additives or additions such as grinding aids are added to cement during manufacture. An addition is a material that is interground or blended during the manufacture of cement. Most concrete used in North America contains at least one admixture. The admixtures are added to improve the quality of concrete in the fresh and hardened state. Publications and patents on admixtures are voluminous. Interest in the development of admixtures is evident from the number of patents taken every year. The total number of patents reported in the years 1985, 1986, 1987, 1988, 1989, 1990, 1991, and 1992 were respectively 44, 64, 99, 133, 242, 57, 53, and 149.

Categorization of admixtures into distinct groups is not easy as one material may belong to more than one category. Broadly, however, they could be divided into the following categories.

1. *Chemical admixtures:* They are generally water-soluble, added mainly to control setting and early hardening of fresh concrete, and to reduce water requirements. Chemical admixtures include accelerators, normal water reducers, superplasticizers, and retarders. The accelerators may contain calcium chloride, alkali hydroxide, calcium formate, calcium nitrate, etc. Examples of retarders are Na, Ca, or NH_4 salts of lignosulfonic acids, hydroxy carboxylic acids, or derivatives of carbohydrates. Normal water reducers may contain salts of refined lignosulfonic acids, hydroxycarboxylic acids, hydroxylated polymers, etc. Superplasticizers may contain sulfonated melamine formaldehyde, sulfonated naphthalene formaldehyde, carboxylated synthetic polymers, etc.

2. *Air-entraining agents:* These are used to improve the durability of concrete. They comprise resins such as neutralized vinsol resin, synthetic detergents, salts of petroleum acids, etc.

3. *Mineral admixtures:* These finely divided materials improve workability, durability and other properties of concrete; examples are fly ash, slag, silica fume, rice husk ash, etc.

4. *Miscellaneous admixtures:* Those other than what have already been described and are known to impart many benefits to concrete. Examples are: latexes, corrosion inhibiting admixtures, alkali-aggregate expansion reducing admixtures, pigments, pumping aids, shotcreting admixtures, etc.

Thermal analysis techniques have been applied to investigate many aspects of concrete science and technology. They include hydration kinetics, mechanism of hydration, deterioration of concrete, examination of the role of newly developed admixtures, admixture detection and estimation, composition of the products, identification of new compounds, etc.

In this chapter an introduction to the role of different types of admixtures on cement and concrete properties is given. This should serve as a useful guide for following the application of thermal techniques to the evaluation of the effect of admixtures on cement and concrete.

2.0 ACCELERATORS

An *accelerating admixture* is a material that is added to concrete for reducing the time of setting and accelerating early strength development. Accelerating admixtures are used in cold water concreting operations and are components of antifreezing admixtures and shotcreting mixes. In cold weather concreting there are other alternatives such as the use of Type III cement, use of higher than normal amount of portland cement Type I or warming of the concrete ingredients. Of the above, the most economical method is the use of Type I cement in conjunction with an accelerator. The advantages include efficient start and finishing operations, reducing the period of curing and protection, earlier removal of forms so that the construction is ready for early service, plugging of leaks and quick setting when used in shotcreting operations.

Many substances are known to act as accelerators for concrete. These include soluble inorganic chlorides, bromides, fluorides, carbonates, thiocyanates, nitrites, nitrates, thiosulfates, silicates, aluminates, alkali hydroxides, and soluble organic compounds such as triethanolamine, calcium formate, calcium acetate, calcium propionate, and calcium butyrate. Some of them are used in combination with water reducers. Quick setting admixtures used in shotcrete applications and which promote setting in a few minutes may contain sodium silicate, sodium aluminate, aluminum chloride, sodium fluoride, strong alkalis, and calcium chloride. Others are solid admixtures such as calcium aluminate, seeds of finely divided portland cement, silicate minerals, finely divided magnesium carbonate, and calcium carbonate. Of these, calcium chloride has been the most widely used because of its ready availability, low cost, predictable performance characteristics, and successful application over several decades.[1][2] In some countries the use of calcium chloride is prohibited, in some others, such as Canada and the USA, the use of calcium chloride is permitted provided certain precautions are taken. Attempts have continued to find an effective alternative to calcium chloride because of some of the problems associated with its use.

2.1 Effect of Calcium Chloride on Calcium Silicates

The silicate phases, C_3S and C_2S together constituting the major portion of the components in portland cement, influence considerably its hydration and strength development. The accelerating influence of $CaCl_2$ on the hydration of C_3S is followed conveniently by estimating at different times the amount of residual unhydrated C_3S, the amount of $Ca(OH)_2$, non-evaporable water content, electrical conductivity, heat liberation, etc.

The amount of $Ca(OH)_2$ formed at different periods may be used to follow the degree of hydration. The rate of hydration of C_3S [determined by estimating unhydrated C_3S or $Ca(OH)_2$] with 2% $CaCl_2$ shows the formation of an increased amount of C-S-H at all times up to 1 month.[1][3]

Calcium chloride accelerates the hydration of C_3S even at higher temperatures and at different W/C ratios.[4]–[6] With the addition of 2% calcium chloride, the degree of hydration at the same period at 22°C and 80°C is 60% and 70%, respectively.[4][5]

Increasing the concentration of $CaCl_2$ even up to 20% with respect to C_3S has been found to accelerate intensely the hydration of the silicate.[1][3] Increasing the amount of $CaCl_2$ not only accelerates the appearance of the conduction calorimetric peak at earlier times, but also intensifies the peak (Fig. 1). The addition of $CaCl_2$ to C_3S not only alters the rate of hydration, but also influences the chemical composition of the C-S-H phase.

Because of the accelerated hydration, the silicate phases show rapid setting characteristics in the presence of $CaCl_2$. In a C_3S:C_2S mixture containing 0, 1, and 2% $CaCl_2$, the setting times have been found to occur at 790, 525, and 105 minutes, respectively.[7]

The addition of a small amount of calcium chloride is capable of influencing the strength of tricalcium silicate. Figure 2 compares the rate of strength development in C_3S pastes with and without $CaCl_2$.[8] At all times up to 28 days, the strength in the paste containing $CaCl_2$ is higher than that hydrated without the chloride. The increased strength is attributable to increased degree of hydration.

Compared to the extensive investigations on the hydration of C_3S in the presence of $CaCl_2$, only meager work has been done on the action of $CaCl_2$ in the hydration of C_2S.[9][10]–[14] Generally, all chlorides accelerate the hydration of C_2S, calcium chloride being more efficient than others at nearly all ages.

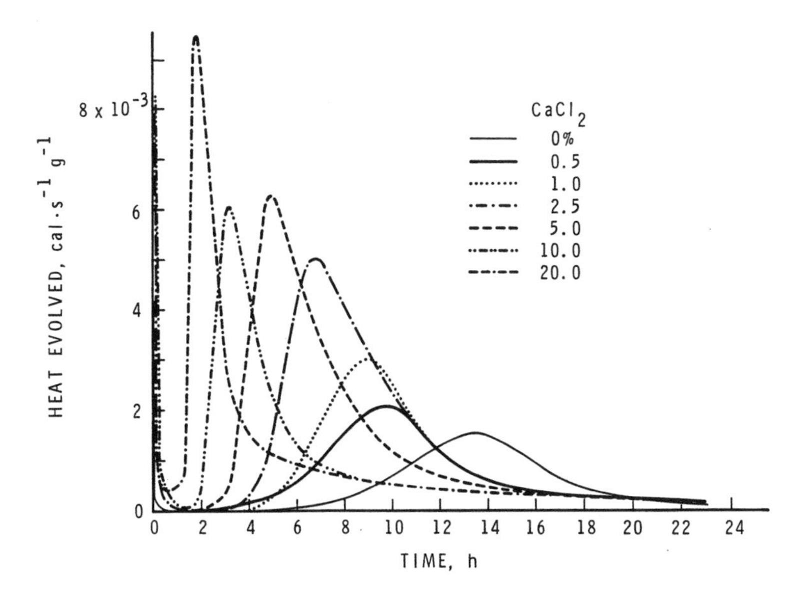

Figure 1. Influence of $CaCl_2$ on the heat evolution characteristics of C_3S.

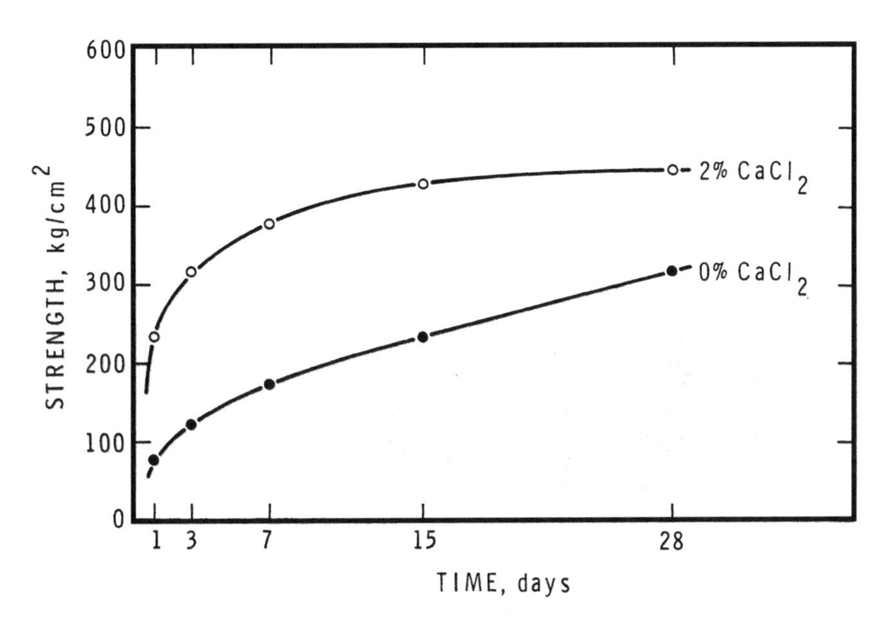

Figure 2. Compressive strength of tricalcium silicate paste containing $CaCl_2$.

The dicalcium silicate phase hydrates at a slower rate than the C_3S phase.[15] The C/S ratio of the C-S-H product in the hydrated C_3S is slightly higher than that formed in the hydrated C_3S. Table 1 shows the effect of different percentages of $CaCl_2$ on the rate of hydration of C_2S.[16] The hydration rate is increased as the amount of chloride is increased.

Addition of $CaCl_2$ to C_2S results in an increase in the rate of strength development. An addition of 3% $CaCl_2$ increases the bending strength of C_2S by 26% at 2 months, about 34% at 3 months, and about 60% at 6 months.[17] Large additions of $CaCl_2$ are, however, detrimental to strength development. The strength development also depends on the fineness of C_2S showing that as the fineness increases, the strength also increases at any particular dosage of $CaCl_2$.[18]

Table 1. Degree of Hydration of β-C_2S with Different Amounts of Calcium Chloride

Sample	Degree of Hydration		
	1 day	7 days	28 days
0.0% $CaCl_2$	16.1	24.3	33.0
0.5% $CaCl_2$	21.3	29.2	47.0
2.0% $CaCl_2$	21.6	34.1	56.1
5.0% $CaCl_2$	26.8	35.9	54.9

Mechanism of Acceleration. Several mechanisms have been suggested for the accelerating influence of calcium chloride on the silicate phases. They include: complex formation, catalytic action, instability of C-S-H phases, nucleation, reduced alkalinity, polymerized silicates, ionic radii effects, chloride diffusion, etc. It appears that no single mechanism can explain all the effects of $CaCl_2$. Possibly, a combination of mechanisms may be operating depending on the experimental conditions and the period of hydration.

2.2 Effect of Calcium Chloride on Calcium Aluminate

Two chloroaluminate hydrates are known, viz., the low form $C_3A•CaCl_2•xH_2O$ and the high form $C_3A•_3CaCl_2•yH_2O$. It is generally believed that the low form is the main reaction product formed under practical conditions of hydration.[19][21][22] The low form crystallizes as hexagonal plates and the high form as needle-shaped crystals. The basal spacing for the low form is 8.1 Å and that for the high form is 10.15 Å. The Differential Thermal Analysis (DTA) technique may be used to differentiate between the two forms of chloroaluminate. Endotherms at about 190°C and 350°C are caused by the monochloroaluminate, and the endotherm at about 160°C is exhibited by the higher chloroaluminate.[20]

In the C_3A-gypsum-$CaCl_2$-H_2O system, substantial amounts of chloride are immobilized within an hour whereas lower amounts of chloride are immobilized by C_3S and cement after longer times.[23] The reaction between C_3A and gypsum is accelerated by calcium chloride. Monochloroaluminate is formed after gypsum is consumed in the reaction with C_3A. Conversion of ettringite to monosulfoaluminate occurs only after all $CaCl_2$ has reacted.[3][5][21][22][24]–[27] The cubic aluminate hydrate (C_3AH_6) also reacts with $CaCl_2$ to form monochloroaluminate hydrate, but the reaction is slower than that with C_3A as the starting material.[28]

In the hydration of C_3A in the presence of $CaCl_2$ at temperatures of 75–100°C, calcium chloroaluminate is identified.[26][29] Using different amounts of gypsum and $CaCl_2$, it has been found that the acceleration of the reaction of gypsum with C_3A is accompanied by the generation of strains and consequent loss in strengths.[25]

In the system C_3S-C_3A-$CaCl_2$-H_2O, more C_4AH_{13} is formed and better strengths attained than that without $CaCl_2$.[30] The main factor may be the accelerated hydration of the C_3S phase.

In the portland cement paste containing different amounts of $CaCl_2$, the hydration of the tricalcium aluminate phase was followed as a function time.[31] The results showed that tricalcium aluminate phase reacted to form compounds such as chloroaluminates, sulfoaluminates, and aluminate hydrates.

It appears that much less work has been done on the effect of $CaCl_2$ on the hydration of the ferrite phase than on other cement minerals. It is generally assumed that the sequence of reactions in the ferrite phase is similar to that of the tricalcium aluminate phase except that the reactions are slower.

2.3 Effect of Calcium Chloride on Cement

The kinetics of hydration of cement may be followed by DTA, TGA, XRD, conduction calorimetry, chemical estimation, etc. The non-evaporable water content is a measure of the amount of hydrated cement formed. In one study, a cement hydrated for 6 hours, 1, 3, 7, and 28 days showed non-evaporable water contents of 7.0, 10.6, 13.6, 14.3, and 15.1% respectively, and the corresponding values with 2% $CaCl_2$ were 10.5, 14.9, 15.1, 15.6, and 16.3%. These results indicate that $CaCl_2$ accelerates the hydration of portland cement.[32]

The hydration of cement, being an exothermic reaction, produces heat, and, if the hydration is accelerated, heat is produced at a faster rate. The position of the peak corresponding to the maximum heat liberation occurs at lower times as the amount of added $CaCl_2$ is increased.[1][33] Calcium chloride also accelerates the rate of hydration of Type II, IV, and V cements, but only marginally that of Type III, a high early strength cement.

Several mechanisms have been suggested for the action of calcium chloride on cement. None of the theories can explain all the effects of $CaCl_2$ in concrete. It is possible that any one mechanism will be able to explain only one or some of the observations. It is likely that an overall mechanism should take into account the amount of the admixture used and the time of hydration and the experimental conditions.

Calcium chloride increases compressive strength of cement pastes especially at earlier times. The most significant effect on compressive strength occurs with portland blast furnace cement and marginally with portland-pozzolan cement. The compressive strength of cement pastes in the presence of 2% $CaCl_2$ improves by about 50, 41, 11, 9, and 8% over the reference at 6 hours, 1, 3, 7, and 28 days, respectively.[34]

Ramachandran and Feldman[35] examined the strength development in portland cement pastes hydrated in the presence of 0, 1, 2, and 3.5% $CaCl_2$. At any particular degree of hydration, the sample with 3.5% $CaCl_2$ had the lowest strength; at lower degrees of hydration, a sample containing 0% $CaCl_2$ was the strongest, although, with time, samples containing 1–2% $CaCl_2$ form stronger bodies than all others. Porosity, density, and bonding affected these results. It is thus evident that addition of $CaCl_2$ not only changes the rate of hydration, but also the intrinsic nature of the hydration products.

2.4 Effect of Calcium Chloride on Concrete

Calcium chloride reduces significantly both the initial and final setting times of concrete. This is useful for concreting operations at low or moderate temperatures. It permits quicker finishing and earlier use of slabs. The setting times are decreased as the amount of $CaCl_2$ is increased. Excessive amounts, e.g., 4–5%, may cause rapid set and should be avoided. Even using the same type of cement, obtained from different sources, shows differences in setting characteristics. There is no direct correlation between the acceleration of setting and subsequent strength development in the hardened state.

It is recognized that $CaCl_2$ increases the early strength of cement paste, mortar, and concrete. The actual values depend on the amount of $CaCl_2$ added, the mixing sequence, temperature, curing conditions, w/c ratios, and the type of cement. Calcium chloride increases the strengths of all types of cements (Types I to V ASTM designation). The time required for concretes to attain a strength of 13.8 MPa using 2% $CaCl_2$ is indicated in Table 2.[36] The optimum dosage suggested varies between 1 and 4%.[37][40] Most practitioners, however, recommend a dosage not exceeding 2% flake $CaCl_2$ or 1.5% anhydrous $CaCl_2$.

The acceleration of strength development is also achieved at temperatures lower than the ambient temperature. Figure 3 shows the relative strengths developed in concrete cured for periods 1, 3, 7, and 28 days and cured at temperatures of -4.0, -4.5, 13, and 23°C.[1][36][40] Although the addition of $CaCl_2$ results in greater strengths at ambient temperatures of curing, the percentage increase in strength is particularly high at lower temperatures of curing.

Table 2. Time Required for Concrete to Attain a Compressive Strength of 13.8 MPa Using 2% $CaCl_2$

Types of Cement	Plain Concrete	Time (days) Concrete Containing Calcium Chloride
I	4	1.5
II	5	2.0
III	1	0.6
IV	10	4.0
V	11	5.5

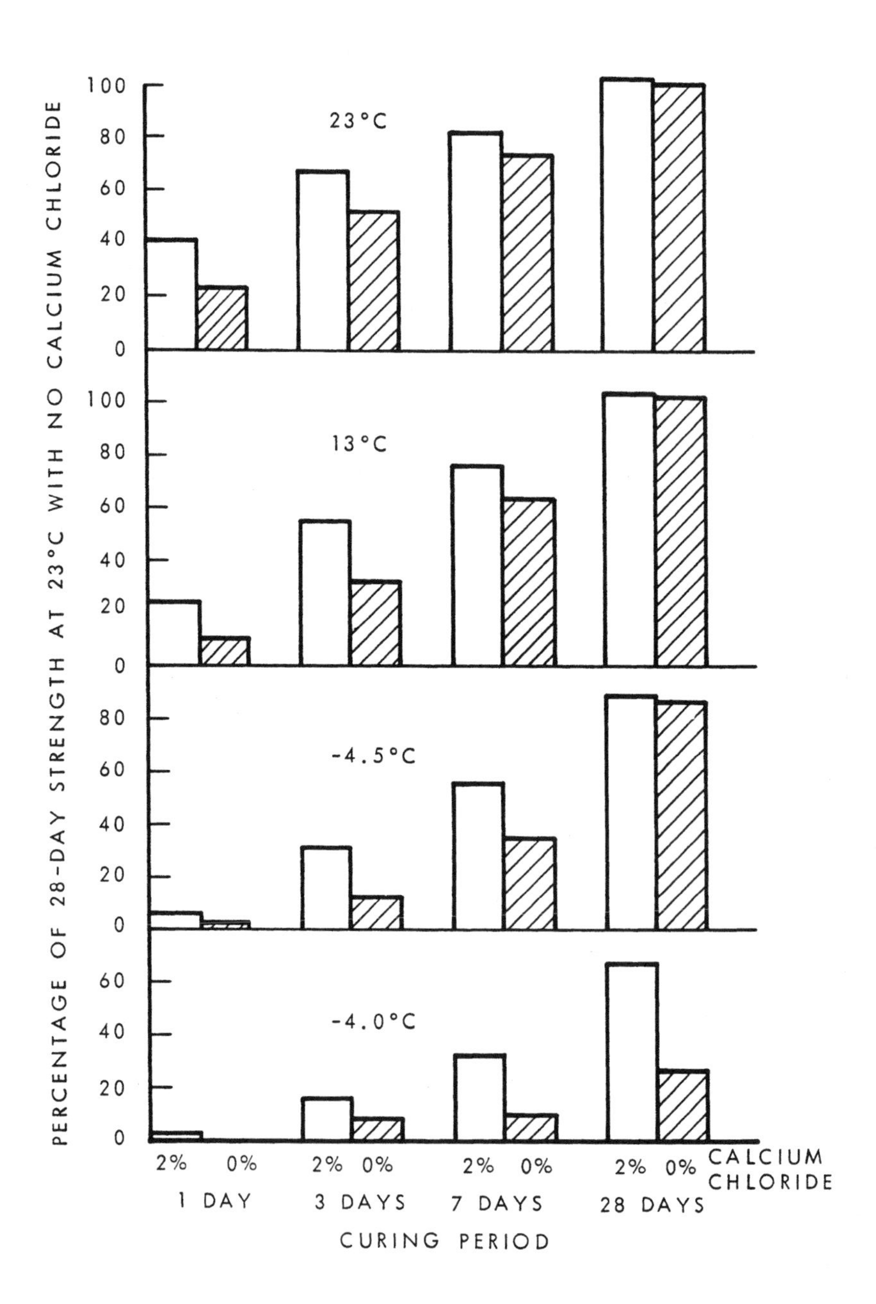

Figure 3. Effect of calcium chloride on strength development in concrete at different temperatures.

The Soviet literature contains references to the use of many complex admixtures containing $CaCl_2$.[38] A study suggests that, of the chlorides of Ca, Ba, Mg, and Fe, 1.6% $BaCl_2$ gives a one day strength equivalent to that obtained with 2% $CaCl_2$.[39]

Calcium chloride also has an effect on the hydration of various other cementitious systems such as pozzolanic cements, slag cements, expansive cements, high alumina cement, gypsum, rapid hardening cement, etc. (See, for example, Refs. 1 and 40.)

One of the limitations to the wider use of calcium chloride in reinforced concrete is that, if present in larger amounts, it promotes corrosion of the reinforcement unless suitable precautions are taken. There is, hence, a continuing attempt to find an alternative to calcium chloride, one equally effective and economical, but without its limitations. A number of organic and inorganic compounds including aluminates, sulfates, formates, thiosulfates, nitrates, silicates, alkali hydroxides, carbonates, halides, nitrites, calcium salts of acetic acid, propionic acid, butyric acid, oxalic acid and lactic acid, urea, glyoxal, triethanolamine, and formaldehyde have been suggested. However, practical experience and research on most of these admixtures is limited.

2.5 Triethanolamine (TEA)

Triethanolamine of formula $N(C_2H_4OH)_3$ is an oily water-soluble liquid having a fishy odor and is produced by reacting ammonia with ethylene oxide. Normally it is used in combination with other chemicals in admixture formulations. Its first use was reported in 1936, and the formulation containing TEA interground with calcium lignosulfonate was used to increase early strengths.[41]

Ramachandran followed the hydration of C_3A (with and without gypsum) containing triethanolamine.[42] It was found that TEA accelerated the hydration of C_3A to the hexagonal aluminate hydrate and its conversion to the cubic aluminate hydrate. The formation of ettringite was also accelerated in the C_3A-gypsum-H_2O system.

In a study of the effect of 0, 0.1, 0.5, and 1.0% triethanolamine on the hydration of C_3S and C_2S, Ramachandran concluded that there was an initial retardation of hydration.[43] At one day, acceleration of hydration took place, thus TEA can be construed as a delayed accelerator.

The retarding effect on tricalcium silicate hydration is almost completely eliminated in the presence of C_3A or NC_3A + gypsum. Adsorption of TEA on ettringite may remove the admixture to facilitate the hydration of tricalcium silicate. Figure 4 shows that the amount of TEA in solution in contact with ettringite decreases with time.[44] Another possibility is that TEA chelates metallic ions in the highly alkaline medium. It can be speculated that the Fe^{+3} in cement, precipitating during the hydration of cement, coats the silicates and aluminates and slows down the reaction.[45] The mono- and diethanolamines also affect the hydration of tricalcium silicate.

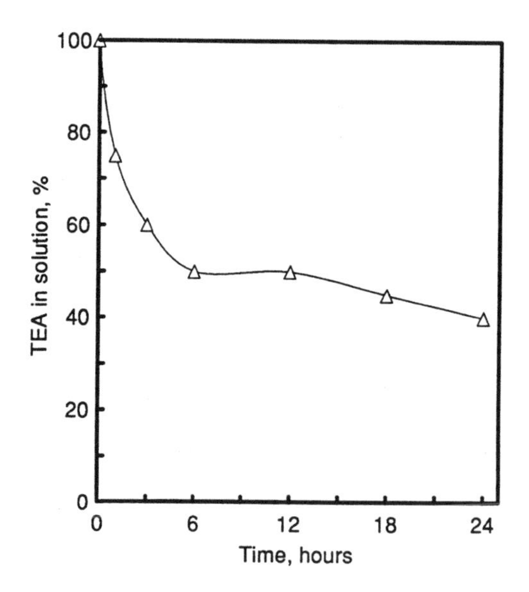

Figure 4. Adsorption of TEA by ettringite.

The initial and final setting characteristics of portland cement treated with 0–0.5% triethanolamine have been reported.[46] With up to 0.05% triethanolamine, the initial setting time is retarded slightly, and there is an accompanying slight extension in the induction period of the hydration of C_3S. At 0.1 and 0.5% triethanolamine, however, rapid setting occurs. The rapid setting may be associated with the accelerated formation of ettringite.

When added to cement, triethanolamine decreases its strength at all ages. Figure 5 shows the strength development in cement pastes containing 0, 0.1, 0.25, 0.35, 0.5, and 1% triethanolamine.[46] Strength decreases as the amount of triethanolamine is increased.

Amine salts are used in combination with other chemicals. Kuroda, et al.,[47] tested the effect of a formulation containing calcium nitrite, calcium rhodonate, and TEA. Both initial and final setting times were accelerated by this admixture combination at 5 and 20°C. The compressive strength of concrete also increased at these temperatures.

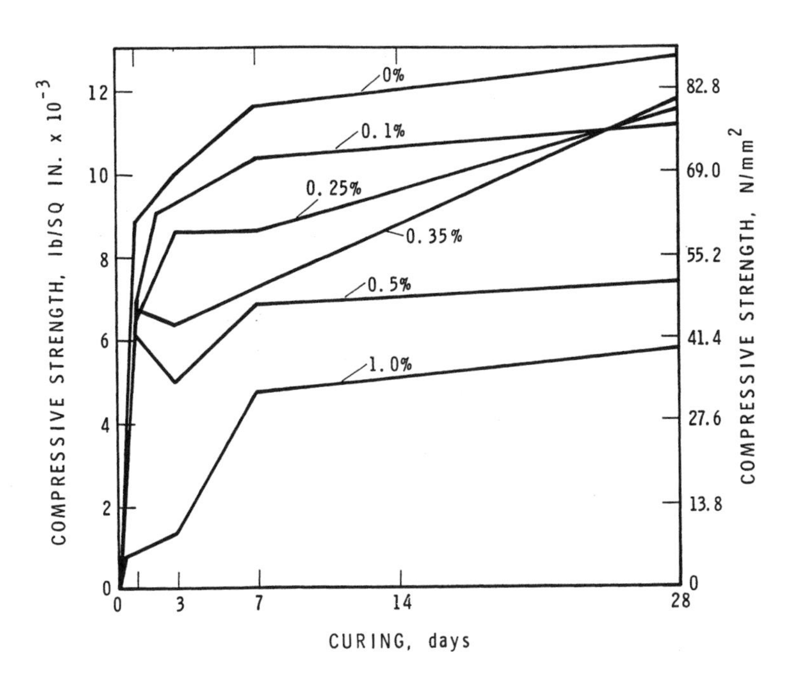

Figure 5. Compressive strengths of cement pastes containing triethanolamine.

Although it has been shown that TEA decreases the strength of portland cement systems, it acts differently with blast furnace cements. In a mortar containing 30% portland cement and 70% slag, at 7 days with 0.5% TEA, the strength increased by about 40%. The total porosity of the paste was lower in the presence of TEA. The TEA promoted acceleration of hydration by increasing the rate of reaction of gypsum with slag in the presence of lime.[48]

2.6 Formates

Calcium formate, of formula $Ca(HCOO)_2$, is a by-product in the manufacture of polyhydric alcohol, pentaerthritol. It is a powder and has a solubility of about 15% in water at room temperature. It is a non-chloride chemical that is used in practice. Many non-chloride accelerating admixture formulations contain formates. Calcium formate is an accelerator for the hydration of C_3S; at equal concentration, however, $CaCl_2$ is more effective (Fig. 6).[49]

The rate of hydration of tricalcium silicate in the presence of different amounts of calcium formate (0.5%–6.0%) has been reported by Singh and Abha.[50] Calcium formate accelerates the tricalcium silicate hydration at 2% addition. It is speculated that in the presence of formate the protective layer on the silicate surface is ruptured, resulting in an acceleratory effect.

The hydration of C_2S is also accelerated by calcium formate. The increase in compressive strength of formate-treated dicalcium silicate samples in relation to the reference sample is evidence of an acceleratory effect (Fig. 7).[51]

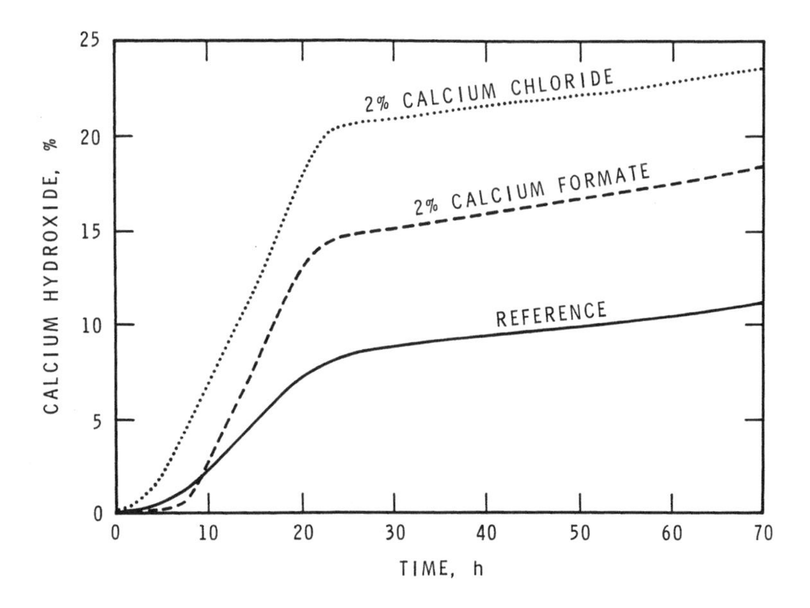

Figure 6. Influence of accelerators on the formation of calcium hydroxide in hydrated $3CaO{\bullet}SiO_2$.

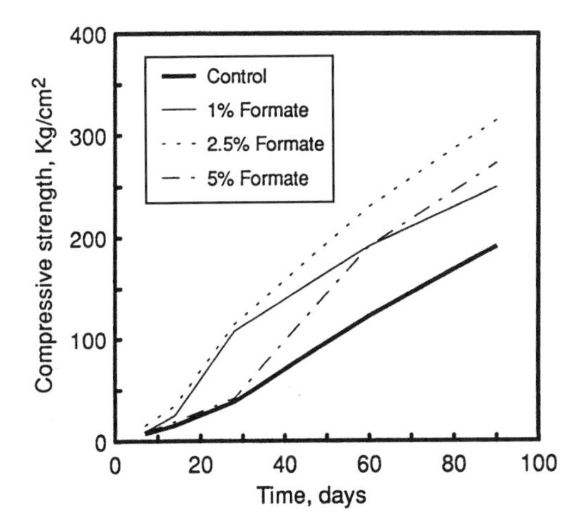

Figure 7. Influence of calcium formate on the strength development in C_2S mortars.

In the hydration of portland cement, although initially larger amounts of heat are developed in the presence of Ca formate, at later stages the heat may be slightly lower or equal to that of the reference material. In the early hydration periods, cement produces more ettringite in the presence of Ca formate than in the presence of $CaCl_2$.[52] However, the amount of C-S-H is higher in the presence of $CaCl_2$.

Ca formate accelerates the setting time of concrete, but comparatively a higher dosage is required to impart the same level of acceleration as that by $CaCl_2$. Table 3 compares the initial and final setting times of cement containing 2% $CaCl_2$ and Ca formate.[52] The effectiveness of Ca formate is also observed at different temperatures.[45]

Table 3. Setting Characteristics of Portland Cement Containing Calcium Chloride or Calcium Formate (2%)

Admixture	Initial Setting (minutes)	Final Setting (minutes)
None	185	225
Calcium chloride	65	75
Calcium formate	80	90

The strength of a cement, mortar, or concrete containing a Ca-formate may depend on the cement composition. Geber[53] found that the early and later strengths in mortar depended on the C_3A/SO_3 ratio of the cement. The compressive strength increase over the control at C_3A/SO_3 ratios (at 1 day and 2% Ca formate) of 2.44, 3.32, 4.34, 5.00, and 7.22 were 3, -14, 25, 88, and 70% respectively. It is suggested that to derive the effectiveness of formate, C_3A/SO_3 ratios higher than 4 should be preferred.

Generally, Ca formate increases the early strength of concrete. In Fig. 8, the results on the effect of 2% and 4% Ca formate indicate that the strengths are increased in the first 24 hours.[54] At lower dosages, the strength development may not be very significant. In combination with sodium nitrite, Ca formate accelerates the early strength development in cement to a significant extent. In comparison with the reference specimen without an admixture, the combination admixture increases strengths by 125, 70, 47, and 23% at 18 hrs, 1 day, 3 days, and 7 days, respectively.[55]

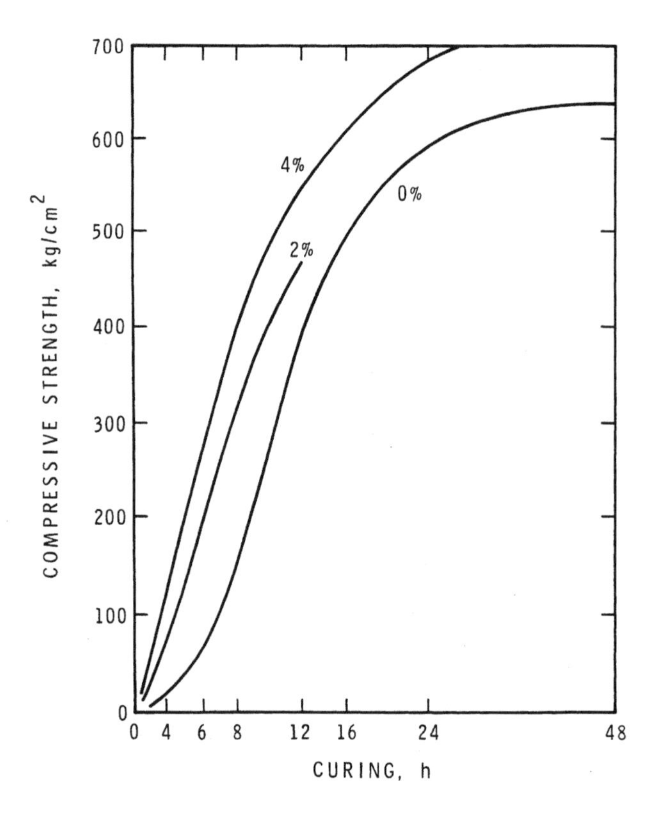

Figure 8. Influence of calcium formate on early strength of concrete.

An extensive study of the strength development in mortars containing Ca formate, Ca formate + Na formate and calcium chloride was carried out by Geber.[53]

2.7 Other Non-Chloride Accelerators

Many other non-chloride accelerators have been suggested which are based on organic and inorganic compounds. Valenti and Sabateli[56] studied the effect of alkali carbonates, viz., Na_2CO_3, K_2CO_3, and Li_2CO_3, on setting times and strengths of cements. At lower dosages, sodium and potassium carbonates retarded the setting times of cement, but at higher dosages (> 0.1%) they accelerated the setting. Lithium carbonate acted as a setting accelerator at all concentrations studied. At concentrations especially higher than 0.1%, the admixture generally increased the compressive strength at 28 days. Addition of Na_2CO_3 is reported to change the intrinsic property of the hydrated cement. In the presence of Na_2CO_3, smaller pores or radius 10–100 Å are decreased and pores of radius > 150 Å are increased slightly.[57] The reduction in the pores may be due to the precipitation of $CaCO_3$.

Calcium nitrate, calcium nitrite, and calcium thiosulfate are also suggested as accelerators. Calcium nitrite accelerates the hydration of cement as evident from the increased amounts of heat developed in its presence. Calcium nitrite also increases the strength and accelerates the setting times, but less effectively than Ca chloride. Table 4 compares the setting times of concrete containing $CaCl_2$ and $Ca(NO_2)_2$.[40][58] Table 5 shows the strength development in concrete containing 2, 3, 4, and 5% calcium nitrite.[40][58] Improvement is less significant at dosages greater than 4%. Calcium thiosulfate increases the strength development of concrete at early ages. For example, at 1 day compared to a compressive strength of 15.8 km/cm^2 for a reference mortar, that containing 1, 2, 3, 5, and 6.5% CaS_2O_3 shows values of 17.1, 18.1, 18.8, 20.1, and 21.2 kg/cm^2 respectively.[59] Sodium thiosulfate also accelerates the setting time, but the compressive strengths are slightly reduced with respect to the reference concrete at dosages of 0.5 and 1.0%.[60] Depending on the dosage of $NaNO_2$, Rosskopf, et al., found that acceleration of setting varied between 0.5 and 2.3 hrs. At dosages of 0.1–1.0% $NaNO_2$, the compressive strength of concrete at 7 days was equal or slightly lower than that of the reference concrete.[60] They also found that formaldehyde and paraformaldehyde in small amounts (0.01–0.25%) reduced the setting time from 9.5 to 6.3 hours.

Table 4. Setting Times of Concrete Containing Calcium Chloride and Calcium Nitrite

Cement Brand	Admixture	Setting Time (hr: min) Initial	Final
A	-----	8:45	12:21
A	Calcium Chloride	4:20	7:30
A	Calcium Nitrite	6:00	10:20
B	-----	8:38	---
B	Calcium Chloride	3:16	5:00
B	Calcium Nitrite	5:24	9:04

(Reprinted with permission from ASTM.)

Table 5. Effect of Calcium Nitrite on Strength Development

Admixture %	Compressive Strength (lbs/in^2) 1 Day	7 Days	28 Days
-----	1299	3401	5032
2	1615	4538	5732
3	1959	4964	5907
4	2294	5337	6383
5	2367	5324	6493

(Reprinted with permission from ASTM.)

The relative acceleratory effects of calcium chloride, calcium nitrate, and sodium thiocyanate on cement have been studied.[61] The total heat is higher in the presence of all accelerators. Calcium chloride addition results in the maximum amount of heat. Sodium thiocyanate accelerates better than calcium nitrate. Calcium nitrate works well at early stages, but sodium thiocyanate is more effective at later ages. Sodium molybdate and trioxide also have an acceleratory effect.[62][63]

Generally, the α hydroxycarboxylic acids are regarded as retarders, but lactic acid behaves differently.[64] In dosages of 0.1 to 0.4%, it reduces both the initial and final setting times of cement. The compressive strength at 28 days is higher than the reference by about 10–15%.

A new accelerator formulation consisting of inorganic salts and organic compounds (one of these being a sulfonate dispersant) is claimed to be effective at a low temperature of -7°C as well at a higher temperature of 35°C. It accelerates setting, increases early and ultimate strengths, decreases shrinkage, and does not initiate corrosion.[65] In Table 6, setting times and strengths of the reference concrete are compared with that containing the above admixture at -7°C. The admixture containing concrete develops better strengths than the reference and can be used in cold weather concreting.

In cements, incorporation of calcium carbonate is permitted in some countries. In Canada, the maximum limit is set at 5%. Calcium carbonate is not an inert filler. It is known to react with calcium aluminate. In a study of the hydration of tricalcium silicate in the presence of finely divided calcium carbonate, Ramachandran observed that the carbonate acted as an accelerator.[66] Ushiyama, et al.,[67] examined the effect of carbonates of Na, K, Li, Cs, and bicarbonates of Na, K, and Li on the hydration of alite. Although small amounts retarded the hydration, larger amounts acted as accelerators.

Table 6. Setting Time and Strength Development in a Newly Formulated Admixture

Admixture	Setting Times, h		Strength, MPa			
	Initial	Final	1 d	3 d	28 d	1 yr
Air Entrained (Reference)	10.67	14.58	1.6	11.7	26.2	36.9
Air Entrained (Accelerator)	6.15	11.50	2.8	9.0	30.8	43.1
Non-Air Entrained + Accelerator	5.50	11.50	3.6	8.7	34.2	46.3

Alkalis, such as NaOH, and Na salts of carbonate, aluminate, and silicate are known to accelerate the hydration of cement and cause early stiffening. In the hydration of tricalcium silicate with NaOH, there was an acceleratory effect even up to 28 days. After 7 days however, the strength of the reference was higher than that containing NaOH.[68] Evidence was also obtained for the incorporation of Na in the hydrated product.

Volatilized silica, obtained as a by-product in the metallic silicon or ferrosilicon industry, is known as silica fume. It has been advocated for incorporation into cement to obtain high strength. It acts as a pozzolan and is also known to accelerate the hydration of tricalcium silicate. Therefore, it can be treated as an accelerator.

There has been continued interest in developing an organic-based accelerator. Ramachandran and coworkers studied the effect of o, m, and p-nitrobenzoic acids on the hydration of tricalcium silicate.[69] The m and p nitrobenzoic acids acted as accelerators. The acceleratory effect was attributed to the complex formation between the organic compound and the C-S-H phase on the surface of the tricalcium silicate phase.

Some work has been carried out using a finely ground portland cement hydrate at a dosage of 2% as a "seeding agent."[70] It is claimed that 2% of seeding is equivalent to 2% $CaCl_2$ in its accelerating effect.

It appears from the published reports that in spite of intensive efforts, few chemicals perform as well as calcium chloride in terms of accelerating the hydration of cement, and developing early strengths when used at equal dosages.

3.0 WATER REDUCERS AND RETARDERS

3.1 Introduction

Most water reducers and retarders are based on similar formulations, and they are treated together in this chapter. These admixtures can be divided into four categories, viz., normal water reducers, water-reducing–retarding admixtures, water-reducing–accelerating admixtures, and retarders.

A normal water reducer lowers the water requirement to attain a given slump. Thus, for the same slump and a constant cement content use of lower w/c ratios, results in general improvement in strength, permeability, and durability. Alternatively, the desired slump is achieved without a change in w/c ratio by lowering the cement content. A water-reducing admixture may also be used to increase the slump to facilitate placements. According to ASTM, water reducers should be able to lower the water requirements by 5% of the control. The water reduction depends on the admixture dosage, fineness of the cement, mix proportions, and the time of addition of admixture. According to the standards, the minimum initial set time should be 1 hour earlier with a maximum of 1½ hours later, with respect to the control. The compressive strength at 3, 7, and 28 days should be at least 110% of the corresponding control mixes.

The water-reducing–retarding admixtures are used for avoiding cold joints and facilitating large pours. They are used particularly in hot weathering operations. These admixtures should increase the initial set time by at least one hour, with a maximum by 3½ hours. The compressive strength should be at least 110% of the control at 3, 7, and 28 days.

The water-reducing–accelerating admixtures may be used in winter concreting as they permit early form removal and enable concrete to be available for service earlier. The addition of these admixtures should result in an initial set one hour earlier, with a maximum at 3½ hours earlier. The compressive strength requirement is 125% of control mix at 3 days and 110% at 28 days.

Set retarders are used to offset the effects of high temperatures, which decrease the set time. They are also applicable when unnecessary delays occur between mixing and placement. In mass concrete, retarders prolong the plasticity of fresh concrete, and that has advantages. Elimination of cold joints is possible with the use of retarders. With retarders, the minimum initial set time should be 1 hour later, with a maximum of 3½ hours. The compressive strength should be a minimum of 90% of the control at 3, 7, and 28 days.

Composition. The materials that are used for water reducers and retarders are generally composed of the following compounds:

1. Lignosulfonic acids and their salts.
2. Modifications and derivatives of lignosulfonic acids and their salts.
3. Hydroxycarboxylic acids and their salts.
4. Modifications and derivatives of hydroxylated carboxylic acids and their salts.
5. Miscellaneous formulations such as inorganic salts including zinc salts, borates, phosphates, chlorides, amines (their derivatives), carbohydrates, polysaccharides, polymeric materials, and sugar acids.

Mode of Action. The common feature of all these admixtures is that they are adsorbed on the solid-water interface. In a cement paste, opposing charges on adjacent particles can exert electrostatic attraction, causing flocculation. A considerable amount of water is imbibed in these agglomerates and adsorbed on the surface, leaving less water for increasing the workability of paste or concrete. In the presence of a water-reducing admixture, the surface charges become the same, hence the particles repel each other and dispersion occurs. Thus, more water is available for reducing the viscosity of the cement paste or concrete.

Several explanations have been offered to account for the retarding action of water reducers. They are based on the following theories:

1. Retarders are adsorbed on the anhydrous cement particles through ionic, hydrogen, or dipole bonding to prevent the attraction of water.

2. An insoluble layer of calcium salt on the hydrating particles is responsible for retardation.

3. The hydrated, rather than the unhydrated constituent, adsorbs the retarder. The surface complex involving the substrate of unhydrated cement and water followed by adsorption is responsible for retardation.

4. The adsorption of retarder on the calcium hydroxide nuclei poisons its future growth; the growth of calcium hydroxide will not proceed until some level of supersaturation is attained. Details of these theories are discussed in a in Ref. 36.

3.2 Retarders

Retarders extend both the initial and final setting times of cement paste and concrete (Table 7).[36] Higher amounts may enhance the retardation process. At equal dosages, sucrose is the most efficient retarder and, hence, an accidental overdose of this admixture could create serious setting problems.

The setting times depend on the type of cement, w/c ratio, temperature, and sequence of addition. Cements with low C_3A and alkali contents are retarded better than those containing large amounts of these constituents. For setting to occur, both C_3A and C_3S phases have to hydrate to some extent. The alkali in the cement is capable of interacting with the retarder and destroying its capability to retard hydration. A small amount of lignosulfonate (0.1%) slightly retards or accelerates the C_3S hydration, depending on the chemical composition and molecular weight of the lignosulfonate.[71][72] Addition of Ca lignosulfonate to C_3S above a particular dosage can indefinitely inhibit its hydration. When a retarder is added to concrete 2–4 minutes after mixing, setting time is delayed 2–4 hours beyond when the admixture is added with mix water.[36]

Table 7. Influence of Some Retarders on the Setting Times of Cement Paste

Admixture	Amount (%)	Initial Set Time (h)	Final Set Time (h)
None	——	4	6.5
Sucrose	0.10	14	24
	0.25	144	360
Citric Acid	0.10	10	14
	0.25	19	44
Ca Lignosulfonate	0.10	4.5	7.5
	0.25	6.5	10

Addition of lignosulfonate retards both the C_3A hydration and the conversion of hexagonal hydrates to the cubic phase.[40] Ca lignosulfonate retards C_3A hydration in the mix of C_3A + gypsum, but a more marked retardation occurs in the conversion of ettringite to monosulfonate.

In the hydration of C_3A or C_3S, the retarding effect appears to be in the decreasing order as follows: sugar acids > carbohydrates > lignosulfonate.

The hydration products of C_4AF with or without gypsum are very similar to those of C_3A and isomorphous with various compounds formed with C_3A hydrate. The rate of hydration of C_4AF is affected to a lesser extent than that of C_3A in the presence of retarders.

Retarders such as salicylic acid retard C_2S hydration especially at higher w/s ratios,[40] but commercial lignosulfonates (at about 0.125%) have a better retarding effect.

In general, retarders are effective in the very early hydration of cement. For a given percentage, Na gluconate, citric acid, or sucrose are better retarders of initial set of a normal low alkali cement than lignosulfonate, salicylic acid, Na heptonate, or boroheptonate.[40]

At a constant w/c, the addition of commercial lignosulfonate results in lower strength cement mortars, but later strengths are higher, as can be seen in Fig. 9.[73] Sometimes lignosulfonate addition may not cause an increase in strength at later ages due to the entrainment of air.

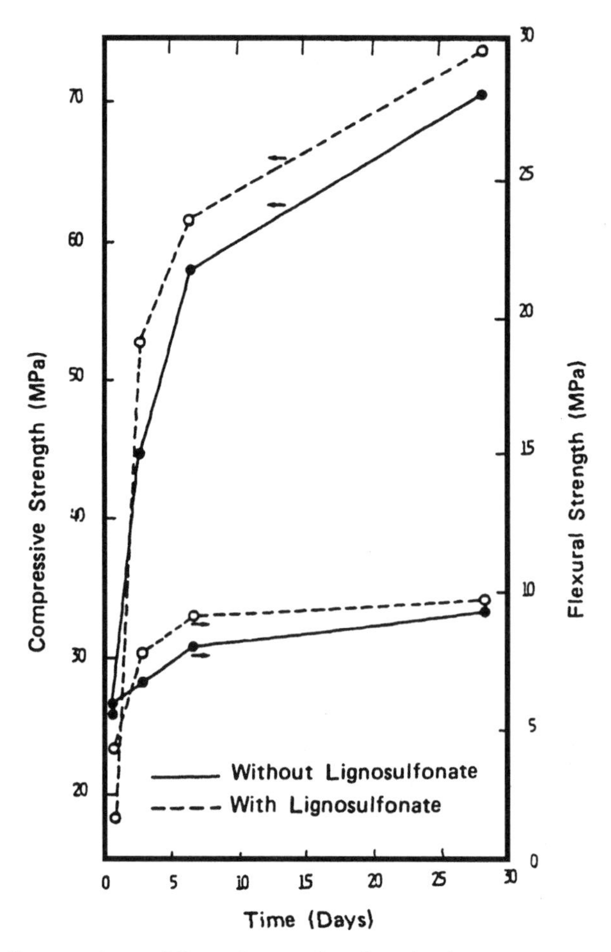

Figure 9. Compressive and flexural strengths of portland cement mortars at a given w/c ratio with lignosulfonate.

3.3 Water Reducers

The effectiveness of water reducers in reducing water requirements for a given consistency depends on the type and amount of water reducer, type of cement, cement content, type of aggregate, aggregate/cement ratio, presence of other admixtures, and the time of addition of the admixture.

A partial explanation of the water reduction by lignosulfonate admixture is its ability to entrain air. It is reported that lignosulfonate promotes higher water reduction than hydroxycarboxylic acid-based admixtures.

There is evidence that water-reducing admixtures are less effective with a high C_3A or alkali content cement. In a cement containing 9.4% C_3A, the water reduction was found to be about 10%, and with 14.7% C_3A, the reduction was 4%.[74]

Table 8. Effect of Water-Reducing Admixtures on Water Reduction

Water Reducer	Nil	Na Gluconate	Glucose	Sugar-free Lignosulfonate
Slump (mm)	95	100	95	100
Water/Cement	0.68	0.61	0.63	0.65
Water Reduction (%)	—	10.3	7.3	4.4

The relative water-reducing effects of three water reducers added at 0.1% dosage are shown in Table 8.[40] Sodium gluconate appears to be more effective than glucose or lignosulfonate.

Water-reducing and retarding admixtures are required to extend the setting time from 1 to 3.25 hours with respect to the reference concrete. Water-reducing and accelerating admixtures accelerate the setting time between 1 and 3.5 hours.

One of the important applications of water reducers is their ability to increase the workability characteristics or slump at a constant w/c ratio, without affecting strengths. The degree to which slump increases depends on the admixture dosage level, cement content, and the type of aggregate. Slump can be increased by 100% by water reducers. The hydroxycarboxylic acid-based water reducers seem to promote higher slumps values than those based on lignosulfonates. Slump loss occurs rapidly in the presence of water reducers. However, water reducers allow longer periods between mixing and placing of concrete without detrimental effects.

Extensive work has been carried out on the effect of lignosulfonate on the hydration of individual compounds as well as on cement itself. The effects are similar to what has already been described under the previous section on "Retarders."

The effect of carboxylic acids such as gluconic acid, and carboxylic acid and sugars such as glucose or sucrose, on portland cement hydration is very similar to that of lignosulfonate, although different percentages of admixtures are required to obtain similar effects.

Generally, water reducers retard both the very early hydration (initial set) and early hydration (final set) of portland cement whereas they increase the degree of hydration at later ages. In some cases, an acceleration of the very early hydration, promoted by carboxylic acids or carbohydrates, causes quick and early stiffening.

At equal dosages, all water reducers are effective in producing concrete of equal or higher compressive strength than that of reference concrete. At equal cement content, air content, and slump, the water reducers increase the 28 day strength by about 10–20%. Table 9 shows the influence of lignosulfonate type admixture on the compressive strength of concrete.[36] Water reduction varies between 5 and 8%.

Although the increase in strength may be explained by water reduction, higher strengths, in many cases, are greater than what would be expected from the reduction in w/c ratio.

Table 9. Effect of Lignosulfonate on the Compressive Strength of Concrete

Admixture, %	W/C	Compressive Strength (% of control)			
		1d	3d	7d	28d
0.0	0.63	100	100	100	100
0.07	0.60	101	104	103	102
0.13	0.60	95	108	111	101
0.26	0.60	107	115	112	115

Water-reducing–accelerating admixtures are particularly effective in promoting early strength development, and should find application in winter concreting.[74] For example, at 2, 5, and 18°C concrete with an admixture shows 3-day strength of 3.6, 5.4, and 15.5 N/mm^2 respectively, but the corresponding concrete containing the admixture shows strength values of 7.2, 8.0, and 19.2 N/mm^2.

4.0 SUPERPLASTICIZERS

Superplasticizing admixtures are capable of reducing water requirements by about 30%, and also impart high fluidity characteristics to concrete. They are variously known as superfluidizers, superfluidifiers, super water reducers, or high range water reducers. They were first introduced in Japan in 1964 and later in Germany in 1972. The basic advantages derived from the use of superplasticizers are as follows: production of concrete having high workability for easy placement without a reduction in cement content or strength; production of high strength concrete with normal workability, but with lower water content; the possibility of making a mix having a combination of better than normal workability and lower than normal amount of water; designing a concrete mix with less cement, but having the normal strength and workability. Because of the significant benefits that superplasticizers confer on concrete, an impressive number of patents have been taken on superplasticizers. In eleven years, 1985–1995, more than three-hundred patents were filed that refer to the development of a superplasticizer or it as an ingredient of a multifunctional admixture.

A variety of organic water-soluble polymers can be used as superplasticizers. The lignosulfonate compound may be considered as a precursor to superplasticizers of today. Polynaphthalene sulfonate formaldehyde (SNF) has been used widely. Another superplasticizer that has been used extensively is sulfonated melamine formaldehyde (SMF). Synthetic organic polymer bearing carboxylic acid groups also act as high dispersants. Polyacrylates are derivatives from a combination of anionic acrylic and substituted acrylic polymers. With other anionic formation groups (phosphonates, phosphates, sulfates) and polar functional groups (hydroxy, ether, amide, amine, etc.), numerous types of polymers can be synthesized for use as superplasticizers.

The mode of action of superplasticizers involves surface adsorption, dispersion of the particles in the aqueous phase assisted by electrostatic repulsion, and steric repulsion. A more detailed interpretation may be needed to explain various physical and chemical phenomena.

Figure 10 shows the effect of a SMF-based superplasticizer on a cement paste. Large irregular agglomerates of cement particles predominate in a water suspension (Fig. 10a). The addition of a superplasticizer causes dispersion into small particles (Fig. 10b). A viscometer is used to measure the effect of a superplasticizer on rheology. Figure 11 plots the viscosity changes in cement containing SNF exposed to different times.[75] At any particular time, the viscosity is lower for the paste containing the admixtures. It has been reported that 1% superplasticizer at a w/c ratio of 0.3 shows the same fluidity as cement paste at a w/c ratio of 0.4 without the admixture.

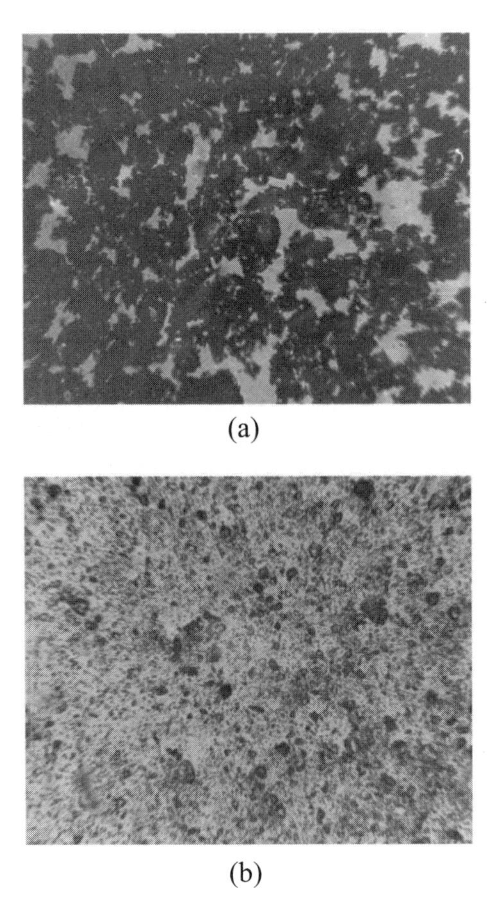

(a)

(b)

Figure 10. Dispersion of cement paste: *a*) without superplasticizer; *b*) with a melamine-based superplasticizer.

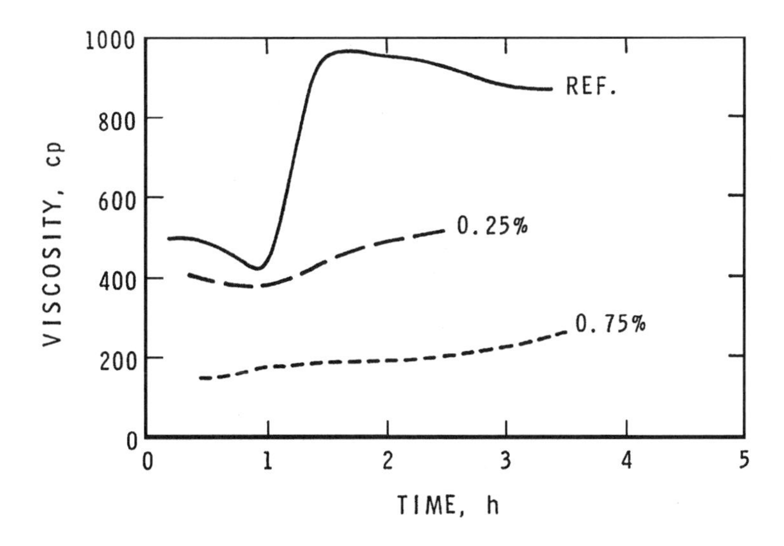

Figure 11. Effect of sulfonated naphthalene formaldehyde on viscosity of cement paste.

In concrete, slump can be increased substantially by the addition of superplasticizers. For example, a slump of 260 mm from an initial slump value of 50 mm is obtained by an addition of 0.6% SMF; the same value is realized with an addition of 0.4% SNF. The time when the superplasticizer is added to concrete affects the slump value. By adding the admixture with the mixing water, the slump is increased, but even higher values result if the admixture is added a few minutes after the concrete is mixed with water.

Higher than normal workability of concrete containing a superplasticizer is maintained for about 30–60 minutes, and then there is a rapid decrease in slump, termed *slump loss.* To control or extend the workability, the superplasticizer should be added at the point of discharge of concrete. Some admixtures are added to superplasticizers to control slump loss. The acrylate-based superplasticizers are claimed to possess good slump retention qualities.

The rate of hydration of cement and cement compounds is influenced by superplasticizers. The reported data on the effect of superplasticizers on the hydration of C_3A and C_3A + gypsum systems are contradictory because of the variations in the materials and conditions of hydration. There is a general consensus that SNF and SMF retard the hydration of

C_3A.[76]–[78] Hydration of C_3S is known to be retarded by the superplasticizer.[79][80] There is also evidence that the C/S ratio of the C-S-H phase is changed. In cements, the hydration of the C_3S phase is retarded, but to a lesser extent than when hydrated in the pure form because part of the admixture is adsorbed by the C_3A phase. Formation of ettringite may be accelerated or retarded by the superplasticizer, depending on the amount of alkali sulfates present in cement.

The amount of water reduction achievable with a particular superplasticizer depends on the dosage and initial slump. Some results are given below (Table 10).[40] There is evidence that beyond a particular dosage, further water reduction is not possible. In all types of cements, water reduction occurs to different extents.

Table 10. Water Reductions in Concrete with an SNF Superplasticizer

Dosage	w/c Ratio	Water Reduction	Slump
Reference	0.60	0	100
Normal	0.57	5	100
Double	0.52	15	100
Triple	0.48	20	100
Reference	0.55	0	50
Normal	0.48	13	55
Double	0.44	20	50
Triple	0.39	28	45

In flowing concrete in the presence of a superplasticizer, the 28 day strength is equal or greater than the corresponding strength of the reference concrete. The strength for flow and water-reduced concrete made with three types of cement is generally commensurate with the values expected from the w/c ratios.[81] Figure 12 gives the strengths developed by concrete containing an SMF superplasticizer.[40][81] Strengths are always higher in the presence of the superplasticizer at any particular w/c ratio. In addition, strengths are higher at lower w/c ratios.

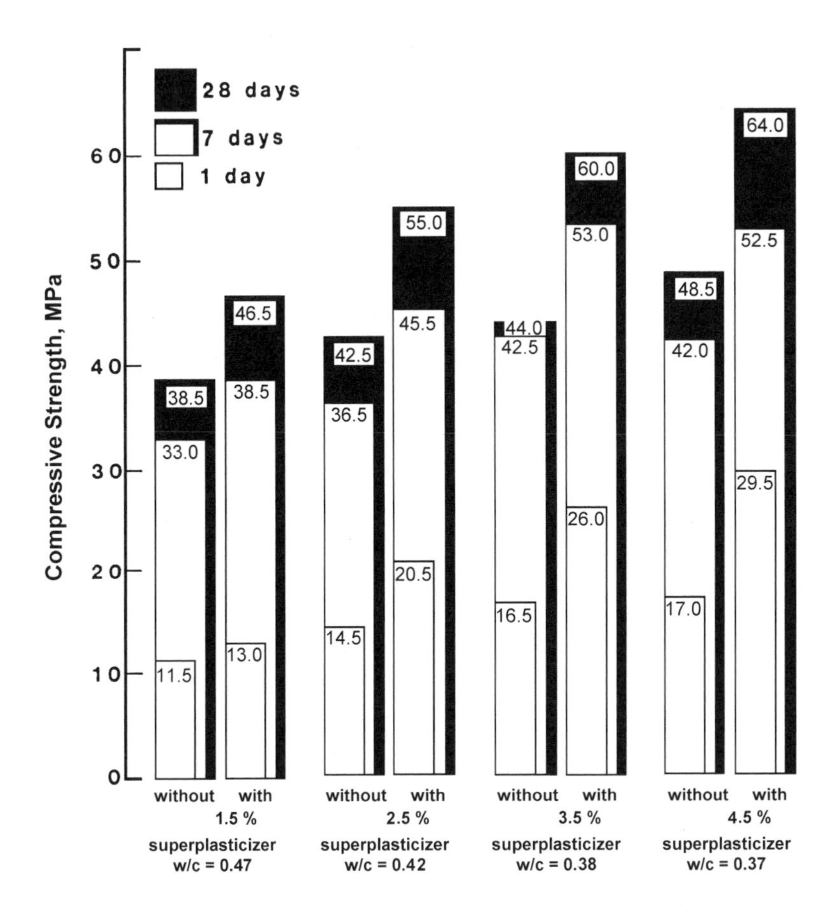

Figure 12. Strength development of high strength flowing concrete in the presence of an SMF-Type superplasticizer.

5.0 AIR-ENTRAINING AGENTS

Air-entrainment is the process whereby many small air bubbles are incorporated into concrete. Air-entrainment is essential for increasing the durability of concrete exposed to freeze thaw conditions. Air-entrainment improves workability of concrete and also decreases bleeding and segregation. Salts of wood resins, synthetic detergents, salts of sulfonated lignin, salts of petroleum acids, salts of proteinaceous materials, fatty and resinous acids and their salts, and organic salts of sulfonated hydrocarbons are examples of air-entraining agents.

The air-entraining action is known to involve adsorption at the air-water or solid-water interfaces. The soluble surfactant ions are adsorbed on cement particles making them hydrophobic, so that as bubbles are generated during mixing, they adhere to the cement. This process stabilizes bubbles, preventing their coalescence. Lowering of surface tension may also stabilize bubbles.

Non-ionic air-entraining agents have no appreciable effect on the rate of cement hydration. Even if some of them possess retardant behavior, the amounts used are such that the effects are not significant. It is known that anionic agents such as sodium vinsol resin react with calcium hydroxide to form a precipitate of calcium salt. When air-entraining agents are used with other admixtures, the interaction between them and the cement has to be considered for compatibility purposes. At higher dosages, the anionic air-entrainers may retard the C_3S hydration.[82]

Many factors influence the amount of entrained air. They include dosage, slump, aggregates, temperature, inclusion of other admixtures, chemical composition of cement, mixing, vibration, etc.

Generally, a strength loss occurs in air-entrained concrete. As it also allows the reduction of w/c ratio, the loss in strength may partly or wholly be offset.

6.0 MINERAL ADMIXTURES

Mineral admixtures are finely divided materials that are added in relatively large amounts as replacement of cement and/or of fine aggregates in concrete. Some of them also possess self-cementitious properties in addition to being pozzolanic. A *pozzolanic material* is one that is able to react with calcium hydroxide forming a cementitious material. Natural pozzolans such as volcanic tuffs, earths, trass, clays, and shales (raw or calcined), and industrial products such as fly ash, slag, silica fume, red mud, and rice husks are the main source of mineral admixtures. Extensive work has been carried out on fly ash, slag, and silica fume.

The mineralogical composition, particle morphology, size, and physical make-up of the mineral admixtures control their reactivity to a relatively larger extent than their chemical composition.

High calcium fly ash and granulated blast furnace slag are cementitious. Low calcium fly ash is termed a *normal pozzolan*. Condensed silica fume and rice husk ash are highly pozzolanic. Slowly cooled blast furnace slag, bottom ash, and field burnt rice husk ash are weak pozzolans.

6.1 Fly Ash

Fly ash contains about 15% crystalline material, the rest being an amorphous material and carbon. The major crystalline materials in fly ash are quartz, mullite, hematite, and magnetite. The reactivity of fly ash depends on the loss on ignition, fineness, and mineralogical and chemical composition. The pozzolanic activity also depends on the mineralogy of coal and on the glassy or non-crystalline structure of fly ash.

Fly ash may be used as a replacement of cement or fine aggregate or as an additional component at the concrete mixing plant. Fly ash has been used up to 60% replacement of cement. Addition of fly ash reduces the water requirement for a particular consistency or flow. Low calcium fly ash acts largely as a fine aggregate initially, but with time will react to pozzolanic compounds. High calcium fly ash participates in the early cementing reactions. Partial replacement of fly ash results in the reduction in temperature rise in fresh concrete due to the reduction in the heat of hydration.

The rate of strength development in fly ash concrete depends on the type of fly ash, temperature, and curing conditions. Concrete containing 50% low calcium fly ash replacement and a superplasticizer is capable of developing 60 MPa compressive strength at 28 days and 20–30 MPa in 3 days. Beyond 7–8 weeks, all concrete mixes containing fly ash show strength comparable to concrete mixes having an equal cement content. At early ages, concretes containing 50% replacement are more permeable than concrete without them. The trend reverses after about 180 days because of the pozzolanic activity of the fly ashes.[83] The sulfate resistance of concrete is enhanced by adding fly ash to concrete.

In 1985, Malhotra and others initiated studies on structural concrete incorporating more than 50% low calcium fly ash.[84][85] This was achieved by using a higher than normal dosage of superplasticizer. The high volume fly ash exhibits a 1-day strength between 5 and 9 MPa. The later age compressive strength reaches 60 MPa at one year. Compressive strength of more than 90 MPa have been reported on test cylinders and drilled cores from various concrete monoliths.[86]

6.2 Slag

Granulated slag is a by-product obtained as a finely divided material from ferrous and nonferrous metal industries. It is used as a component of blended cement or as a supplementary cementing material. Slags have similar chemical composition to that of normal cement. They are essentially glassy with glass contents around 75%. The chemical composition corresponds to melilite phase, a solid solution phase between gehlenite (C_2AS) and akermanite (C_3MS_2). It may take about three days before cementing properties of slags to be noticeable. Some hydration takes place immediately after water addition and a protective layer is formed on the surface of the slag. Alkalis, gypsum, and lime serve as activators or accelerators for slag hydration. Alkali-activated slags have found commercial uses in some countries. Up to 65–80% of slag replacement in slag concretes is possible.

The resistance of slag concrete to sulfate is increased if more than 65% cement is replaced by slag at a constant gypsum addition of 5%. The high resistance to sulfate attack can be attributed partly to lower permeability of concrete.

6.3 Silica Fume

Silica fume is a by-product from the silicon metal or ferro-silicon industry. Its particle size ranges from 0.1 μm to 0.2 μm, the surface area being 20–23 m^2/g. It is highly pozzolanic and improves the properties of concrete in fresh and hardened states.

Addition of silica fume accelerates the hydration of C_3S. The CaO/SiO_2 ratio of the C-S-H product is also changed. For example, at a w/s ratio of 10, C_3S yields a C-S-H product with CaO/SiO_2 of 0.8 at 24 hours, and this ratio decreases to 0.33 and 0.36 at C_3S/SiO_2 ratios of 1 and 0.4, respectively. In the hydration of cement, the calcium hydroxide formed as a product is consumed by silica fume. Silica fume in cement increases the overall rate of reaction.[87] Figure 13 gives the strength results of concrete containing silica fume. Silica fume was added as an addition. The increase in strength depends on the amount of silica fume added.[87]

Durability of silica fume concrete to freeze thaw action is generally satisfactory especially at silica fume content of less than 20%.[88] Incorporation of silica fume in concrete increases its resistance to sulfate attack. Reduced availability of lime and increased impermeability are the main reasons for sulfate resistance.

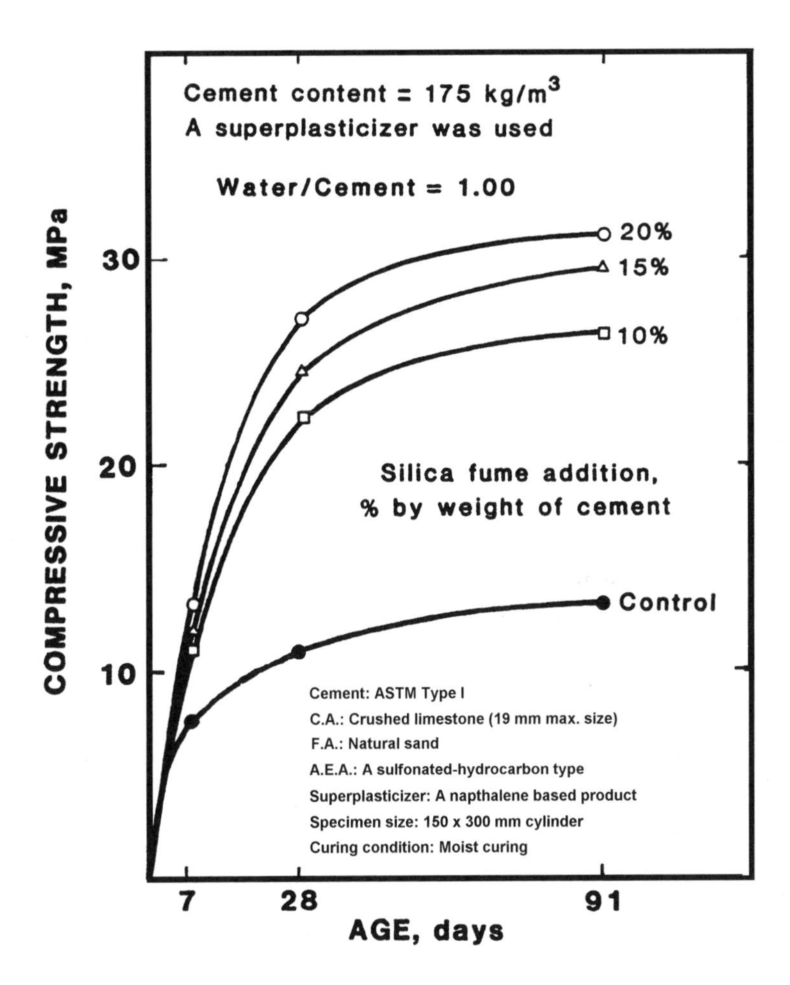

Figure 13. Relation between compressive strength and age for concrete incorporating various percentages of silica fume as an addition to cement.

7.0 MISCELLANEOUS ADMIXTURES

In addition to admixtures already described, many others are used to produce special modification of concrete and mortar. Thermal Analysis Techniques have not been applied significantly to examine the role of these admixtures in concrete, although they have potential. A thorough treatment of these admixtures is presented in Ref. 89. A brief description of some of these admixtures follows.

7.1 Expansion Producers

These admixtures control settlement and provide expansion in the plastic state. In the plastic state, metallic Al, Mg, or Zn produce gases. Admixtures that produce expansion both in plastic and hardened states are calcium sulfoaluminate and lime based materials. Those which produce expansion only in the hardened state are granulated iron filings and chemicals that promote the oxidation of iron and offset volume changes due to shrinkage.

7.2 Pigments

Concrete may be colored by integral addition or by the application of color in the hardened state. Integral colored compounds include natural and synthetic products, for example, chromium oxides, cobalt oxide, and organic pigments based on Cu complexes of phthalocyanine.

7.3 Dampproofing and Waterproofing Admixtures

They are applied on roofs, slabs on ground, basements, water-retaining structures, concrete blocks, and clay bricks. Waterproofing admixtures reduce the permeability of concrete. The dampproofing admixtures impart water repellency and reduce moisture migration by a capillary action. Examples of these admixtures are soaps and fatty acids which react with cement, conventional water reducers, methyl siliconates, etc.

7.4 Pumping Aids

They are used to pump marginally pumpable concrete. Materials that are used for this purpose are water-soluble synthetic and natural organic compounds such a cellulose ether, polyethylene oxide, alginates, organic water-soluble flocculants, emulsion of organic materials, fine clays, kaolin, diatomaceous earth, and calcined pozzolans.

7.5 Flocculating Admixtures

The chief function of these admixtures is to reduce the destructive type of bleeding to a more gradual seepage in mixes. The materials include maleic anhydride copolymer, Na salt of styrene, polyacrylamide, polymethacrylic acid, etc.

7.6 Bacterial, Fungicidal, and Insecticidal Admixtures

Many organisms contribute to concrete degradation by establishing and maintaining large colonies on or within concrete. Bactericidal admixtures consist of polyhalogenated phenols, Na benzoate, benzalkonium chloride, and copper compounds. Fungicidal admixtures include polyhalogenated phenols, copper acetoarsenite, and copper arsenate. Termite-proofing admixtures contain emulsion of the chemical dieldrin and is used at a dosage of 0.5% by weight.

7.7 Shotcreting Admixtures

The shotcreting process involves quick setting (3–12 minutes) of the mix on a surface and should also provide a thick layer without sloughing off, and a rapid strength of the order of 500–1000 psi. Materials such as air-entraining agents, strong alkalis, accelerators, organic bases, and sodium aluminate are used for this purpose.

7.8 Antiwashout Admixtures

These are cohesion-inducing admixtures that render concrete cohesive enough to allow limited exposure to water with little loss of cement. Such admixtures allow placement of concrete under water without the use of conventional tremies. The active ingredient in such admixtures is a water-soluble cellulose ether or water-soluble acrylic type polymer. Other auxiliary agents are also added.

7.9 Corrosion Inhibiting Admixtures

The use of de-icing salts on concrete pavements has a deleterious effect on reinforced concrete. Many bridges and parking garages with cast-in-place concrete and reinforced steel develop cracks by allowing corrosive chemicals to react with the steel. In normal concrete, the water in the pores is highly alkaline and a protective iron oxide layer forms on the steel. This protective layer may be severely affected under salt exposure conditions. One way to improve the resistance of concrete is to enhance its impermeability by incorporation of silica fume, fly ash, slag, or polymer modified mortar. Some chemicals known as corrosion inhibitors have been advocated to counter corrosion. They are Ca/Na nitrites, Na benzoate, Na/K chromates, Na salts of silicates and phosphates, stannous chloride, hydrazine hydrates, etc. Extensive work has been carried out on the effect of nitrite-based compounds, which have shown promise as efficient corrosion inhibitors.

7.10 Alkali-Aggregate Expansion Reducing Admixtures

The alkali-aggregate expansion involves chemical interaction between the alkali hydroxides derived from the cement and reactive components of aggregates. In the alkali-silica reaction, hydrolysis of reactive silica and alkali forms alkali-silica gel. The alkali-silicate reaction is the result of aggregate containing greywackes. The alkali-carbonate reaction is different from others as the affected concrete does not contain any significant amounts of silica gel. The effective methods of counteracting this expansion are: use of silica fume, fly ash, slag, air-entrainment, and chemical admixtures. Of the chemicals studied lithium salts, Cu sulfate, Al powder, and some proteins were found to be effective in decreasing the alkali-aggregate expansion reaction.[90]

7.11 Polymer-Modified Mortars/Concrete

These admixtures are used to increase the bond strength in repair applications, to decrease shrinkage, increase tensile strength, etc. The polymers used include latexes, redispersible polymer powders, water-soluble polymers, liquid resins, and monomers. In practical applications, styrene butadiene rubber, polyacrylic ester and polyvinylidene chloride-vinyl chloride, methylcellulose, etc., have been used.

7.12 Admixtures for Oil Well Cements

Oil well cement needs to have the properties such as low permeability, satisfactory bond with the rock or bore holes. Depending on conditions, cement slurries with densities from 1 to 2.4 kg/lit at pumping times 2–6 hs, and exposed to temperatures from freezing to 200°C and pressures up to 140 MPa are utilized. The types of admixtures used are many and they include retarders (lignosulfonate-based), polyhydroxy organic acids of sugar origin, carboxymethyl/hydroxyethyl cellulose, etc. Accelerators are utilized for lowering thickening times and increasing early strengths when cement surface casing and conductor strings below 40°C. Accelerators are: calcium chloride, sodium chloride, sodium silicate, sea water and calcium sulfate hemihydrate, sodium aluminate, etc.

Strength retrogression is countered by the addition of active silica sand or flour, natural pozzolans, or fly ash. Fluid loss controllers are used to maintain constant w/s ratio. Polymers such as polyacrylamide or polyethylene amines are also used. Many other types of admixtures are used for various purposes such as gas migration controllers, weighting agents, extenders, dispersants, thixotropic agents, defoamers, foamers, coloring agents, etc. Details of these admixtures are described in a chapter of Ref. 91.

7.13 Antifreezing Admixtures

Antifreezing admixtures are capable of depressing the freezing point of water in concrete considerably. Antifreezing admixtures have been used in Russia for concreting at temperatures of -30°C. These admixtures allow earlier stripping and use of forms, earlier completion of construction projects and occupancy. Many admixtures such as Na nitrite, Na chloride, weak electrolytes, and organic compounds (high molecular weight alcohols and carbamide) have been advocated. In Table 11, strength gain in concrete cured at -5°C and containing antifreezing admixtures is given.[40] The binary admixture at lower dosages imparts higher strengths than when used individually. Some novel formulations have been proposed that contain superplasticizers. Much more information has to be obtained before these admixtures can be used in practice. Antifreezing admixtures are expensive, especially as they are used in high dosages. Compatibility with other admixtures and long term durability of concrete has yet to be established for each mixture and curing regime.

Table 11. Strength Gain in Concrete Containing Antifreezing Admixtures

Admixture	Amount	% of Strength (28 d) With Respect to Normally Cured Specimens		
		7 d	28 d	90 d
Sodium Nitrite	4–6	30	70	90
Ca Nitrite + Urea	3–5	35	80	100
Ca Chloride + Na Chloride	2–3	35	80	100

REFERENCES

1. Ramachandran, V. S., *Calcium Chloride in Concrete-Science and Technology*, p. 216, Applied Science Publishers, London (1981)
2. Collepardi, M., Rossi, G., and Spiga, M. C., Hydration of Tricalcium Silicate in the Presence of Electrolytes, *Ann. di Chimica*, 61:137–148 (1971)
3. Tanoutasse, N., The Hydration Mechanism of C_3A and C_3S in the Presence of Calcium Chloride and Calcium Sulphate, *5th Int. Symp. Chem. Cements,* Part 2, pp. 372–378, Tokyo (1968)
4. Mukherji, K., and Ludwig, U., Influence of Addition of Calcium Chloride and Calcium Sulphate on the Rate of Hydration of C_3S and β-C_2S, *Tonind. Ztg.,* 97:211–216 (1973)
5. Skalny, J., and Maycock, J. N., Mechanisms of Acceleration by Calcium Chloride—A Review, 77th Annual Meeting, ASTM, Washington, DC, (June 1974)
6. Collepardi, M., and Massida, L., Hydration of Tricalcium Silicate in Suspension, *Ann. di Chimica*, 61:160–168 (1971)
7. Ramachandran, V. S., Possible States of Chloride in the Hydration of Tricalcium Silicate in the Presence of Calcium Chloride, *Materiaux et Constr.*, 4:3–12 (1971)
8. Rio, A., Celani, A., and Saini, A., New Investigation on the Action Mechanism of Gypsum and Calcium Chloride and Their Influence on the Structural and Mechanical Characteristics of the Hydrosilicates Produced by the Hydration of C_3S, *Il Cemento*, 67:17–26 (1970)

9. Odler, I., and Skalny, J., Pore Structure of Hydrated Calcium Silicates. II-Influence of Calcium Chloride on the Pore Structure of β-Dicalcium Silicate, *J. Colld. Int. Sci.*, 36:293–297 (1971)

10. Andreeva, E. P., Segalova, E. E., and Kormanovskaya, G. N., Effect of Calcium Chloride on the Metastable Solubility of β-Dicalcium Silicate and Tricalcium Silicate, *Kolld Zh.*, 26:04–408 (1964)

11. Collepardi, M., and Massida, L., Hydration of Beta Dicalcium Silicate Alone and in the Presences of $CaCl_2$ or C_2H_5OH, *J. Am. Ceram. Soc.*, 56:181–183 (1973)

12. Collepardi, M., Massida, L., and Usai, G., Sull'idratazione del ß-Silicato bi Calcico-Nota I. Idratazione in Pasta, *Ann. di Chimica*, 62:321–328 (1972)

13. Collepardi, M., Massida, L., and Merlo, E. G., Sull'idratazione Del ß-Silicato Bicalcico, Nota III - Idratazione in Mulino a Sfere, *Ann. di Chimica*, 62:337–344 (1972)

14. Collepardi, M., Marcialis, A., and Massida, L., Sull'idratazione Del β-Silicato Bicalcico, Nota III-Idratazione in Mulino a Sfere, *Ann. di Chimica*, 62:37–344 (1972)

15. Kantro, D. L., Weise, C. H., and Brunauer, S., Paste Hydration of ß-Dicalcium Silicate, Tricalcium Silicate and Alite, *Symp. Structure of Portland Cement Paste and Concrete,* pp. 309–327, Highway Res. Board (1966)

16. Teoreanu, I., and Muntean, M., The Kinetics of the Hydration Process of the Silicate Constituents of Portland Cement Clinker Under the Influence of Electrolytes, *Silicates Ind.*, 39:49–54 (1974)

17. Ragozina, T. A., Hydrolysis and Hydration of ß-Calcium Orthosilicate in Solutions of Salts, *Zh. Prik. Khim*, 22:545–552 (1949)

18. Balazs, Gy., and Boros, M., Effect of Calcium Chloride and Citric Acid on the Hydration of C_3S and β-C_2S Pastes, *Periodica Polytechnica* 22:29–35 (1978)

19. Taylor, H. F. W., (ed.), *The Chemistry of Cements,* Vol. I, Academic Press Inc., London (1964)

20. Stukalova, N. P., and Andreeva, E. P., Chemical Interactions and Structure Formation During Hydration of Tricalcium Aluminate in Calcium Chloride Solutions Saturated with Calcium Hydroxide, *Kolld. Zh.*, 31:446–450 (1969)

21. Gupta, P., Chatterji, S., and Jeffery, J. W., Studies of the Effect of Various Additives on the Hydration Reaction of Tricalcium Aluminate, Part III, *Cement Tech.*, 3:146–153 (1972)

22. Gupta, P., Chatterji, S., and Jeffery, J. W., Studies of the Effects of Various Additives on the Hydration Reaction of Tricalcium Aluminate, Part I: *Cement. Tech.*, 1:3–10 (1970), Part II: *Cement Tech.*, 3:21–26 (1970), Part IV: *Cement Tech.*, 4:63–68 (1973), Part V: *Cement Tech.*, 4:146–149 (1973)

23. Ramachandran, V. S., Seeley, R. C., and Polomark, G. M., Free and Combined Chloride in Hydrating Cement and Cement Components, *Mater. Struct.*, 17:285–289 (1984)

24. Schwiete, H. E., Ludwig, V., and Albeck, T., Combining of Calcium Chloride and Calcium Sulphate in Hydration of Aluminate-Ferrite Clinker Constituents, *Zement Kalk Gips*, 22:225–234 (1969)

25. Andreeva, E. P., and Segalova, E. E., Contribution to the Investigation of Process of Crystallization Structure Formation in Aqueous Suspensions of Tricalcium Aluminate in the Presence of Electrolytes, *Kolld. Zh.*, 23:129–133 (1961)

26. Traetteberg, A., and Sereda, P. J., Strength of C_3A Paste Containing Gypsum and $CaCl_2$, *6th Int. Congr. Chem. of Cement,* Suppl. Paper, Sec. II, Moscow (1974); *Cem. and Concr. Res.*, 6:461–474 (1976)

27. Janosne, B., and Gyorgy, B., Hydration in the System C_3A-$CaSO_4$-$CaCl_2$-H_2O, *Kozlekedesi Dokumentacios Vallalat,* Budapest (1972)

28. Andreeva, E. P., Chemical Nature of the New Substances Formed in Aqueous Suspension of Tricalcium Aluminate with Additions of Calcium Chloride, *Zh. Prik. Khim*, 33:1042–1048 (1960); See also Andreeva, E. P., Segalova, E. E., and Volynets, E. E., *Dok. Akad. Nauk.*, SSSR, 123:1052–1055 (1958)

29. Liberman, G. V., and Kireev, V. A., Reaction of Tricalcium Aluminate with Water in the Presence of the Chlorides of Calcium, Sodium and Potassium at Elevated Temperatures, *Zh. Prik. Khim.*, 37:194–196 (1964)

30. Kurczyk, H. G., and Schwiete, H. E., Electron Microscopic and Thermochemical Investigations on the Hydration of the Calcium Silicates $3CaO{\bullet}SiO_2$ and β-$2CaO{\bullet}SiO_2$ and the Effects of Calcium Chloride and Gypsum on the Process of Hydration, *Tonind. Ztg.*, 84:585–598 (1960)

31. Odler, I., and Maulana, A., Effect of Chemical Admixtures on Portland Cement Hydration, *Cem. and Concr. Res.*, 9:38–43 (1987)

32. Skalny, J., and Odler, I., The Effect of Chlorides Upon the Hydration of Portland Cement and Upon Clinker Minerals, *Mag. Concr. Res.*, 19:203–210 (1967)

33. Lerch, W., *The Effect of Added Materials on the Rate of Hydration of Portland Cement Pastes at Early Ages,* Portland Cem. Assn. Fellowship, p. 44 (1944)

34. Odler, I., and Skalny, J., The Effect of Chlorides Upon the Hydration of Portland Cement and Upon Some Clinker Minerals, *Mag. Concr. Res.*, 19:203–210 (1967)

35. Ramachandran, V. S., and Feldman, R. F., Time-Dependent and Intrinsic Characteristics of Portland Cement Hydrated in the Presence of Calcium Chloride, *Il Cemento*, 75:311–322 (1978)

36. Ramachandran, V. S., Feldman, R. F., and Beaudoin, J. J., *Concr. Sci.*, Heyden & Son, London (1981)

37. Shideler, J. J., Calcium Chloride in Concrete, *J. Amer. Concr. Inst.*, 23:537–559 (1952)

38. Machedlov-Petrosyan, O. P., Ol'ginski, A. G., and Doroshenko, Y. M., Influence of Certain Combined Chemical Additives on Hydration and Mechanical Properties of Clinker Cement, *Zh. Prik. Khim.*, 51:1493–1498 (1977)

39. Ranga Rao, M. V., Investigation of Admixtures for High Early Strength Development, *Indian Concr. J.*, 50:279–289 (1976)

40. Ramachandran, V. S., *Concrete Admixtures Handbook,* p. 1153, Noyes Publications, NJ (1995)

41. Tucker, G. R., Kennedy, H. L., and Renner, H. L., Concrete and Hydraulic Cement, US Patent, 2,031,621, (Feb. 25, 1936)

42. Ramachandran, V. S., Influence of Triethanolamine on the Hydration of Tricalcium Aluminate, *Cement Concr. Res.*, 3:41–54 (1973)

43. Ramachandran. V. S., Influence of Triethanolamine on the Hydration Characteristics of Tricalcium Silicate, *J. Appl. Chem. Biotechnol.*, 22:1125–1138 (1972)

44. Pauri, S, Monosi, S., Moriconi, G., and Collepardi, M., Effect of Triethanolamine on the Tricalcium Silicate Hydration, *8th Int. Congr. Cement. Chem.,* Rio De Jeneiro, Brazil, 3(2):121–129 (1986)

45. Dodson, V., *Concrete Admixtures,* p. 211, Van Nostrand Reinhold, New York (1990)

46. Ramachandran, V. S., Hydration of Cement - Role of Triethanolamine, *Cem. Concr. Res.* 6:623–631 (1976)

47. Kuroda, T., Goto, T., and Kobayashi, S., Chloride-Free and Alkaline Metal Accelerator, *Annual Review, Cement Association of Japan,* pp. 194–197, (1986)

48. Tachihato, S., Kotani, H., and Loe, Y., The Effect of Triethanolamine on the Hydration Mechanism and Strength Development of Slag Cement, *Silicates Ind.*, 49:107–112 (1984)

49. Ramachandran, V. S., Investigation of the Role of Chemical Admixtures in Cements—A Differential Thermal Approach, *7th Int. Conf. Therm. Analysis,* pp. 1296–1302 (1982)

50. Singh, N. B., and Abha, K., Effect of Calcium Formate on the Hydration of Tricalcium Silicate, *Cem. Concr. Res.*, 13:619–625 (1983)

51. Valenti, G. L., Rinaldi, D., and Sabateli, V., Effect of Some Acceleratory Admixtures on the Hardening of β-Dicalcium Silicate, *J. Mater. Sci. Lett.*, 3:821–824 (1984)

52. Bensted, J., Early Hydration Behaviour of Portland Cement in Water, Calcium Chloride and Calcium Formate Solutions, *Silic. Ind.*, 45:67–69 (1980)

53. Geber, S., Evaluation of Calcium Formate and Sodium Formate as Accelerating Admixtures for Portland Cement Concrete, *ACI J., Proc.* 80:439–444 (1983)

54. Massazza, F., and Testolin, M., Latest Development in the Use of Admixtures for Cement and Concrete, *Il Cemento*, 77:73–146 (1980)

55. Diamond, S., Accelerating Admixtures, *Proc. Int. Congr. Admixtures,* p. 192, London (1980)

56. Valenti, G. L., and Sabateli, V., The Influence of Alkali Carbonates on the Setting and Hardening of Portland and Pozzolanic Cements, *Silic. Ind.*, 45:237–242 (1980)

57. Sereda, P. J., Feldman, R. F., and Ramachandran, V. S., Influence of Admixtures on the Structure and Strength Development, *7th Int. Congr. Chem. Cement,* Part VI, pp. 32–44, Paris (1980)

58. Rosenberg, A. M., Gaidis, J. M., Kossivas, T. G., and Previte, R. W., *A Corrosion Inhibitor Formulated with Calcium Nitrite for Use in Reinforced Concrete,* ASTM STP-629:89–99 (1977)

59. Ghosh, S. N. (ed.), *Advances in Cement Technology,* Pergamon Press, Oxford (1983)

60. Rosskopf, P. A., Linton, F. J., and Peppler, R. B., Effect of Various Accelerating Admixtures on Setting and Strength Development of Concrete, *J. Testing Evaln.*, 3:322–330 (1975)

61. Abdelrazig, B. E., Bonner, D. G., Nowell, D. V., Dransfield, J. M., and Egan, P. J., *Effects of Accelerating Admixtures on Cement Hydration, Admixtures for Concrete,* (E. Vazquez, ed.), p. 586, Chapman and Hall, London (1990)

62. Thomas, N. L., and Egan, P. J., Assessment of Sodium Molybdate as an Accelerator of Cement Hydration and Setting, *Advances Cem Res.*, 2:89–98 (1989)

63. Fischer, H. C., Molybdenum Trioxide—An Accelerator of Portland Cement Hydration, *Cem. Concr. Agg.*, 12:53–55 (1990)

64. Singh, N. B., Prabha Singh, S., and Singh, A. K., Effect of Lactic Acid on the Hydration of Portland Cement, *Cem. Conc. Res.*, 16:545–553 (1986)

65. Brook, J. W., and Ryan, R. J., A Year Round Accelerating Admixture, *3rd Int. Conf., Superplasticizers and Other Chemical Admixtures in Concrete,* ACI SP-119, p. 665 (1989)

66. Ramachandran, V. S., Thermal Analysis of Cement Components Hydrated in the Presence of Calcium Carbonate, *Thermochimica Acta*, 127:385–394 (1988)

67. Ushiyama, H., Kawano, Y., and Kamegai, N., Effect of Carbonates on Early Stage Hydration of Alite, *8th Int. Cong. Chem Cements,* Rio de Jeneiro, Brazil, III(2):154–166 (1986)

68. Ramachandran, V. S., Beaudoin, J. J., Sarkar, S. L., and Aimin, X., Physico-Chemical and Microstructural Investigation of the Effect of NaOH on the Hydration of $3CaO•SiO_2$, *Il Cemento*, 90:73–84 (1993)

69. Ramachandran, V. S., Beaudoin, J. J., and Paroli, R. M., The Effect of Nitrobenzoic and Aminobenzoic Acids on the Hydration of Tricalcium Silicate: A Conduction Calorimetric Study, *Thermochimica Acta*, 190:325–333 (1991)

70. Vazquez, E., (ed.), *Admixtures for Concrete,* pp. 197–208, Chapman and Hall, London (1990)

71. Ramachandran, V. S., Effect of Sugar-Free Lignosulphonates on Cement Hydration, *Zement-Kalk-Gips*, 31:206–210 (1978)

72. Milestone, N. B., Hydration of Tricalcium Silicate in the Presence of Lignosulphonates, Glucose, and Sodium Gluconate, *J. Am. Ceram. Soc.*, 62:321–324 (1979)

73. Collepardi, M., Marcialis, A., and Solinas, V, V., The Influence of Calcium Lignosulphonate on the Hydration of Cements, *Il Cemento*, 70:3–14 (1973)

74. Rixom, M. R., *Chemical Admixtures for Concrete,* p. 234, E.& F. N. Spon Ltd., London (1978)

75. Roy, D. M., and Asaga, K., Rheological Properties of Cement Pastes, *Cement Concr. Res.*, 10:387–394 (1980)

76. Ramachandran, V. S., Adsorption and Hydration Behavior of Tricalcium Aluminate-Water and Tricalcium Aluminate–Gypsum-Water Systems in the Presence of Superplasticizers, *J. Am. Concr. Inst.*, 80:235–241 (1983)

77. Massazza, F., Costa, U., and Corbella, E., Influence of Beta Naphthalene Sulfonate Formaldehyde Condensate Superplasticizing Admixture on C_3A Hydration in Reaction of Aluminates During Setting of Cements, CEMBUREAU Rept., p. 3 (1977)

78. Slanicka, S., Influence of Water-soluble Melamine Formaldehyde Resin on the Hydration of C_3S, C_3A+$CaSO_4 \cdot 2H_2O$ Mixes and Cement Pastes, *7th Int. Congr. Chem. Cements, Paris,* II:161–166 (1980)
79. Ramachandran, V. S., Influence of Superplasticizers on the Hydration of Cement, *3rd Int. Congr. Polymers in Concrete, Koriyama, Japan,* pp. 1071–1081, (1981)
80. Lukas, W., The Influence of Melment on the Hydration of Clinker Phases, *5th Int. Melment Symp., Munich,* pp. 17–21, (1979)
81. Ghosh, R. S., and Malhotra, V. M., Use of Superplasticizers as Water Reducers, CANMET Division Report MRP/MRL 78–189(J) CANMET, Canada Energy Mines and Resources, Ottawa (1978)
82. Ramachandran, V. S., Admixture and Addition Interactions in the Cement-Water System, *Il Cemento,* 83:13–38 (1988)
83. Ghosh, S. N., *Mineral Admixtures in Cement and Concrete,* p. 565, ABI Book Ltd., New Delhi (1993)
84. Malhotra, V. M., Superplasticized Fly Ash Concrete for Structural Applications, *ACI Concr. Int.,* 8:28–31 (1986)
85. Malhotra, V. M., High Volume Fly Ash and Slag Concrete, *Concrete Admixtures Handbook,* (V. S. Ramachandran, ed.), Ch. 12, Noyes Publications, NJ (1995)
86. Langley, W. S., Carette, G. G., and Malhotra, V. M., Strength Development of High Volume Fly Ash as Determined by Drilled Cores and Standard Moist Cured Cylinders, Division Report MSL 90-24 (OP & J) (1990)
87. Malhotra, V. M., Ramachandran, V. S., Feldman, R. F., and Aitcin, P. C., *Condensed Silica Fume in Concrete,* p. 231, CRC Press (1987)
88. Malhotra, V. M., Carrette, G. G., and Aitcin, P. C., Mechanical Properties of Portland Cement Concrete Incorporating Blast Furnace Slag and Condensed Silica Fume, *Proc. RILEM-ACI Symp. Technol., Monterrey, Mexico,* p. 395 (1985)
89. Mailvaganam, N. P., *Concrete Admixtures Handbook* (V. S. Ramachandran, ed.), pp. 939–1024, Noyes Publications, NJ, (1995)
90. McCoy, W. J., and Caldwell, A. G., A New Approach to Inhibiting Alkali-Aggregate Expansion, *J. Am. Concr. Inst.,* 47:693–706 (1951)
91. Bensted, J., Admixtures for Oil Well Cements, *Concrete Admixtures Handbook,* (V. S. Ramachandran, ed.), pp. 1077–1111, Noyes Publications, NJ (1995)

5

Accelerating Admixtures

1.0 INTRODUCTION

Admixtures can influence the properties of concrete from the moment water comes into contact with the concrete ingredients. In the fresh state, the water requirement, workability, bleeding, segregation, rate of hydration, setting time, air content, pumpability, etc., and in the hardened state, compressive, flexural and tensile strengths, creep, shrinkage, permeability, and durability are affected to different extents, depending on the type and amount of admixture added. The general types of admixtures used in concrete and their applications have already been described in Ch. 4 of this book.

Thermal analysis techniques have been applied widely for the investigation of the role of admixtures, especially that related to the hydration of cement and cement components. Application of thermal analysis permits determination of the heat of reaction, mechanism of reaction, kinetics of reactions, compatibility of admixtures with cements, prediction of some properties, durability problems, material characterization and selection, development of new admixtures, quick assessment of some physical properties, etc. In some instances, they yield results that are not possible to obtain with the use of other techniques.

In this chapter, typical examples of the application of various thermal techniques to the study of the effect of admixtures on the hydration of cement and cement compounds is emphasized. Where relevant, the results obtained with these techniques are compared with those derived from other tools.

2.0 CALCIUM CHLORIDE

Calcium chloride is a unique accelerator in the sense that, of the various cation-anion combinations, the Ca^{+2} and Cl^- combination ranks as the best accelerator for cements. The accelerating influence of $CaCl_2$ on the hydration of calcium silicates was observed more than sixty years ago by Sloane, et al.,[1] and Haegerman,[2] and has been confirmed by subsequent work.[3] The accelerating influence of calcium chloride on the hydration of C_3S is conveniently followed by adopting any of the following methods, viz., estimating at different times the amount of residual unhydrated C_3S, the amount of $Ca(OH)_2$, loss on ignition, electrical conductivity, heat liberation, etc.

The techniques of DTA, TG, and Conduction Calorimetry have proven to be valuable to follow the hydration of C_3S in the presence of calcium chloride.[3] Typical results of the application of DTA are illustrated in Figs. 1, 2, and 3.[4] Thermograms of C_3S hydrated to different times in the presence of 0, 1, or 4% $CaCl_2$ enable a study of the progress of hydration, identification, and estimation of some of the products. Unhydrated C_3S exhibits endothermal effects at about 680, 930, and 970–980°C representing crystalline transitions (Fig. 1). Onset of hydration is indicated by the endothermal effects below 300°C. They are caused by expulsion of loosely, as well as firmly, bound water from the C-S-H gel. The increase in the intensity of this effect with time is indicative of the formation of larger amounts of C-S-H product. A very small endothermal effect at about 480°C, which appears within a few minutes becoming more evident at one hour and after, may be attributed to the dehydration effect of $Ca(OH)_2$. In the first eight hours, the amount of $Ca(OH)_2$ produced is about 25% of all the calcium hydroxide that is formed in thirty days.

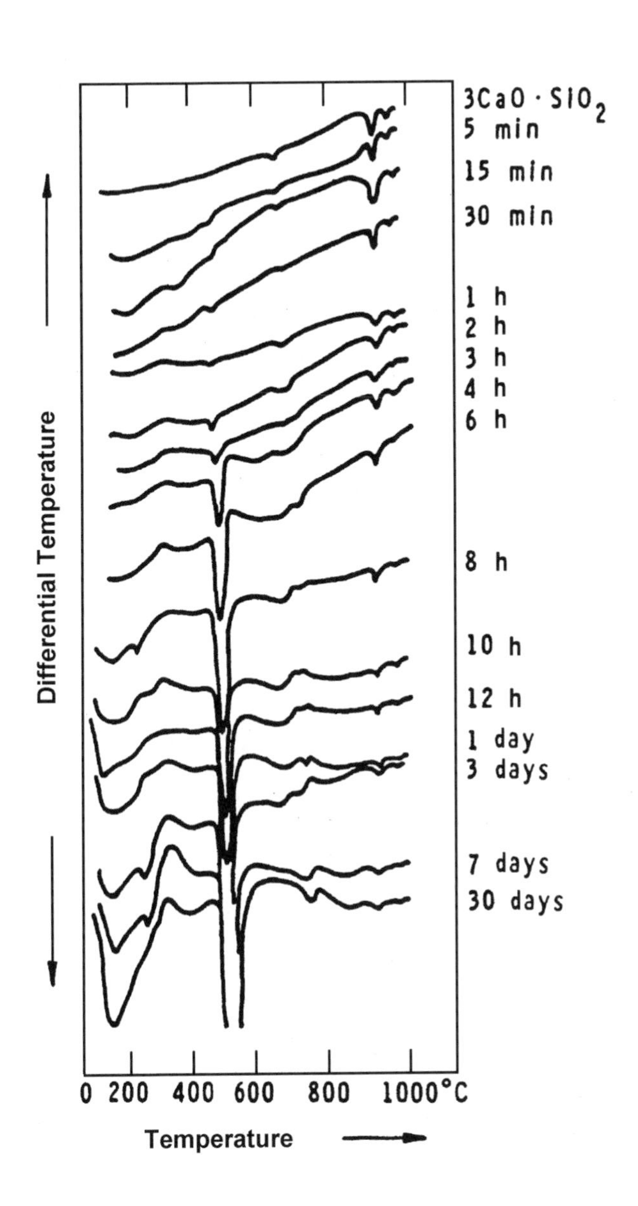

Figure 1. Rate of hydration of $3CaO{\bullet}SiO_2$ by DTA.

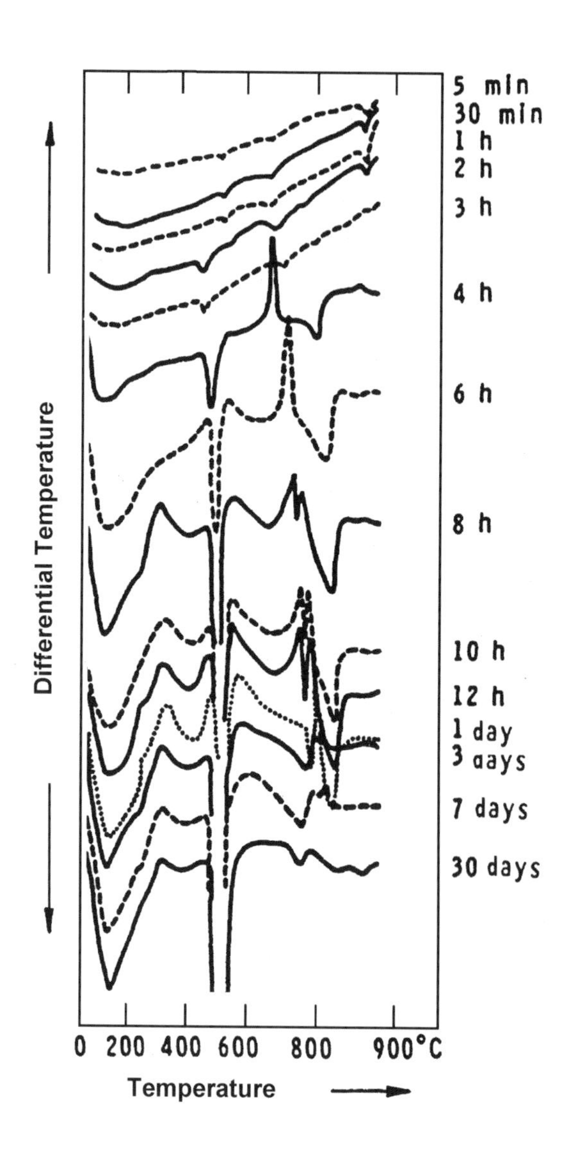

Figure 2. Rate of hydration of $3CaO{\bullet}SiO_2$ in the presence of 1% $CaCl_2$ by DTA.

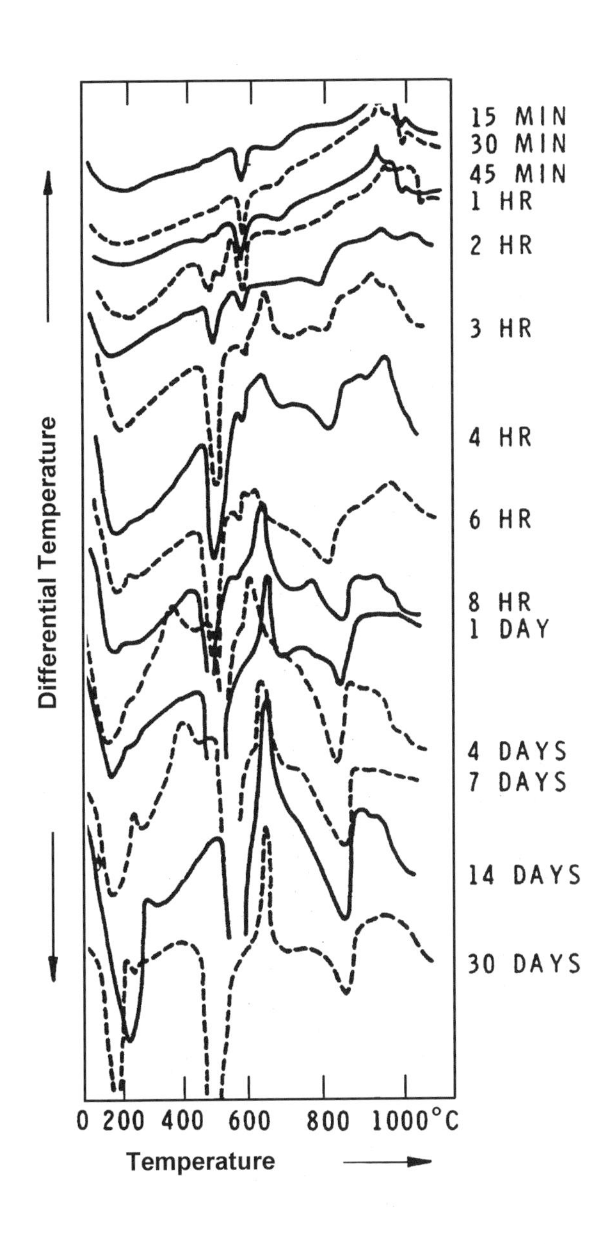

Figure 3. Rate of hydration of $3CaO{\bullet}SiO_2$ in the presence of 4% $CaCl_2$.

In the presence of 1% $CaCl_2$ the thermograms of hydrating C_3S show significant differences from those hydrated without calcium chloride addition (Fig. 2). The endothermal effects below 300°C in the presence of $CaCl_2$ are much larger than those obtained in samples without the addition of calcium chloride (Figs. 1 and 2). An endotherm at 550°C appearing up to two hours in the presence of 1% $CaCl_2$ is absent in C_3S hydrated without $CaCl_2$. There is also evidence that the endothermal effect due to $Ca(OH)_2$ is more intense in samples containing 1% $CaCl_2$ than without it. Of the total amount of $Ca(OH)_2$ formed at 30 days, 33% is formed within 8 hours of hydration. A remarkable feature of these thermograms is the onset of an intense exothermic peak at four hours at a temperature of 690°C. This peak is always followed by a large endothermal dip at about 800–840°C. There is some evidence[5] that it may be caused by the chemisorbed chloride on the C-S-H surface and possibly also by chloride ions in the interlayer positions.

In the presence of 4% $CaCl_2$, some thermal effects become more intense at earlier times than the corresponding ones hydrated in the presence of 1% $CaCl_2$ (Fig. 3).[4] Exothermal peaks are also evident at temperatures above 600°C at three hours and beyond. The possibility of a surface complex of chloride on the hydrating silicate phase is suggested by an endothermal effect in the range 570–590°C. A larger effect at 810–850°C following an exothermal effect is present from three hours to thirty days.

If the rate of hydration of C_3S is determined in terms of the amount of $Ca(OH)_2$ formed at different times, at six hours the sample containing 4% chloride will have the largest amounts of calcium hydroxide. At 24 hours and 30 days, the sample containing 1% will have higher amounts of calcium hydroxide. If the hydration is determined by the disappearance of C_3S, then at 30 days C_3S with 4% $CaCl_2$ is hydrated to the maximum extent followed by that containing 1% $CaCl_2$. The apparent discrepancy is due the differences in the CaO/SiO_2 ratios of the C-S-H products formed during the hydration.

Calcium chloride accelerates the hydration of C_3S even at higher temperatures. The effect is particularly greater at earlier periods of hydration. Heat evolution curves show that at temperatures of 25, 35, and 45°C, the addition of 2% $CaCl_2$ not only influences the total heat developed at early periods but also the time at which the maximum heat evolution peak occurs.[6] Increasing the concentration of $CaCl_2$ up to 20% with respect to C_3S has been found to influence the conduction calorimetric curves.[7] In Fig. 4, conduction calorimetric curves of C_3S containing 0–20% calcium chloride are given. The sample containing no chloride shows a hump with a peak effect at about 13–14 hours. This peak occurs at lower temperatures and is also sharper as the amount of added chloride is increased. At 20% $CaCl_2$, a sharp peak occurs at about two hours.

By applying thermal analysis, XRD, and chemical methods, it has been concluded that calcium chloride may exist in different states in the system tricalcium silicate-calcium chloride-water.[5] The chloride may be in the free state, as a complex on the surface of the silicate during the dormant period, as a chemisorbed layer on the hydrate surface, in the interlayer spaces, and in the lattice of the hydrate. Figure 5 gives the estimate of the states of chloride in the silicate hydrated for different periods.[5] The results show that the amount of free chloride drops to about 12% within 4 hours, becoming almost nil in about 7 days. At 4 hours, the amount of chloride existing in the chemisorbed and/or interlayer positions rises sharply and reaches about 75%. Very strongly held chloride that cannot be leached, even with water, occurs to an extent of about 20% of the initially added chloride. Since this will not be in a soluble state in water, it would not be available for corrosion processes. The formation of complexes may explain effects such as the acceleration of hydration, the increase in surface area, morphological changes, and the inhibition of formation of afwillite (a crystalline form of calcium silicate hydrate) in the presence of calcium chloride.

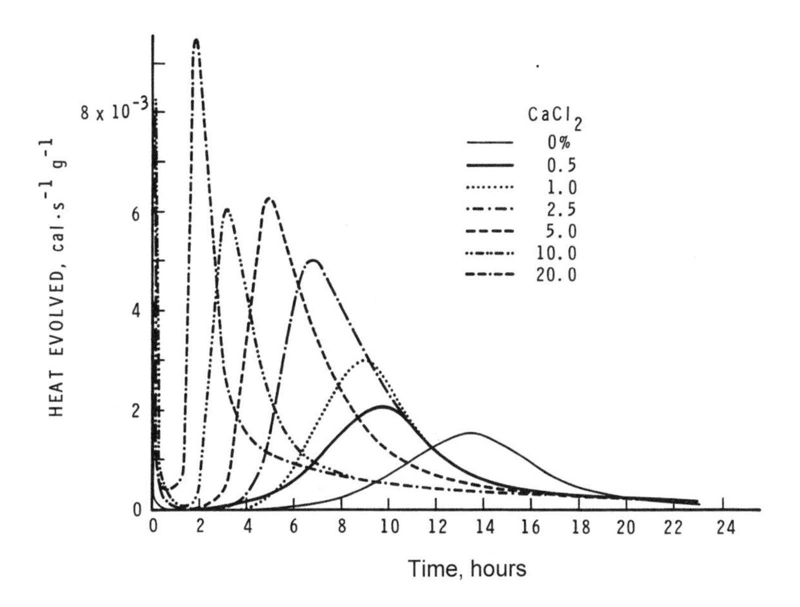

Figure 4. Influence of $CaCl_2$ on the heat evolution characteristics of hydrating C_3S.

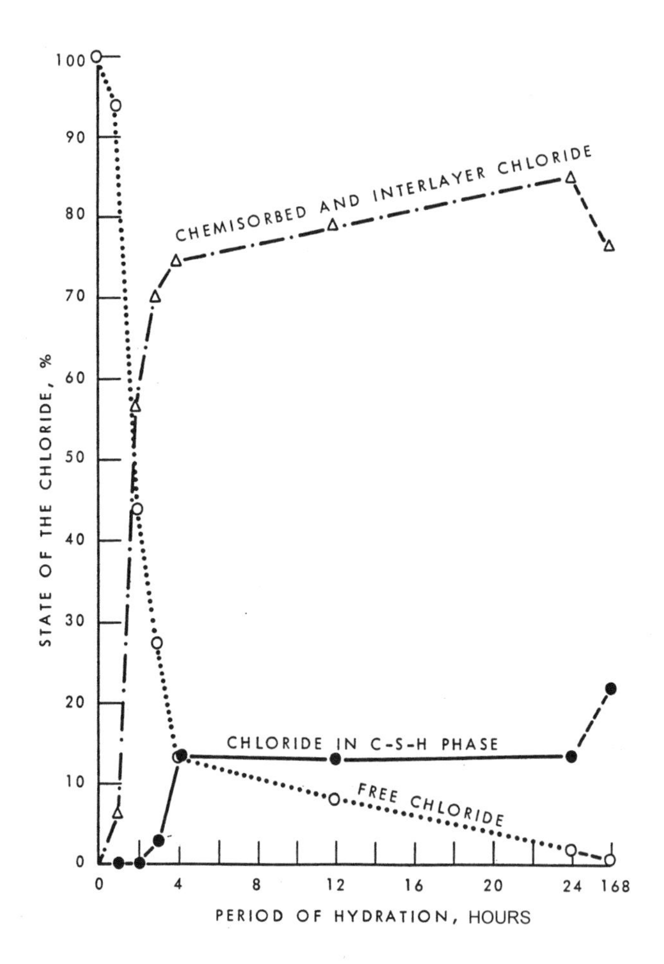

Figure 5. Possible states of chloride in tricalcium silicate hydrated for different periods.

Collepardi, et al., studied the hydrates formed during the hydration of C_3S in the presence of calcium chloride.[8][9] In an autoclaved sample containing $CaCl_2$, they detected an exothermic peak in the thermogram that was attributed to the chemisorption of chloride on the C-S-H surface and its penetration into the molecular layers of C-S-H.

Compared to the extensive thermal investigations of C_3S hydration in the presence of $CaCl_2$, only a limited amount of work has been reported on the effect of calcium chloride on the hydration of C_2S. Calcium chloride accelerates the hydration of dicalcium silicate. In the thermograms of C_2S paste hydrated for 1–3 months, lower amounts of calcium hydroxide are formed in the presence of calcium chloride, compared to that hydrated without the chloride.[10] This would imply that a higher C/S ratio C-S-H product results in the presence of calcium chloride. There is also evidence that in the hydration of C_2S some chloride is bound rigidly.

The reaction of C_3A with calcium chloride results in the formation of high and low forms of tricalcium chloroaluminates. Under normal conditions of hydration, the low form, viz., $C_3A \cdot CaCl_2 \cdot XH_2O$ is obtained. The DTA technique may be used to differentiate between the two forms. Endothermal effects at about 190 and 350°C are caused by the low form and the endotherm at about 160°C is exhibited by the high form. In the system C_3A-CaO-$CaCl_2$-H_2O, at higher concentrations of calcium chloride, calcium hydroxychloride is formed that is identified by peaks at 130, 145, and 485°C.[11]

Calcium chloride influences the rate of hydration of C_3A + gypsum mixtures. In Fig. 6, the conduction calorimetric curves of the mixtures C_3A + 20% gypsum + 12.5% $CaCl_2$ are given along with the identified compounds at different times.[3][7] A comparison of this curve with that obtained with C_3A + gypsum (G) or C_3A + $CaCl_2$ would lead to the following conclusions. The reaction between C_3A and gypsum is accelerated by calcium chloride. Monochloroaluminate (MCA) is formed after gypsum is consumed in the reaction with C_3A. Conversion of ettringite (TSA) to monosulfoaluminate occurs only after all $CaCl_2$ has reacted.[12]

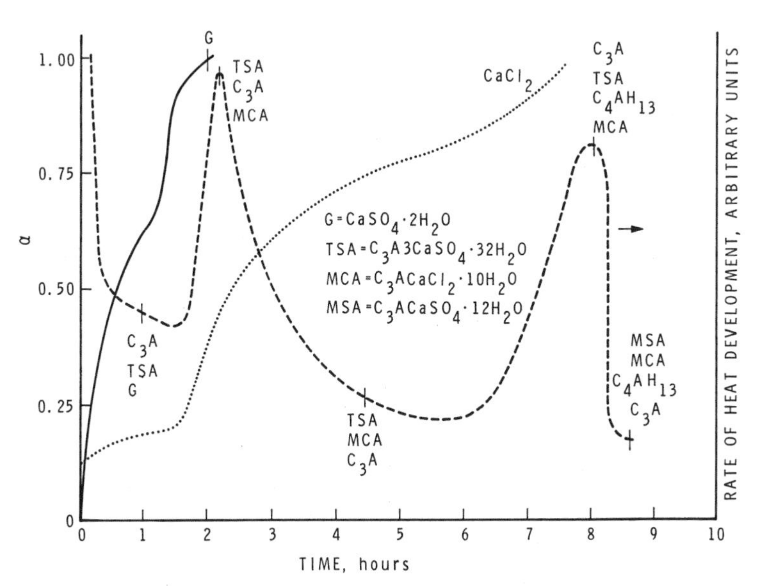

Figure 6. The rate of consumption of various components in the C_3A-gypsum-$CaCl_2$-H_2O system.

The effect of calcium chloride on the hydration of portland cement has been studied by various techniques such as XRD, TG, conduction calorimetry, chemical analysis, etc. Thermograms of portland cement hydrated with or without 2% $CaCl_2$ are shown in Fig. 7.[3] The endothermal effect at about 450–475°C is indicative of the presence of $Ca(OH)_2$. Compared to the DTA of plain cement paste, that hydrated in the presence of calcium chloride indicates substantial differences in the form and size of the endotherm appearing in the region 150–200°C. The broad endothermal valley is caused mainly by the presence of a high form of sulfoaluminate and the C-S-H phase. The relatively larger intensity of these effects in the presence of the chloride demonstrates the increased formation of these products. Particularly significant is the increased intensity of the $Ca(OH)_2$ peak between 2 and 4 hours, in the presence of the chloride. The small endothermal effect around 350°C observed only in the presence of calcium chloride is probably due to calcium chloroaluminate hydrate or to its solid solution with the hexagonal calcium aluminate hydrate.

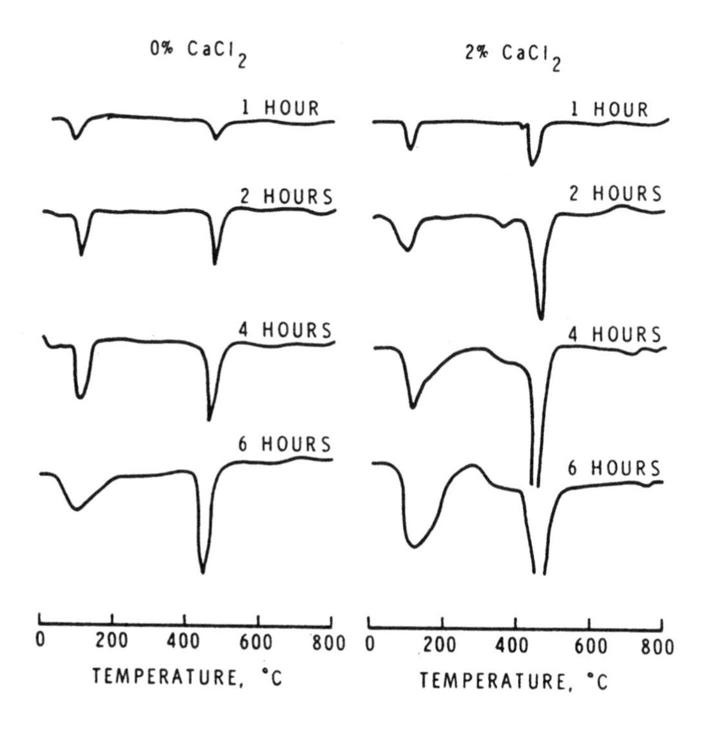

Figure 7. Differential thermograms of portland cement hydrated with 2% $CaCl_2$.

The thermogravimetric method may also be used to estimate the amount of $Ca(OH)_2$ formed by the hydration of cement. Determination of the weight loss between 450 and 550°C caused by the decomposition of $Ca(OH)_2$, and that between 550 and 900°C caused by the decomposition of $CaCO_3$ permits the estimation of the amount of lime formed during the hydration of cement with 0, 1, and 2% $CaCl_2$. Table 1 gives the results for cement samples hydrated by adding the chloride to the mixing water or by adding it a few minutes after the cement is mixed with water.[13] The lime formed is reported in terms of CaO. The amount of CaO at 1 day in the presence of calcium chloride is almost twice that formed in the absence of the admixture. The degree of hydration is generally higher in samples hydrated with 2% $CaCl_2$. It is also observed that even at 90 days the samples with calcium chloride are hydrated to a greater extent than those without it. Hydration is promoted by delayed addition of the chloride. This is explained as follows—in the absence of chloride, the reaction that occurs initially involves the aluminate and ferrite phases, and gypsum. In the presence of $CaCl_2$, where the aluminate and ferrite phases preferentially react with the chloride, less chloride is available for accelerating the hydration of C_3S. If, however, chloride is added a few minutes after the cement is mixed with water, the sulfate would have already reacted with the aluminate and ferrite phases, and there is more chloride available to accelerate the silicate reaction.

Table 1. Effect of $CaCl_2$ on the Amount of CaO Formed in Portland Cement Pastes

Calcium Chloride Addition	Method of Addition of $CaCl_2$	Percentage CaO formed at (days)				
		1	3	9	28	90
0%	———	3.5	6.8	8.0	7.9	8.1
1%	with water	6.7	7.0	8.0	8.3	8.4
1%	2 minutes later	6.9	7.7	8.1	8.4	8.5
1%	4 minutes later	7.0	7.7	8.3	8.4	8.5
1%	8 minutes later	6.5	7.2	8.1	8.0	8.4
2%	with water	7.1	7.7	8.2	8.5	8.6
2%	2 minutes later	7.8	8.0	8.3	8.6	8.8
2%	4 minutes later	7.9	8.0	8.3	8.6	8.8
2%	8 minutes later	7.3	7.8	7.9	8.0	8.3

The setting and hardening of concrete are accelerated in the presence of calcium chloride. These are related to the effect of $CaCl_2$ on the rate of hydration of cement in concrete. The hydration of cement is exothermic, resulting in the production of heat, and if the heat is produced at a faster rate, larger amounts of the hydrates are formed at earlier times in the presence of accelerators. This is particularly significant in the first 10–12 hours. The influence of different amounts of calcium chloride on the rate of heat development is depicted by conduction calorimetric curves (Fig. 8).[14] The position of the peak corresponding to the maximum liberation of heat moves towards shorter times as the amount of $CaCl_2$ is increased. It occurs at about eight hours for the reference cement and at about three hours with 2% $CaCl_2$. Calcium chloride also increases the amount of heat liberated during the first few hours. Their reaction can also be accelerated by hydration at higher temperatures. The addition of 2% $CaCl_2$ was found to have the same accelerating effect on hydration as an increase of 11°C in the hydration temperature.[14]

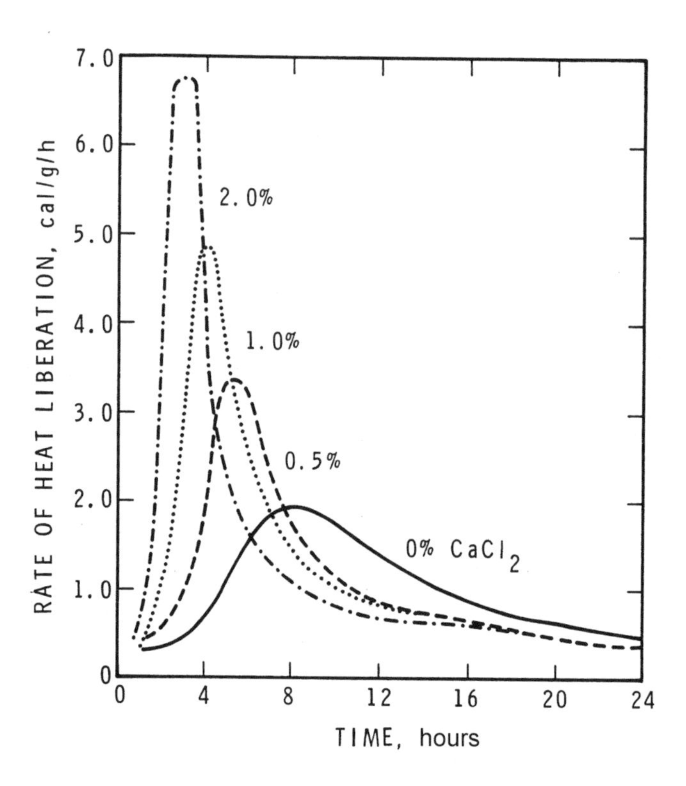

Figure 8. The effect of $CaCl_2$ upon the rate of heat development in cement.

The development of heat produced by the addition of calcium chloride on portland cement Type I from different plants shows significant variations (Fig. 9). This is due to the differences in the fineness and chemical composition of cement.

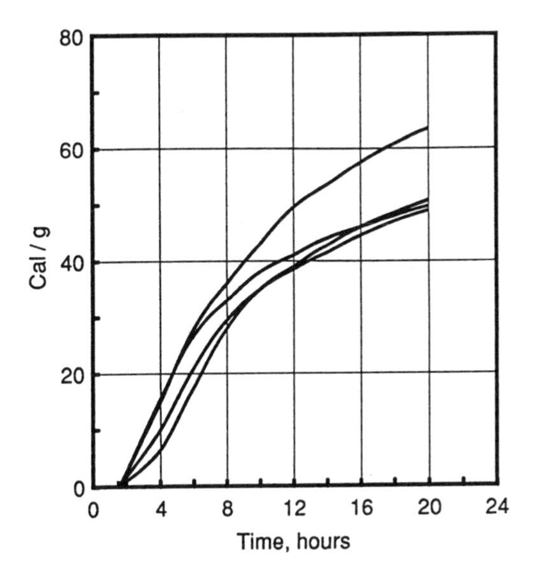

Figure 9. The effect of $CaCl_2$ on the heat development in four Type I cements.

Ramachandran and Feldman[15] compared the rate of hydration of cement in the presence of calcium chloride by following the amount of $Ca(OH)_2$, non-evaporable water, or heat development. Calcium chloride, in amounts of 0, 1, 2, and 3.5%, was added to cement at w/s ratios of 0.25 and 0.40. Hydration was followed from a few minutes to thirty days. There was good correlation between the non-evaporable water content and heat development. It was also observed that cement containing 3.5% hydrated to the greatest extent in the early periods, but lower amounts of calcium hydroxide were formed at this dosage. This was explained by the formation of C-S-H with a higher ratio in such a cement. It was also concluded that the hydration kinetics cannot be accurately studied by the estimation of lime, especially at a lower w/c ratio.

Calcium chloride also accelerates the rate of hydration of Type II, IV, and V cements but only marginally that of Type II, a high early strength cement.

3.0 NON-CHLORIDE ACCELERATORS

One of the limitations to the wider use of calcium chloride in reinforced concrete is that, if present in larger amounts, it promotes corrosion of the reinforcement unless suitable precautions are taken. The use of calcium chloride is banned in many countries. There is, hence, a continuing attempt to find an alternative to calcium chloride, one equally effective and economical, but without its limitations. A number of organic and inorganic compounds including aluminates, sulfates, formates, thiosulfates, nitrates, silicates, alkali hydroxides, carbonates, halides, nitrites, calcium salts of acetic acid, propionic acid, butyric acid, oxalic acid and lactic acid, urea, glyoxal, triethanolamine, and formaldehyde have been suggested. However, practical experience and research on these admixtures are limited. The effect of these compounds on the hydration of individual cement compounds and cement has been widely studied by thermal analysis techniques.

Triethanolamine of formula $N(C_2H_4OH)_3$ (TEA for short) is an oily water-soluble liquid having a fishy odor and is produced by reacting ammonia with ethylene oxide. Normally it is used in combination with other chemicals in admixture formulations.

Ramachandran[16] followed the hydration of C_3A (with and without gypsum) containing triethanolamine. Figure 10 refers to thermal curves of C_3A hydrated for 0, 1, 5, 10, 30, and 60 minutes in the presence of 0, 0.5, 1.0, and 5% TEA. At 1 minute, the C_3A sample containing no TEA exhibits endothermal peaks at 150 and 250°C typical of a mixture of hexagonal aluminate hydrates. At about 5 minutes, all samples show endotherms between 300 and 400°C. They are typical of cubic tricalcium aluminate hexahydrate. At 1 minute, C_3A samples containing 0.5 and 1% TEA show the formation of larger amounts of hexagonal hydrates, as evidenced by the greater intensity of the peak at about 150°C. At 5% TEA, the existence of a larger amount of C_3AH_6 even at 1 minute indicates that the hydration of C_3A is enhanced by increased amounts of TEA. An exotherm (400–500°C) in samples containing 5% TEA is caused by a complex that forms between hydrated aluminate and TEA.

DTA results of C_3A + 25% gypsum mixtures hydrated with 0% and 1% TEA are shown in Fig. 11. The unhydrated mixture exhibits an endothermal doublet between 100 and 150°C due to a stepwise dehydration of gypsum. Hydration seems to start from the first minute of contact of water with the sample. At 5 minutes, a larger amount of gypsum disappears from

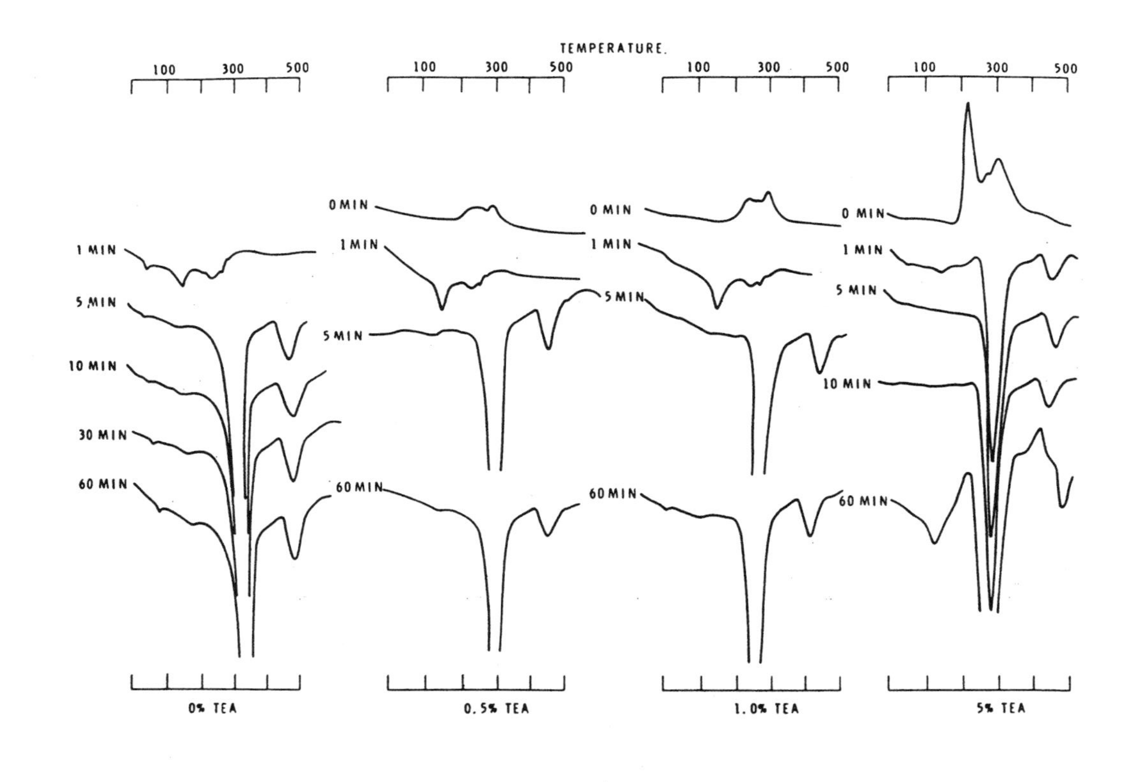

Figure 10. Hydration of tricalcium aluminate with different amounts of triethanolamine.

the sample containing TEA. At 10 and 30 minutes, the ettringite peak occurring between 150 and 200°C is more intense in the sample with TEA. It appears that the formation of ettringite is accelerated between 5 and 10 minutes in the presence of TEA whereas in the untreated sample ettringite is detected between 10 and 30 minutes. The accelerating effect of TEA in the C_3A + gypsum mixture is also reflected in a greater amount of heat developed at earlier times in conduction calorimetric curves. Scanning electron microscopic investigation has indicated the formation of more densified ettringite needles in samples with TEA.

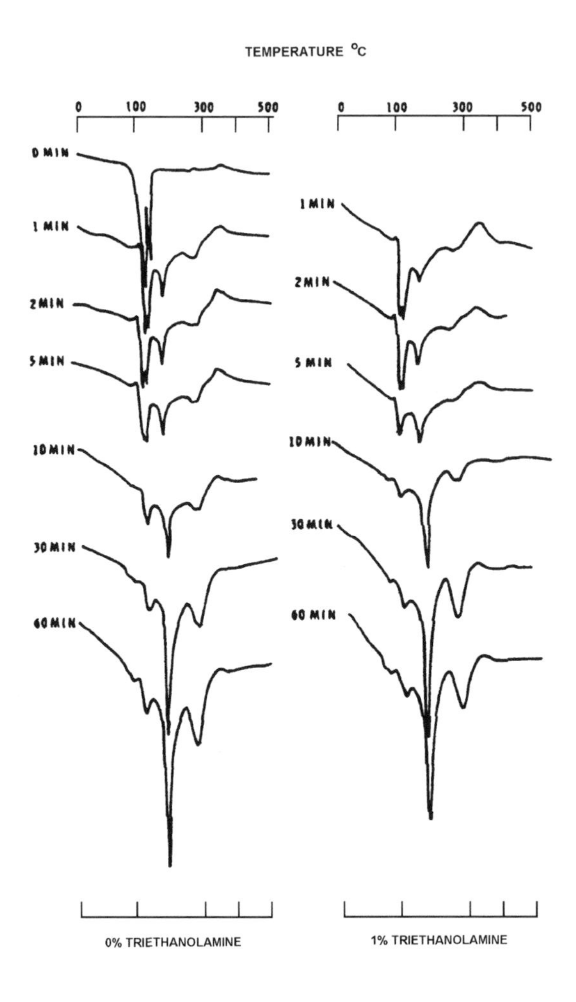

Figure 11. Hydration of C_3A + 25% gypsum with triethanolamine.

Triethanolamine influences the hydration of the C_3S phase. The effect of 0.5% TEA on the hydration of C_3S is illustrated by DTA curves in Fig. 12.[17] Figure 12a refers to the DTA behavior of C_3S hydrated at a w/s ratio of 0.5 from 5 minutes to 28 days. A small endothermal effect at about 480–500°C appearing at 1 hour and increasing subsequently at later ages is due to $Ca(OH)_2$ dehydration. In the thermograms of C_3S, hydrated in the presence of triethanolamine (Fig. 12b), a very small endothermal effect representing the formation of $Ca(OH)_2$ appears only after 10 hours. The curves indicate that TEA initially retards the rate of hydration of the silicate. The sharp exothermal effect appearing at 6 hours is attributed to the complex of TEA formed on the surface of the hydrating C_3S. This complex may be responsible for the retardation effect. TG results are in conformity with those of DTA.

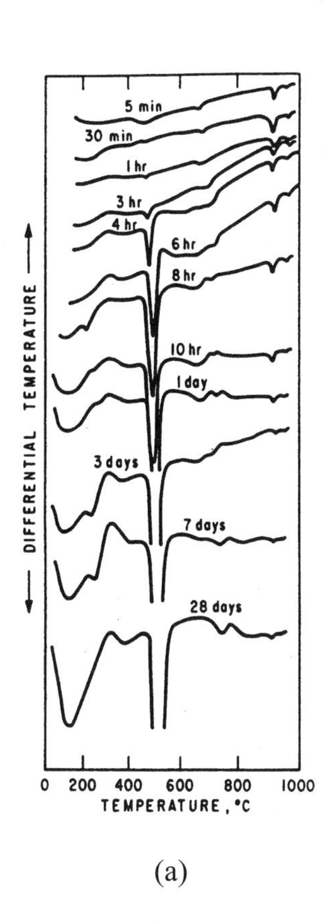

(a)

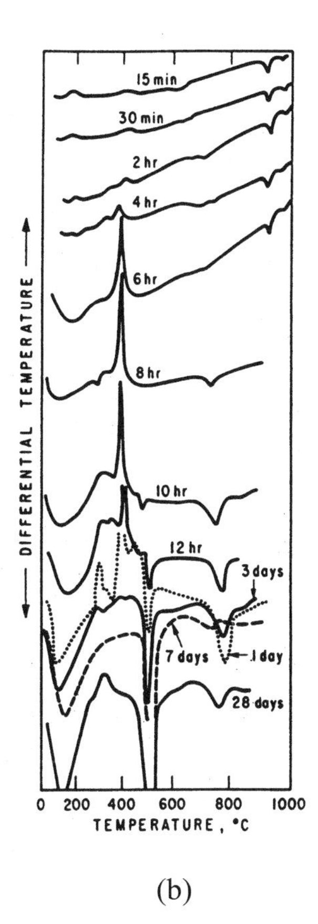

(b)

Figure 12. Thermograms of C_3S hydrated to different periods: *(a)* C_3S + 0% TEA; *(b)* C_3S + 0.5% TEA.

Isothermal conduction calorimetric curves of C_3S hydrated with different amounts of TEA are shown in Fig. 13.[17] In the C_3S sample hydrated without TEA, there is little heat development in the first few hours, representing the dormant period. The curve starts to rise steeply at about 2.5 to 3 hours denoting the beginning of the acceleratory period. Maximum heat develops between 7 and 8 hrs. By the addition of TEA, the induction period is extended. The C_3S hydration peak occurs later, but with a higher intensity so that TEA acts like a delayed accelerator. The mono- and diethanolamines also affect the hydration of C_3S, similar to TEA.

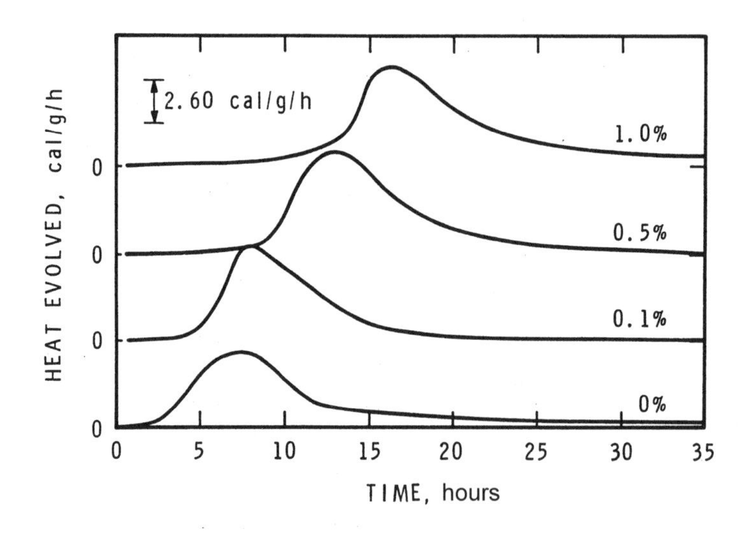

Figure 13. Conduction calorimetric curves of 3CaO•SiO_2 with different amounts of added triethanolamine.

The hydration of C_2S with 0.5% TEA has been followed by DTA (Fig. 14).[3][17] Generally, endothermal peaks due to $Ca(OH)_2$ decomposition (450–480°C) are of lower intensity in samples containing TEA. An additional endotherm appears in these samples and has been attributed to the presence of less crystalline $Ca(OH)_2$.

In the DTA and TG studies of cement containing different amounts of TEA, evidence was obtained for the formation of lower amounts of $Ca(OH)_2$ in the presence of the admixture. In Fig. 15 the amount of lime (DTA estimation) formed in the cement paste in the presence of TEA is given.[18] Comparison of the results at the same degree of hydration has revealed that TEA promoted the formation of C-S-H with a higher C/S ratio.

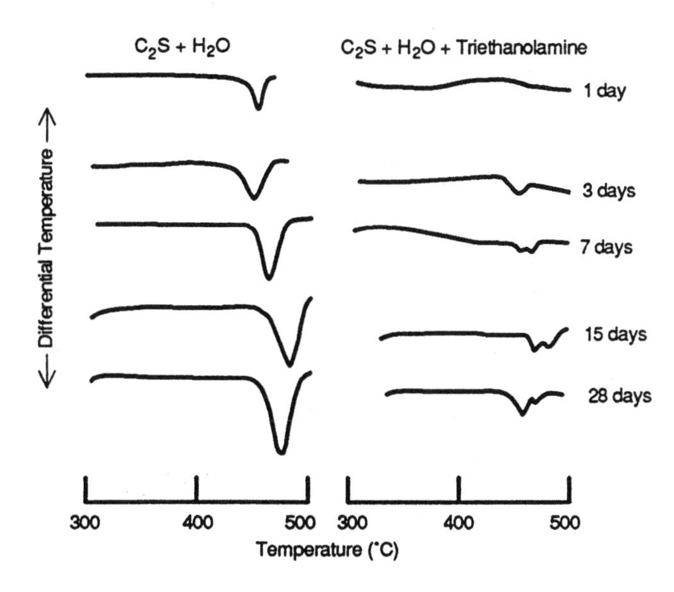

Figure 14. Thermograms of $2CaO{\bullet}SiO_2$ hydrated alone and in the presence of triethanolamine.

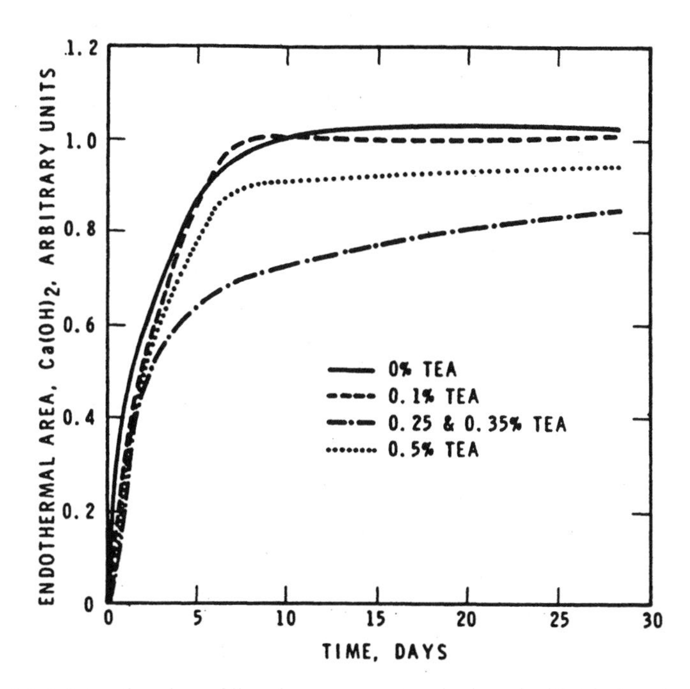

Figure 15. DTA estimation of lime in cement pastes hydrated with triethanolamine.

Conduction calorimetric curves of cement hydrated in the presence of different amounts of TEA have also been obtained.[18] In Fig. 16, conduction calorimetric curves of cement hydrated in the presence of 0, 0.1, 0.25, 0.35, and 0.5% TEA are given. Immediately on contact with water all the samples evolved heat (not shown) that is caused by the heat of wetting, the hydration of free lime, and the reaction of C_3A to form ettringite. The magnitude of this peak increased with the amount of TEA added. This indicates that TEA accelerates the formation of ettringite. The second peak, occurring after 9–10 hours in cement without TEA is caused by the hydration of C_3S. A small hump at about 22 hours denotes the formation of low sulfoaluminate from the reaction between ettringite and excess C_3A. In the curves of samples containing TEA dual peaks appear between 10 and 15 hours. One of them represents the ettringite-monosulfate reaction that is accelerated by TEA. The other peak represents the hydration of C_3S, triggered by the aluminate-gypsum reaction. This peak occurs at a later period compared to that registered for C_3S (with no TEA). DTA curves indicated that at 20 hours lower amounts of lime were formed in samples with TEA. The setting times were in the range of 4.3–4.8 hours in samples containing less than 0.05% TEA, but a rapid setting occurs (2–6 minutes) at 0.1–0.5% TEA levels. This may be associated with the accelerated formation of ettringite.

Calcium formate of formula $Ca(HCOO)_2$ is a by-product in the manufacture of polyhydric alcohol, pentaerthritol. It is a powder and has a low solubility of about 15% in water at room temperature. It is a popular non-chloride chemical that is advocated in practice. Many non-chloride accelerating admixture formulations contain formates. Calcium formate is an accelerator for the hydration of C_3S; at equal concentration, however, $CaCl_2$ is more effective in accelerating the hydration of C_3S (Fig. 17).[19]

The hydration of C_2S is also accelerated by calcium formate. The increase in compressive strength of the formate-treated dicalcium silicate samples in relation to the reference sample is evidence of the acceleratory effect.

In the hydration of portland cement, although initially larger amounts of heat are developed in the presence of Ca formate, at later ages the heat may be slightly lower or equal to that of the reference material. Calcium formate accelerates the hydration of all types of cement. Figure 18 gives the relative amounts of heat produced by adding 0.18 molar calcium chloride, calcium nitrite, and calcium formate to Type V cement. Calcium chloride is the best accelerator followed by calcium formate.[12]

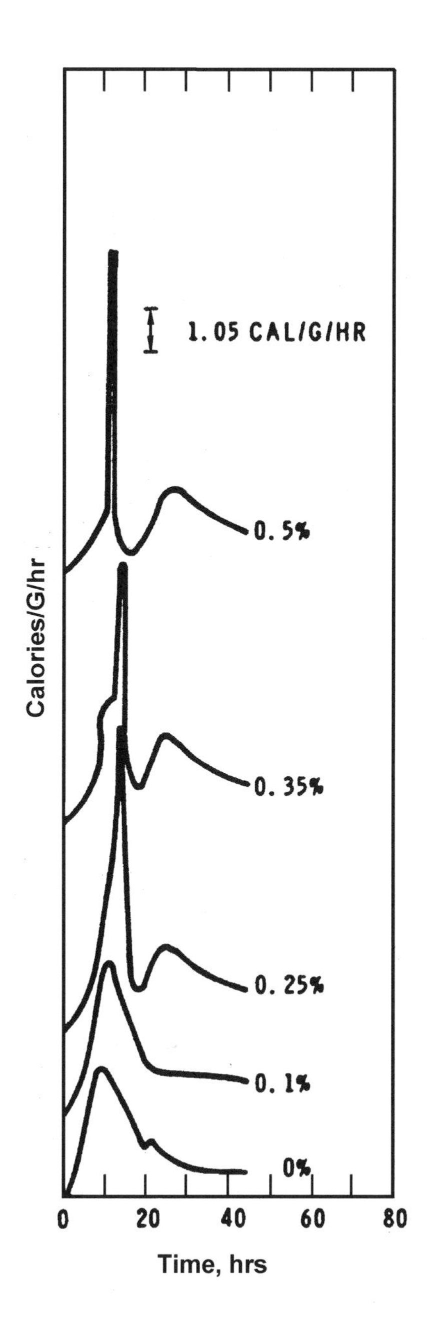

Figure 16. Conduction calorimetric curves of cement hydrated in the presence of triethanolamine.

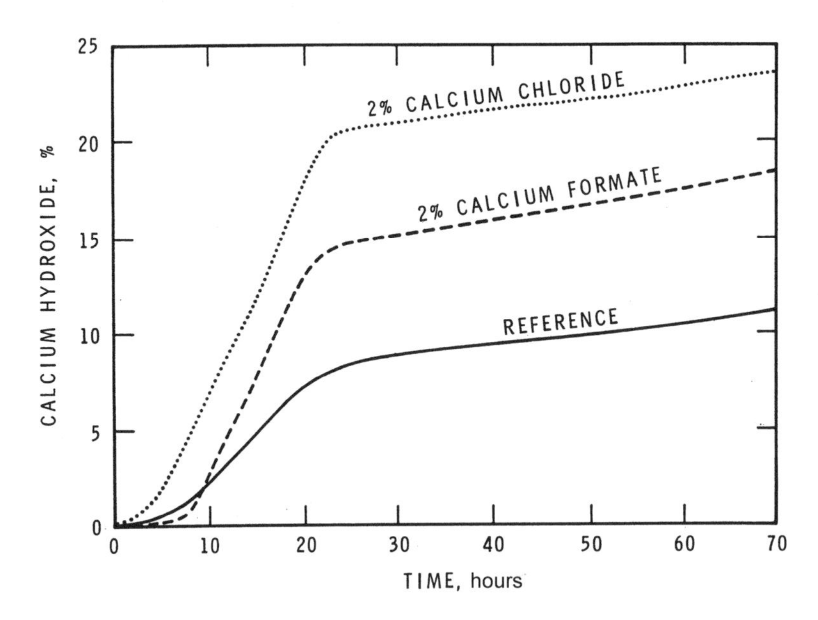

Figure 17. Influence of accelerators on the formation of calcium hydroxide in hydrated $3CaO \cdot SiO_2$.

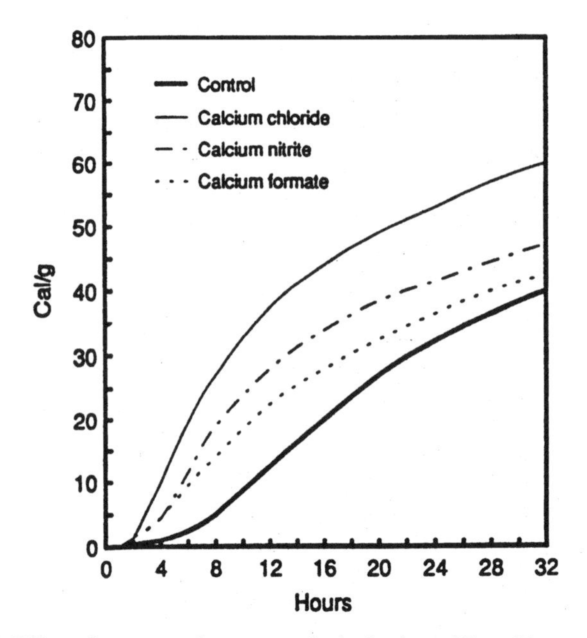

Figure 18. Effect of some accelerators on the hydration of Type V cement.

Many inorganic and organic salts have been examined for their action on the hydration of cement and cement compounds utilizing thermal techniques. They include sodium and calcium salts of chloride, bromide, nitrite, thiosulfate, thiocyanate, iodide, nitrate, hydroxide, carbonate, hydroxide, etc. A few typical examples are given illustrating the application of thermal techniques in the investigation of these compounds on cements and cement components.

Kantro[20] carried out an extensive investigation on the effect of various salts on the hydration of C_3S using conduction calorimetry. In Fig. 19, the rate of heat development occurring in C_3S hydrated in the presence of soluble salts calcium, viz., Br, Cl, SCN, I, NO_3, and ClO_4, is compared with that containing no additive. All the salts promote production of larger amounts of heats and at earlier times. The chloride and bromide compounds however, exhibit the best acceleration effect.

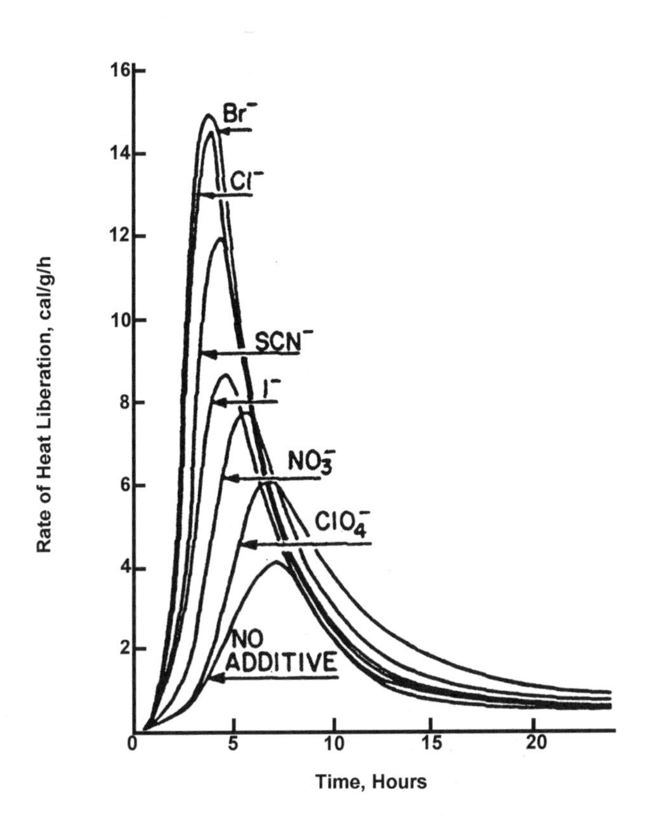

Figure 19. Rate of heat liberation vs. time for C_3S containing various salts.

A comparison of the DTA curves of C_3S containing Ca nitrate, thiosulfate, and chloride shows that in terms of $Ca(OH)_2$ formed at 7 hours all the compounds act as accelerators; calcium chloride, however, is the best accelerator (Fig. 20).[21]

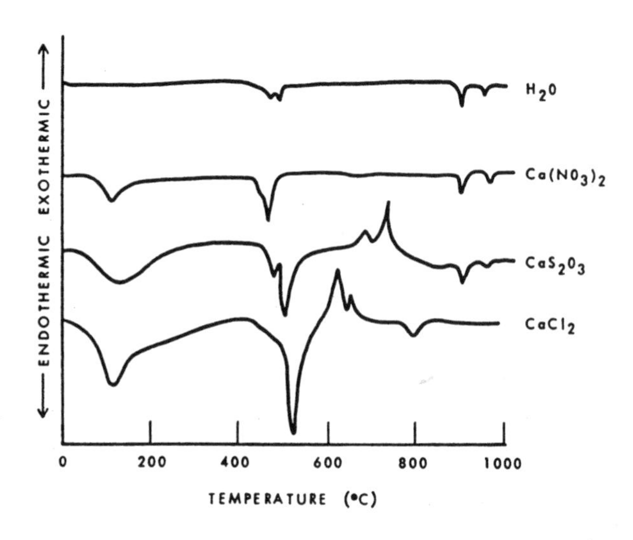

Figure 20. DTA curves of the hydration products of C_3S in the presence of calcium salts.

In cold weather concreting, the rate of hydration of cement is very slow. Attempts have been made to develop so-called antifreezing admixtures in which accelerators and a freezing point depressant are used. Thiocyanates are potential accelerators that may be used in these formulations. A systematic work has been carried out to determine the effect of various thiocyanates such as Na, K, NH_4, Ca, and Li thiocyanates at dosages of 1.5 and 3%, on the hydration of cement cured at 20, 0, and -5°C. Both conduction calorimetry and TG techniques were used.[22] All the thiocyanates increased the rate of hydration of cement at all temperatures. The most efficient early acceleration, as well as strength development, occurs with 3% Ca thiocyanate. A linear relationship exists between the amount of lime formed and strength within a range of values (Fig. 21). The lines for the paste cured at -5°C can be broadly divided into three groups. Pastes containing 1.5% Na, Li, and K thiocyanate exhibit better strengths than the reference material at lime contents greater than 3%. Best strengths are obtained with 1.5–3.0% Ca thiocyanate and 3% Na, K, or Li thiocyanates. These data imply that the differences in the microstructure are responsible for the variation in the intrinsic strengths of the pastes when compared at equal degrees of hydration (equal calcium hydroxide contents).

Applying DSC, Abdelrazig, et al.,[23] compared the amount of lime formed when portland cement was hydrated from 1 hour to 3 days in the presence of calcium chloride, calcium nitrate, sodium thiocyanate, and calcium thiocyanate. At 3 hours, all salts acted as accelerators, but calcium chloride appeared to be the best accelerator. At about 6 hours, more $Ca(OH)_2$ was formed in the presence of calcium thiocyanates. At 3 days, the chloride and the thiocyanates produced about the same amount calcium hydroxide. Calcium nitrate was found to be the least efficient accelerator.

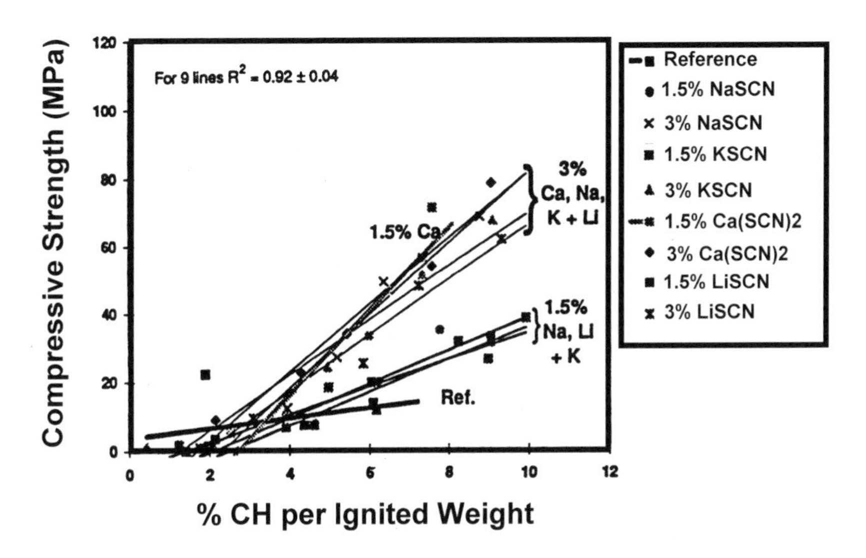

Figure 21. Relationship between strength and calcium hydroxide content in cement paste hydrated at -5°C

A new accelerator formulation consisting of inorganic salts and organic compounds (one of these being a sulfonate dispersant) is claimed to be effective at as low a temperature as -7°C as well at a higher temperature of 35°C. It accelerates setting, increases early and ultimate strengths, decreases shrinkage, and does not initiate corrosion.[24] Conduction calorimetric curves, of the sample (at -7°C) containing this admixture, exhibit exothermal heat effects that cannot be detected in the reference sample. In Table 2, the setting times and strengths of the reference concrete (-7°C) are compared with that containing the above admixture. The admixture develops better strengths than the reference and can be used in cold weather concreting.

Table 2. Setting Time and Strength Development in a Newly Formulated Admixture

Admixture	Setting Times, Hrs		Strength, MPa			
	Initial	Final	1 d	3 d	28 d	1 yr
Air Entrained (Reference)	10.67	14.58	1.6	11.7	26.2	36.9
Air Entrained (Accelerator)	6.15	11.50	2.8	9.0	30.8	43.1
Non-Air Entrained + Accelerator	5.50	11.50	3.6	8.7	34.2	46.3

Calcium nitrite has been used as a corrosion inhibitor in concrete. It accelerates the hydration of cement. In Fig. 18, its acceleration effect on Type V cement is obvious from the larger amount of heat developed compared to that by the reference.[12] It appears to be a better accelerator than calcium formate.

In cements, incorporation of calcium carbonate is permitted in some countries. In Canada, the maximum limit is set at 5%. Calcium carbonate is not an inert filler. It is known to react with calcium aluminate. In a study of the hydration of tricalcium silicate in the presence of finely divided calcium carbonate, Ramachandran[25] observed that the carbonate acted as an accelerator. In Fig. 22, the conduction calorimetric curves clearly show the accelerating influence of calcium carbonate. There was evidence of formation of a complex of the carbonate with the hydrated silicate.[25]

Alkalis such as NaOH, and Na salts of carbonate, aluminate, and silicate are known to accelerate the hydration of cement and cause early stiffening. In the hydration of tricalcium silicate with NaOH, there is an acceleratory effect, as is evident from the conduction calorimetric curves

(Fig. 23).[26] The peak occurs earlier and the amount of heat developed in the sample with NaOH is higher, even up to 28 days. The acceleratory effect does not reflect in increased strengths. The lower strengths in NaOH treated samples is attributable to the plate-like and reticular morphology of the hydration product with poor bonding characteristics.

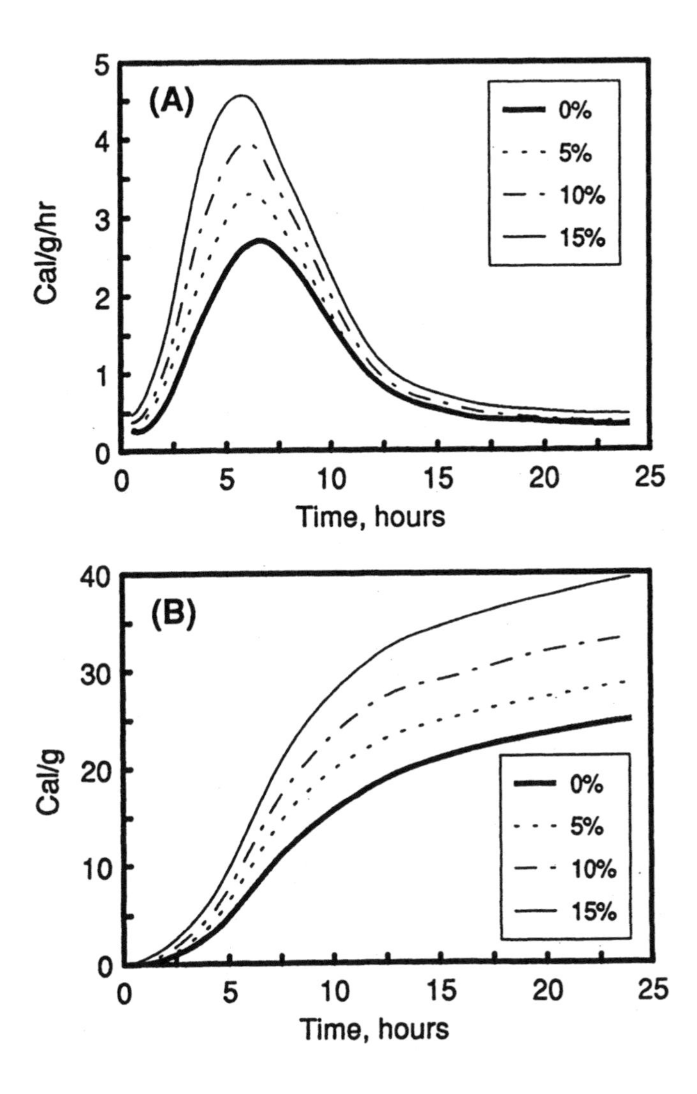

Fig 22. Conduction calorimetric curves of C_3S in the presence of calcium carbonate.

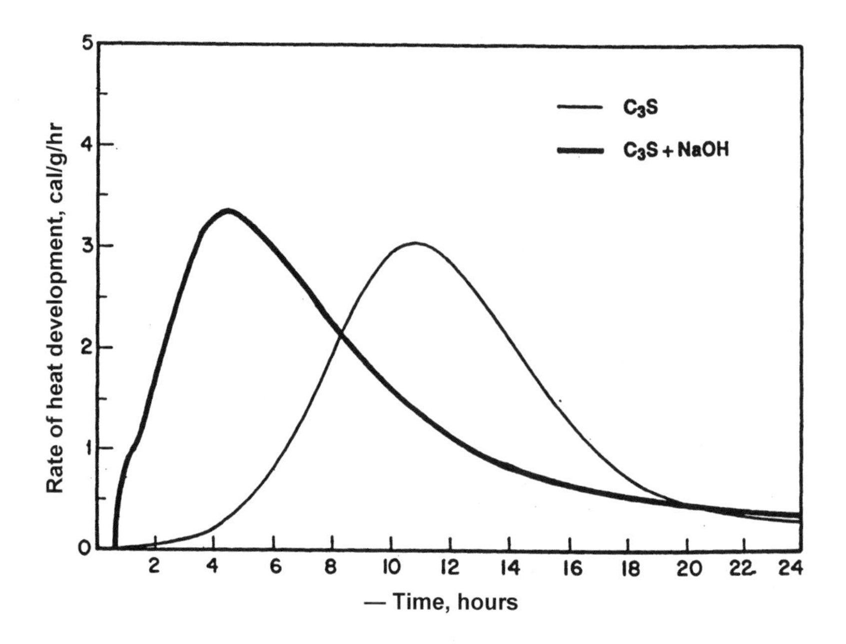

Figure 23. Rate of heat development of C_3S hydrated in the presence of NaOH.

There has been continued interest in developing an organic-based accelerator. Ramachandran and coworkers studied the effect of *o, m,* and *p*-nitrobenzoic acids on the hydration of tricalcium silicate.[27] The *m* and *p* nitrobenzoic acids acted as accelerators (Fig. 24). Paranitrobenzoic acid appears to be the best accelerator. The acceleratory effect was attributed to the complex formation between the organic compound and the C-S-H phase on the surface of the tricalcium silicate phase.

In addition to studies on C_3S,[25] Ramachandran and Zhang-Chunmei[28]–[31] have also carried out extensive investigations of the effect of calcium carbonate on the hydration of C_3A, C_3A + gypsum and cement utilizing DSC and TG techniques. Calcium carbonate was found to suppress the formation of calcium aluminate hydrate in the hydration of C_3A. It accelerates the formation of ettringite and the conversion of ettringite to monosulfoaluminate in the hydration of C_3A + gypsum. In Fig. 25, DSC curves for C_3A containing 12.5% gypsum and 12.5% calcium carbonate hydrated for 5 minutes to 3 days are given. In the C_3A-gypsum mixture, gypsum is consumed in 30 minutes as evidenced by the decreasing intensity of the peak at 130°C. After 30 minutes, the intensity of the endothermic peak at 165–180°C increases substantially. This peak represents monosulfoaluminate hydrate. XRD data confirm the presence of monosulfoaluminate at 1 hour. In the presence of 12.5% gypsum and 12.5% carbonate, the rate of disappearance of gypsum is accelerated. Gypsum is consumed in 20 minutes. Formation of ettringite is accelerated by calcium carbonated (peak at about 100°C). In the presence of calcium carbonate, even at 10 minutes, ettringite lines appear in XRD. DSC results also demonstrate that the ettringite peak occurs within 5 minutes in the presence of the carbonate, but not in the sample with gypsum only.

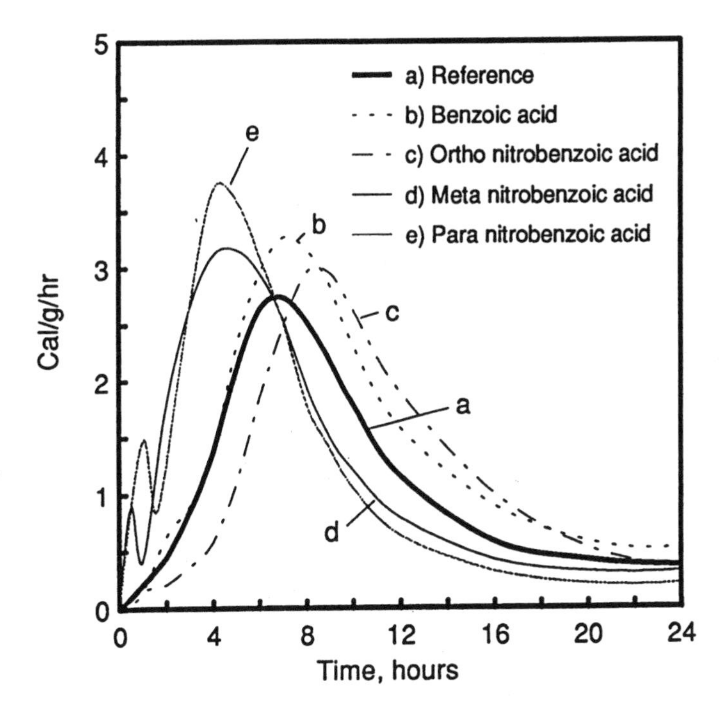

Figure 24. Conduction calorimetric curves of C_3S hydrated in the presence of nitrobenzoic acids.

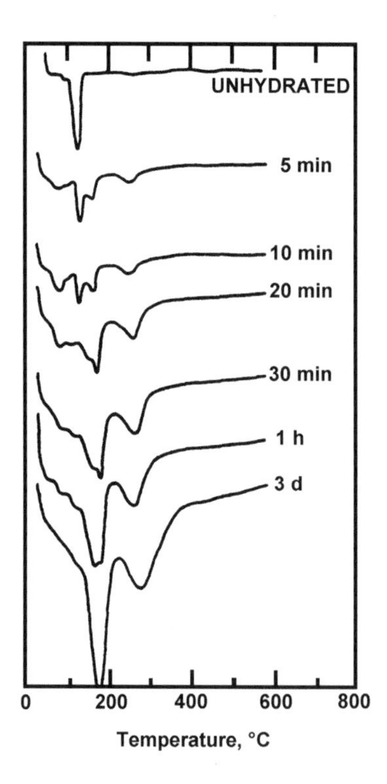

Figure 25. DSC curves of C_3A hydrated for different periods in the presence of 12.5% gypsum + 12.5% $CaCO_3$.

REFERENCES

1. Sloane, R. C., McCaughey, W. D., Foster, W. D., and Shreve, C., Effect of Calcium Chloride as an Admixture in Cement Paste, Eng. Expt. Station, Ohio State Bull., p. 61 (1931)
2. Haegerman, G., (See Forsen, L., The Chemistry of Retarders and Accelerators, *Symp. Chem. Cements*), Stockholm, pp. 298–363 (1938)
3. Ramachandran, V. S., *Calcium Chloride in Concrete*, p. 216, Applied Science Publishers, London (1976)
4. Ramachandran, V. S., Kinetics of the Hydration of Tricalcium Silicate in the Presence of Calcium Chloride by Thermal Methods, *Thermochimica Acta*, 2:3–12 (1971)

5. Ramachandran, V. S., Possible States of Chloride in the Hydration of Tricalcium Silicate in the Presence of Calcium Chloride, *Mater. and Construction*, 4:3–12 (1971)
6. Skalny, J., and Maycock, J. N., Mechanisms of Acceleration of Calcium Chloride—A Review, *77th Annual Meeting ASTM, Washington, DC* (1974)
7. Tenoutasse, N., The Hydration Mechanism of C_3A and C_3S in the Presence of Calcium Chloride and Calcium Sulphate, *5th Int. Symp. Chem Cements, Tokyo,* 2:372–378 (1968)
8. Collepardi, M., Rossi, G., and Spiga, M. C., Hydration of Tricalcium Silicate in the Presence of Electrolytes, *Annali di Chimica*, 61:137–148 (1971)
9. Chichio, G., and Collepardi, M., The Influence of $CaCl_2$ on the Crystallization of Calcium Silicates in the Hydration of C_3S, *Cem. Concr. Res.*, 4:861–868 (1974)
10. Kurczyk, H. C., and Schwiete, H. E., Electron Microscopic and Thermochemical Investigation on the Hydration of Calcium Silicates $3CaO{\bullet}SiO_2$ and β-$2CaO{\bullet}SiO_2$ and the Effect of Calcium Chloride and Gypsum on the Process of Hydration, *Tonindustr. Ztg.*, 84:585–598 (1960)
11. Stukalova, N. P., and Andreeva, E. P., Chemical Interactions and Structure Formation During Hydration of Tricalcium Aluminate in Calcium Chloride Solutions Saturated with Calcium Hydroxide, *Koll. Zh.*, 31:446–450 (1969)
12. Ramachandran, V. S., *Concrete Admixtures Handbook*, p. 1152, Noyes Publications, NJ (1995)
13. Tamas, F. D., Acceleration and Retardation of Portland Cement Hydration by Additives, *Symp. Struct. Portland Cem. Paste and Concrete,* Special Report 90, pp. 392–397, Highway Res. Board, Washington, DC (1966)
14. Lerch, W., *The Effect of Added Materials on the Rate of Hydration of Portland Cement Pastes at Early Ages*, p. 44, Portland Cement Assoc. Fellowship (1944)
15. Ramachandran, V. S., and Feldman, R. F., Time-Dependent and Intrinsic Characteristics of Portland Cement Hydrated in the Presence of Calcium Chloride, *Il Cemento*, 75:311–322 (1978)
16 Ramachandran, V. S., Action of Triethanolamine on the Hydration of Tricalcium Aluminate, *Cem. Conr. Res.*, 3:41–54 (1973)
17. Ramachandran, V. S., Influence of Triethanolamine on the Hydration of Characteristics of Tricalcium Silicate, *J. App. Chem. Biotech.*, 22:1125–1138 (1972)
18. Ramachandran, V. S., Hydration of Cement—Role of Triethanolamine, *Cem. Concr. Res.*, 6:623–632 (1976)

19. Ramachandran, V. S., Investigation of the Role of Chemical Admixtures in Cements—A Differential Thermal Approach, *7th Int. Conf. Thermal Anal., Kingston, Canada,* pp. 1296–1302 (1982)

20. Kantro, D. L., Tricalcium Silicate Hydration in the Presence of Various Salts, *J. Testing and Evaluation*, 3:312–321 (1975)

21 Murakami, K., and Tanaka, H., see: Ramachandran, V. S., Elucidation of the Role of Chemical Admixtures in Hydrating Cements by DTA Technique, *Thermochimica Acta,* 3:343–366 (1972)

22. Wise, T., Ramachandran, V. S., and Polomark, G. M., The Effect of Thiocyanates on the Hydration of Portland Cement at Low Temperatures, *Thermochimica Acta*, 264:157–171 (1995)

23. Abdelrazig, B. E. I., Bonner, D. G., and Nowell, D. V., Estimation of the Degree of Hydration in Modified Ordinary Portland Cement Pastes by Differential Scanning Calorimetry, *Thermochimica Acta*, 168:291–295 (1990)

24. Brook, J. W., and Ryan, R. J., A Year Round Accelerating Admixture, *3rd Int. Conf. Superplasticizers and Other Admixtures in Concrete,* ACI SP-119, p. 665 (1989)

25. Ramachandran, V. S., Thermal Analysis of Cement Components Hydrated in the Presence of Calcium Carbonate, *Thermochimica Acta*, 127:385–394 (1988)

26. Ramachandran, V. S., Beaudoin, J. J., Sarkar, S. L., and Xu, A., Physico-Chemical and Microstructural Investigations of the Effect of NaOH on the Hydration of $3CaO{\bullet}SiO_2$, *Il Cemento*, 90:73–84 (1993)

27. Ramachandran, V. S., Beaudoin, J. J., and Paroli, R. M., The Effect of Nitrobenzoic and Aminibenzoic Acids on the Hydration of Tricalcium Silicate: A Conduction Calorimetric Study, *Thermochimica Acta*, 190:325–333 (1991)

28. Ramachandran, V. S., and Zhang Chun-Mei, Dependence of Fineness of Calcium Carbonate on the Hydration Behavior of Tricalcium Silicate, *Durability of Building Materials,* 4:45–66 (1986)

29. Ramachandran, V. S., and Zhang Chun-Mei, Thermal Analysis of the $3CaO{\bullet}Al_2O_3$-$CaSO_4{\bullet}2H_2O$-$CaCO_3$-H_2O System, *Thermochimica Acta*, 106:273–282 (1986)

30. Ramachandran, V. S., and Zhang Chun-Mei, Hydration Kinetics and Microstructural Development in the $3CaO{\bullet}Al_2O_3$-$CaSO_4{\bullet}2H_2O$-$CaCO_3$-H_2O System, *Materiaux et Constructions*, 19:437–444 (1986)

31. Ramachandran, V. S., and Zhang Chun-Mei, Influence of $CaCO_3$ on the Hydration and Microstructural Characteristics of Tricalcium Silicate, *Il Cemento*, 83:129–152 (1986)

6

Retarding and Water Reducing Admixtures

1.0 INTRODUCTION

Lignosulfonate-based admixtures form a major percentage of the retarder/water reducer type of admixtures used in the concrete industry. Several techniques have been used to obtain an understanding of the action of retarders/water reducers, such as the mechanism of their action, rate of hydration, setting times, microstructure, etc. The techniques that have yielded important results include DTA, TG, DSC, scanning electron microscopy, XRD, chemical shrinkage measurements, isotherms, loss on ignition, IR, etc. These techniques are generally applied on samples that are hydrated for certain periods of time. The conduction calorimetry, however, follows the instantaneous evolution of heat as a function of time. It provides a method of quickly assessing the relative rates of hydration in the presence of different amounts/types of admixtures. The time of termination of the induction period gives information on the relative retarding action of various types of retarders.

2.0 LIGNOSULFONATES

The lignosulfonate-based admixtures have been used more widely than other water reducers. They are capable of reducing water requirements and retarding the setting times of concrete. They influence the dispersion and the hydration rate of the individual cement compounds, and, thus, the cement itself. Techniques such as XRD, DTA, DSC, TG, DTG, and conduction calorimetry have been used extensively to follow the hydration of cement and cement compounds containing different types and amounts of lignosulfonates (LS).

2.1 Tricalcium Aluminate

Tricalcium aluminate hydrates to form, initially, hexagonal phases identifiable by endothermal effects in DTA at 150–200°C and 200–280°C. These are converted to a cubic phase of formula C_3AH_6 which exhibits endothermal effects at 300–350°C and 450–500°C. The addition of lignosulfonate influences the rate of formation of these phases and their interconversions to the cubic phase. Depending on the amount of lignosulfonate, the hexagonal phase may be stabilized even up to fourteen days or more with lignosulfonate, but in that hydrated without the admixture, the cubic form may appear at six hours or earlier (Fig. 1).[1]

The mechanism by which the hexagonal phase of calcium aluminate is stabilized by calcium lignosulfonate (CLS) can be examined by treating this phase with lignosulfonate and subjecting it to DTA studies. Adsorption-desorption studies have indicated that the hexagonal phase adsorbs lignosulfonate irreversibly. There is evidence that lignosulfonate enters the interlayer positions of the hexagonal phase. Figure 2 shows the thermograms of the hexagonal hydrate containing CLS and that of CLS itself.[2] An intense exothermic peak at about 360°C is common to CLS and the hexagonal phase containing CLS, but is reduced to a small hump in vacuum DTA. The important difference is that an exothermic peak appearing at 730°C in the sample treated with CLS is absent in the CLS sample. This exothermic peak signifies oxidation of the strongly bound CLS in the interlayer positions. The weight loss curves obtained through TG investigation also show evidence of the formation of a complex between the hexagonal phase and CLS.

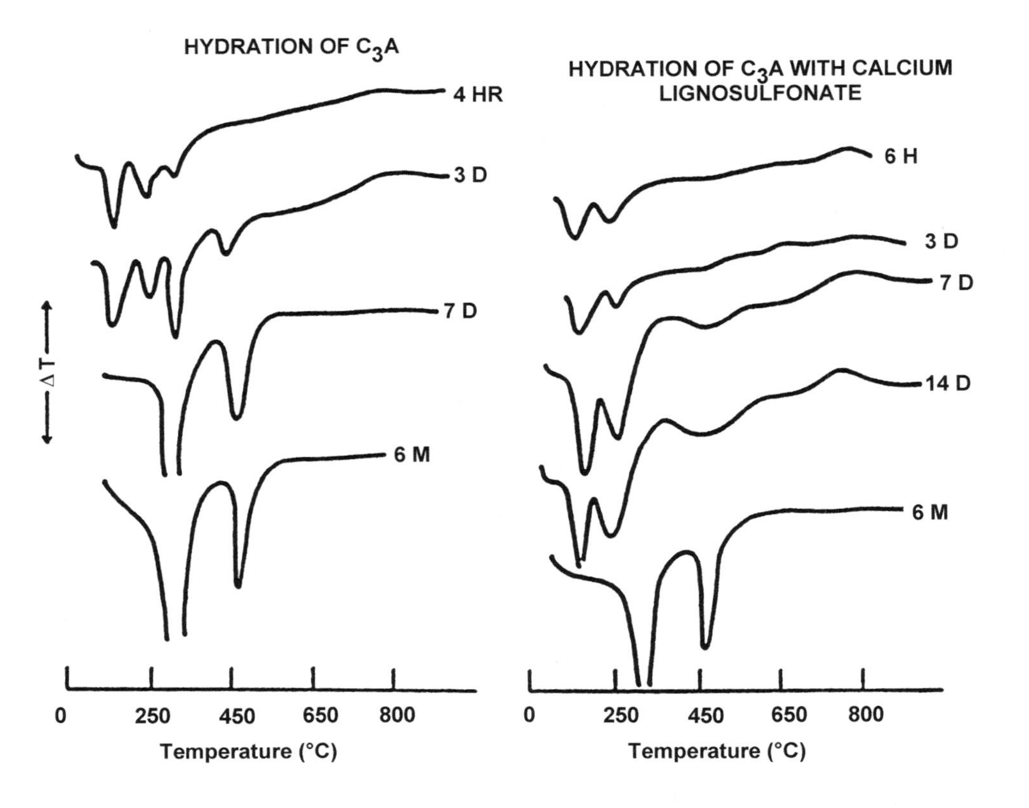

Figure 1. The influence of calcium lignosulfonate on the hydration of C_3A.

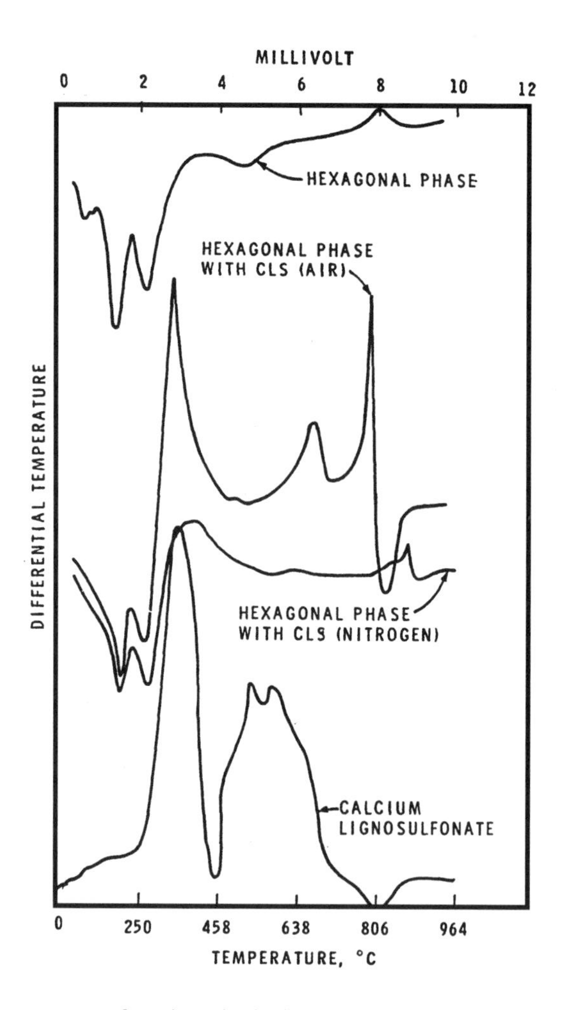

Figure 2. Thermograms of products in the hexagonal phase-CLS-H_2O system.

2.2 Tricalcium Aluminate-Gypsum-Calcium Lignosulfonate-Water

In the hydration of C_3A in the presence of gypsum and Na lignosulfonate (NLS), the conduction calorimetric curve indicates that, in the absence of lignosulfonate, ettringite is formed rapidly and converted to the low sulfoaluminate form. In the presence of NLS, the conversion of ettringite to the low sulfoaluminate form is retarded.[3]

DTA curves of C_3A-gypsum(10%)-$Ca(OH)_2$(5%)-H_2O with or without lignosulfonate (LS) show differences (Fig. 3).[1][4] The formation of high sulfoaluminate is recognized by an endothermic peak at about 150°C. The conversion of this phase to the low sulfoaluminate form or a solid solution is indicated by the emergence of an endothermic peak at about 200°C. Thermograms show that it takes up to three days for the high sulfoaluminate to convert to the low form in the presence of LS, whereas in its absence, the conversion occurs in three hours.[4]

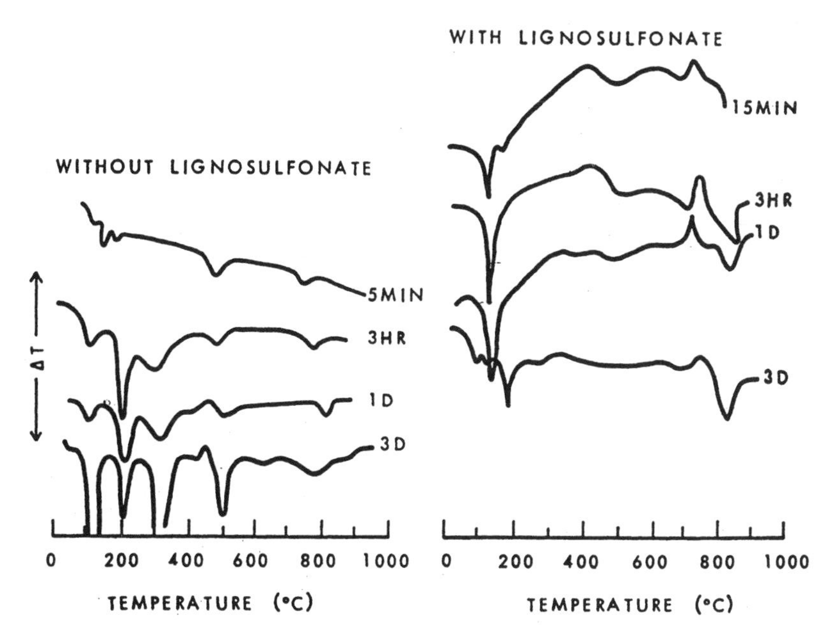

Figure 3. Thermograms of C_3A hydrated in the presence of gypsum, lime or lignosulfonate.

2.3 Tetracalcium Aluminoferrite-Calcium Lignosulfonate-Water

Tetracalcium aluminoferrite adsorbs less retarder than the C_3A phase. The hydration products of C_4AF being amorphous may be detected by DTA and DTG.[5] The effect of sugar-free NLS and Na_2CO_3 on the hydration of C_4AF was studied by Collepardi, et al., using thermogravimetric analysis.[6] Larger amounts of lignosulfonate (2.5%) retarded the formation of $C_3(A,F)H_6$ up to 7 days. At lower dosages (0.625%), lignosulfonate appeared to act as a slight accelerator. Lorprayon and Rossington who studied the hydration of C_4AF with lignosulfonate containing sugar found that C_4AF hydration was retarded even at a dosage of 0.125% lignosulfonate.[7] In the presence of a mixture of sugar-free LS (2.5%) + Na_2CO_3 (2.5%), C_4AF hydration was inhibited up to about 1 day.[6]

2.4 Tricalcium Silicate-Lignosulfonate-Water

Conduction calorimetric curves of C_3S hydrated in the presence of 0.3% calcium lignosulfonate show that the induction period is extended from seven hours to about twenty hours. At 0.5% addition, the hydration of C_3S is completely inhibited.[8] Other studies have shown that addition of small amounts of different types of CLS (0.5%) to C_3S pastes retard setting indefinitely, although some of them contained calcium chloride. XRD showed absence of calcium hydroxide in these pastes.[9][10]

The inhibitive action of CLS on the hydration of C_3S does not seem to be a simple function of the percentage of admixture added. The concentration in the aqueous phase has to be taken into account. In Fig. 4,[11] calorimetric curves A, B, C, and D represent those of C_3S treated with 0, 0.125, 0.25, and 1% solution of CLS at a water/solid ratio of 2. The samples B and C show reduced peaks occurring at longer periods indicating retardation. The sample D, containing 1% CLS, completely inhibits hydration. If C_3S is treated with 1% CLS solution at a water/solid ratio 0.5, the percentage of CLS based on C_3S would be 0.5, the same as sample C. The resulting curve, however, would be similar to D. These results suggest that C_3S is stabilized in contact with a CLS solution of concentration above 0.25%.

The TG and DTA techniques have been used to study the effect of 0.5% CLS on C_3S hydration. Free lime and combined water contents were determined at different times of hydration. The initial rate of hydration of C_3S was retarded. At a dosage of 0.8%, the hydration is inhibited up to thirty days.[11][12]

DTA was used by Ramachandran to explain the adsorption-desorption of calcium lignosulfonate in the C_3S-H_2O system.[11][13] The adsorption and desorption curves for C_3S containing CLS are given in Fig. 5. Lignosulfonate does not adsorb on the anhydrous C_3S in a nonaqueous system. A steep increase occurs in the CLS uptake up to an initial concentration of about 1 g/L. This is due to the formation of a large amount of C-S-H (of high specific surface) formed in the presence of low amounts of lignosulfonate. DTA curves show large endothermal effects below 200°C and one between 400 and 500°C for the formation of lime. At CLS concentrations between 1 and 3 g/L there is a decrease in the uptake of CLS as the degree of hydration of C_3S is reduced. At a concentration above 3 g/L, there is again a gradual increase in the amount of CLS uptake, although the hydration of C_3S is minimal. Therefore, the gradual increase in the adsorption of CLS above a concentration of 3 g/L, may either be due to a better dispersion of C_3S particles which increases its surface area and/or to the penetration of CLS molecules between the interlayer position of C-S-H phase. On the desorption branch, at low concentration levels, there is complete irreversibility of adsorption, indicating that CLS is strongly adsorbed on the hydrated C_3S as complex.

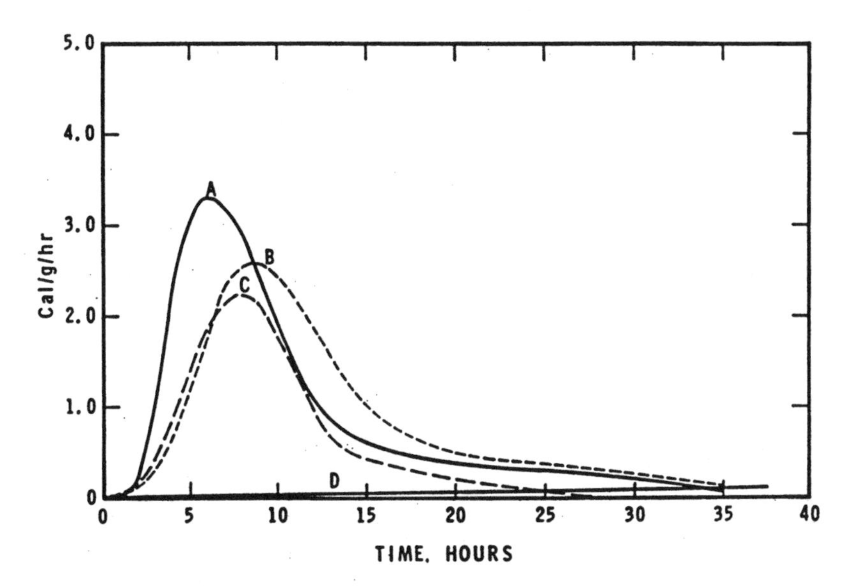

Figure 4. Rate of hydration of C_3S as a function of the concentration of CLS by conduction calorimetry.

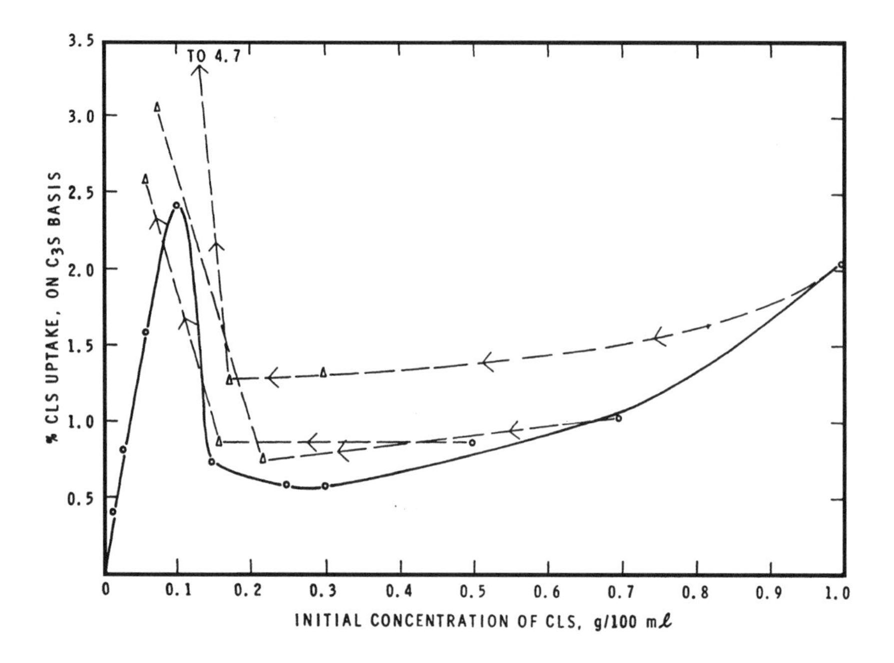

Figure 5. Adsorption of calcium lignosulfonate on C_3S in an aqueous medium.

In Fig. 6,[11][13] the adsorption-desorption of LS on a sample of completely hydrated C_3S is shown. There is a rapid initial adsorption ofCLS followed by a much slower rate at concentration levels higher than 0.5 g/L. The scanning desorption isotherms do not follow the adsorption isotherm, indicating the formation of increasing amounts of irreversibly adsorbed lignosulfonate as its concentration increases. This is due to the fact that water disperses the C-S-H and CH particles and promotes the penetration of LS into the interlayer spaces within the C-S-H phase. That CLS not only chemisorbs on the C-S-H, but also enters the interlayer position, is confirmed by an exothermic effect at about 800°C in DTA. A partial chemical reaction with CH is also indicated by an exothermal effect.

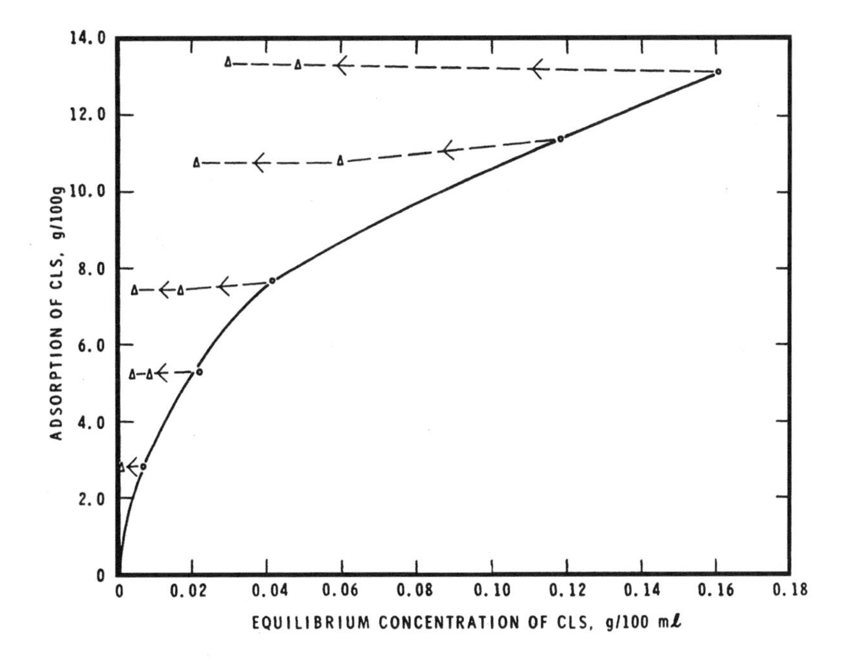

Figure 6. Adsorption-desorption isotherms of calcium lignosulfonate in completely hydrated C_3S in an aqueous medium.

2.5 Dicalcium Silicate-Lignosulfonate-Water System

Not much work has been reported on the influence of lignosulfonate on the hydration of dicalcium silicate. The rate of hydration of dicalcium silicate is slow, and it is further retarded by the addition of lignosulfonate. Evidence of this is the lower amounts of lime formed in the hydrated product in the presence of 0.3% lignosulfonate. It has been reported that sugar-free Na-lignosulfonate (0.3%) retards the hydration of C_2S, but to a lower extent than the commercial lignosulfonate.[14]

2.6 Tricalcium Silicate-Tricalcium Aluminate-Lignosulfonate-Water System

Adsorption of lignosulfonate on C_3A during hydration is much higher than that on the hydrating C_3S. In the system C_3A-C_3S-H_2O, C_3A may act as a sink for lignosulfonate.[14] Ramachandran[16] studied the effect of CLS on the hydration of C_3S containing C_3A, hexagonal aluminate hydrate, and cubic aluminate hydrate. The degree of hydration of C_3S was monitored by determining the DTA peak area due to $Ca(OH)_2$ decomposition.

The effect of the addition of CLS on the hydration of C_3S in the system C_3S-C_3A-H_2O can be demonstrated with reference to Fig. 7.[15][16] The rate of hydration is estimated by the amount of $Ca(OH)_2$ formation. Curve 1 refers to the hydration of C_3S with water. By the addition of 0.8% calcium lignosulfonate (curve 2), the hydration is completely inhibited even up to thirty days, as evident from the absence of $Ca(OH)_2$. When 5% C_3A is added to the mixture of C_3S and lignosulfonate, curve 3 results. The hydration of C_3S is retarded owing to the CLS, but not completely inhibited as in curve 2 because most of the admixture is consumed by the hydrating C_3A. Curve 4, in which C_3S and C_3A are hydrated for five minutes prior to the addition of CLS, shows a much higher retardation of the C_3S hydration compared to curve 3. This shows that the pre-hydrated mixtures adsorb less lignosulfonate and allow greater adsorption to take place on C_3S. At larger additions of the retarder (curve 5), even though C_3A adsorbs initially a large amount of the admixture, larger concentrations are still present to effectively retard C_3S hydration for a long time (compare curves 5 and 3).

The combined effect of Na-lignosulfonate (0.3%) and Na_2CO_3 in the hydration of C_3S (80%) + C_3A (20%) is approximately similar to that in the presence of lignosulfonate except for complete blocking of C_3S and C_3A hydration for at least thirty minutes (Fig. 8).[14][17] At higher dosages of both lignosulfonate and carbonate (0.9%), an induction period of one day occurs.[17] The influence of other organic water reducers, such as glucose and sodium gluconate at additions of 0.1–0.3% on the C_3S (80%) + C_3A (20%) system appears to be similar to that of lignosulfonate except for a better retarding action of Na-gluconate.[14]

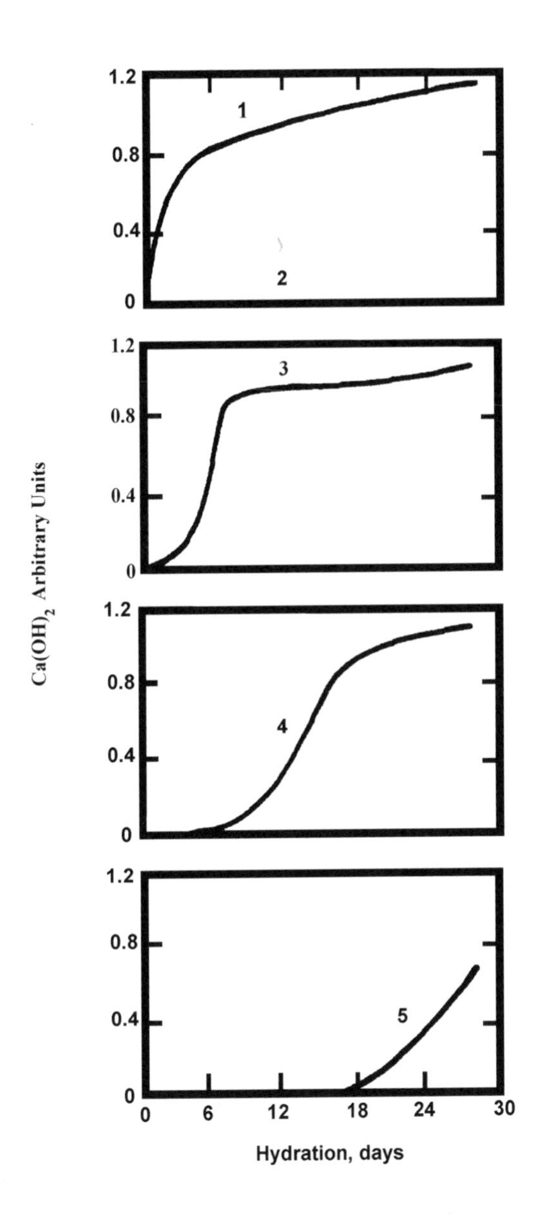

Figure 7. Effect of lignosulfonate on the hydration of C_3S in the system C_3S-C_3A-H_2O. (*1*) $C_3S + H_2O$; (*2*) C_3S + CLS (0.8%) + H_2O; (*3*) $C_3S + C_3A$ + CLS (0.8%) + H_2O; (*4*) C_3S + C_3A (hydrated 5 minutes) + CLS (0.8%) + H_2O; (*5*) $C_3S + C_3A$ + CLS (3.2%) + H_2O.

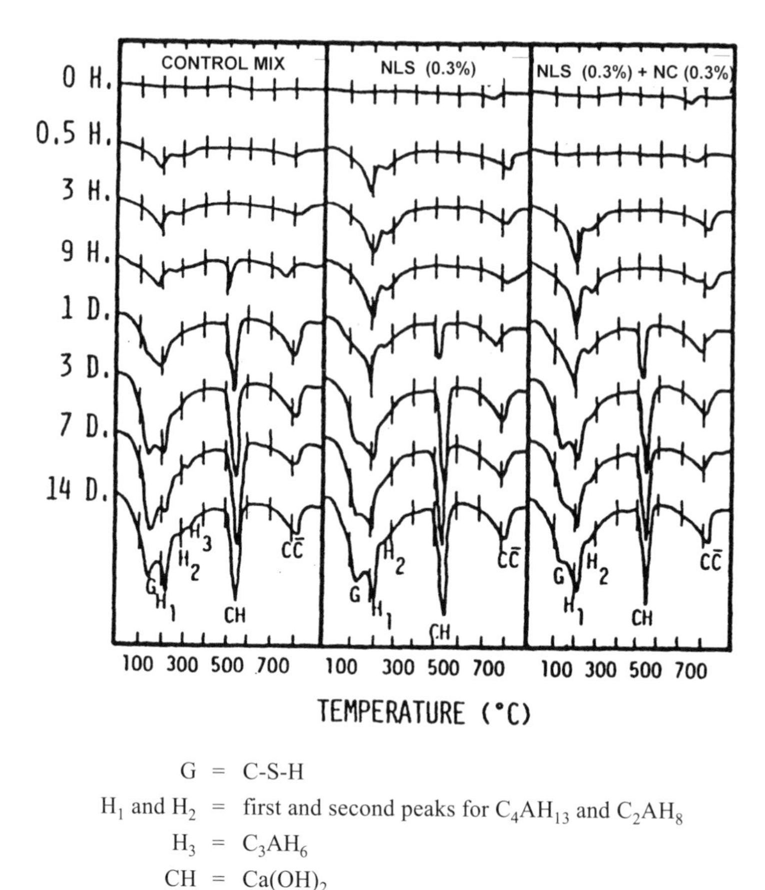

G = C-S-H

H_1 and H_2 = first and second peaks for C_4AH_{13} and C_2AH_8

H_3 = C_3AH_6

CH = $Ca(OH)_2$

CC = $CaCO_3$

Figure 8. Influence of sodium carbonate (NLS) and sodium lignosulfonate + sodium carbonate (NC) on the paste hydration of C_3S (80%)-C_3A (20%).

2.7 Cement-Lignosulfonate-Water System

Thermal investigations have shown that the rate of hydration of C_3A and the inter-conversion of ettringite to the low monosulfoaluminate in the C_3A-$CaSO_4$•$2H_2O$-CLS-H_2O system, and hydration of individual cement phases C_4AF, C_3S, and C_2S are all retarded by lignosulfonates. Thus, it is expected that the rate of hydration of cement should also be retarded by the addition of lignosulfonates.

An addition of 0.1% CLS may extend the initial and final setting times of cement mortar by two and three hours, respectively. The influence of 0.3% CLS on the hydration of cement is shown in Fig. 9.[8] Thermograms indicate that the reference cement containing no admixture exhibits a broad endothermal peak below 200°C, representing the formation of both ettringite and C-S-H phase. These peaks increase in intensity as the hydration period is increased. The effect between 450 and 500°C is caused by the dehydration of $Ca(OH)_2$, and its intensity indicates the extent to which the hydration of the C_3S component has progressed. The cement hydration, in the presence of lignosulfonate, is retarded as seen by the lower intensity of the $Ca(OH)_2$ decomposition peak. The low temperature effect below 300°C in the presence of CLS is not sharp as that obtained in the reference sample.

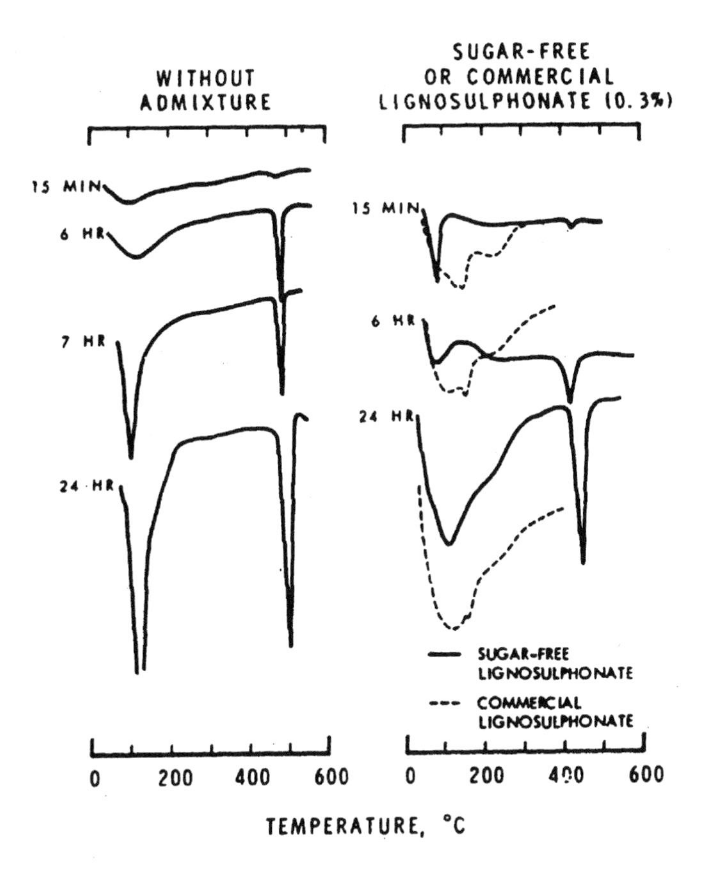

Figure 9. DTA of cement hydrated in the presence of lignosulfonates.

Conduction calorimetric curves also demonstrate the retarding effect of lignosulfonate on cement.[18] Conduction calorimetric curves of cement containing 0, 0.1, 0.3, and 0.5% CLSshow that the cement containing CLS delays the appearance of the peaks caused by the C_3S phase.

Odler and Becker followed the DTA behavior of high early strength cement and a laboratory-made cement containing 0.5% lignosulfonate._They were hydrated for different periods.[12] Thermograms indicated that the formation of ettringite was accelerated in the commercial cement while it was retarded in the laboratory cement. The rate of formation of $Ca(OH)_2$ was significantly lower in both cements containing the lignosulfonate admixture.

The conduction calorimetric technique has been applied to follow the hydration of Type I and Type V cements containing 0.28% CLS. Table 1 gives the heat of hydration in two cements containing CLS.[19] Generally, the amount of cumulative heat is lower in the sample with lignosulfonate. The retarding action of the admixture at later times is compensated for as the time of hydration increases.

Table 1. Effect of CLS-Based Admixture on Cumulative Heat of Hydration (cals/g) *(Reprinted with permission from Elsevier Science)*

Age	Type I Cement		Type V Cement	
	No Admix.	Normal Dosage	No Admix.	Normal Dosage
5 mins	2.80	2.20	0.65	0.17
10 mins	3.70	3.66	1.20	0.54
30 mins	4.67	5.65	2.18	2.27
2 hrs	5.50	6.80	2.80	4.40
4 hrs	7.10	7.50	4.00	5.10
12 hrs	29.2	13.5	21.1	8.90
24 hrs	50.0	43.1	40.9	32.5
48 hrs	63.7	59.5	55.5	49.2
72 hrs	71.5	67.9	61.0	54.5

The influence of lignosulfonate on cements containing different amounts of SO_3 has been studied by conduction calorimetry.[20] The hydration of the C_3S component was retarded. In the presence of CLS, the rate of heat evolution in the first few minutes is sustained longer suggesting the immediate retarding effect of gypsum on C_3A has not occurred, but an acceleration of C_3A hydration has resulted. In the first twenty-four hours, for Type I cement containing 1.65% SO_3 the total heat was reduced by 26% compared to that in the control sample. For the same cement with 2.15% SO_3, the corresponding reduction was only 14%. It was found that CLS increased the optimum gypsum content. The cement low in C_3A was more strongly retarded and gave lower strengths.

The hydration of cement undergoes a significant modification in the presence of a mixture of lignosulfonate and alkali carbonate. The heat evolution peaks occur separated by two induction peaks. The possibility of a highly anionic complex between lignosulfonate and carbonate ions has been proposed to explain its more effective dispersing effect than lignosulfonate. The first induction period is due to the competitive interaction of one or more of the admixture with C_3A.[21]

3.0 SUGAR-FREE LIGNOSULFONATE

Lignosulfonate is a complex molecule of high molecular weight. It is derived from substituted phenyl propane containing hydroxy, methoxy, phenolic, and sulfonate groups. The commercial lignosulfonates, not being pure, contain varying amounts of sugars such as mannose, glucose, xylose, galactose, arabinose, and fructose. Sugars are known to be good retarders of setting of cement, hence, the major part of the retarding action of a commercial lignosulfonate is believed to be due to the sugars contained in it.[8]

Conduction calorimetric curves of C_3S hydrated up to 55 hours show that the exothermic peak at about 7–8 hours found in the reference C_3S is retarded or completely annulled at dosages of about 0.3% sugar-free Na-lignosulfonate.[8] Work of Monosi, et al.,[13] also has shown the effectiveness of sugar-free lignosulfonate on C_3S hydration. In the DTG of C_3S hydrated with or without 0.3% sugar-free lignosulfonate, the endothermal peak at about 500°C for $Ca(OH)_2$ decomposition is absent in the sample containing sugar-free lignosulfonate (Fig. 10).[13][22]

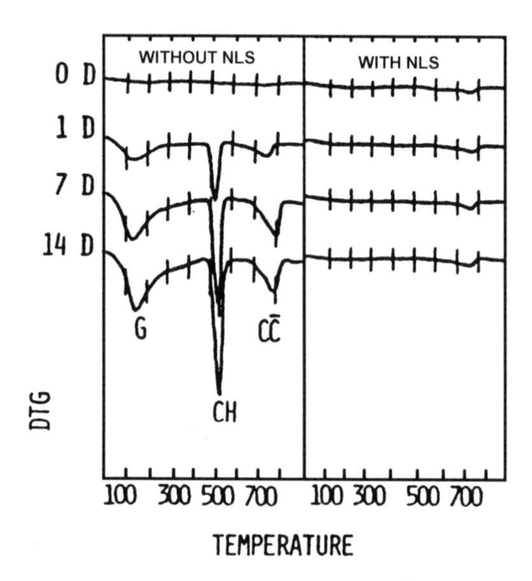

Figure 10. DTG curves for anhydrous and hydrated C_3S with or without 0.3% sugar-free sodium lignosulfonate.

Chatterji's work[23] on C_3A hydration shows that sugar-free lignosulfonate is an inferior retarder in comparison with commercial lignosulfonate. Disagreement amongst researchers is due to the differences of methods adopted for separation of sugars, variability of materials, and molecular weights and methods adopted for following the hydration. Based on XRD results, Pauri, et al.,[24] found that sugar-free lignosulfonate retards the hydration of C_3A. The retarding effect was less significant on C_3A-Na_2O solid solution than that on pure C_3A.

Studies have indicated that calcium lignosulfonate is an active admixture in concrete, interacting with the hydrating cement minerals.[14] Ramachandran[8] examined the effect of sugar-free Na lignosulfonate prepared by continuous diffusion (using a Dowex 50WX2 resin bed) on the hydration of portland cement. The effect of 0.1–0.3% sugar-free lignosulfonate on DTA thermograms of C_3A hydrated from 1 day to 60 days are shown in Fig. 11. The peak at about 100°C is due to the expulsion of adsorbed water. The endothermal effects at about 200° and 250–280°C are ascribed to the dehydration of hexagonal phases C_2AH_8-C_4AH_{13}. The effects at 300° and 300–500°C represent dehydration effects of the cubic C_3AH_6 phase. In the absence of lignosulfonate, even at 15 minutes, a large endothermal peak occurs at 300°C due to the presence of the cubic phase. In the presence of 0.1% sugar-free lignosulfonate, there is a slight retardation effect in the early periods, but at 1 day the cubic phase predominates. At 0.5% addition, there is a significant retardation effect as evidenced from the presence of a large amount of the hexagonal phase. Complete conversion of the hexagonal to the cubic phase occurs only at 28 days. Further retardation is observed at 1–3% additions of Na lignosulfonate. Thermogravimetric analysis and XRD investigations confirmed the results obtained from DTA. It also appears that sugar-free Ca and Na lignosulfonates are nearly as effective as the commercial lignosulfonate in delaying the setting characteristics of cement (Table 2).[15]

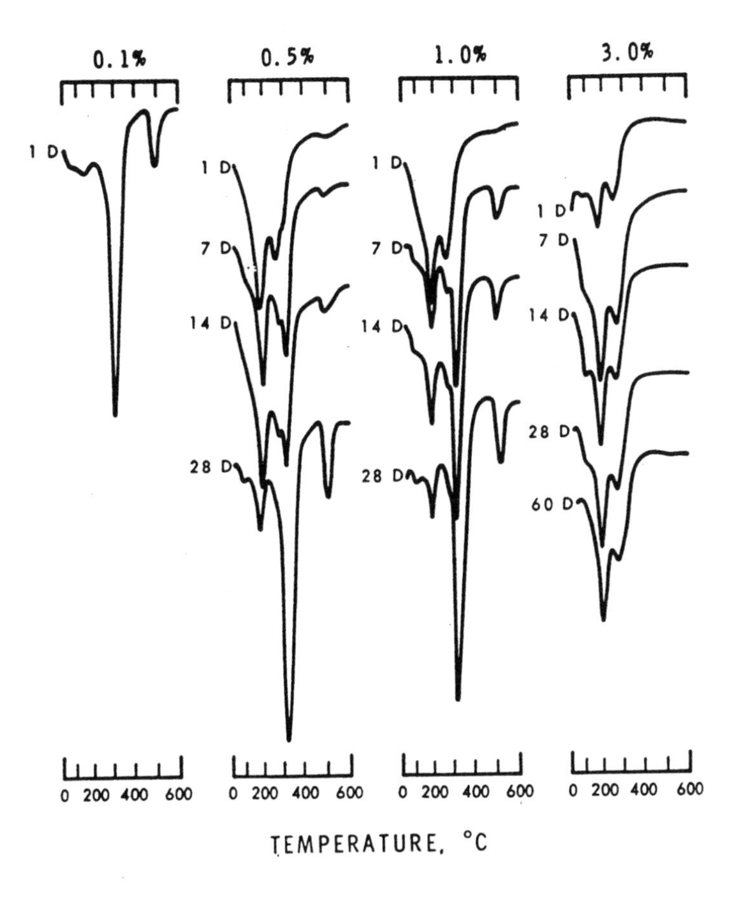

Figure 11. DTA of C_3A hydrated in the presence of sugar-free Na lignosulfonate.

Table 2. Water Reduction and Setting Characteristics of Mortar Containing Lignosulfonates

Admixture Type	Amount (%)	W/C	Initial Set (Hrs)	Final Set (Hrs)
None	Nil	0.550	5	9
Commercial CLS	0.1	0.540	7	12
Sugar-Free NLS	0.1	0.540	7	12
Sugar-Free CLS	0.1	0.540	7	12
Commercial CLS	0.3	0.440	Quick Set	13.5
Sugar-Free NLS	0.3	0.440	12	15
Sugar-Free CLS	0.3	0.440	14	16.5
Commercial CLS	0.5	0.425	Quick Set	22
Sugar-Free NLS	0.5	0.425	23	28
Sugar-Free CLS	0.5	0.425	22	27.5

Conduction calorimetric analysis has also been carried out on the effect of sugar-free lignosulfonate on the hydration of cement. It was found that sugar-free lignosulfonate was nearly as efficient as commercial lignosulfonate in retarding rate of reaction and setting of cement.[8]

4.0 HYDROXYCARBOXYLIC ACIDS

Hydroxycarboxylic acids have several (OH) groups and either one or two terminal carboxylic acid (OH) groups attached to a relatively short carbon chain. Typical compounds considered for use in concrete include gluconic, citric, tartaric, mucic, malic, salicylic, heptonic, saccharic, and tannic acids. They are generally used as aqueous solutions of sodium salt.

These admixtures retard the hydration of the C_3A phase by delaying the conversion of the hexagonal phase to the cubic phase. By applying DTA and XRD, Lorprayon and Rossington showed that salicylic acid retards the hydration of C_3A.[7] It is also expected that this chemical would retard the hydration of the silicate phases.

A complex results when a cement is hydrated in the presence of salicylic acid. Such a product, when added to cement, slows down the hydration. TG and other methods have been used to identify this complex.[25][26]

5.0 SUGARS

The retarding action of sugars and their oxidation products on concrete have been well documented. All saccharides with the exception of trehalose strongly retard the set of cement. The five-membered rings, sucrose and raffinose, are the best retarders. The reducing sugar glucose, maltose, and lactose are moderate retarders. The strong retarding action of C_3A hydrated with 1% sucrose as opposed to the weak retarding action of 1% trehalose is well substantiated by DTA curves (Fig. 12).[1][27] Three peaks in the range of 100–200°C represent dehydration of a mixture of hexagonal aluminate hydrates. The peaks in the range 300–350°C and 500°C are mainly due to the dehydration effect of the cubic aluminate hydrate. The hydration of C_3A without the sugar shows a strong endothermic peak at about 550°C, as early as 3 hours. In the presence of sucrose, a large amount of the hexagonal phase persists up to 90 days, as indicated by an endothermal effect below 250°C. However, C_3A with trehalose shows an intense endothermal effect for the presence of the cubic phase after a relatively short time.

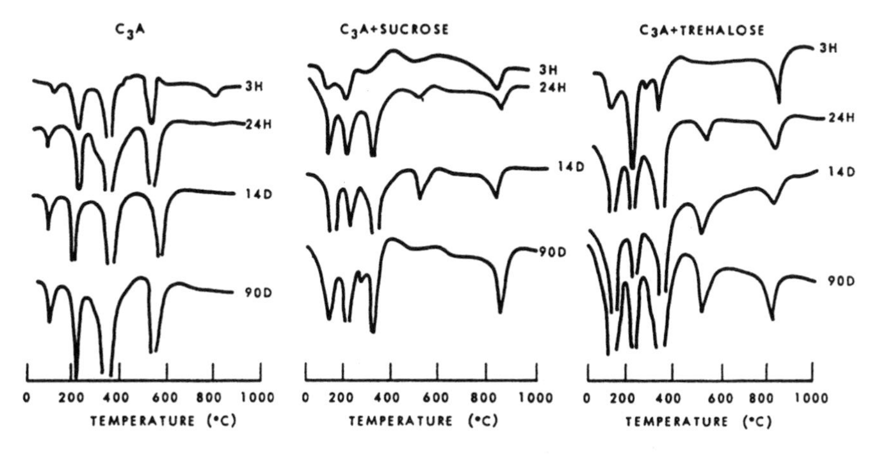

Figure 12. Influence of sucrose and trehalose on the hydration of C_3A.

The retarding action of sugars is attributed to the formation of a complex between the hexagonal phase and sugars. The formation of an interlayer complex in the presence of Ca or Na gluconate is confirmed by an endothermal peak at 80°C in the DTA thermogram.[28][29] This peak cannot be assigned to the C_4AH_{19} because, under the drying conditions that were adopted, it would be transformed to C_4AH_{13}. Retardation of the formation of the ettringite phase in the presence of sugars is also evident from DTA studies. For example, at 14 days of hydration the mixture containing sucrose has 25% ettringite and that without it, 52%. Thus, thermograms also reveal that the conversion of ettringite to the low sulfoaluminate form is retarded in the presence of sugars.[27]

It has also been observed that a sugar, in certain concentrations, may also act as an accelerator for the hydration of the tricalcium aluminate systems. Thermograms indicate that at 0.025–0.05% sucrose dosage, peaks due to cubic phases are intensified or those due to the hexagonal phases are decreased. This would show that sucrose is an accelerator. At concentrations of 2–5%, however, the number of hexagonal phases formed is diminished indicating that the hydration of C_3A is retarded.[1]

The DTG, DTA, and XRD techniques have been applied to study the retarding effect of sucrose on the hydration of C_3S. In Fig. 13,[10] the influence of 0.1% glucose (G), sodium gluconate (NG), and lignosulfonate high in sugar (LSA) or low in sugar (LS) on C_3S hydration is compared using XRD. The addition of 0.1% glucose causes an initial delay in the hydration of C_3S followed by a slight acceleration, as was found for lignosulfonate high in sugar acids. The induction period, however, is longer with glucose than when lignosulfonate was used. The induction period is longest with sodium gluconate (50 days). In the presence of 0.1% glucose or 0.01% gluconate, amorphous calcium hydroxide was detected by DTA at 7 days. At later ages, the crystalline form of calcium hydroxide was identified by both XRD and DTA.

6.0 PHOSPHONATES

Phosphonic acid-based chemicals are known to form complexes with many inorganic species, and the action of many retarders is based on complex formation. Several phosphonic acids such as amino trimethylene phosphonic acid (ATMP), 1-hydroxyethylidene-1, 1-diphosphonic acid

(HEDP), diethylenenitramine pentamethylene phosphonic acid (DTMP), and their corresponding sodium salts were added to cement and cement compounds in small amounts of 0.03–0.09% in order to study their retarding action. Conduction calorimetry and DSC techniques were applied.[30]

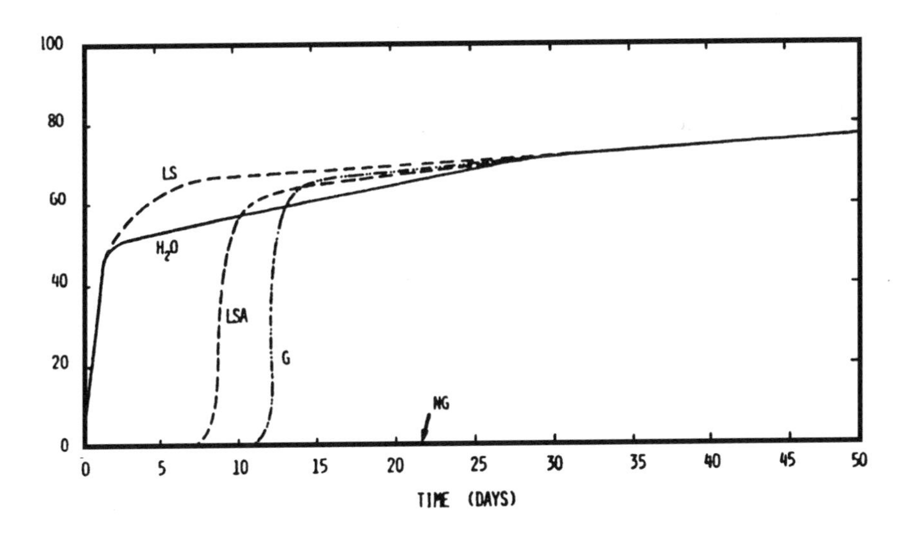

Figure 13. Influence of 0.1% glucose (G), sodium gluconate (NG), and lignosulfonate high or poor in sugar acids (LSA and LS respectively) on C_3S hydration.

A phosphonate-based admixture, ATMP (amino trimethylene phosphonic acid) in amounts of 0.03–0.05%, was added to C_3S and C_3A at a w/s ratio of 1.0 and the heat, developed with time, was followed by conduction calorimetry.[31] Phosphonic acid affects the hydration of C_3S (Fig. 14). The reference C_3S produced a maximum exothermic peak with a heat equivalent to 5.32 cals/g/h at 8.5 hours. Addition of phosphonic acid reduced the amplitude of the peak, delayed the occurrence of the peak, and also modified the appearance of the peak. About 1.79 cals/g /h of heat was developed at 11.5 hours with the addition of 0.03% phosphonic acid, and, at a dosage of 0.04%, an exothermal peak occurred at 31 hours with a heat production of 1.41 cals/g/h. At 0.05% addition, there was practically no heat produced even up to 72 hrs.

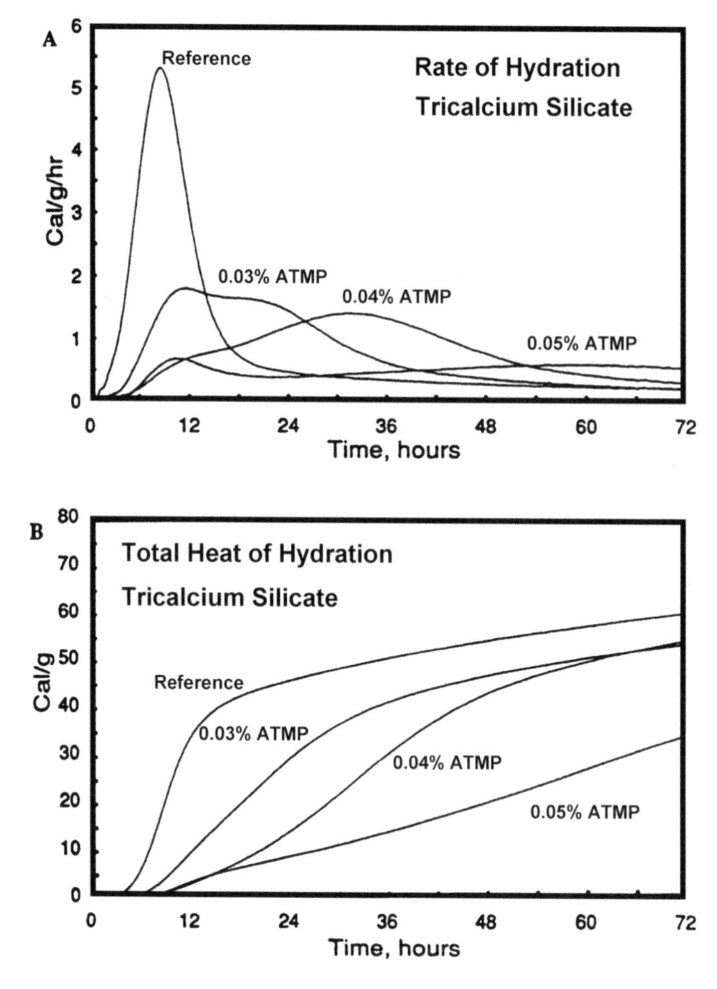

Figure 14. Calorimetric curves showing the effect of phosphonic acid on the rate of hydration of tricalcium silicate.

Phosphonic acid had a retarding effect on C_3A hydration (Fig. 15). The reference C_3A produced the maximum heat effect at 3.6 minutes with a value of 113.7 cals/g/h. At an addition of 0.03% acid, the peak occurred at 19.2 minutes producing 76 cals/g/h.

DSC investigations showed that the addition of phosphonic acid to the C_3A + gypsum mixture, in the presence of water, retards the hydration and inter-conversions in the system.[31]

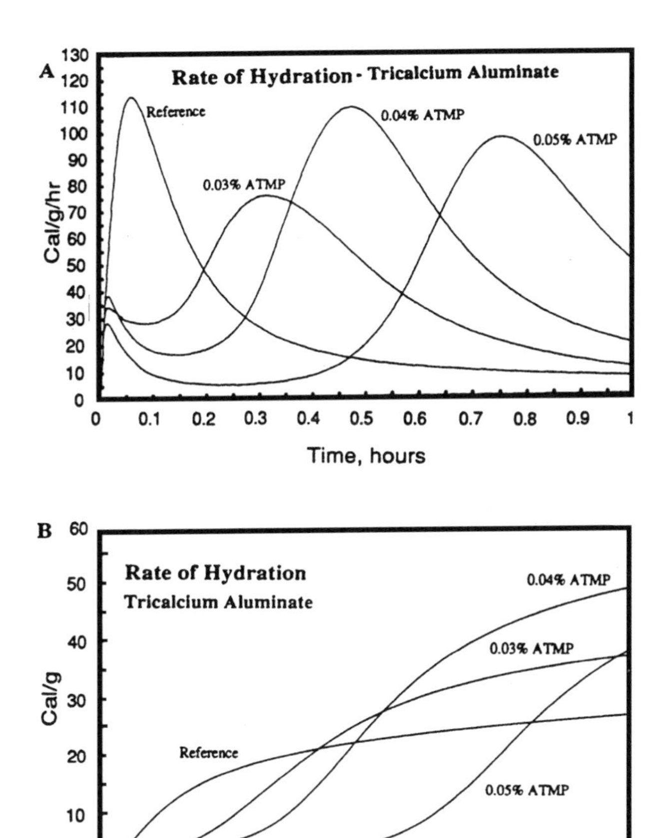

Figure 15. Calorimetric curves showing the effect of a phosphonic acid on the rate of hydration of the tricalcium aluminate component of cement.

Phosphonic acid-based chemicals such as ATMP, HEDP, DTMP, and their sodium salts were added to portland cement in dosages of 0.03–0.09% at a constant w/c ratio of 0.35 and their conduction calorimetric and DSC curves were compared.[30] The conduction calorimetric curves of cement containing sodium salt of ATMP are shown in Fig. 16.[30] Addition of this salt results in the retardation of cement hydration in terms of extended induction period and the late appearance of the exothermic peak. At 72 hours, however, the addition of the salt increases total heat development.

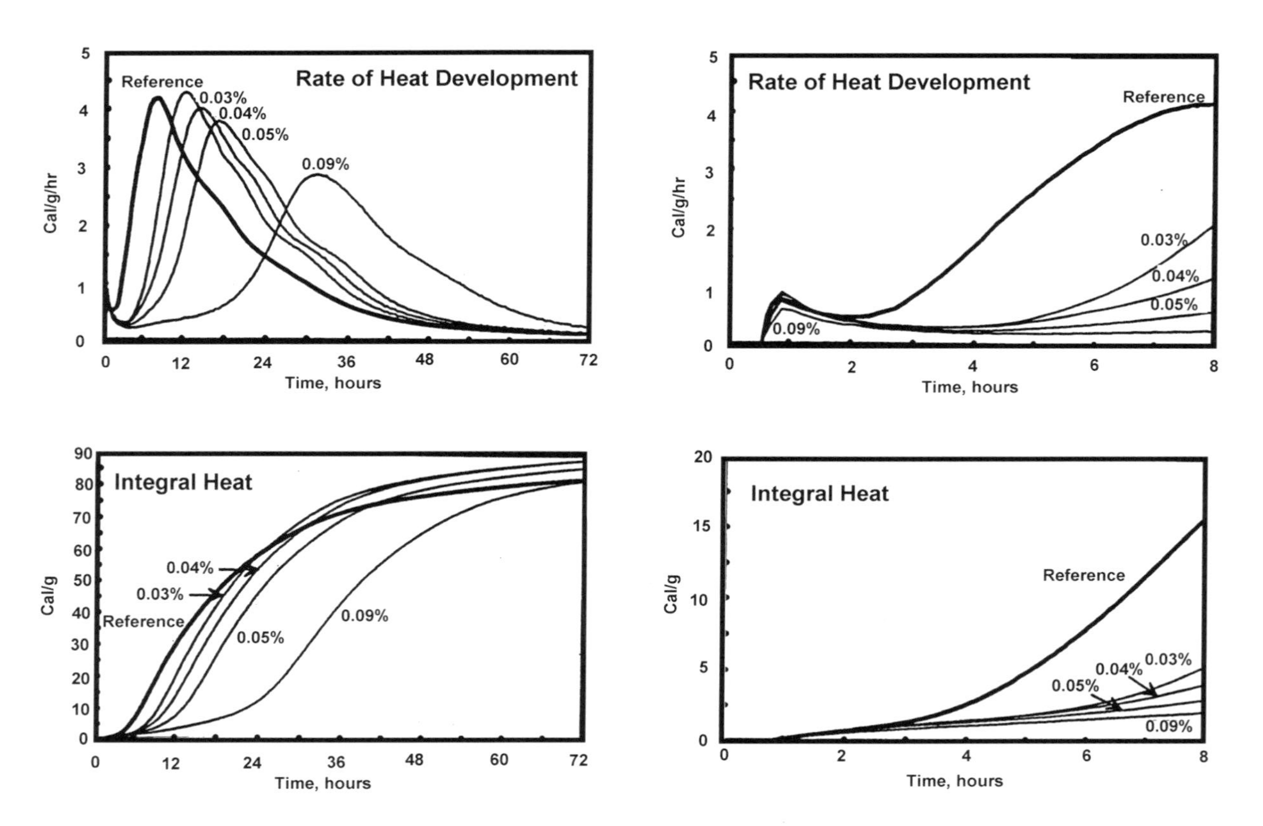

Figure 16. Conduction calorimetric curves of portland cement in the presence of phosphonate systems.

The relative effects of phosphonates can be evaluated by comparing the time at which the induction period is terminated at different dosages of a chemical. In Fig. 17, the time of termination of the induction period is plotted as a function of dosage.[30] Among the acids, the most efficient retarder is DTPMP which requires 0.05% for an induction period of 21 hours, compared to periods of 13 and 10 hours respectively for acids ATMP and HEDP at the same dosage. Among the salts, Na_6 DTMP is the most effective retarder.

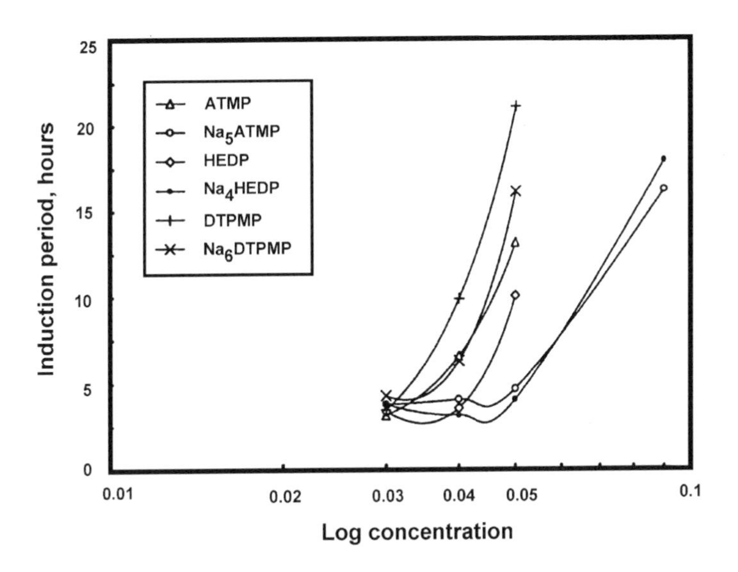

Figure 17. The effect of various concentrations of phosphonate compounds on the induction period of portland cement.

7.0 CONDUCTION CALORIMETRIC ASSESSMENT OF RETARDERS

Several techniques have been applied to obtain an understanding of the rate of hydration, types of products, complex formation, inter-conversions, mechanism of hydration, etc., in the hydration of cements and their individual phases. The common techniques used for these investigations include DTA, DSC, DTG, TG, chemical shrinkage measurements, XRD, loss on ignition, IR, electron microscopy, and others. Most of these techniques are applied to products that are hydrated to a specific length of time. In the conduction calorimetric technique, the heat developed is followed from the time water comes into contact with the cement or its phase. The time at which the exothermic peak appears may be used to assess the relative retarding actions of retarders. The time of termination of the induction period provides an idea of the relative times of setting of cements containing various admixtures. The total amount of heat produced at different times may be used for determining the rate of hydration.

Ramachandran and Lowery[32] applied conduction calorimetry to the study of the relative effects of various retarders on the hydration of cement. The retarders used were calcium gluconate, glucose, glycolic acid, molasses, sodium borate, sodium citrate, sodium heptonate, sodium hexametaphosphate, sodium pyrophosphate, sugar-free lignosulfonate, and sucrose. The dosage of the chemicals varied between 0.025 and 1.2%.

Gluconate was found to be a good retarder. Its action is possibly related to its poisoning action on the hydration products of cement. It has also been advocated for controlling slump loss in concrete. Conduction calorimetric curves of cement with gluconate show large hump effects for the hydration of the silicate phase (Fig. 18a).[32] At a dosage of 0.15%, the hydration is retarded up to 54 hours. Compared to gluconate, sodium borate is a mild retarder. Even at 0.7% the retardation effect lasts only for 24 hours (Fig. 18b).

The most effective retarders, that need a dosage of about 0.15% or less to achieve an induction period of 40 hours, include Ca gluconate, sodium heptonate, and sucrose. The least effective retarders that require more than 0.5% dosage for extending the induction period by 40 hours are sugar-free lignosulfonate, Na pyrophosphate, Na hexametaphosphate, Na borate, and glycolic acid. Glucose, molasses, and Na citrate are termed moderate retarders requiring a dosage between 0.15 and 0.5%.

Conduction calorimetry has been used to obtain the induction periods of cement containing different amounts of retarders. Table 3[32] provides data on the minimum concentration of a retarder required to achieve different induction periods.

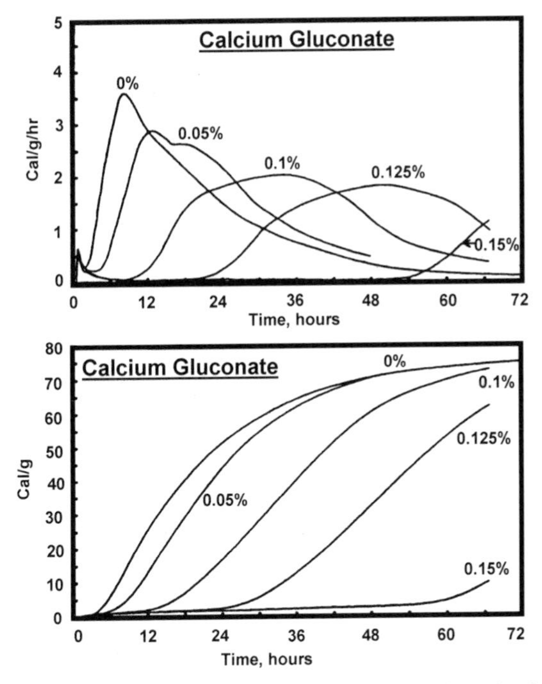

Figure 18. The effect of calcium gluconate on the hydration of portland cement. *(a)* rate of heat development *(b)* integral heat development.

Table 3. Minimum Concentrations of Various Retarders (%) Required to Achieve Desired Induction Periods

Retarding Agent	Induction Periods (Hrs)			
	10	20	30	40
Ca Gluconate	0.10	0.12	0.13	0.14
Glucose	<0.10	0.15	0.23	>0.30
Glycolic Acid	<0.50	0.55	0.64	0.79
Molasses	0.10	0.13	0.20	>0.20
Na Borate	0.51	0.63	>0.70	<0.10
Na Citrate	0.21	0.32	0.39	<0.50
Na Heptonate	<0.10	0.12	0.14	>0.15
Na Hexametaphosphate	<0.20	1.00	>1.00	——
Na Pyrophosphate	<0.20	0.38	1.00	>1.20
S.F. Ca Lignosulfonate	0.51	0.52	0.54	0.58
Sucrose	0.04	0.06	0.073	>0.075

In studies above, the time of termination of the induction period is taken to evaluate the relative setting times of cements containing retarders. The setting times of cement mortars are determined according to ASTM C403 using the Proctor needle. It is an arbitrary method and the time atwhich the setting occurs may involve both physical and chemical phenomena. Ramachandran[33] obtained the DSC curves of set mortars containing different types of retarders with or without a superplasticizer(Fig. 19). All of them exhibited endothermal peaks at about 100° and 450–500°C indicating the formation of $Ca(OH)_2$ at the time of set. Thus, it is evident that at the time of set some hydration of C_3S takes place. The degree of hydration of C_3S for the samples is different, depending on the admixture used. It, therefore, appears that the morphological and other factors also play a part in the setting phenomenon. It appears that the setting time occurs at some point after the induction period has terminated, but before the time at which the exothermic peak appears in the conduction calorimetry.

8.0 SLUMP LOSS

Addition of a superplasticizer to concrete results in significantly increased slump values. This increase is transient, however, and is generally not maintained beyond a period of 30–60 minutes. One method of controlling this loss is to add a retarder in the superplasticized concrete. Conduction calorimetric and different thermal methods may be adopted to examine the relative retardation action of admixtures. Of the several retarders studied, calcium gluconate was found to retard the hydration to a maximum extent. In Fig. 20,[33] the conduction calorimetric curves show that by the addition of 0.1–0.2% gluconate there is retardation of hydration of cements. However, in combination with 0.1–0.2% sulfonated melamine formaldehyde, there is increased retardation. It was found that by the addition of 0.05–0.1% sodium gluconate, the slump loss can be decreased. For example, with only gluconate added to concrete, the slump value is about 40 mm at 2 hours, but, in combination with SMF, the slump is as high a 140 mm (Fig. 21).[33]

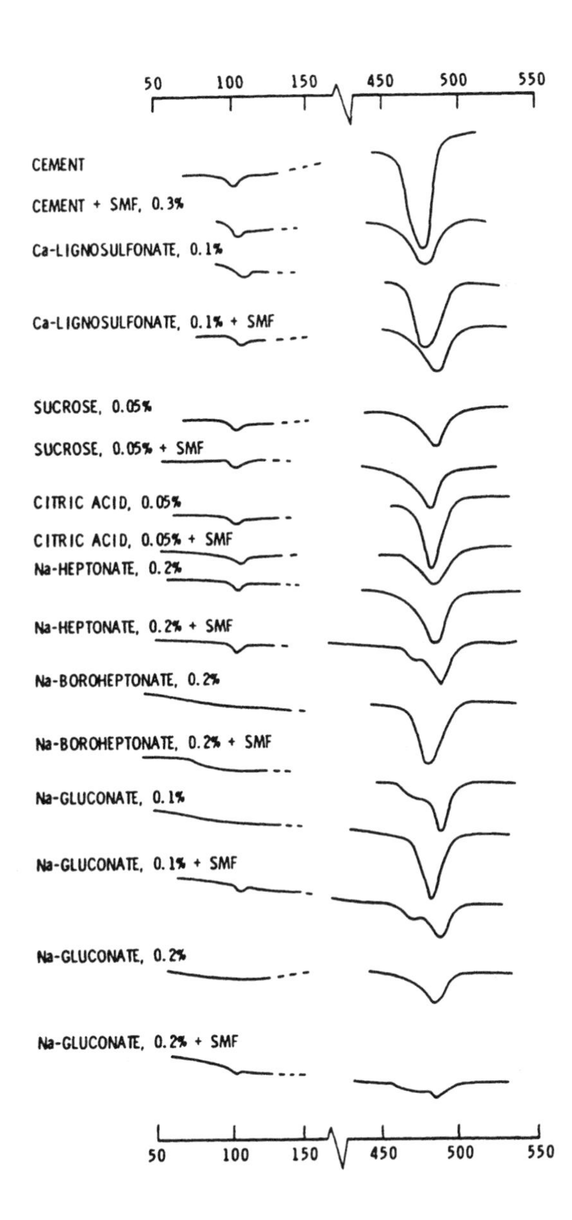

Figure 19. DSC curves (at times of set) of cement mortars containing various admixtures.

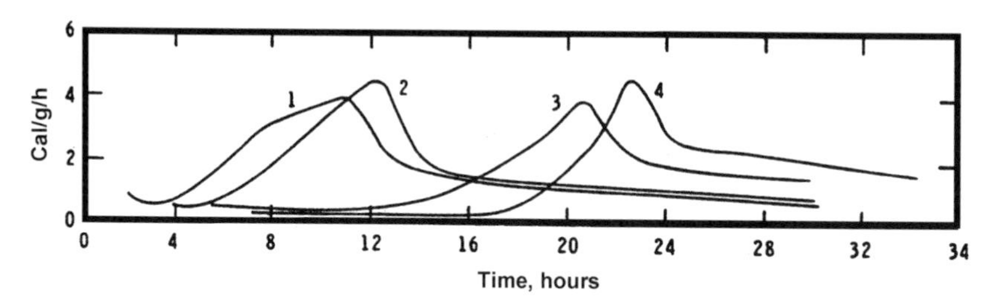

Figure 20. Conduction calorimetric curves of cement mortars containing Na gluconate and sulfonated melamine formaldehyde (SMF).

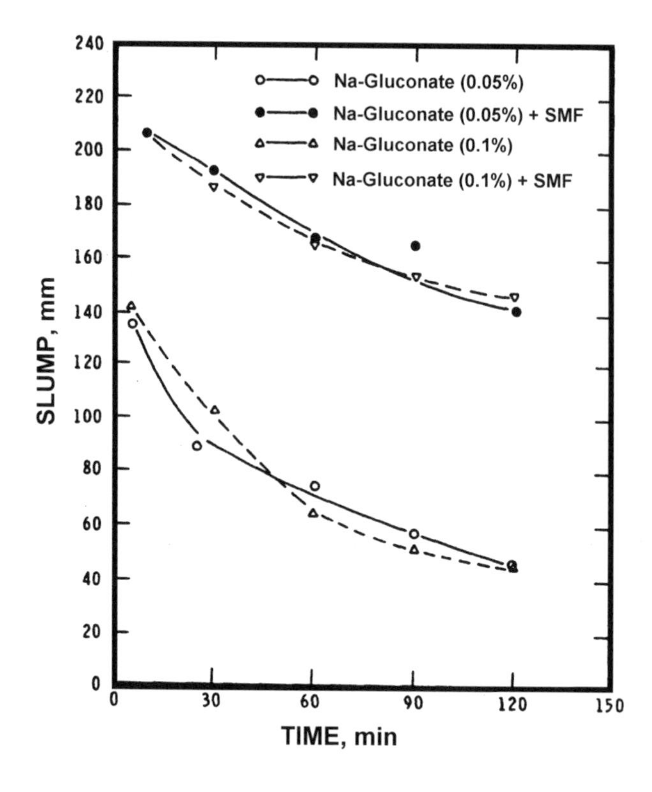

Figure 21. Effect of gluconate and its mixtures with SMF on slump loss.

9.0 ABNORMAL SETTING

Several instances of quick setting of cements have been encountered involving the use of admixtures. Greene[34] found that a particular white cement containing 1.54% SO_3 set abnormally when used in the presence of a carbohydrate-type of water reducer. In Fig. 22, DTA of the unhydrated cement shows peak doublets below 200°C for the gypsum hydrate and another at about 500°C for calcium hydroxide.[34] Comparison of the samples with and without the admixture at 4 hours indicates differences. The endothermal peaks at about 200°C indicate low sulfoaluminate formation in the presence of the admixture. Even at 1 hour, low sulfoaluminate hydrate had formed (not shown in the figure). By the addition of 0.4% SO_3, there was no formation of low sulfoaluminate at 4 hours. These results suggest that, in the presence of the admixture, SO_3 consumption was accelerated, and an insufficient amount of SO_3 was present in the solution to control C_3A hydration, leading to the quick setting.

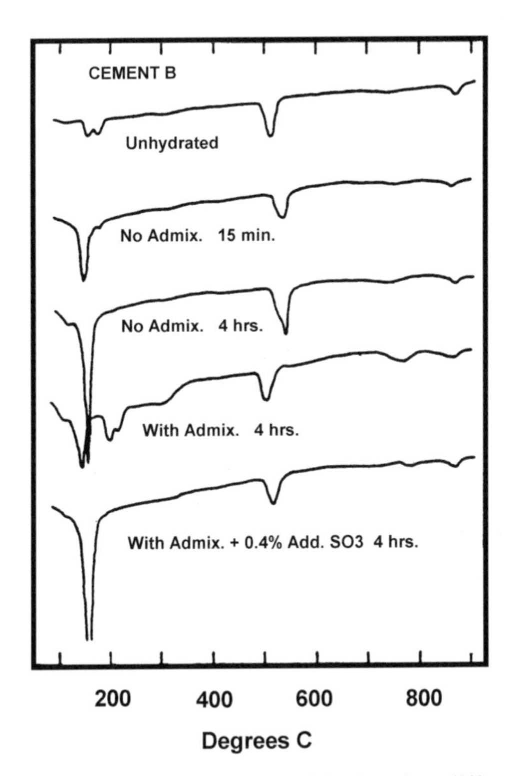

Figure 22. DTA curves of cement, unhydrated and hydrated, at different times with or without the water-reducing admixtures.

In another example, conduction calorimetry was use to explain a flash set in an oil cement slurry containing a lignosulfonate admixture.[22] Conduction calorimetric curves were obtained for systems such as portland cement, C_3A-gypsum-lime, and C_4AF-lime-gypsum systems containing refined or unrefined lignosulfonates. The results showed that unrefined lignosulfonate destabilizes the hydration process of the C_4AF-gypsum-lime mixture causing a flash set. In the conduction calorimetry, the C_4AF-10% gypsum-5% lime with 1–2% unrefined lignosulfonate (UL) dramatically increases the intensity of the very early ettringite peak and reduces the peak intensity for C_3S hydration at about 9–10 hours (Fig. 23). The induction period is practically absent. It was concluded that the xylose present in lignosulfonate was responsible for the flash set.

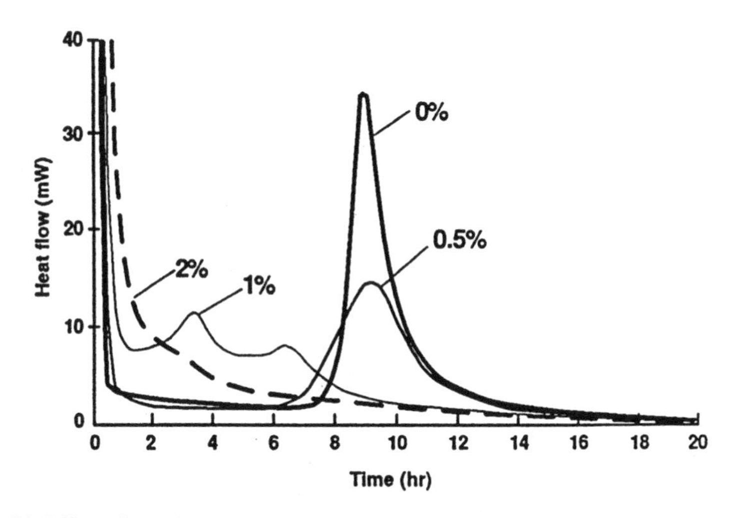

Figure 23. Effect of unrefined lignosulfonate on the hydration of C_4AF-10% gypsum- 5% lime system.

10.0 READY-MIX CONCRETE

In ready-mix concrete trucks, the unused concrete is washed out and this poses environmental, waste, and other problems. A novel approach consists of adding a retarder-based admixture, such as carboxylic acid or phosphorous-based organic acids and salts, to the concrete mix in the truck.[35] This stabilizes the concrete from further hydration. When an activator is added, subsequently normal hydration occurs.

In Fig. 24, calorimetric investigation shows the effect of the stabilizer.[35] Normal cement (cell 1) shows an exothermal peak at about 10 hours. Cell 2 refers to the mix to which the stabilizer is added with the mix water. The hydration is retarded and the peak shifts to 30–40 hours. Addition of a stabilizer, 2 hours after the mix was prepared, shows that hydration is almost stopped (cell 3). The sample in cell 4 was stabilized at 6 hours after the initial injection of mix water. The hydration is retarded for about 24 hours. When the activator is added at 24 hours reaction was initiated.

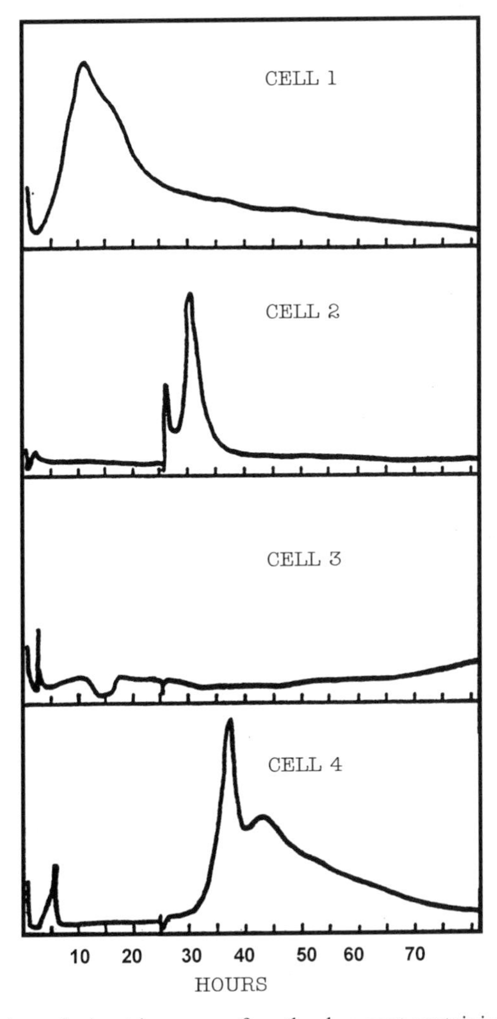

Figure 24. Conduction calorimetric curves of portland cement containing a stabilizer and an activator.

11.0 OTHER ADMIXTURES

Many chemicals and waste products may act as retarders. Thermoanalytical techniques are very useful to assess the effect of these products on the hydration of cements. A few examples are described below.

Copper refining wastes containing Cr and Sb were added to cements and the rate of hydration was followed by DTA, TG, and DTG. There was a significant decrease in the amount of $Ca(OH)_2$ in pastes hydrated with the copper refining waste, suggesting that it is a good retarder.[36]

The metal plating waste products from a zinc plating process were added to portland cement and the hydration was followed by conduction calorimetry. At an addition level of 3–10%, the rate of hydration was greatly reduced.[37]

In an investigation of various polymers, it was found that sodium carboxy methyl cellulose and versicol acted as strong retarders. For example, with the reference portland cement pastes, the peak in the conduction calorimetry occurred at 11 hours, and with versicol, the peak shifted to 20, 29, and 34 hours at dosages of 0.25, 0.50, and 0.75% respectively.[38]

Zinc oxide, in small amounts, retards the setting of cement very effectively. Calorimteric investigation has shown that, at an addition of 0.25% ZnO, hydration is almost nil up to 12 hours, and, at 1% addition, hydration is not initiated even up to 2 days.[39]

Thermoanalytical investigations have shown that potassium chromate retards the hydration of C_3A.[39]

12.0 IDENTIFICATION OF WATER REDUCERS/ RETARDERS

Admixtures, heated in the presence or absence of air, yield peaks that may be used to characterize them. Even within the same admixture group, thermograms may show differences that could be applied for quality control and identification purposes.

In a study of various water reducers, Gupta and Smith[40] applied DTA and TG techniques. They obtained characteristic peaks depending on the type of water reducer. Lignosulfonates decomposed at around 700°C in oxygen. Three stage decomposition occurred in TG curves. Most lignosulfonates gave a common peak in the region of 400–500°C. DTA showed one endothermic peak and three exothermic peaks.

Sodium salts of gluconic acid and α-glucoheptonic acid indicate different thermal behaviors from lignosulfonate.[40] An exothermic peak of large intensity was evident at 710–850°C for gluconate whereas Na α-glucoheptonate gave an exothermic peak at 800°C.

An extensive investigation was carried out by Milestone on various types of admixtures by subjecting them to DTA in the presence of oxygen.[41] DTA of admixtures based on lignosulfonates is given in Fig. 25.[41] Although the overall pattern of the thermograms is similar, no two samples had exactly the same pattern. Hence, it was concluded that the pattern serves as a fingerprint for identification of these admixtures. The temperature of the exotherm, number of exotherms, and the intensity of the exotherms can be used for differentiating one lignosulfonate from the other.

Hydroxycarboxylic acids and their salts also show characteristic peaks in the thermograms. It was found that the high temperature exothermal peaks (500–700°C) were common for many hydroxycarboxylic acids salts, but they are absent in pure acids. It was also observed that two admixtures, supposedly identical, in fact, had different thermal patterns.[41]

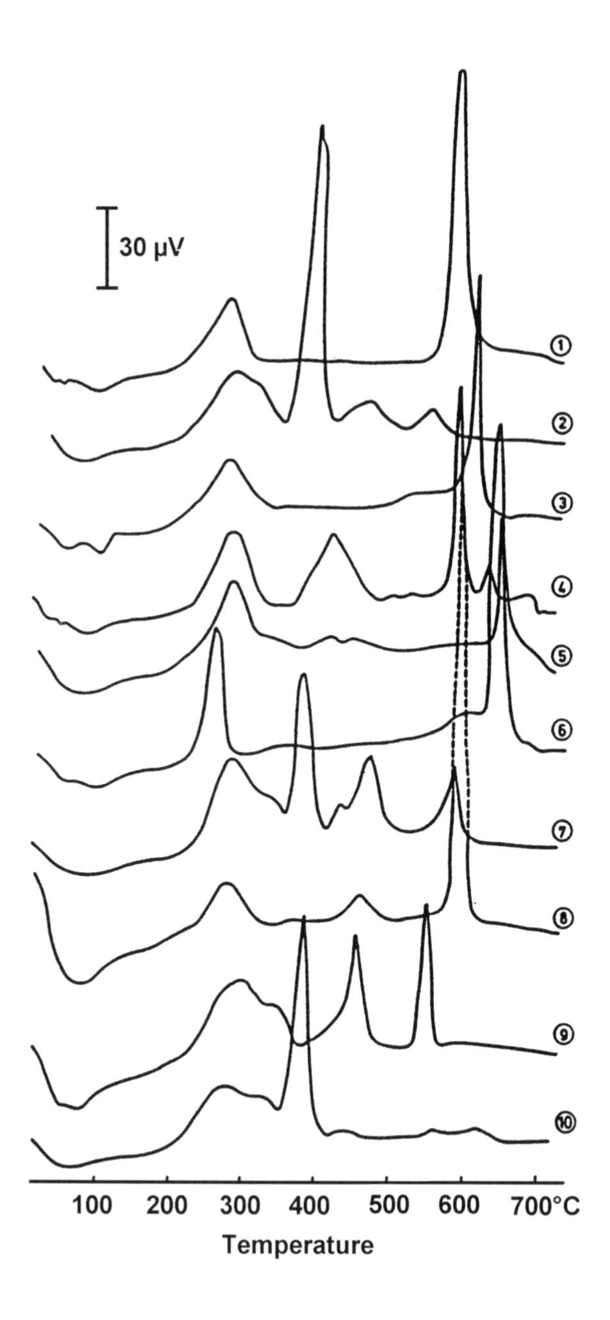

Figure 25. DTA curves of various lignosulfonates.

REFERENCES

1. Ramachandran, V. S., Elucidation of the Role of Chemical Admixtures in Hydrating Cements by DTA Technique, *Thermochimica Acta,* 3:343–366 (1972)
2. Ramachandran, V. S., and Feldman, R. F., Effect of Calcium Lignosulfonate on the Tricalcium Aluminate and Its Hydration Products, *Mater. Const.*, 5:67–76 (1972)
3. Massazza, E., and Costa, U., Effect of Superplasticizers on C_3A Hydration, *7th Int. Symp. Chem. Cements, Paris,* 4:529–534 (1980)
4. Young, J. F., Hydration of Tricalcium Aluminate with Lignosulfonate Additives, *Mag. Concr. Res.*, 14:137–142 (1962)
5. Bhatty, J. I., A Review of the Application of Thermal Analysis to Cement-Admixture Systems, *Thermochimica Acta*, 189:313–350 (1991)
6. Collepardi, M., Manosi, M., and Moriconi, G., Combined Effect of Lignosulfonate and Carbonate on Pure Portland Cement Clinker Compound Hydration, I: Tetracalcium Aluminoferrite Hydrate, *Cement Concr. Res.,* 10:455–462 (1980)
7. Lorprayon, V., and Rossington, D. R., Early Hydration of Cement Constituents with Organic Admixtures, *Cement Concr. Res.,* 11:267–277 (1981)
8. Ramachandran, V. S., Effect of Sugar-Free Lignosulfonates on Cement Hydration, *Zement-Kalk-Gips,* 31:206–210 (1978)
9. Seligman, P., and Greening, N. R., Studies of Early Hydration Reactions of Portland Cement by X-ray Diffraction, *Highway Res. Rec.,* 62:80–105 (1964)
10. Milestone, N. B., Hydration of Tricalcium Silicate in the Presence of Lignosulfonates, Glucose and Sodium Gluconate, *J. Am. Ceram. Soc.,* 62:321–324 (1979)
11. Ramachandran, V. S., Interaction of Calcium Lignosulfonate with Tricalcium Silicate, Hydrated Tricalcium Silicate and Calcium Hydroxide, *Cement Concr. Res.,* 2:179–194 (1972)
12. Odler, I., and Becker, T., Effect of Some Liquefying Agents on the Properties and Hydration of Portland Cement and Tricalcium Silicate, *Cement Concr. Res.,* 10:321–331 (1980)
13. Collepardi, M., Manosi, S., Moriconi, G., and Pauri, M., *Concr. Admixtures Handbook* (V. S. Ramachandran, ed.), pp. 326, 329, 331, Noyes Publ., NJ (1995)
14. Ramachandran, V. S., *Concr. Admixtures Handbook,* pp. 334, 337, 338, Noyes Publ., NJ (1995)

15. Ramachandran, V. S., Feldman, R. F., and Beaudoin, J. J., *Concr. Science,* p. 427, Heyden & Sons, London (1981)
16. Ramachandran, V. S., Differential Thermal Investigation of the System Tricalcium Silicate-Calcium Lignosulfonate-Water in the Presence of Tricalcium Aluminate and its Hydrates, *3rd Int. Conf. Therm. Analysis, Davos, Switzerland,* pp. 255–267 (1971)
17. Monosi, S., Moriconi, G., and Collepardi, M., Combined Effect of Lignosulfonate and Carbonate on Pure Portland Clinker Compounds Hydration, III: Hydration of Tricalcium Silicate Alone and in the Presence of Tricalcium Aluminate, *Cement Concr. Res.,* 12:415–424 (1982)
18. Rixom, M. R., *Chemical Admixtures for Concrete,* E. & F. N. Spon Ltd., London (1978)
19. Khalil, S. M., and Ward, M. A., Influence of Lignin-Based Admixture on the Hydration of Portland Cement, *Cement Concr. Res.,* 3:677–688 (1973)
20. Khalil, S. M., and Ward, M. A., Influence of SO_3 and C_3A on the Early Reaction Rates of Portland Cement in the Presence of Calcium Lignosulfonate, *Am. Ceram. Soc. Bull.,* 57:1116–1122 (1978)
21. Jawed, I., Klemm, W. A., and Skalny, J., Hydration of Cement-Lignosulfonate-Alkali Carbonate System, *J. Am. Ceram. Soc.,* 62:461–464 (1979)
22. Ramachandran, V. S., and Beaudoin, J. J., *Handbook of Analytical Techniques in Concrete Science and Technology,* William Andrew Publishing, Norwich, New York, p. 964 (2001); Michaux, M., and Nelson, E. B., Flash Set Behavior of Oil Well Cement Slurries Containing Lignosulfonates, *9th Int. Congr. Cements, New Delhi, India,* 4:584–590 (1992)
23. Chatterji, S., Electron-Optical and X-Ray Diffraction Investigation of the Effect of Lignosulfonates on the Hydration of C_3A, *Indian Concr. J.,* 41:151–160 (1967)
24. Pauri, M., Ferrari, G., and Collepardi, M., Combined Effect of Lignosulfonate and Carbonate on Pure Portland Clinker Compound Hydration, IV: Hydration of Tricalcium Aluminate–Sodium Hydroxide Solid Solution, *Cement Concr. Res.,* 13:61–68 (1983)
25. Dunster, A. M., Kendrick, D. P., and Personage, J. R., The Mechanism of Hardening and Hydration of White Portland Cement Admixed with Salicylic Aldehyde, *Cement Concr. Res.,* 24:542–550 (1994)
26. Diamond, S., Salicylic Acid Retardation of C_3A, *J. Am. Ceram. Soc.,* 55:177–180 (1972)
27. Young, J. F., The Influence of Sugars on the Hydration of Tricalcium Aluminate, *Proc. 5th Int. Symp. on the Chem. Cements, Tokyo,* 2:256–267 (1968)

28. Milestone, N. B., The Effect of Glucose and Some Glucose Oxidation Products on the Hydration of Tricalcium Aluminate, *Cement Concr. Res.*, 7:45–52 (1967)
29. Sarsale, R., Sabetelli, V., and Valenti., G. L., Influence of Some Retarders on the Hydration at Early Ages of Tricalcium Aluminate, *7th Int. Symp. Chem. of Cements, Paris,* 4:546–551 (1980)
30. Ramachandran, V. S., Lowery, M. S., Wise, T., and Polomark, G. M., The Role of Phosphonates in the Hydration of Portland Cement, *Mater. Struct.,* 26:425–432 (1993)
31. Ramachandran, V. S., and Lowery, M. S., Effect of Phosphonate-Based Compound on the Hydration of Cement and Cement Compounds, *4th CANMET/ACI Int. Conf. on Superplasticizers and Other Chem. Admixtures in Concr., Montreal, Canada,* SP-148:131–151 (1994)
32. Ramachandran, V. S., and Lowery, M. S., Conduction Calorimetric Investigation of the Effect of Retarders on the Hydration of Portland Cement, *Thermochimica Acta,* 195:373–387 (1994)
33. Ramachandran, V. S., Effect of Retarders/ Water Reducers on Slump Loss in Superplasticized Concrete, Developments in the Use of Superplasticizers, *Am. Concr. Inst. Special Publ.,* 68:393–407 (1981)
34. Greene, K. T., A Setting Problem Involving White Cement and Admixture, *Transportation Res. Rec.,* 564:21–26 (1976)
35. Kinney, F. D., Re-Use of Returned Concrete by Hydration Control: Characterization of a New Concept, *3rd CANMET/ACI Int. Conf. of Superplasticizers and Other Admixtures in Concr.,* SP-119:19–40, Am. Concr. Res. Inst. Special Publ. (1989)
36. Zivica, V., Stabilization of Copper Refining Waste in Cement Matrix Using Thermal Analysis, *J. Therm. Analysis,* 46:611–617 (1996)
37. Hills, C. D., Koe, L., Collars, C. J., and Perry, R., Early Heat of Hydration During Solidification of a Metal Plating Sludge, *Cement Concr. Res.,* 22:822–832 (1992)
38. Olmez, H., Dollimore, D., Gamlen, G. A., and Mangabhai, R. J., The Mechanical Properties of Polymer-Modified OPC/PFA Paste, *Proc. CANMET/ACI 1st Int. Conf. on Use of Fly Ash, Silica Fume, Slag and Other Mineral By-products in Concr.,* ACI-SP-79:623 (1983)
39. Ghosh, S. N., *Advances in Cement Technology*, p. 804, Pergamon Press, Oxford (1983)
40. Gupta, J. P., and Smith, J. I. H., Thermoanalytical Investigation of Some Cement Admixtures, *2nd European Symp. Therm. Analysis, Aberdeen,* pp. 483–484, Heyden Press, London (1981)
41. Milestone, N. B., Identification of Concrete Admixtures by Differential Thermal Analysis, *Cement Concr. Res.,* 14:207–214 (1984)

7

Superplasticizing Admixtures

1.0 INTRODUCTION

The advantages derived from the use of superplasticizing admixtures include production of concrete having high workability and production of high strength concrete having normal workability and lower water content. Another possible application is the fabrication of concrete with less cement, having normal strength and workability. Superplasticizers have also been used with advantage in concretes containing fly ashes, silica fume, rice husk ash, slags, etc.

A study of the hydration of cement and cement compounds in the presence of superplasticizers is useful for theoretical and practical considerations. Many types of thermal techniques including DTA, DSC, TG, DTG, Conduction Calorimetry, and EGA have been used for such studies. They have yielded important results that could be correlated with physical and mechanical characteristics of cement systems. Investigations have included the measurement of heat of hydration, the mechanism of reactions, strength development, microstructure, permeability, durability aspects, compatibility problems between cement and superplasticizers, the prediction of some properties, material characterization and selection, mathematical modeling of hydration, development of test methods, and cement-superplasticizer interactions.

2.0 TRICALCIUM ALUMINATE

There is general agreement that the superplasticizers based on sulfonated melamine formaldehyde (SMF) and sulfonated naphthalene formaldehyde (SNF) retard the hydration of tricalcium aluminate.[1]–[3] The influence of a superplasticizer depends on its dosage, molecular weight, the type of cation associated with it, the water/cement ratio, and the temperature. Figure 1 compares the conduction calorimetric curves of C_3A hydrated in the presence of 2% SMF and that without the admixture. The results show that, within a few seconds of contact with water, a rapid rate of heat development occurs with a peak at about 8–9 minutes for the sample with no admixture.[4][5] In the specimen containing SMF, the total amount of heat generated in the first 30 minutes and the amplitude of the peak are lower in comparison with that without SMF.

DSC curves demonstrate that C_3A hydration is retarded in the presence of SMF. It is also seen from the DSC curves (Fig. 2) that the hydration of C_3A to the hexagonal phases, as well as the conversion of the hexagonal to the cubic phase, are retarded at SMF dosages of 1, 2, and 4%.[1] At 1 day, C_3A containing 0 or 4% SMF shows a large endothermal peak at about 320–340°C for the presence of the cubic phase. The corresponding peak for the samples containing 2 or 4% SMF is of lower amplitude. The endothermic peak at about 170–185°C is indicative of the presence of the hexagonal phase. It is evident that higher amounts of SMF retard the hydration of C_3A and the conversion of the hexagonal to the cubic phase more efficiently than the lower dosages of this admixture.[6] There has been some evidence that the SNF admixture does not retard the hydration of C_3A as efficiently as SMF.

Collepardi, et al.,[7] applied DTA and XRD to the study of the hydration of $C_3A + Ca(OH)_2$ mixtures containing 0.6% SNF. There were no substantial changes in the rate of hydration of this mixture in the presence of SNF.

The adsorption and dispersion effects of superplasticizers on the aluminate phase have been studied.[1][5] Adsorption of SMF on C_3A occurs as soon as the solution comes into contact with the solid. The rate and the amount of adsorption on C_3A far exceeds that on the C_3S phase or cement (Fig. 3).[1][5] Even the hexagonal phase adsorbs large amounts of SMF. Adsorption is irreversible, indicating that a chemical interaction occurs between the hydrating C_3A and SMF. Thus, the retardation of C_3A may be explained by adsorption of SMF on the hydrating C_3A surface. Similar

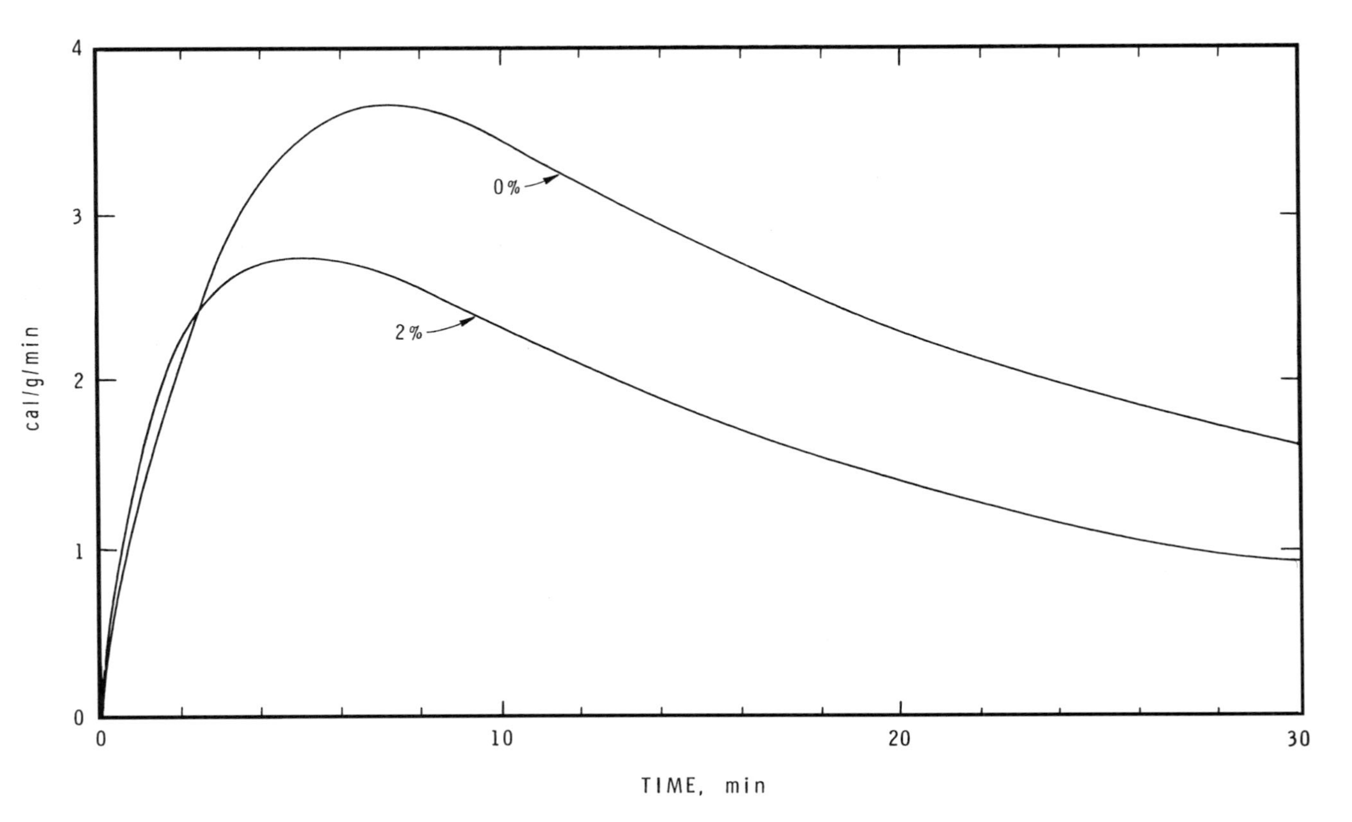

Figure 1. Conduction calorimetric curves of C_3A hydrated in the presence of SMF.

conclusions have been drawn from the thermal analysis data on the water-reducing admixture-C_3A interactions. The retarding effect of dicarboxylic acid ester, lignin modified melamine formaldehyde condensate, and SNF have also been investigated.[8]

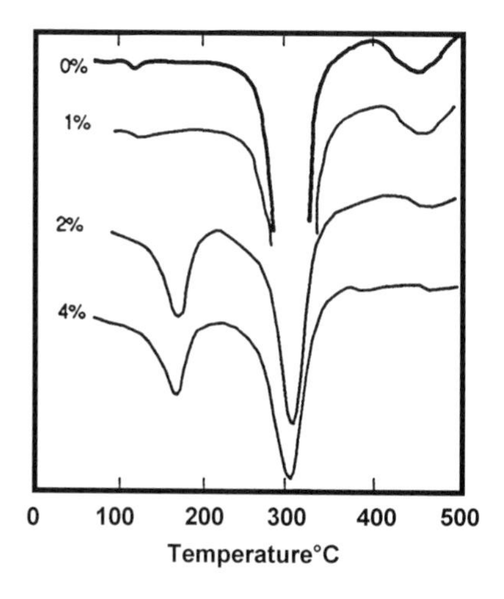

Figure 2. DSC curves of C_3A hydrated in the presence of SMF.

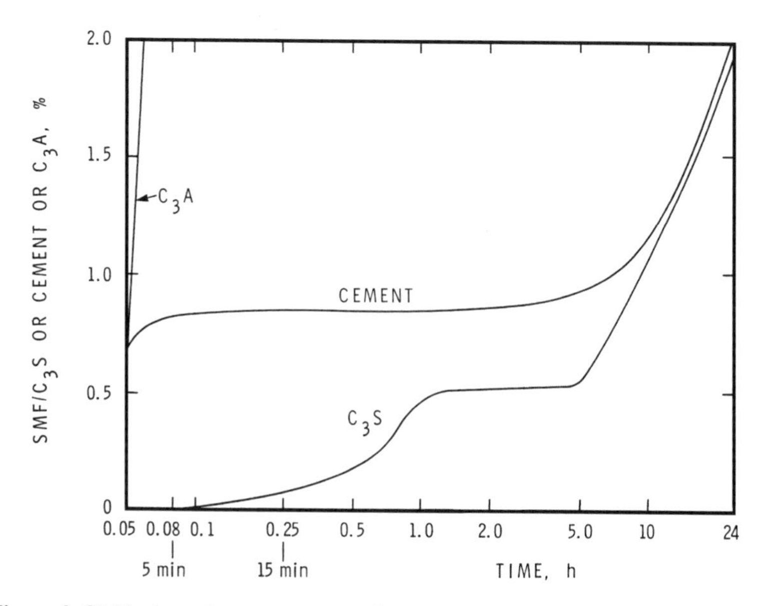

Figure 3. SMF adsorption on cement and cement compounds during hydration.

The rate of hydration of C_4AF is retarded in the presence of superplasticizers.[8] In an investigation involving conduction calorimetry, Gagne[10] found that, in the presence of SNF and SNF + SO_4^{2-}, the hydration of tetracalcium aluminoferrite was retarded during the 60 minutes of hydration.

Monocalcium aluminate, a major component of high alumina cement, hydrates to metastable products CAH_{10}, C_4AH_{13}, and C_2AH_8 which eventually convert to the stable cubic aluminum hydrate, C_3AH_6. In the presence of 2–4% SMF, SNF, or modified lignosulfonate, the DTA results have shown that the degree of conversion of CAH_{10} and C_2AH_8 phases to the cubic phase is marginally retarded.[9]

3.0 TRICALCIUM ALUMINATE-GYPSUM SYSTEM

The role of the superplasticizer on the rate of hydration of C_3A-gypsum mixtures is not clear. All the possibilities, viz., retardation, acceleration, and ineffectiveness have been reported. The variation in results reported from different sources may be due to the differing conditions of hydration such as the w/s ratio, the type of superplasticizer, its molecular weight, the type of cation associated with it, the proportion and particle size of the C_3A and gypsum, the type of sulfate, and the temperature. There is, however, general agreement that the conversion of ettringite to monosulfate is retarded in the presence of a superplasticizer.

The effect of 0, 1, 2, and 4% SMF on the hydration of C_3A + 25% gypsum may be followed by conduction calorimetry (Fig. 4).[1] An exothermic peak appears at 20 hours with 1% SMF, at about 23 hours with 2% SMF, and at about 15 hours with 4% SMF. A continuous hump is registered for the sample containing no admixture. The retardation effect of C_3A-ettringite reaction to produce low sulfoaluminate may be caused by the adsorption of SMF on the hydrating C_3A surface.

Slanicka[3] used XRD and DTA techniques to follow the hydration of C_3A-gypsum mixtures containing 1–4% SMF. The mixture was hydrated for 2 hours to 24 hours, and the products were estimated by DTA and XRD. Table 1 shows the amounts of ettringite and monosulfate hydrate formed at different times of hydration in the presence of 4% SMF. It is obvious that the formation of ettringite is almost nil at 2 hours in the

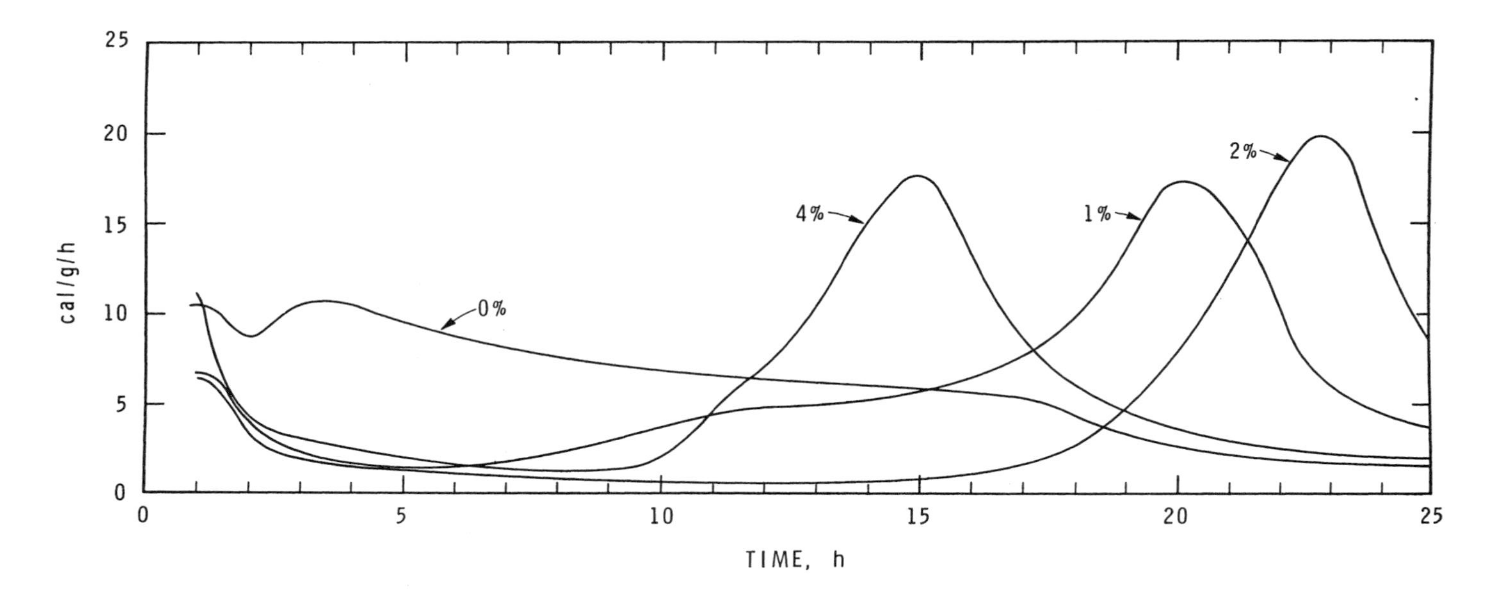

Figure 4. Conduction calorimetric curves of C_3A + gypsum + H_2O containing SMF.

presence of SMF, but, at 6 hours, there is more ettringite in this mixture than in the reference sample. More monosulfate is found at 6 hours in the reference than in the sample with 4% SMF. The rates of these reactions are dependent on the dosages of the admixture. The lower amounts of monosulfate formed at 6 hours and 48 hours in the presence of SMF versus reference samples suggest that the conversion of ettringite to the monosulfate form is retarded by SMF.

Table 1. Amounts of Ettringite and Monosulfate Hydrate Formed in the C_3A-$CaSO_4$•$2H_2O$-H_2O System in the Presence of SMF Superplasticizer

Hydration Time (Hrs)	Ettringite		Monosulfate Hydrate	
	Reference	SMF (4%)	Reference	SMF (4%)
2	28.3	—	0	—
6	17.6	24.4	5.5	3.5
24	12.3	—	13.2	—
48	10.0	18.8	15.0	9.9

Massazza and Costa[11] applied the DSC, XRD, and conduction calorimetric techniques to investigate the effect of a SNF type of superplasticizer on the hydration of C_3A-gypsum mixtures. The results indicated that SNF retards the formation of ettringite and its conversion to the monosulfate form.

DTG technique has been adopted for the investigation of the C_3A-$Ca(OH)_2$-gypsum system containing SNF hydrated for different periods.[7] It was found that at a dosage of 0.6% SNF by weight of C_3A, the rate of ettringite formation and its conversion to monosulfate was not affected (Fig. 5). The peak at 130°C is attributed to ettringite and that at 160°C to gypsum. It can be seen that the intensities of these peaks are of the same magnitude in the reference as well as in the mixtures containing SNF. The conversion of ettringite to monosulfate is identified by a peak at 220–270°C. The intensity of the peak remains the same in the reference and that containing the superplasticizer. Conduction calorimetric studies confirmed these observations.

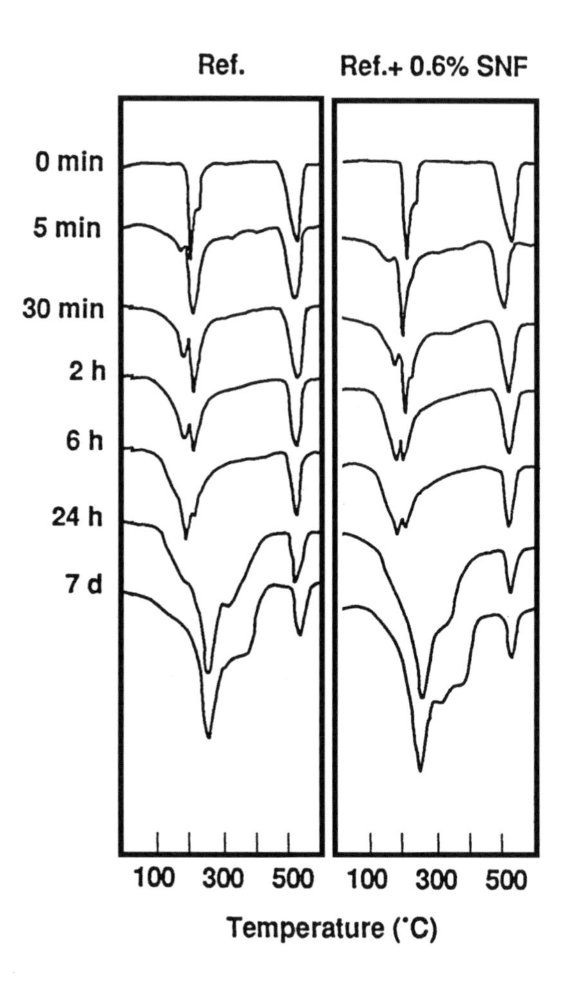

Figure 5. DTG of the mixture C_3A-$Ca(OH)_2$-gypsum hydrated for different periods in the presence of SNF admixture.

The conduction calorimetric work and adsorption studies indicate that the retarding action of SMF and SNF in the C_3A-gypsum system is related to the preferential adsorption of superplasticizer molecules on the C_3A and the competition of SO_4^{-2} ions in the early stages of hydration.[1][10]

4.0 TRICALCIUM SILICATE

Superplasticizers affect the hydration of tricalcium silicate. It is generally known that superplasticizers retard the hydration of C_3S.[4][12]–[15] The effect of different amounts of SMF on the hydration of C_3S can be illustrated by conduction calorimetric curves (Fig. 6).[4][5] By the addition of 1% SMF, the induction period and the exothermic peak are shifted to higher times suggesting that substantial retardation has taken place. At 2 and 4% SMF, further retardation of hydration occurs with a very low rate of heat development. There is evidence that the CaO/SiO_2 ratio of the C-S-H product is changed by SMF.

The effect of superplasticizers on the hydration of C_3S has also been studied by estimating calcium hydroxide formed at different times by applying DTA and TG.[16] Contrary to many other investigations, it is reported in this work that superplasticizer does not affect the hydration of C_3S. It is not easy to compare the results obtained by different investigators as the rate of hydration depends on the type of superplasticizer, the cation associated with it, the molecular weight, the amount and hydration conditions, and the time at which the hydration degree is compared.

The TG technique has been used to follow the hydration of alite in the presence of 2–4% SMF.[16] Calcium hydroxide and the non-evaporable water contents were determined and not much difference in the hydration rates was observed between the reference alite sample and that containing the superplasticizer.

The effect of 0, 0.3, 1.0, and 3% SMF on the hydration of C_3S has been investigated using the XRD and Franke methods of determining lime.[15] At 12 hours, the degree of hydration, in terms of C_3S reacted, was only 8% in the presence of the superplasticizer whereas for the reference material the value was 34%. Thus, a significant amount of retardation seems to have taken place. Thermal methods have also been applied to determine the amount of lime formed at different times in the presence of 0, 0.32, and 1.0% SMF. The results are summarized in Table 2. At 1 day, there is retardation in the formation of lime, and a severe retardation is evident with 1% SMF. At 3 days and beyond, a slight acceleration seems to occur with SMF.

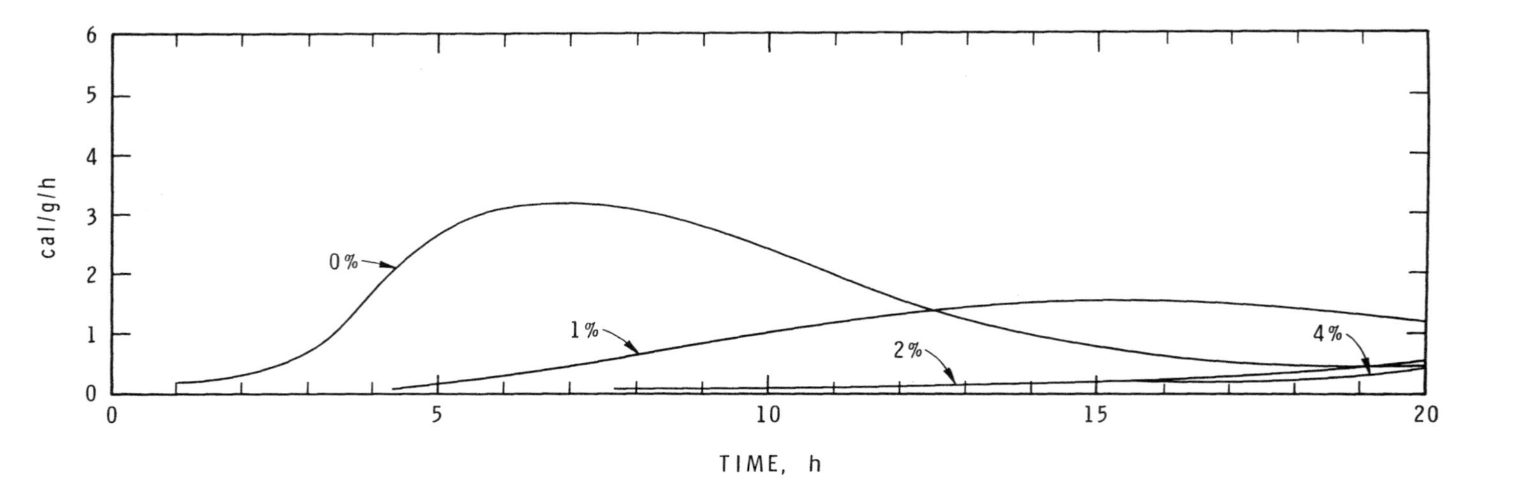

Figure 6. The influence of SMF on the conduction calorimetric curves of C_3S hydration.

Table 2. Amount of Lime Formed in the Hydration of C_3S containing SMF

Hydration Time, Days	0% SMF	0.32% SMF	1% SMF
1	13.0	11.0	1.4
3	16.5	17.2	17.2
7	19.9	21.0	20.9
28	21.9	22.0	22.3

The effect of sodium SNF of different polycondensation degrees (n) on C_3S hydration has been examined by the microcalorimetric technique. Typical results are shown in Table 3.[17] The first peak refers to that occurring during the pre-induction period and the second to that occurring during the acceleratory period. The rate of heat development (dQ/dt) (1st peak) is increased at $n > 2$. The second peak is of lower intensity and appears at longer times as n is increased. This retardation is due to the adsorption effect of the superplasticizer. The minimal effect for the first peak is attributed to the ineffectiveness of SNF towards the rate of interaction between C_3S and water. Although there is significant retardation in the appearance of the second peak, the total heat corresponding to this peak is higher in all samples with $n>2$. The DTG results confirm those of the calorimeter.

Table 3. Effect of Condensation Degree of SNF on the Thermal Characteristics of C_3S Hydration

n	Rate of Heat (dQ/dt)	Time for the Max Peak Effect		Heat Developed (Main Peak) kJ/kg (Hrs)
		1st Peak	2nd Peak	
1	1.7	4.3	18	71
2	1.6	4.7	15	51
5	4.3	2.9	155	127
7	4.3	1.25	278	180
17	4.5	0.72	350	317

The relative effects of superplasticizers on the hydration of C_3S has been compared in the presence of 1% SMF, 1% SNF, and 0.25% LS (lignosulfonate). The rate of hydration was determined by TG technique (Fig. 7).[18] The amount of calcium hydroxide formed in pastes with these admixtures was distinctly lower than those in reference samples, indicating significant retardation effects. It was also found that the C-S-H formed in the presence of the admixtures had a higher C/S ratio. It is evident from the figure that the retardation effect of even 0.25% LS is significant.

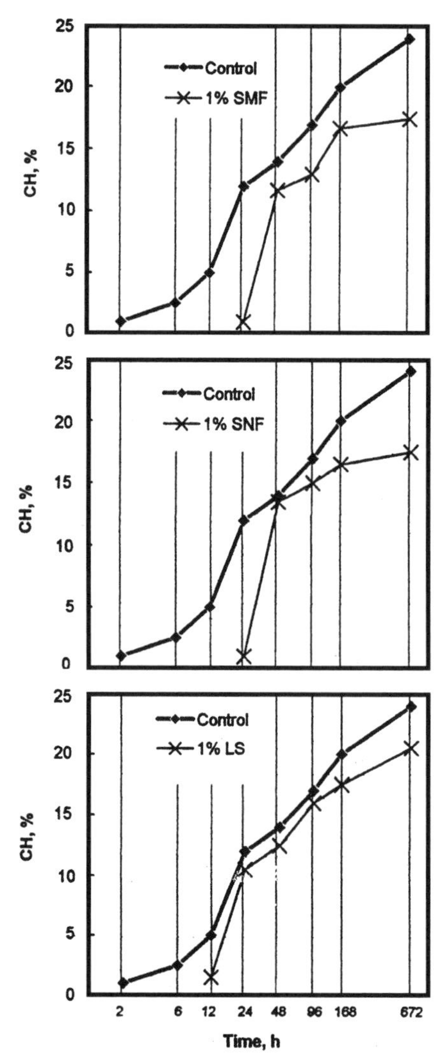

Figure 7. Effect of SMF, SNF, and LS on the amount of $Ca(OH)_2$ formed in the hydration of C_3S.

Not much work has been reported on the effect of superplasticizers on the hydration of C_2S. It has been reported that substantial retardation occurs in the hydration of C_2S at all dosages of SMF.[13]

5.0 CEMENT

The characteristics of cement pastes or concrete from the initial slump to strength and durability are directly or indirectly related to the type, the rate formation, and the amount of the hydration products. Several factors have to be taken into consideration in the elucidation of the role of superplasticizers on the hydration of cement. Some of the factors that affect the rate of hydration include the type of cement, the characteristics of the ions in the solution (sulfate, OH, and alkalis), the particle size/surface area of cement, the relative amounts of the individual phases of cement, the dosage of the superplasticizer, the cation associated with the superplasticizer, the type of superplasticizer, its molecular weight, the dosage, the temperature of hydration, and the water:cement ratio.

In cements, the hydration of C_3S is retarded, but not to the same extent as when it is used in a pure form. In cement, part of the superplasticizer is adsorbed by the C_3A phase so that only smaller amounts are available to retard the hydration of the silicate phase. The rate of hydration of cement in the presence of superplasticizers has been studied by many thermal techniques. The effect of SMF on the hydration of cement can be followed by conduction calorimetry (Fig. 8).[4][5] The peak occurring at 5 hours in the reference sample is reduced and appears at later times. Retardation of the appearance of the peak is increased with the amount of SMF added. At 4% SMF, the peak effect becomes very broad and extends over a longer period.

The influence of superplasticizers on the hydration of cement may also be determined by TG by estimating the amount of calcium hydroxide formed at different times. In Table 4, the amounts of calcium hydroxide formed in the presence of 1 or 2% SMF are given.[5][19] At earlier times, the hydration is significantly reduced by 2% SMF up to 24 hours. After 24 hours, the difference in values is marginal between the reference sample and that containing SMF. The retardation effects of superplasticizers have also been confirmed by XRD, hydration degree, setting times, and conductivity investigations.[20]–[24]

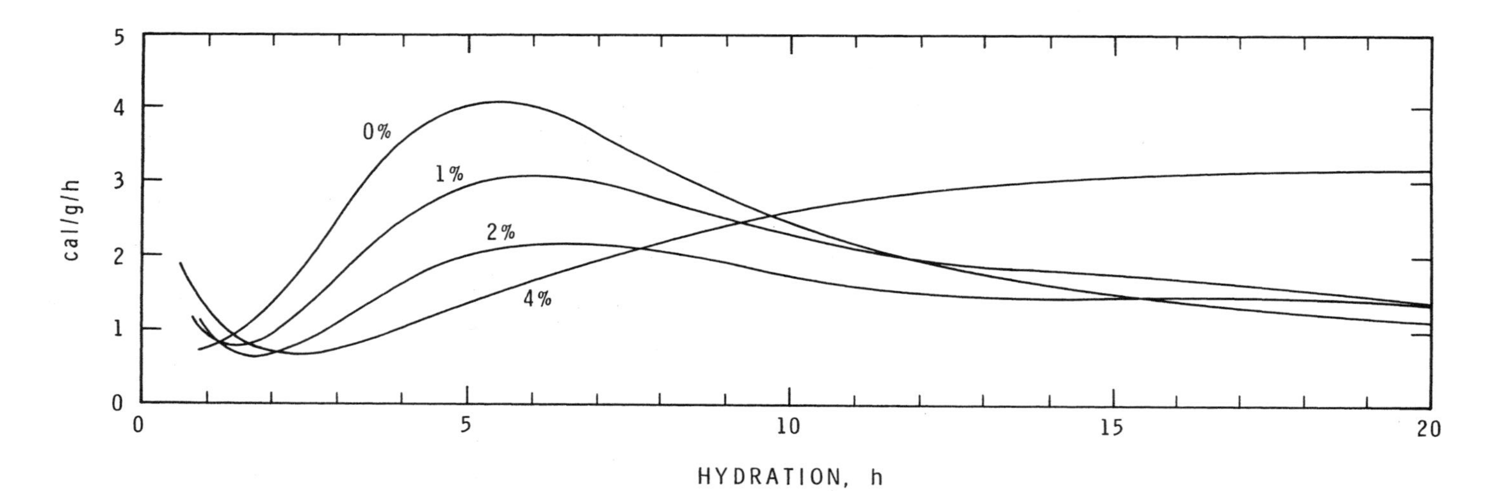

Figure 8. Conduction calorimetric curves of portland cement hydrated in the presence of SMF.

Table 4. Amounts of Calcium Hydroxide Formed in a Cement Paste Hydrated in the Presence of SMF Superplasticizer

Mixture	3 h	6 h	9 h	24 h	3 d	7 d	28 d
Reference	2.3	6.8	8.0	13.7	17.8	19.3	19.8
Cem. + 1% SMF	2.1	6.4	7.3	13.1	17.6	19.4	19.6
Cem. + 2% SMF	1.9	3.3	4.0	12.9	17.1	18.8	19.4

The influence of different amounts of SNF on the hydration of cement has also been studied by conduction calorimetry (Fig. 9).[20] At 0.5% SNF, the peaks appear at earlier times. Improved dispersion is thought to be responsible for this acceleration effect. As the concentration of SNF increases, the retardation effect becomes evident by the appearance of peaks at later times. Uchikawa,[25] however, found only minor changes in the hydration kinetics when SNF admixtures were used.

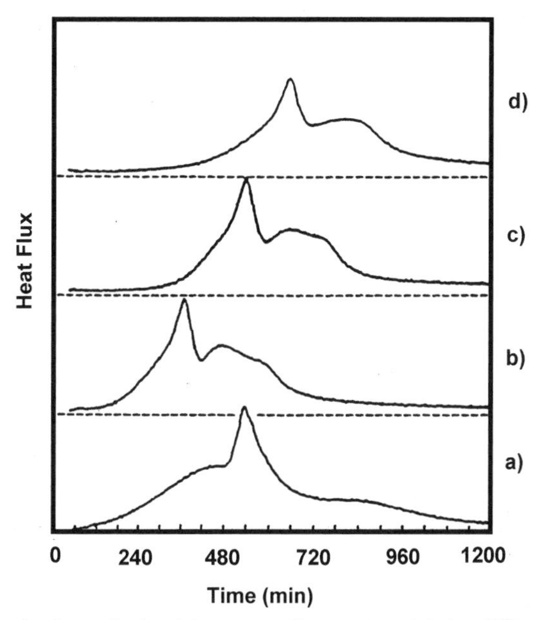

Figure 9. Conduction calorimetric curves of cement containing different amounts of SNF admixture.

The TG and impedance spectroscopic techniques have been applied to study the relative effects of three types of superplasticizers on the hydration of cement.[26] The amounts of calcium hydroxide formed were determined by TG in samples containing 1.5% of sodium-SNF, calcium-SNF, and sodium-SMF. The results are tabulated in Table 5. At 2 hours, there is a slight acceleration effect, but at 7–24 hours, there is a definite decrease in the amounts of $Ca(OH)_2$ formed in samples containing the superplasticizers. Of the admixtures studied sodium-SMF seems to retard the hydration to a greater extent than others.

Table 5. The Percentage of Lime Formed in the Presence of Superplasticizers

Time (Hrs)	Reference	Na-SNF	Ca-SNF	Na-SMF
2	1.09	1.12	1.13	1.01
7	3.76	2.06	1.65	1.10
24	10.06	7.87	9.28	5.46

(Reprinted with permission from Elsevier Science.)

Cabrera and Lynsdale[27] investigated the effect of superplasticizers based on modified lignosulfonate, SNF, SMF, and acrylic acid-hydroxypropyl methacrylate on the hydration of cement. The DTG/TG techniques were adopted. There was a slight acceleration of hydration in the presence of admixtures at 1 day, but the long-term hydration was not affected.

DTA can be used to investigate the role of different types of superplasticizers on cement hydration at early periods. In Fig. 10, the DTA curves of cement containing SMF, SNF, and modified lignosulfonate hydrated for different periods are shown.[6][18] Curve (*a*) is for the reference, curve (*b*) denotes the paste containing 1% SMF, curve (*c*) refers to that with 1% SNF, and curve (*d*), to that with 0.5% lignosulfonate. The ettringite peak occurs at 100°C. In the reference sample, the ettringite peak is sharp and increases with time. The corresponding peak, at 30 minutes for the pastes containing the superplasticizer, is stronger in intensity. This is evidence that these admixtures accelerate the formation of ettringite. In addition to the dispersion effect, the acceleration may also be related to the higher amounts of alkali sulfates present in the cement. In another cement containing a negligible amount of sulfate, the formation of ettringite was retarded.

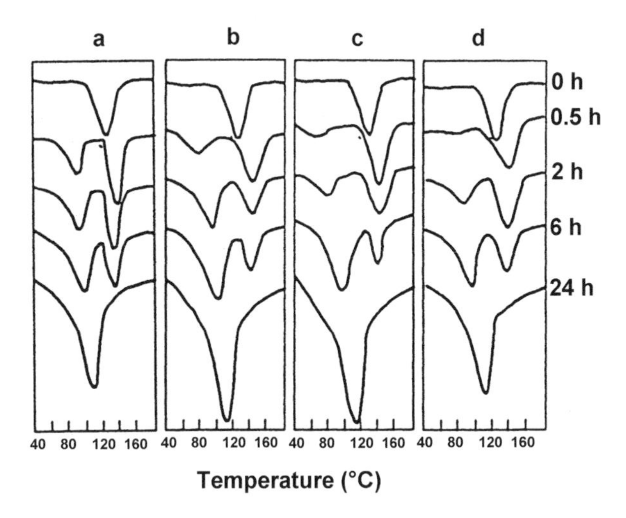

Figure 10. DTA curves for cement hydrated for different periods in the presence of superplasticizers.

An extensive work was carried out by Uchikawa[28] on the conduction calorimetry of superplasticizers, such as SNF (NS in the figure), lignosulfonate (LS), a co-polymer of acrylic acid with acrylic ester (PC), and a three dimensional polycondensate product of aromatic aminosulfonic acid with trimethyl phenol (AS) (Fig. 11). The first peak in the calorimetry corresponds to the heat of dissolution of alite, the heat of formation of the AFt phase, and the calcium hydroxide formation from free lime. The second peak corresponds to the heat of hydration of alite. The admixtures were found to accelerate the formation of the ettringite phase. At w/c ratios of 0.3 and 0.5 and a later addition of the admixture, the appearance of the second peak was significantly delayed and the peaks were of lower intensity. Most retardation occurred with polycarboxylic acid and amninosulfonic acid-based admixtures (Fig. 11). DSC was used to determine the amount of lime formed at different times. The DSC results show that the addition of admixtures at different w/c ratios generally decreases the amounts of lime in the presence of superplasticizers (Fig. 12).

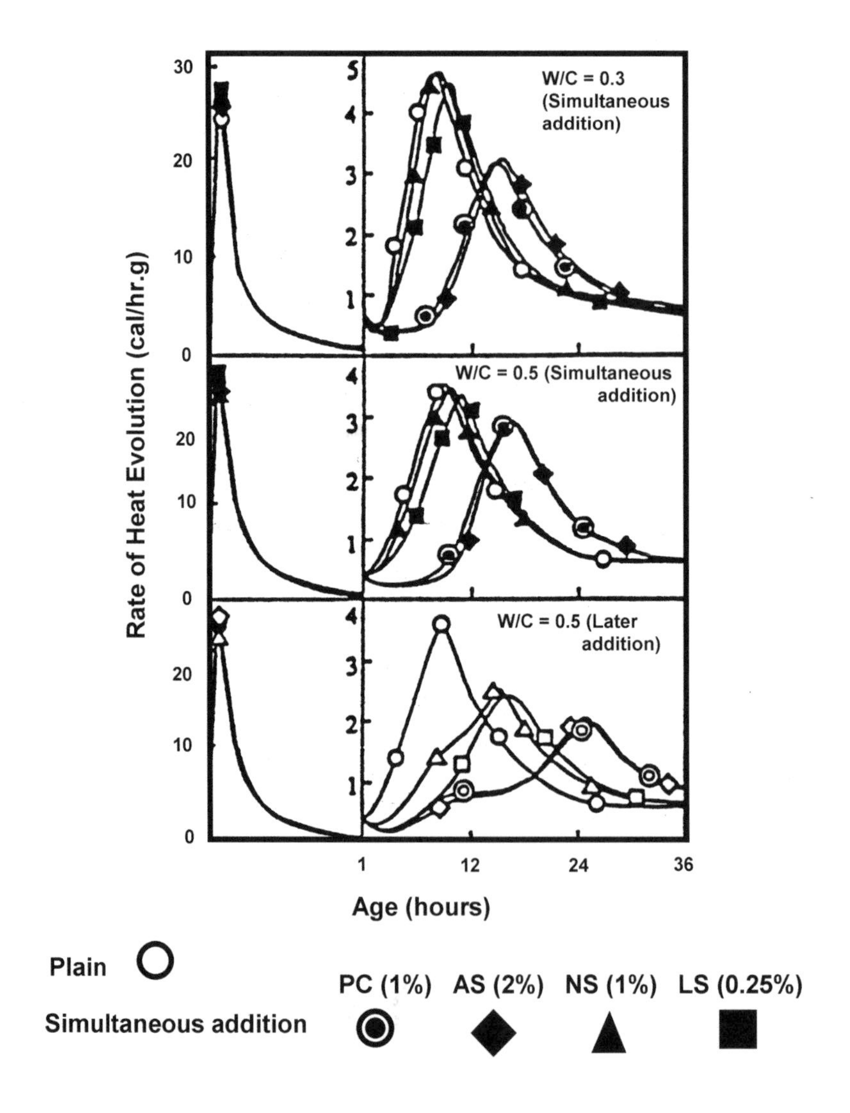

Figure 11. The influence of late addition on the heat effects of cement containing superplasticizers.

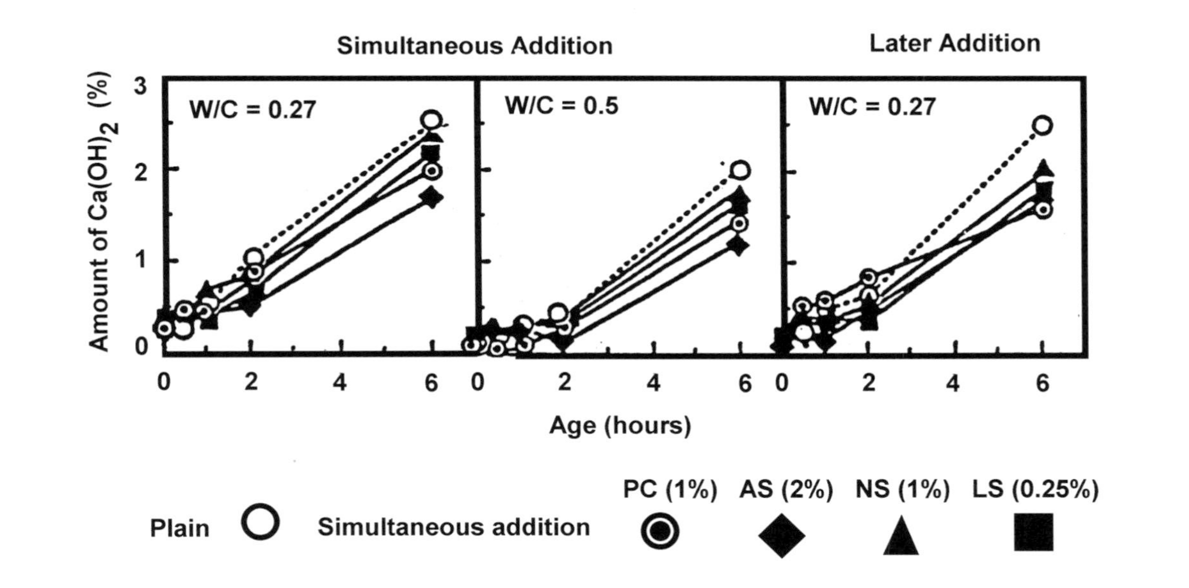

Figure 12. The amount of calcium hydroxide formed at different w/c ratios in cements containing superplasticizers.

In gypsum-less cement, a conduction calorimetric study showed that a strong retardation of hydration occurred in the presence of a lignosulfonate-based superplasticizer containing sodium carbonate.[29]

The rate of hydration of cement is influenced by the temperature to which it is exposed. In the presence of superplasticizers, the hydration rate is modified depending on the temperature. The rate of heat development as well as the cumulative heat effect in cement containing SMF and hydrated at 20, 40, and 55°C are compared in Fig. 13.[6][14][19] At all temperatures, the peaks are shifted to higher times, showing that the hydration is retarded by SMF. This effect seems to increase at 40°C and decreases at a higher temperature. In TG/DTG and DTA, the decomposition of the AFm phase and calcium hydroxide, migration of calcium hydroxide to the C-S-H phase, and the interaction of $Ca(OH)_2$ with the superplasticizer have to be taken into account for the interpretation of the curve.[30]

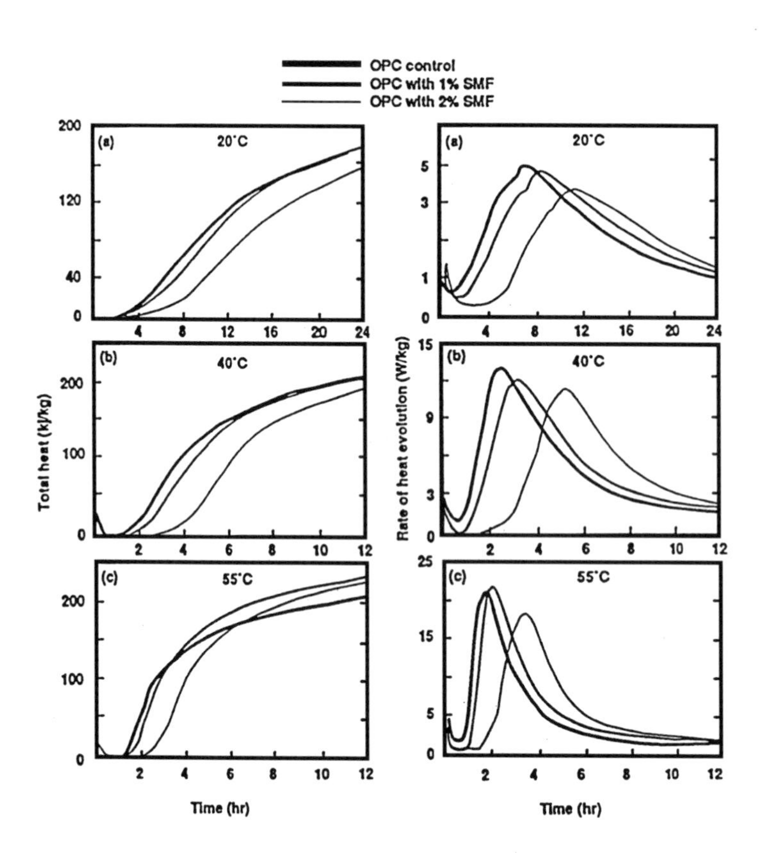

Figure 13. Conduction calorimetric curves of portland cement hydrated at different temperatures in the presence of SMF superplasticizer.

The rate of hydration of cement also depends on the particle size of cement. The effect of SNF on the hydration of Type III cement has been studied.[31] Three fractions of cement, viz., 72–30 μm (coarse), 30–4 μm (medium), and 4 μm (fine) were hydrated in the presence of SNF. Heat evolution occurring at different times is compared in Fig. 14. The reference sample exhibits a hump at about 12 hours, but, in the presence of SNF, it is shifted to about 20 hours. The fine fraction evolving the most heat is the least retarding of all the samples. Only minimal heat is evolved by the coarse fraction. The presence of higher amounts of alkali and SO_3 in the finer fraction explains why it shows the least retardation. The alkali content in the medium fraction is 0.86% and that in the fine fraction is 1.01%. The SO_3 contents in the fine and medium fractions are 12.3 and 2.6% respectively.

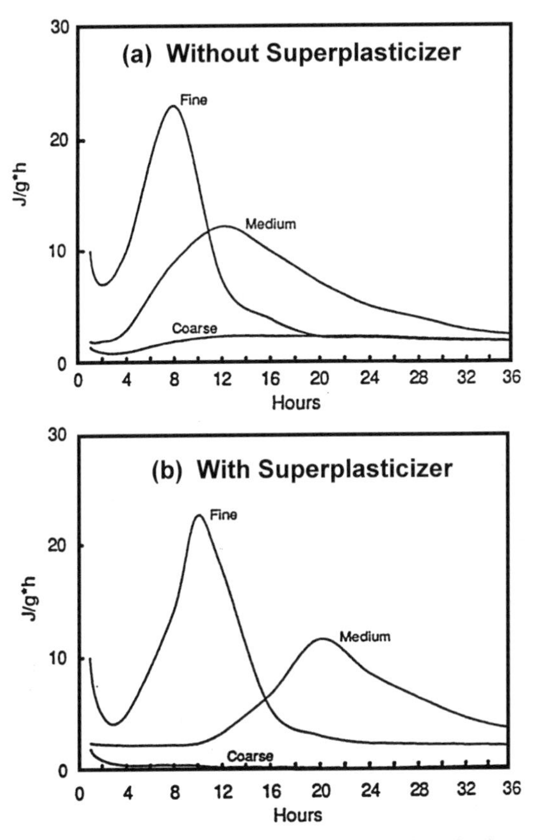

Figure 14. Calorimetric curves of cements of different fractions in the presence of SNF superplasticizer.

Surface area and composition of cement are other factors that determine its rate of hydration. The effect of a superplasticizer (SNF type) on four different cements is illustrated in Fig. 15.[6][20] The heat developed by the four cements designated *A, B, C,* and *D* contained different amounts of C_3S, C_2S, and C_3A. The surface areas of cements *A, B, C,* and *D* were respectively 460, 375, 383, and 600 m^2/kg. The sharp inflection in the curves is attributable to C_3S hydration. Cement *A* exhibits the peak at earlier times than cement B although they are of similar composition. This can be explained by the differences in the surface area of these two cements. Also, cement *A* has a higher amount of C_3S. Cements *B* and *C* of comparable surface areas behave differently. Cement *C* contains more C_3A and gypsum than cement *B*. Thus, there is an acceleration of hydration of C_3S by gypsum. In addition, the greater absorption of the superplasticizer by the hydrating C_3A results in a lower amount of superplasticizer in the solution phase. Consequently, a superplasticizer has less influence on C_3S hydration. Cement D, with the largest surface area, is retarded the most. The main reason is that it has the lowest amount of C_3A, hence there is a lower amount of interaction between the C_3A phase and the superplasticizer resulting in a larger amount of the superplasticizer remaining in the liquid phase. This leads to a better retardation of hydration. Cement *D,* with 12.6% C_4AF, does not appear to be interacting with the superplasticizer to any significant extent.

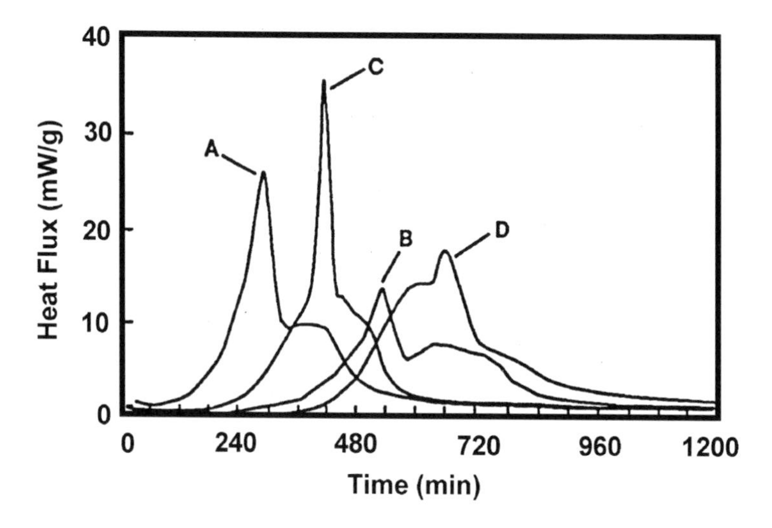

Figure 15. Calorimetric curves of cements of different compositions and size fractions hydrated in the presence of SNF.

In order to study the effect of SO_3 on slump loss in superplasticized cement, Khalil and Ward[32] adopted the conduction calorimetric technique. The heat effects were used to explain the workability loss. It was concluded that for cements cured at 25°C, the optimum SO_3 content was higher in the presence of the superplasticizer.

The rate of hydration of the C_3S phase in cement (containing a superplasticizer) is influenced by the type of sulfate and alkali present in the cement. In Figs. 16 and 17, the effect of different types of sulfates on the hydration of low and high alkali cements is given.[5][33] In the cement containing high sulfate, the retardation is significant at 1% SNF when dihydrate or anhydrite are used, but is negligible when hemihydrate is used.

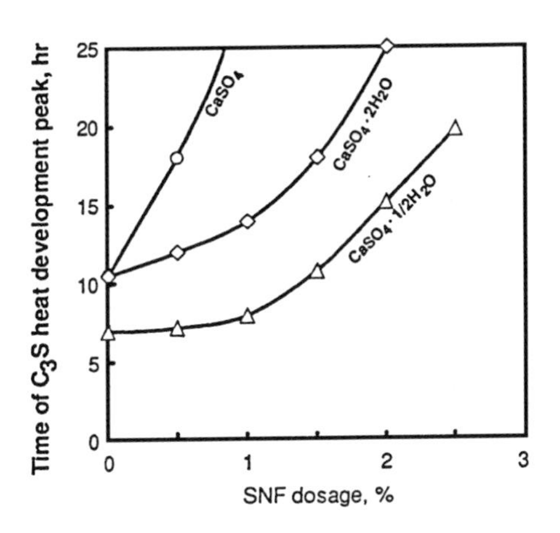

Figure 16. Effect of SNF on the hydration of tricalcium silicate phase in a high alkali cement.

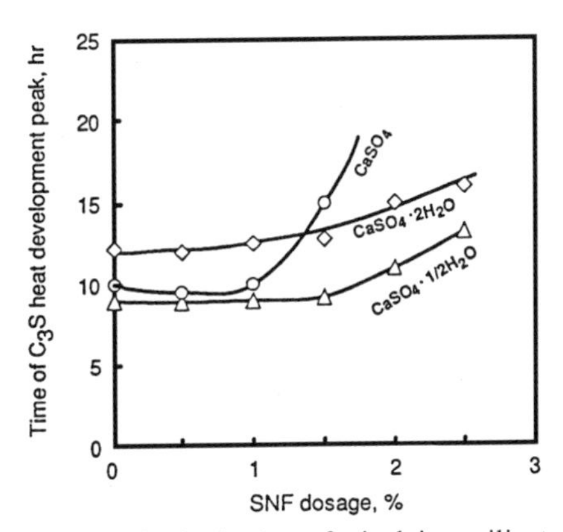

Figure 17. Effect of SNF on the hydration of tricalcium silicate phase in a low alkali cement.

The effectiveness of a superplasticizer may also depend on its degree of poly-condensation. Vovk, et al., prepared a sodium SNF type superplasticizer of polycondensation degree (*n*) from 2 to 17 using liquid chromatography.[17] Table 6 gives the heat developments in cements containing a polymer with *n* varying between 2 and 17. The higher the degree of polycondensation, the greater the extent to which it will combine with the hydration products of Al-containing products. In Table 6, as *n* increases from 2 to 10, the value of the main exothermic peak intensity (C_3S hydration peak) decreases, and it occurs at later times. With further increase in the value of *n* (17), the thermal peak is enhanced, and the peak appears much earlier.

Table 6. Heat Evolution in Cement Containing SNF of Different Degrees of Polycondensation

n	*dQ/dt*	Induction Period (hrs)	Peak Appearance (hrs)
-	1.6	3	12
2	1.0	6	20
5	0.9	10	35
10	0.7	35	65
17	1.2	26	40

In another investigation, Basile, et al.,[37] added 0.4% SNF of average molecular weights 260, 290, 480, and 640 to cement and followed by DTG the amounts of calcium hydroxide formed at 8 hours and beyond. The results are tabulated in Table 7. The minislump results showed that the fluidizing effect increased as the molecular weight of SNF was increased. The superplasticizers retarded the hydration from 8 hours to 1 day. At 7 and 28 days, the amounts of $Ca(OH)_2$ formed increased in the presence of superplasticizers. At higher molecular weights, more $Ca(OH)_2$ seems to have formed.

Table 7. Influence of Molecular Weight of SNF Superplasticizer on the Hydration of Cement in Terms % $Ca(OH)_2$ Formed

Mol. Wt. Average	8 hours	1 day	7 days	28 days
Reference	2.0	7.3	9.0	10.50
260	1.8	6.6	9.5	10.7
290	1.7	6.5	10.5	11.2
480	1.6	6.7	11.0	12.1
640	1.5	6.8	12.0	15.2

The type of cation associated with the superplasticizer may influence the rate of hydration of cement. For example, the time to attain the maximum heat effect in Type I cement with SNF, having cations such as NH_4, Co, Mn, Li, and Ni is 12.8, 11.5, 10.5, 10.3, and 9.3 respectively. The exact mechanism for these variations is not understood.[34]

Type I portland cement concrete, made at a particular w/c ratio, with or without the superplasticizer (SMF or SNF) does not show differences in strength values. It has been found, however, that superplasticized Type V cement concrete develops lower strengths from early ages to 90 days.[5] For example, at a w/c ratio of 4, the 1, 3, 7, 28, and 90 day strengths of the reference concrete are 15.2, 24.5, 30.4, 40.6, and 52.1 MPa respectively, and the corresponding values for the concrete containing 0.6% SNF superplasticizer are 15.9, 22.3, 27.3, 36.0, and 44.6 MPa. The conduction calorimetric curves for Type I and Type V cement pastes treated with a 0.6% SMF superplasticizer showed differences (Fig. 18).[5][35] A more efficient retardation occurs in Type V cement. The causes leading to lower strengths in Type V cement are explained by differences in porosity values of the products and lower rates of hydration.

Ramachandran[36] found that by the addition of calcium gluconate slump loss in superplasticized concrete could be reduced. As the addition of gluconate also leads to a retardation of set, a quick method was needed to assess the maximum amount of gluconate that can be added without causing undue retardation of set. Conduction calorimetry was used for this purpose. Figure 19 gives the calorimetric curves for four samples containing gluconate and SMF. Gluconate (curve 4), in excess of 0.1%, leads to the peaks shifting to longer times, hence causing unacceptable retardation.

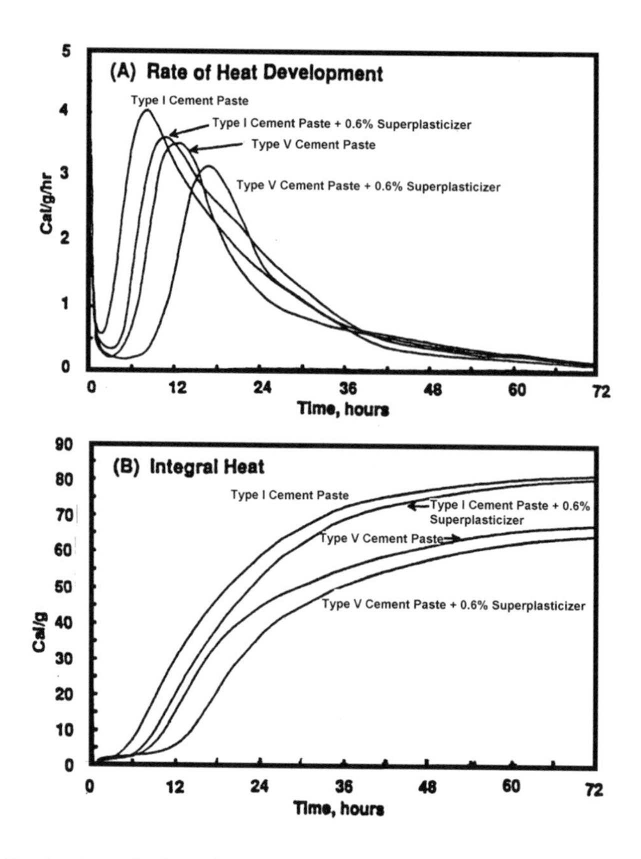

Figure 18. Conduction calorimetric curves for Type I and Type V cements containing SMF superplasticizer.

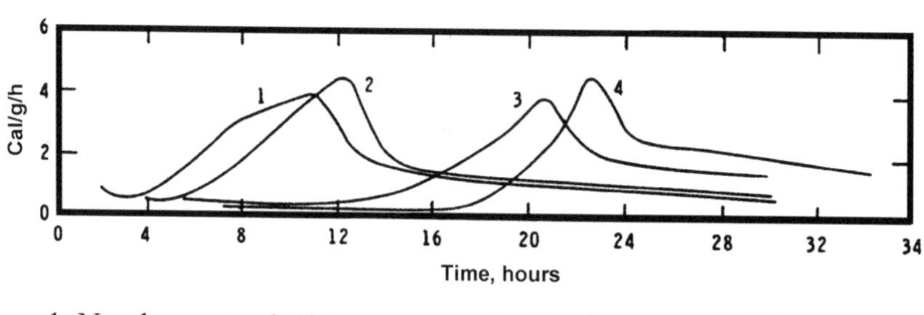

1. Na gluconate, 0.1%
2. Na gluconate, 0.1% + SMF
3. Na gluconate, 0.2%
4. Na gluconate, 0.2% + SMF

Figure 19. Conduction calorimetric curves of cement containing SMF and Na gluconate.

The physico-chemical phenomena associated with the setting of cement is complex. The time of setting is defined in ASTM in terms of the penetration of a needle to a specific depth. Ramachandran[36] examined by DSC several cement samples containing water reducers, retarders, and superplasticizers that had just set. In all the mixes, an endothermal peak typical of $Ca(OH)_2$ was detected, suggesting that at the time of set at least some C_3S must hydrate. It was also observed that the superplasticized cement paste had hydrated to a lesser extent than the reference cement paste at the time of setting. This indicates that physical forces are also involved in the setting phenomenon.

6.0 THERMAL ANALYSIS OF SUPERPLASTICIZERS

The analysis of superplasticizers is not easy as the manufactured product may not be pure. Techniques such as IR and UV spectroscopy yield useful, but not adequate, data so that analysis by other tools becomes necessary. Thermal analysis methods have been applied with some success to evaluate superplasticizers. The results could also be used for quality control purposes.

Milestone[38] carried out DTA of superplasticizers and other admixtures in an atmosphere of oxygen and obtained several thermal inflections that could be used for differentiation of superplasticizers of similar or dissimilar types. In one example, two superplasticizers, thought to be identical, yielded different DTA patterns. In Fig. 20, two superplasticizers, apparently identical, show different patterns.

DTG/TG has been used to differentiate various types of admixtures including superplasticizers.[39] It was found that maximum weight loss occurred at 400°C for the sodium SMF admixture. Sodium SNF showed maximum weight loss at about 900°C. Six industrial superplasticizers (5 SNF and 1 SMF type) were subjected to DTG/TG investigations and thermal stabilities were found to be different so that the data could be used as finger prints for identification purposes.[40]

Evolved Gas Analysis (EGA) technique has also been applied to investigate superplasticizers.[41] In this technique, the admixture is heated at a constant rate and the evolved gas, such as water, is measured. In water-containing compounds, water is adsorbed and measured coulometrically. In

Fig. 21, EGA curves of six superplasticizers are given. The intensity of the peak and the temperature at which it occurs can be used for differentiating them.

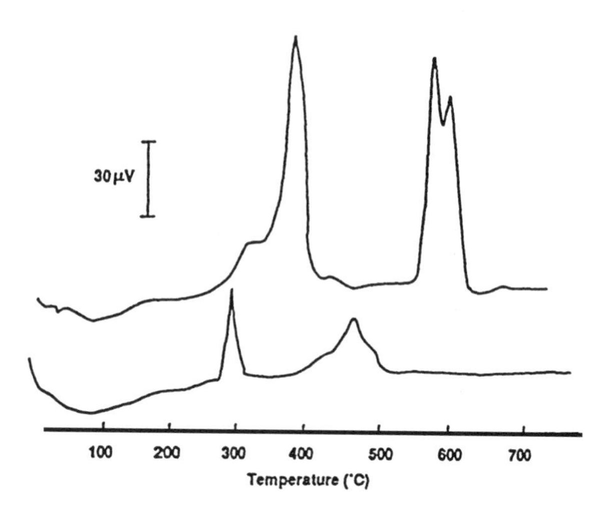

Figure 20. DTA of two superplasticizers.

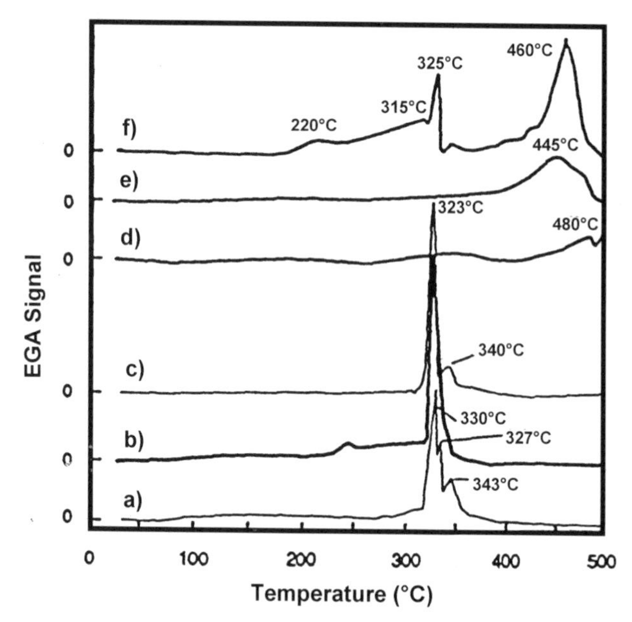

Figure 21. Evolved gas analysis of superplasticizers.

REFERENCES

1. Ramachandran, V. S., Adsorption and Hydration Behavior of Tricalcium Aluminate-Water and Tricalcium Aluminate-Gypsum-Water Systems in the Presence of Superplasticizers, *J. Am. Concr. Inst.*, 80:235–241 (1983)
2. Massazza, F., Costa, U., and Corbella, E., Influence of Beta Naphthalene Sulfonate Formaldehyde Condensate Superplasticizing Admixture on C_3A Hydration in Reaction of Aluminate During Setting of Cements, CEMBUREAU Rept., p. 3 (1977)
3. Slanicka, S., Influence of Water Soluble Melamine Formaldehyde Resin on the Hydration of C_3S, $C_3A + CaSO_4{\bullet}2H_2O$ Mixes and Cement Pastes, *7th Int. Congr. Chem. Cements, Paris,* Part II:161–166 (1980)
4. Ramachandran, V. S., Influence of Superplasticizers on the Hydration of Cement, *3rd Int. Congr. Polymers in Concr., Koriyama, Japan,* pp. 1071–1081 (1981)
5. Ramachandran, V. S. (ed.), *Concrete Admixtures Handbook*, 2nd Ed., Noyes Publ., NJ (1995)
6. Ramachandran, V. S., Malhotra, V. M., Jolicoeur, C., and Spiritos, N., *Superplasticizers, Properties and Applications in Concrete*, Natural Resources Canada, Ottawa (1998)
7. Collepardi, M., Corradi, M., and Baldini, G., Hydration of C_3A in the Presence of Lignosulfonate-Carbonate System or Sulphonated Naphthalene Polymer, *7th Int. Congr. Chem. Cements, Paris,* IV:524–528 (1980)
8. Henning, O., and Goretzki, I., Effect of Plasticizers on the Degree of Hydration, *RILEM Int. Conf. Concrete at Early Ages*, I:151–155, Assn. Amicales des Ingineurs, Paris (1982)
9. Quon, D. H. J., and Malhotra, V. M., Effect of Superplasticizers on Slump and Degree of Conversion of High Alumina Concrete, Developments in the Use of Superplasticizers, (V. M. Malhotra, ed.), SP-68:173–187, *Am. Concr. Res.*, Detroit (1981)
10. Gagne, P., Chaleur d'Hydratation Initiale de Ciments en Presence de Divers Adjuvanta, M. S. Thesis, p. 98, University of Sherbrooke, Canada (1993)
11. Massazza, F., and Costa, U., Effect of Superplasticizers on the Hydration of C_3A, *7th Int. Congr. Chem. Cements, Paris,* IV:529–535 (1980)
12. Hayek, E., and Dierkes, P., X-Ray Examination into the Effect of Melment on the Hydration of Gypsum and Cement Phases C_3A and C_3S, *Melment Symp., Trostberg,* p. 14 (1973)
13. Lukas, W., The Influence of Melment on the Hydration of Clinker Phases in Cement, *5th Int. Melment Symp., Munich,* pp. 17–21 (1979)

14. Singh, N. B., and Prabha Singh, S., Superplasticizers in Cement and Concrete, *J. Sci. Ind. Res.*, 52:661–675 (1993)
15. Odler, I., and Abdul-Maula, S., Effect of Chemical Admixtures on Portland Cement Hydration, *Cement Concr. Aggregates*, 9:38–43 (1987)
16. El-Hemaly, S. A. S., El-Sheikh, R., Mosolamy, F. H., and El-Didamony, H., The Hydration of Alite in the Presence of Concrete Admixtures, *Thermochimica Acta*, 78:219–225 (1984)
17. Vovk, A. I., Vovk, G. A., and Usherov-Marshak, A. V., Regularities of Hydration and Structure Formation of Cement Pastes in the Presence of Superplasticizers with Different Molecular Mass, *5th CANMET/ACI Int. Conf. on Superplasticizers and Other Chemical Admixtures in Concr.*, SP-173:763–780 (1997)
18. Odler, I., and Becker, T., Effect of Some Liquefying Agents on Properties and Hydration of Portland Cement and Tricalcium Silicate Pastes, *Cement Concr. Res.*, 10:321–331 (1980)
19. Yilmaz, V. T., and Glasser, F. P., Influence of Sulphonated Melamine Formaldehyde Superplasticizer on Cement Hydration and Microstructure, *Advances Cements Res.*, 2:111–119 (1989)
20. Simard, M. A., Nkinamubanzi, P. C., Jolicoeur, C., Parraton, D., and Aitcin, P. C., Calorimetry, Rheology and Compressive Strength of Superplasticized Cement Paste, *Cem. Concr. Res.*, 23:939–950 (1993)
21. Singh, N. B., and Singh, A. K., Effect of Melamine on the Hydration of White Portland Cement, *Cem. Concr. Res.*, 19:547–553 (1989)
22. Popescu, G., Montean, M., Horia, B., Stelien, I., Dan-Florin, A., and Bujor, A., Effect of Superplasticizers on Portland Cement Mortars and Pastes, *Il Cemento*, 70:107–114 (1982)
23. Singh, N. B., Sarvahi, R., and Singh, P., Effect of Superplasticizer on the Hydration of Portland Cement, *9th Int. Congr. Chem. Cements, New Delhi,* pp. 564–569 (1992)
24. Chen, Y., Wu, Z., and Yu, K., Effect of Mechanism of Concrete Admixtures Based Aromatic Hydrocarbon Sulphonate, *9th Int. Congr. Chem. Cements, New Delhi,* IV:550–556 (1992)
25. Uchikawa, H., Hanehara, S., Shirasaka, T., and Sawaki, D., Effect of Admixture on Hydration of Cement, Adsorptive Behavior of Admixture and Fluidity and Setting of Fresh Cement Paste, *Cem. Concr. Res.*, 22:1115–1129 (1992)
26. Gu, P., Xie, P., Beaudoin, J. J., and Jolicoeur, C., Investigation of the Retarding Effect of Superplasticizers on Cement Hydration by Impedance Spectroscopy and Other Methods, *Cem. Concr. Res.*, 24:433–442 (1994)

27. Cabrera, J. G., and Lynsdale, C. J., The Effect of Superplasticizers on the Hydration of Normal Portland Cement, *2nd ACI/CANMET Int. Symp. on Advances in Concr. Technol., Las Vegas,* Suppl. Papers, pp. 741–758 (1995)
28. Uchikawa, H., Hydration of Cement and Structure Formation and Properties of Cement Paste in the Presence of Organic Admixture, *Regourd Symp. on High Performance Concr., Sherbrooke, Canada,* (P. C. Aitcin, ed.), pp. 49–102 (1994)
29. Hrazdira, J., Role of Superplasticizers in Gypsum-less Portland Cement, *4th CANMET/ACI Int. Conf. on Superplasticizers and Other Admixtures in Concr.*, Montreal, Canada, (V. M. Malhotra, ed.), pp. 407–414 (1994)
30. Singh, S. B., Sarvahi, R., Singh, N. P., and Shukla, A. K., Effect of Temperature on the Hydration of Ordinary Portland Cement in the Presence of Superplasticizer, *Thermochimica Acta*, 247:381–388 (1994)
31. Aitcin, P. C., Sarkar, S. L., Regour, M., and Volant, D., Retardation Effect of Superplasticizers on Different Cement Fractions, *Cement Concr. Res.*, 17:995–999 (1987)
32. Khalil, S. M., and Ward, M. A., Effect of Sulfate Content of Cement Upon Heat Evolution and Slump Loss of Concrete Containing High Range Water Reducers, *Mag. Concr. Res.*, 32:28–38 (1980)
33. Nawa, T., Eguchi, H., and Fukuya, Y., Effect of Alkali Sulfate on the Rheological Behavior of Cement Paste Containing Superplasticizer, *Am. Concr. Res. Inst., SP-119:405–424 (1989)*
34. Sinka, J., Fleming, J., and Villa, J., Condensates on the Properties of Concrete, *1st Int. Symp. Superplasticizers, CANMET, Ottawa, Canada,* p. 22 (1978)
35. Ramachandran, V. S., Lowery, M. S., and Malhotra, V. M., Behavior of ASTM Type V Cement Hydrated in the Presence of Sulfonated Melamine Formaldehyde, *Mater. Const.*, 28:133–138 (1995)
36. Ramachandran, V. S., *Effect of Retarders/Water Reducers on Slump Loss in Superplasticized Concrete, Developments in the Use of Superplasticizers*, (V. M. Malhotra, ed.), ACI., SP-68:393–407 (1981)
37. Basile, F., Biagni, S., Ferrari, G., and Collepardi, M., Influence of Different Sulfonated Naphthalene Polymers in the Fluidity of Cement Paste, *3rd CANMET/ACI Int. Conf. on Superplasticizers and Other Chemical Admixtures in Concr., Ottawa, Canada,* (V. M. Malhotra, ed.), SP-119:209–220 (1989)
38. Milestone, N. B., Identification of Concrete Admixtures by Differential Thermal Analysis, *Cem. Concr. Res.*, 14:207–214 (1984)

39. Gupta, J. P., and Smith, J. I. H., Thermoanalytical Investigations of Some Concrete Admixtures, *2nd European Symp. Therm. Anal., London, UK,* pp. 483–484 (1981)
40. Khorami, J., and Aitcin, P. C., Physico-Chemical Characterization of Superplasticizers, *3rd CANMET/ACI Int. Conf. on Superplasticizers and Other Admixtures in Concr.*, (V. M. Malhotra, ed.), SP-119:117–131 (1989)
41. Gal, T., Pokol, G., Leisztner, L., and Gal, S., Investigation of Concrete Additives by Evolved Gas Analysis, *J. Therm. Anal.*, 33:1147–1151 (1998)

8

Supplementary Cementing Materials and Other Additions

1.0 INTRODUCTION

Chemical admixtures such as accelerators, retarders, water reducers, and superplasticizers are generally water soluble and are added in small amounts. They significantly affect the properties of concrete in the fresh and hardened states through physico-chemical and surface interactions. In contrast to chemical admixtures, materials such as fly ash, silica fume, blast furnace slag, natural pozzolans, and others are added in substantial amounts to concrete. Most of them react significantly with the components of the cement paste yielding higher strengths and better durability characteristics. They are used as cement replacement materials. Some of them also possess self-cementitious properties. Utilization of these materials confers several benefits such as energy savings, reduction of the heat of hydration, conservation of resources, environmental protection, and improved durability to various types of chemical attack.

These materials are known by various names. The European concrete standards refer to them as additions.[1] The ASTM classifies them as mineral admixtures. In the literature, they are also described as supplementary cementing materials, cement replacement materials,[2] or composite materials.[3] In this chapter, these materials are referred to as supplementary cementing materials. Also included in this chapter is a discussion on other additions that are not treated in other chapters.

The main methods of investigating the effect of supplementary materials and other additions on cement hydration include XRD, SEM, NMR, Mossbauer spectroscopy, IR, and thermal analysis. Poorly crystallized products that form in these materials are advantageously investigated by TG, DTG, DTA, DSC, and conduction calorimetry. Thermal methods have also been suggested for characterization of the supplementary and other materials.

2.0 FLY ASH

Hydration of fly ash cement differs from pure cement in terms of the hydration rates of the clinker phases, amount of calcium hydroxide formed, composition of the clinker hydration products, and additional hydration products from the reaction of the fly ash.[3] Lower amounts of lime are formed in the presence of fly ash because of the pozzolanic reaction between the fly ash and lime formed in cement hydration. Fly ash generally retards the reaction of alite in the early stages and accelerates the middle stage reaction. The accelerated reaction is attributed to the existence of nucleation sites on fly ash particles. The aluminate and ferrite phases hydrate more rapidly in the presence of fly ash, and also there is a significant difference in the hydration rate of the belite phase up to 28 days. Table 1 gives the relative hydration rates of cement compounds in the presence of fly ash as derived from conduction calorimetry.[4] It can be seen that the earlier rates of hydration are generally retarded, and the later stage hydration rates are accelerated.

The reactivity of fly ash in cement depends on the glass content, RH, temperature, alkali content of cement, particle size, and sulfate ion concentration. High calcium fly ash is mainly glassy with some crystalline phases and has a self-hardening property. In a three day old fly ash cement paste, the composition of the hydrate formed from alite or belite is similar to that formed from a normal portland cement paste, but at later stages the CaO/SiO_2 ratio is lower and Al_2O_3/CaO ratio, higher in a fly ash cement paste. The silicate anion condensation occurs more rapidly in C_3S-fly ash paste than in C_3S paste.[4]

Table 1. Relative Hydration Rates of Cement Compounds in the Presence of Fly Ash

Synthesized Compounds or Clinker Minerals	Early Stage Hydration I	II	III	Later Stage Hydration
Tricalcium Silicate	R	R	A	A
Alite	R	R	A	A
Tricalcium Aluminate	R	R	A	A
Tricalcium Aluminate + Ca Sulfate	R	R	A	A
Tricalcium Aluminate + Ca Sulfate+				
Calcium Hydroxide	R*	R*	A	A
Interstitial Phases	R*	R*	A	A

A = Acceleration

R = Retardation

R* = Some fly ashes having high Ca adsorption capacity and high pozzolanic activity will accelerate the hydration at these stages

The addition of fly ash to cement results in the formation of decreased amounts of calcium hydroxide in the hydration product. This is attributable to the dilution effect and to the consumption of calcium hydroxide by the pozzolanic reaction with the fly ash. In Fig. 1, the amount of calcium hydroxide formed at different times of hydration in cement containing fly ash is given.[5] The amount of $Ca(OH)_2$ estimated by TG was found to be lower in samples containing fly ash. With the increase in the amount of fly ash, less calcium hydroxide was formed because of the pozzolanic reaction and dilution effect. Even at 60% fly ash, some lime was present in the mortar, and the pH was found to be 13.5. At this pH value, the passivity of steel is assured. It can also be observed that there is more lime at 60% fly ash than at 75% slag addition.

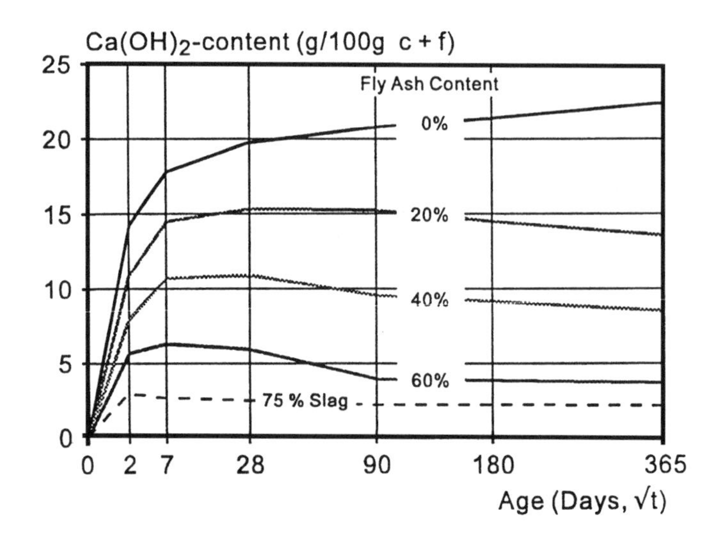

Figure 1. Amount of lime formed in the fly ash mortar containing different percentages of fly ash.

Studies involving DTG, TG, and DTA show that the composition and character of hydration products formed in the portland cement and portland cement-fly ash pastes are not very different while the relative proportions of the products vary.[6]

Conduction calorimetric curves for cement with 40% ordinary fly ash are different from that containing 40% high calcium fly ash (Fig. 2).[4] Both ordinary and high calcium fly ashes delay the appearance of the peak effect due to alite hydration. The appearance of the peak, however, is delayed to a greater extent by high calcium fly ash. Fly ash with a low calcium adsorption capacity, higher contents of Ca^{2+}, dissolved alkalis, and unburned carbon retards the hydration of alite, by hindering the saturation rate of $Ca(OH)_2$ in the liquid phase.

It has been shown by conduction calorimetry that Type 10 (Type I ASTM) cement containing 25% fly ash exhibits similar behavior to Type 40 (Type IV ASTM) low heat cement containing no fly ash.[7] Both these cements have lower heats of hydration compared to pure Type 10, normal portland cement (Fig. 3).

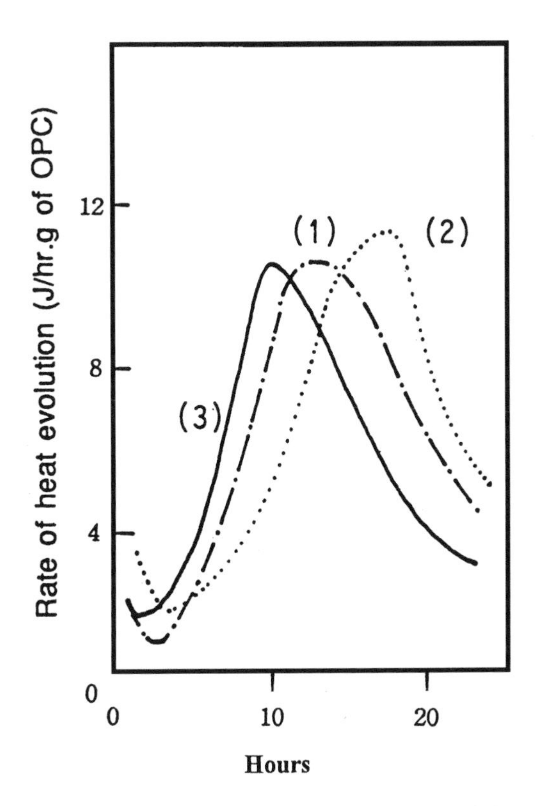

Figure 2. Conduction calorimetric curves of cement pastes containing ordinary fly ash and high calcium fly ash. 1. Ordinary fly ash. 2. High calcium fly ash. 3. Ordinary portland cement.

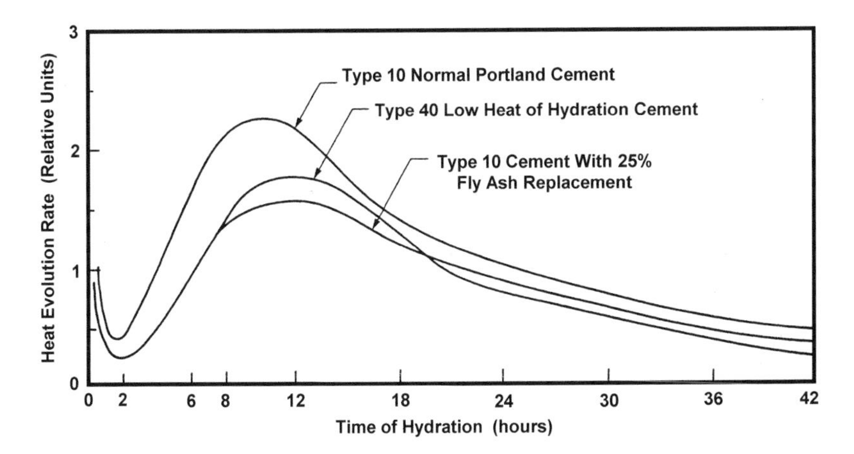

Figure 3. Conduction calorimeteric curves of normal portland cement and low heat cement containing fly ash.

The rate of hydration of cement-fly ash also depends on the particle size of fly ash. Conduction calorimetric curves have been obtained for cements containing 30% fly ash of surface areas of 200, 450, and 650 m^2/kg. These mixtures are designated PFA-0, PFA-1, and PFA-2 respectively in Fig. 4.[8] The fly ash cements exhibited sharper peaks at earlier times at 11–12 hours compared to the reference cement. These peaks become sharper as the surface area of fly ash increases. The peak effects are associated with the hydration of C_3A after the depletion of gypsum in the system. The initial exothermal effect due to alite hydration is retarded by the fly ashes.

Grinding of high calcium fly ashes enhances their reactivity. The DTA studies indicate that the intensity of peaks at about 150°C is increased with highly ground fly ash additions. This is evidence that higher amounts of ettringite are formed in the presence of finely ground fly ash.[9]

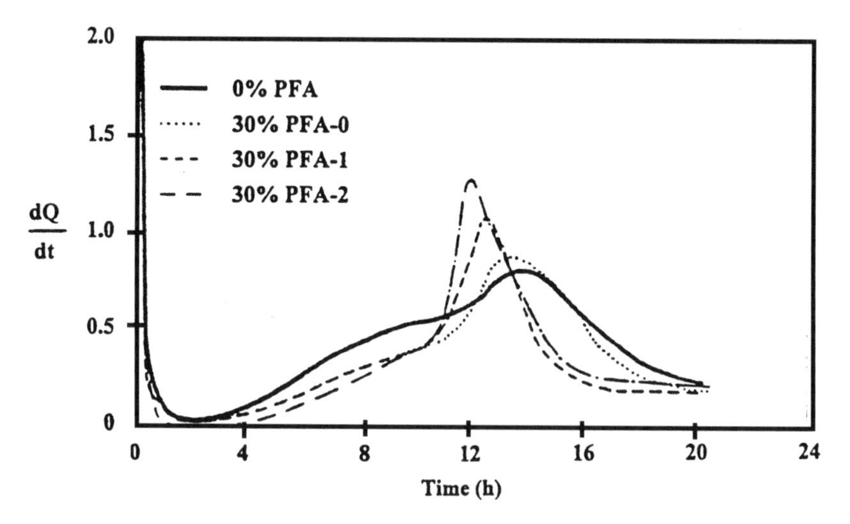

Figure 4. The influence of particle size of fly ash on the hydration of cement.

The Canada Center for Mineral and Energy Technology (CANMET) developed a structural concrete incorporating high volumes (>50%) of low calcium fly ash. Such concretes yield adequate early strengths, workability, low temperature rise, and high later strengths.[10] The mechanism of strength development in high volume fly ash concretes was investigated by Feldman, et al., applying several techniques such as electron microscopy, permeability, pore size distribution, and thermal analysis.[11] Fly ash cement pastes containing different amounts of fly ash were hydrated, and the hydration products were subjected to TG analysis. The amounts of non-evaporable water and $Ca(OH)_2$ were estimated up to 90 days. Up to 7 days, $Ca(OH)_2$ contents in high volume fly ash cements started to decrease. In untreated cements, the $Ca(OH)_2$ was about 11–12% compared to only 2–3% in fly ash cements. These results and other tests indicated that in high volume fly ash cements a relatively more homogeneous product resulted with low lime contents, and the formation of a low C/S ratio C-S-H product produced a stronger body.

Many fly ashes cannot be used as supplementary materials or mineral admixtures because they possess low pozzolanic activity. Some activators may be added to make them hydraulic. The addition of Na_2SiO_3 and $Ca(OH)_2$ to fly ash cement mixtures results in the acceleration of the hydration of cement as evidenced by increased intensities of the endothermal peak (calcium hydroxide decomposition) in thermal analysis techniques. Strengths are also increased.[12] Fly ash can also be activated by mixing it with phosphogypsum and wet hydrated lime. The estimation of ettringite and lime may be carried out by DTA, and the results may be used to explain the strength development in such systems.[13]

Bonding occurs when fly ash and lime are hydrothermally treated. The products formed in such mixtures are advantageously identified by DTA. Thermal analysis indicates that initially C-S-H (I) forms and is then converted to 11 Å tobermorite. Carbonation also can be followed by DTA.[6] By a hydrothermal process, a cementing material was synthesized from a mixture of fly ash, lime, and water. The resulting product was analyzed by DTA, TG, and XRD.[14]

Atmospheric fluidized bed coal combustion (AFBC) at a temperature of about 850°C with calcite as a sorbent, produces solid residue of different chemical composition from that produced at higher temperatures. The AFBC residue possesses cementing properties. It consists of anhydrite, free lime, quartz, and aluminosilicates. DTA and TG techniques have been utilized to determine melting intervals, oxidation of unburnt coal, limestone, and anhydrite.[15]

The presence of unburnt carbon in fly ashes has many adverse effects on concrete including the appearance of grayish blackness in mortars and concrete, need for higher w/c+f ratio to obtain required consistency, and decreased air entertainment. Thermogravimetric analysis has been developed to estimate hydrated lime, calcium carbonate, and unburned carbon in fly ashes.[16] The methodology is as follows:

a) The material is heated up to 100°C in a N_2 atmosphere and purged by N_2 at 100°C.

b) It is heated in an inert atmosphere from 100° to 750°C, cooled from 750° to 100°C with a flow of nitrogen.

c) It is then subjected to isothermal treatment at 100°C for 5 minutes with air flow.

d) It is heated in a reactive atmosphere from 100° to 1000°C in air.

Weight loss at 400–450°C in an inert atmosphere was ascribed to water loss from $Ca(OH)_2$, and that at 600–750°C (inert atmosphere) was caused by the decomposition of $CaCO_3$. The loss of weight in the range of 500–740°C in an oxidizing atmosphere signified oxidation of carbon. Table 2 gives the amounts of lime, calcium carbonate, and unburned carbon determined in six fly ashes by TG.[16]

Table 2. TG Estimation of Calcium Carbonate, Calcium Hydroxide, and Unburned Carbon in Fly Ashes

Fly Ash No.	$Ca(OH)_2$, %	$CaCO_3$, %	Unburned Carbon, %
1	0.48	0.40	0.09
2	0.16	0.81	2.83
3	—	0.37	2.39
4	—	0.75	4.86
5	0.76	2.26	1.01
6	1.37	3.19	0.70

(Reprinted with permission from Elsevier Science.)

3.0 SILICA FUME

Silica fume is available in a powdered, condensed, or slurry form. Powdered silica fume is not easy to use in practice because of dust and other problems. Other forms of silica that have potential for use in concrete are synthetic silica, silica gel, colloidal silica, and fumed silica. Condensed silica acts as a filler and is also a good pozzolan. The specific surface area of silica fume may be as high as 20,000 m^2/kg, which is almost an order of magnitude higher than that of normal portland cement. Fineness, amorphous nature, and high silica contents make silica fume a very reactive material. Depending on the amount and nature of silica fume and w/c ratio, the concrete strength at 28 days could reach a value of 100 MPa. The action of silica fume on the hydration of cement and cement components can be followed by DTA, DSC, TG, and conduction calorimetry. Variability in the hydration rates of cement in the presence of silica fume reported by different investigators is expected because silica fumes differ in their physical and chemical nature. Most concretes made with silica fume contain a superplasticizing admixture. Conduction calorimetric studies indicate that all the stages of hydration in pure components of cement, such as C_3S, C_3A, $C_3A + CaSO_4$, $C_3A + CaSO_4 + Ca(OH)_2$, alite and the interstitial phase, are accelerated with a few exceptions.[4]

Addition of even a small amount of active silica accelerates the hydration of tricalcium silicate.[17]–[21] The effect of different percentages of finely divided SiO_2 (Aerosil) (S) on the hydration of tricalcium silicate has been studied by conduction calorimetry (Fig. 5).[17] The addition of even a small amount of silica accelerates the hydration of C_3S by lowering the time at which the exothermic peak appears. The peak appears earlier as the dosage of SiO_2 increases. The degree of hydration of tricalcium silicate has been determined by estimating the amount of calcium hydroxide formed at different times, using TG. Table 3 gives the amount of lime formed at 3, 7, and 24 hours.[17] At all times, the amount of lime is higher in samples containing SiO_2. Larger amounts of lime formed, especially at higher dosages, are indicative of acceleration of C_3S.

The C/S ratio of the C-S-H product decreases by the addition of silica. It is reported that in the absence of SiO_2 at a w/s ratio of 10, C_3S yields a C-S-H product with a C/S ratio of 0.8 at 24 hours, and this ratio decreases to 0.33, and 0.36 at w/s ratios of 1 and 0.4 respectively.[19] The acceleration of early hydration of C_3S is ascribed to the fineness of the particles that would provide preferential nucleation sites. The pozzolanic particles also offer a surface for the precipitation of C-S-H from the pore solution. Thus, the removal of the Ca and silicate ions would result in a reduction in the thickness of the relatively impermeable C-S-H growing on the C_3S grains.[1]

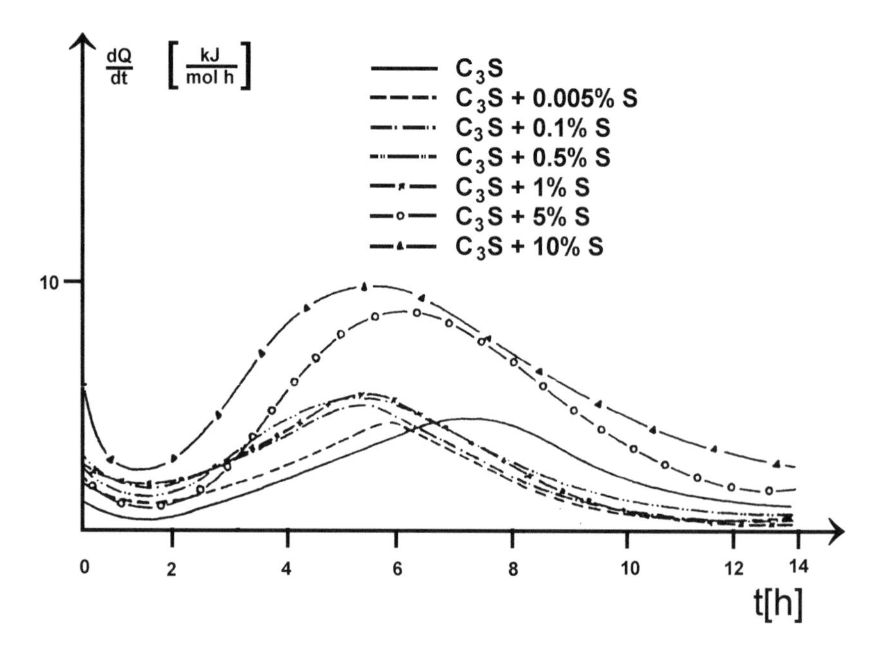

Figure 5. Conduction calorimetry of C_3S containing finely divided SiO_2.

Table 3. Calcium Hydroxide Content in C_3S Pastes in the Presence of SiO_2

Sample	Calcium Hydroxide (%), at Different Times (Hrs)		
	3	7	24
Tricalcium Silicate	1.2	2.4	10.0
Tricalcium Silicate + 0.0% SiO_2	1.2	3.2	10.4
Tricalcium Silicate + 0.5% SiO_2	1.6	2.8	11.2
Tricalcium Silicate + 1.0% SiO_2	2.4	5.6	12.4
Tricalcium Silicate + 5.0% SiO_2	2.4	6.4	12.8
Tricalcium Silicate + 10% SiO_2	4.0	7.6	16.6

The influence of particle size of silica fume on the hydration of C_3S can be studied by calorimetry.[20] Figure 6 shows that the induction period decreases from 180 minutes to 90 minutes, and the maximum exothermal effect from 660 minutes to 280 minutes. The height of the second peak initially increases as the fineness increases, but then decreases at a higher specific surface area of SiO_2.

In the hydration of cement and tricalcium silicate, calcium hydroxide formed as a reaction product reacts with silica fume. In this context, some work has been carried out on the reaction occurring in the system $Ca(OH)_2$-SiO_2-H_2O.[21] The heat effect that occurs in this system is mainly attributed to the reaction between $Ca(OH)_2$ and SiO_2. The rate of disappearance of CH from the system containing SiO_2, of surface areas of 20,000 m^2/kg and 90,000 m^2/kg, is compared in Table 4.[21] The results indicate that the reaction rate depends on the surface area of SiO_2. Lime was consumed almost completely within one day when the high surface area silica was used.

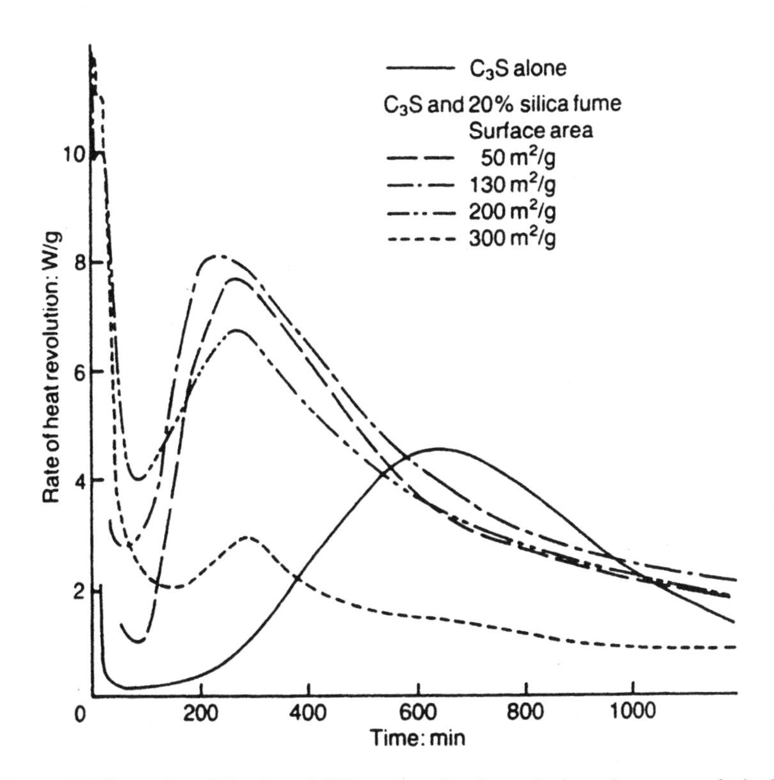

Figure 6. Effect of particle size of SiO_2 on conduction calorimetric curves of tricalcium silicate.

Table 4. Amount of $Ca(OH)_2$ Remaining in $Ca(OH)_2$-SiO_2 Mixtures at Different Times

Surface Area of SiO_2	$Ca(OH)_2$(%), Formed at Different Times (Hrs)							
	0	1	3	5	12	24	72	168
20,000 m^2/kg	55	-	-	53.9	-	50	40.2	27.7
90,000 m^2/kg	55	27.3	27.5	20	4.1	1.4	0	0

Thermal techniques have been applied extensively to monitor the influence of silica fume on the hydration of cement.[19][22]–[32] The rate of heat development in cement pastes with 0, 10, 20, and 30% silica fume is plotted in Fig. 7.[19][22] The intense peak below 1 hour may be attributed to the aluminate reaction. The second effect occurring in the cement paste (reference) with a peak effect at about 6–7 hours is caused by the C_3S hydration. Addition of silica fume alters the profile of the curve. A peak appears at about 5 hours and another at about 6–10 hours. The peaks increase in amplitude and become sharper as the amount of silica is increased. The peaks appear at earlier times in pastes containing silica fume signifying an acceleration effect on C_3S hydration. The more prominent second exothermal effect in the presence of silica fume may be caused by the aluminate reaction. It is also possible that it is owing to the reaction of calcium hydroxide (hydration product) with silica fume. The total amount of heat developed in the reaction is increased in the silica fume-containing pastes.

The pozzolanic activity of silica fume can be demonstrated in the systems Cement-Silica Fume-H_2O by estimating the amount of lime in the hydrated product at different times by thermogravimetric analysis.[25][26] Figure 8 shows the amount of calcium hydroxide formed in cement-silica fume pastes. At all curing periods, lime content is less in portland cement pastes incorporating silica fume. Extrapolation of the curve shows that, at about 24% silica, no lime would be left in the pastes. This figure may vary depending on the materials and method of estimation.

The amount of lime formed as a function of water/cement + silica fume ratio has also been studied by TG.[26]–[29] Figure 9 indicates that, as a consequence of the pozzolanic reaction of silica fume with lime, the amount of lime is decreased; higher amounts of silica fume are more effective.[26]

The XRD of these pastes at longer times shows no indication of $Ca(OH)_2$. It can be concluded that the lime is present in a poorly crystallized form in these pastes. The non-evaporable water contents are also lower in portland cement pastes containing silica fume.[32]

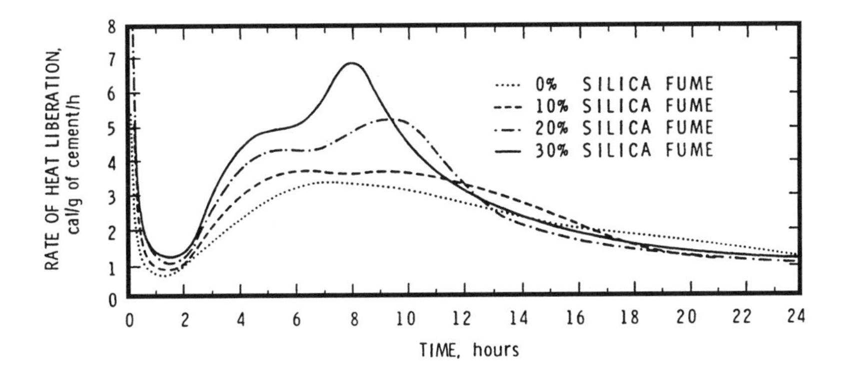

Figure 7. Conduction calorimetric curves of cement pastes containing different amounts of silica fume.

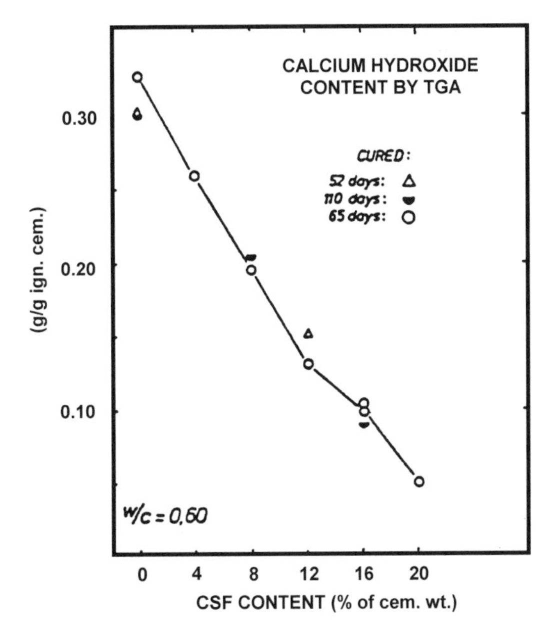

Figure 8. Amount of calcium hydroxide formed in cement-silica fume pastes hydrated for different times.

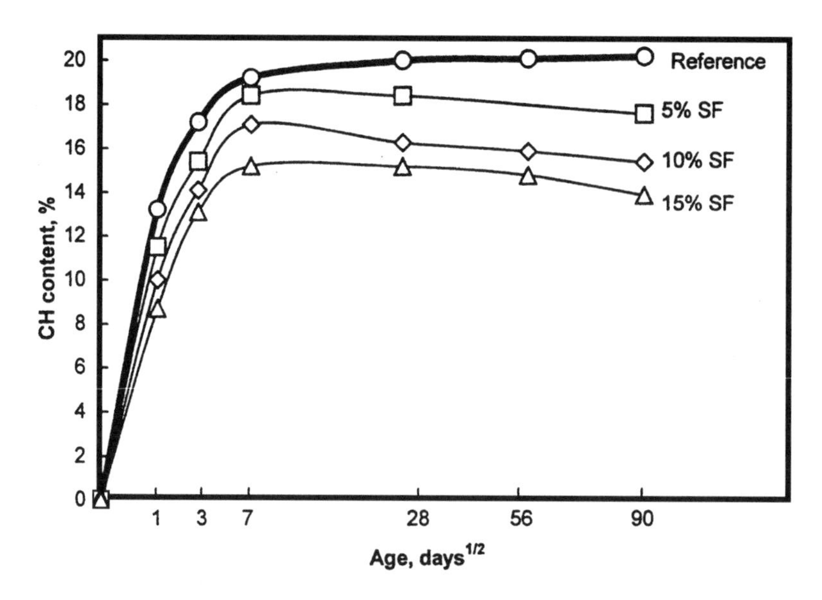

Figure 9. Effect of time of hydration on calcium hydroxide contents in cement pastes.

In normal portland cement, hydration leads to the formation of ettringite. In many instances, the formation of ettringite leads to expansion. To prevent the formation of ettringite, a gypsum-based material with 75% hemihydrate, 20% portland cement, 5% silica fume, and a superplasticizer was fabricated. The pastes were cured in water for 1 to 10 minutes and subjected to DTA.[33] In Fig. 10, thermograms show the stepwise dehydration endothermal peaks (shown upwards) at 150° and 200°C as is typical of gypsum. There was no indication of ettringite or monosulfate, which normally are identified by peaks at 125–130° and 190–195°C.

It is generally thought that the product resulting from the hydration of cement in the presence of silica fume consists of C-S-H (I) type calcium silicate hydrate. The C/S ratio of this hydrate has values between 0.9 and 1.3.[19] Differential thermal technique is capable of identifying the low C/S ratio C-S-H product by an exothermal peak in the vicinity of 900°C and beyond. In Fig. 11, thermograms of cement containing 30% silica fume exhibit a typical exothermal effect for the presence of C-S-H (I) in all the pastes cured from 1 day to 6 months.[19][34]

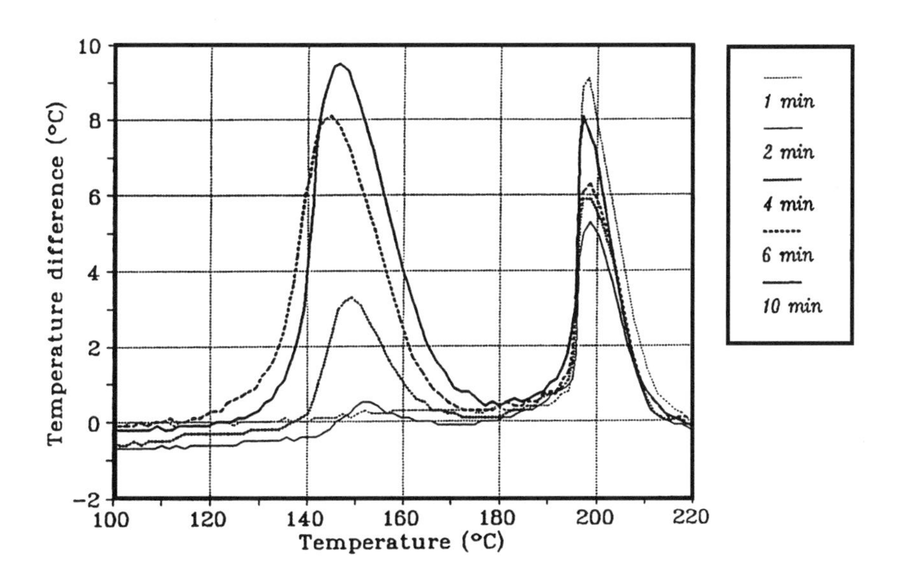

Figure 10. Differential thermograms of gypsum-portland cement-SiO_2 mixes cured for different periods.

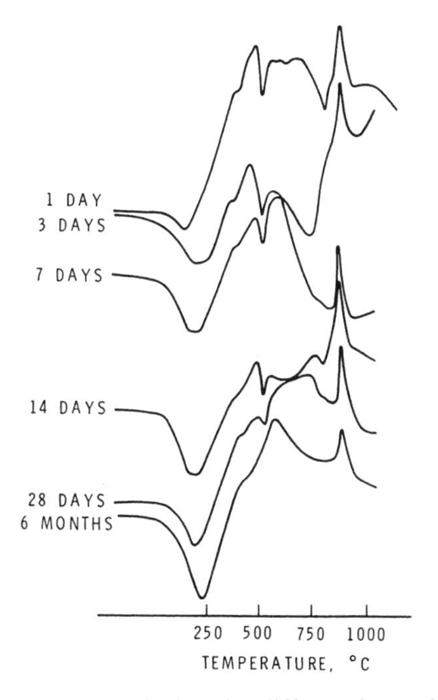

Figure 11. DTA of cement pastes hydrated to different times with 30% silica fume.

4.0 SLAGS

Blast furnace slag (referred to as *slag* in this chapter) is formed as a liquid in the manufacture of iron, and when cooled slowly, crystallizes and has virtually no cementing properties. If cooled rapidly below 800°C, it forms a glass that has hydraulic properties and is called granulated slag. It is blended or mixed with portland cement in amounts up to 80%. The reactivity of slag depends on its bulk composition, glass content, and fineness.

The glass content in slag can be determined by light microscopy or from the difference shown in XRD determination of crystalline phases. Some success has been achieved in the estimation of glass content by DTA.[35]

The principle hydration products of slag cements are essentially similar to those found in portland cement pastes. The microstructure of slag cement pastes is also similar to that of portland cement pastes. X-ray microanalysis has, however, shown that the C/S ratio of C-S-H product in hydrated slag cement is lower than that found in portland cement paste.

Slags can be activated by calcium sulfate and other compounds such as NaOH, lime, water glass, phosphogypsum, etc. When calcium sulfate is added, the precipitation of ettringite provides a sink for Ca^{2+} and $Al(OH)_4^-$ ions released from the slag. The nature of products resulting from the activation of slags is similar to that in normal portland cement pastes (Table 5).[1]

Supersulfated cement is made by grinding together a mixture of 80% slag, 5% portland cement, and about 15% gypsum or anhydrite. Supersulfated cement does not contain lime and is capable of resisting sulfate attack. In the supersulfated cement, the main products are C-S-H and ettringite. The amount of ettringite reaches a limit in about 3 days, the period during which all sulfate will have reacted.

Thermal techniques such as DTA and conduction calorimetry have been applied extensively for a study of the mechanism of hydration of slag, divitrification in slags, the rate of hydration of slags under different conditions, the identification of compounds, and the effect of various activators.[37]

Estimation of the glassy phase in slags and slag-cements is needed for control purposes. The glass content of a slag has a strong influence on the strength of cement-slag systems. Schwiete and Dolbar[36][37]determined the glass contents of several slags and found a good correlation between the glass contents and compressive strengths (Fig. 12).

Table 5. Reaction Products of Slag Treated with Various Activators

Activator	Crystalline Phases	Comments
NaOH, Na_2CO_3, Na silicate	C-S-H, C_4AH_{13}, C_2AH_8, $Mg(OH)_2$	Some Si is incorporated in C_4AH_{13}; C/S ratio of C-S-H is less than that in cement paste
$Ca(OH)_2$	C-S-H, C_4AH_{13}	C_2AH_8 is absent
Gypsum, Hemihydrate, Phosphogypsum	C-S-H, AFt, $Al(OH)_3$	SO_4 in slag may act as an internal activator
Cement	C-S-H, AFm, AFt, Hydrogarnet, Hydrotalcite-like phase, Vicatite ($C_3S_2H_3$)	Any one paste may not have all these phases

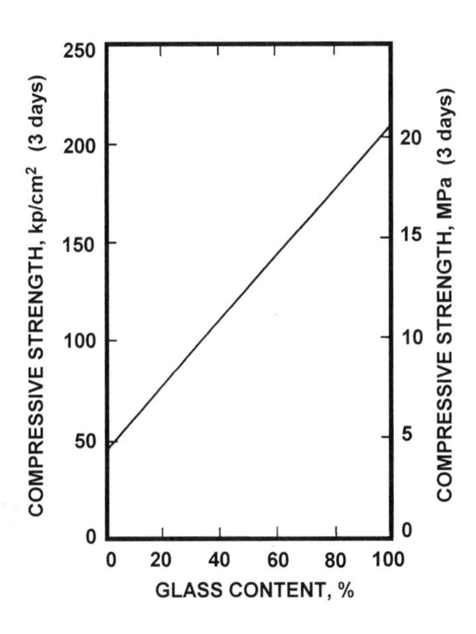

Figure 12. Effect of glass content on compressive strength in slag cements.

A granulated slag heated to a temperature of 800–900°C divitrifies resulting in the appearance of one or more exothermal peaks in DTA. The temperature, the number, and the intensity of the peaks differ from one slag to the other. The area of the exothermic peak is indicative of the amount of the glassy phase. The ground portland cement clinker does not exhibit any peak in this temperature range. Figure 13 illustrates the differential thermograms of three slags of high magnesia contents.[38] Kruger found a good relationship between the peak areas and slag content for three slags (Fig. 14).[37][38] A straight line relationship is obtained with a correlation coefficient higher than 0.99.

The conduction calorimetric technique has been applied extensively to study the hydration of slag cement. The influence of slag on the hydration rate of individual cement compounds is summarized in Table 6.[4] At stage I for cement and C_3A + gypsum systems, ettringite is formed. In pure C_3A, formation of aluminate hydrates and, in the C_3S phase, formation of surface reactions may cause a heat effect at stage I. In stage II, the peak effect is due to the rapid hydration of alite or C_3S and the conversion of ettringite to monosulfate hydrate. During stage III, hydration continues at a lower rate through the diffusion process. Acceleration occurs for all compounds at stage III. Except for $C_3A + CaSO_4 + Ca(OH)_2$, all others exhibit retardation effects at stages I and II.

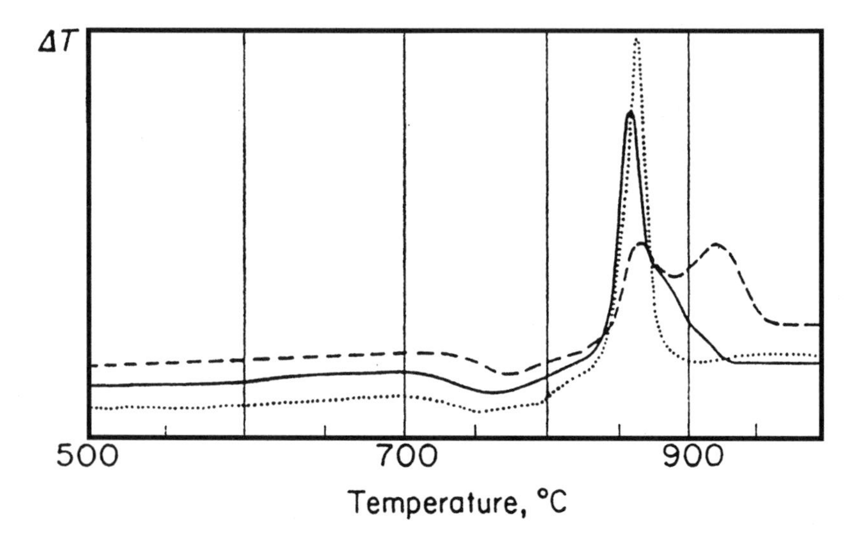

Figure 13. DTA curves of three slags.

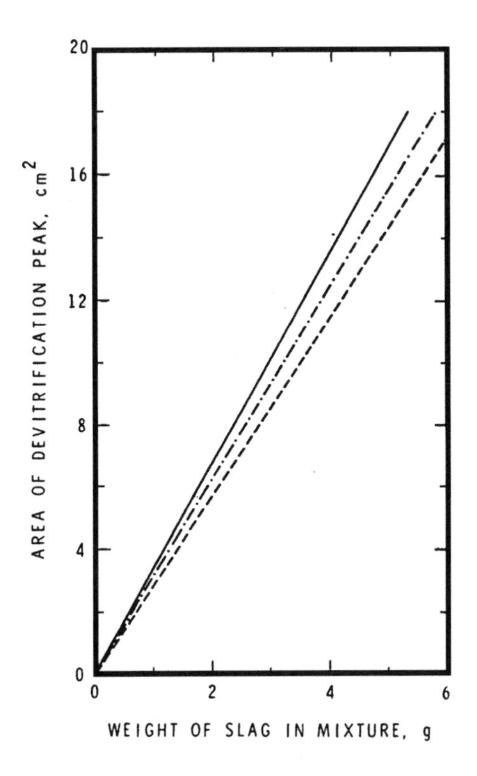

Figure 14. Relationship between the area of the devitrification peak and slag content.

Table 6. Hydration Rate of Cement Compounds Containing Slag

Compound	Stages		
	I	II	III
C_3A	Retardation	Retardation	Acceleration
Alite	Retardation*	Retardation*	Acceleration
C_3S	Retardation	Retardation	Acceleration
$C_3A + CaSO_4$	Retardation	Retardation	Acceleration
$C_3A + CaSO_4 + Ca(OH)_2$	Acceleration	Acceleration	Acceleration

* Under certain conditions, alite may be accelerated when high surface area slag is used.

The rate of hydration of slag cements has been followed by conduction calorimetry, TG, and DTA.[39]–[42] The temperature of curing and the dosage of slag determine the hydration characteristics. Figure 15 shows the conduction calorimetric curves of slag cements containing 0, 35, 55, and 70% slag hydrated at temperatures of 20, 30, and 40°C.[41] The amplitude of the peaks is higher as the temperature is increased at a constant dosage of slag. The amplitude of the peaks is lower in all the mixes containing slag compared to that of the reference paste. The peak for the portland cement paste is more symmetrical than that for slag cement. The C_3S peaks appear later in slag cements, signifying the retardation effect. In slag cements, the time of appearance of the peak at 20, 30, and 40°C is respectively 14, 10, and 7 hours. Conduction calorimetric curves also have shown that the onset of the peak is increased with the SO_3 content. The data derived from conduction calorimetry suggests that the total heat liberated depends on the quality and amount of slag in slag cement.[1] In slag cement containing 30 to 70% slag, the amount of heat developed at 108 hours varied between 174 and 341 J/g.

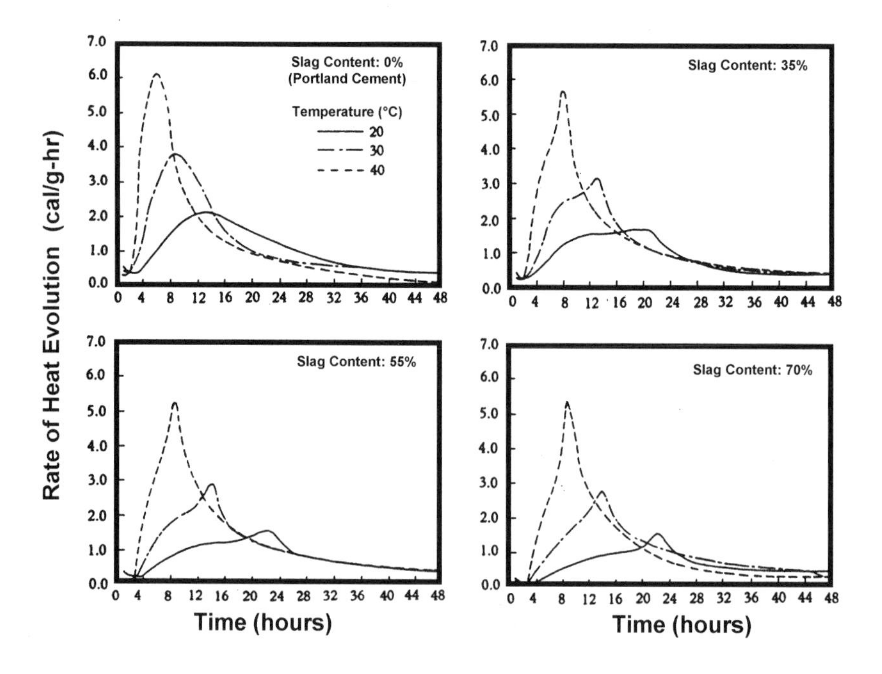

Figure 15. Conduction calorimetric curves of slag cements hydrated at different temperatures and dosages of slag.

Differential Thermal Analysis may be used to monitor hydration products such as calcium hydroxide in slag cements cured at different curing temperatures and times.[27] Some results are shown in Fig. 16.[41] The amount of $Ca(OH)_2$ is decreased with increasing temperature in slag cements. The $Ca(OH)_2$ content decreases as the percentage of slag in the mixture is increased. This is due to the consumption of lime by slag during hydration. This is a dilution effect. There is evidence that the C/S ratio of the C-S-H phase in the pastes is lower than that in portland cement paste. DTA also provides information on the type of C-S-H that is produced at different conditions of curing.

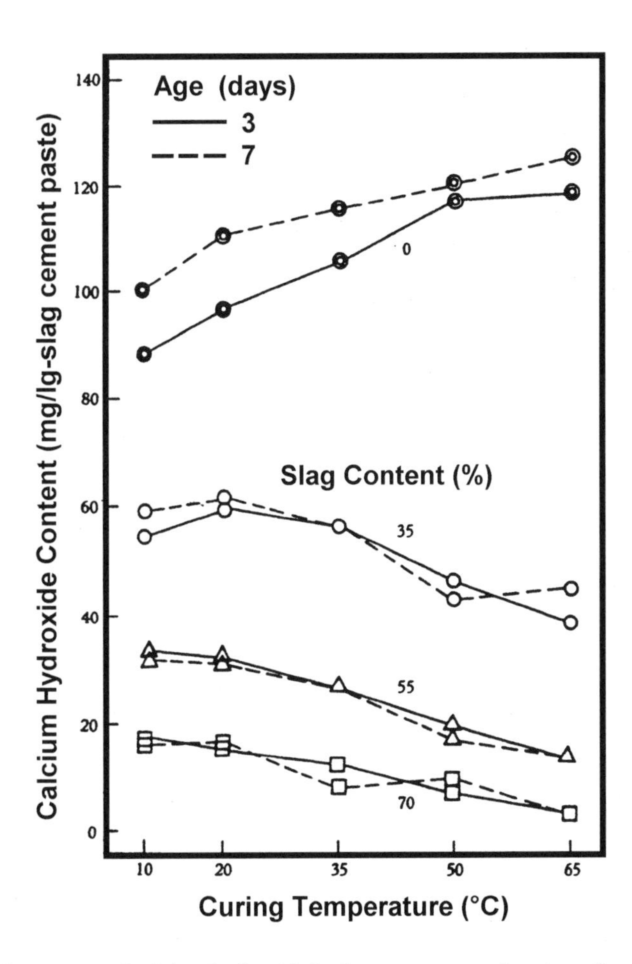

Figure 16. The amount of calcium hydroxide in slag cements as a function of temperature.

Thermogravimetric Analysis has also been carried out to estimate $Ca(OH)_2$ in slag cements. TG yields similar results to those obtained by DTA.

The hydration behavior of slag cement also depends on its surface area. Addition of finer slag promotes acceleration of hydration of the alite phase. For example, a slag cement (40% slag), containing slag having a specific surface area of 4000 cm^2/g, exhibits an exothermic peak at about 11–12 hours in conduction calorimetry. However, in the presence of slag with a specific surface area of 5920 cm^2/g, a more intense peak appears 1–2 hours earlier.[4]

Superplasticizing admixtures are used widely in high performance concrete production. Not only do they influence the rheological parameters, but also the setting characteristics. These effects depend on the type and dosage of the admixture. In Fig. 17, the role of three types of superplasticizers, viz., 0.5% Ca-SNF, commercial SMF, or Na-SNF on the hydration of slag cements is examined.[43] Addition of the superplasticizer results in the retardation in terms of the time of appearance of the exothermal peak and also a decrease of the peak intensity. Na-SNF retards most of the superplasticizers studied.

Steel slag, a by-product obtained in the manufacture of steel, has been used as an aggregate. Conduction calorimetric investigations have shown that steel slag hydrates slowly compared to iron slag, as evidenced by a decrease in the intensity of the exothermic peak.[44]

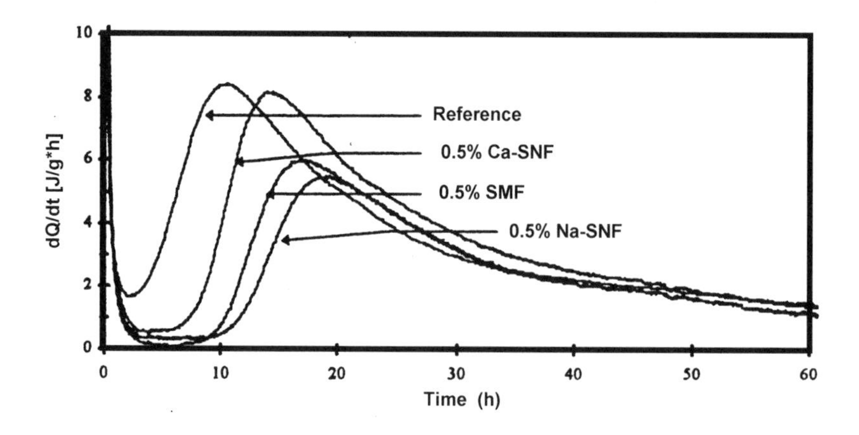

Figure 17. The influence of superplasticizers on the rate of heat development in slag cements.

Slags derived from the manufacture of nickel, copper, and lead are not used as widely as those from iron production. Douglas, et al.,[45] carried out physico-mechanical investigations, including thermal analysis on several non-ferrous slags. Non-evaporable water and $Ca(OH)_2$ contents were determined by TG/DTG up to 90 days. Results for a period up to 28 days are tabulated in Table 7. For most slags, there is a general decrease in the lime content compared to the reference. This would indicate that the slags are pozzolanic. In some slags (B, F, and G), there is a gradual decrease in lime content while for others (A, C, and D) there is either a progressive increase or a more erratic distribution from 7 days to 90 days. Addition of lead slag does not result in the setting of cement as indicated by the absence of lime. It would also appear that slags B, F, and J with low lime contents accelerate the early hydration of portland cement. Correlation was obtained between lime content and the corresponding glass content (in slags) or compressive strength.

Table 7. The Amount of Lime Formed at Different Times in Non-Ferrous Slag Cements

Type of Slag	Calcium Hydroxide, % (Portland Cement Basis)			
	1 day	3 days	7 days	28 days
Reference	13.1	20.0	20.9	21.1
Cu Slag A	9.6	13.1	16.0	17.4
Cu Slag B	8.2	24.5	14.4	15.3
Cu Slag E	8.4	15.4	13.1	16.8
Cu Slag F	7.6	23.1	13.4	13.7
Cu Slag I	10.1	14.9	16.0	19.1
Cu Slag J	8.4	22.0	11.5	12.6
Cu Slag K	9.3	13.6	18.7	14.2
Ni Slag C	12.4	11.9	15.1	14.7
Ni Slag D	10.6	15.4	14.9	17.6
Pb Slag G	0	10.6	23.3	15.4
Pb Slag H	0	11.6	9.6	14.5

Many massive concrete elements are made from slag cements. There is a possibility of early age thermal cracking in such structures, hence, a general hydration model was developed based on conduction calorimetric studies.[46] In this model, the heat production is calculated as a function of degree of hydration or temperature. This model enables simulation of the field temperature in massive concrete elements.

The DTA and X-ray methods have proven useful in the examination of set supersulfated cements.[47] Thermograms of supersulfated cement hydrated to different times are plotted in Fig. 18. The cement exhibits an endothermal effect at 150–170°C that is mainly attributed to ettringite. A large amount of ettringite is present even at 28 days and is lower at later periods. The increased strength in cement at 28 days is due mainly to the C-S-H phase and alumina gel. A quantitative estimation of products of hydration of supersulfated cement has also been carried out by thermal methods.[47]

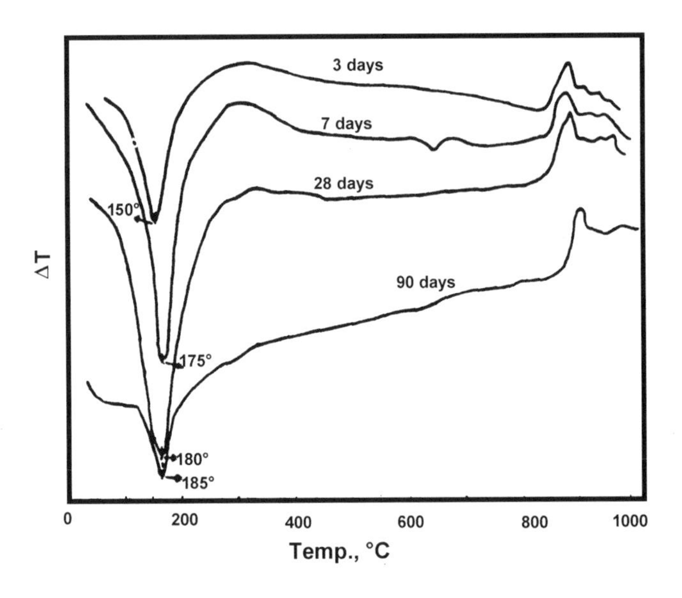

Figure 18. Differential thermograms of supersulfated cement.

Autoclave treatment of slag with lime results in an increased strength. Kalousek and Chopra[47] studied by DTA the autoclave products of slag-lime mixes and tried to investigate the causes leading to strength decreases in certain instances. The increase in strength corresponded to an increase in the amount of the C-S-H phase. In certain mixes containing larger amounts of $Ca(OH)_2$, the formation of α-C_2SH resulted in low strengths. The CH, C-S-H, and α-C_2SH (endothermal peak at 500°C) phases could be identified by DTA.

The long-term behavior of slag concrete exposed to carbonation is a subject of controversy. Thermal methods are suitable to monitor the depth of carbonation by estimating $CaCO_3$, as well as the $Ca(OH)_2$ contents. In a systematic study, Litvan and Meyer[48] examined core samples taken from different depths from slag concrete walls that were exposed to natural elements for 20 years. The amounts of carbonation and lime at various depths were determined by TG from the inside and outside of the building. Figure 19 compares the data for slag cement concrete (GBFSC) with those of normal cement concrete (OPC). Very low amounts of $Ca(OH)_2$ were found at all depths in slag concrete. In normal concrete, however, there was a significant amount of $Ca(OH)_2$. On the exposed surface, the rate of carbonation is higher in slag concrete than in normal concrete. While normal concrete contains a significant amount of lime even in carbonated areas, slag concrete is devoid of it. This means the pH in slag concrete will be low, and this condition will promote corrosion of reinforced steel.

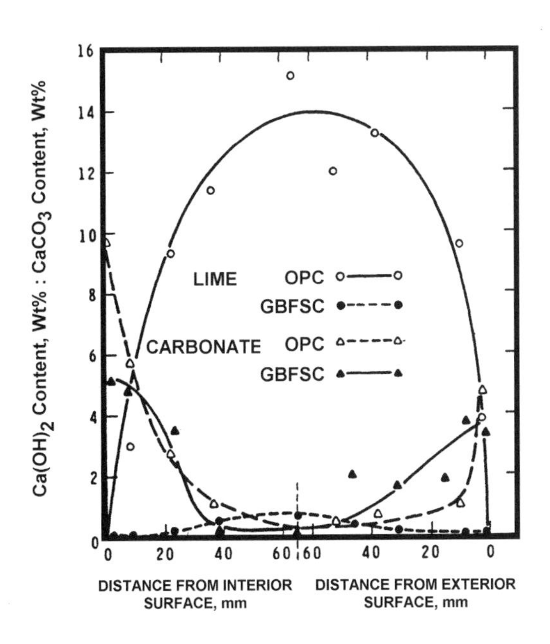

Figure 19. Concentration of $CaCO_3$ and $Ca(OH)_2$ at various depths for normal and slag concretes.

The slag by itself is not hydraulic, but treating it with an activator enables it to react with water. Conduction calorimetry and DTA techniques are suited to evaluate the efficiency of activators for slags. Thermal techniques have been applied to study the effectiveness of activators, such as NaOH, Na_2CO_3, and Na_2SiO_3,[49] NaBr, NaI, Na_2SO_4,[50] etc., and water glass.[51] The DTA of the slag-gypsum anhydrite mixture activated by small amounts of sodium sulfate, calcium hydroxide, and ferrous sulfate is shown in Fig. 20.[49] The endotherms at 145–150°C and 200–220°C are due to ettringite and gypsum, respectively. Ettringite is present in larger amounts at earlier times. The endotherm at 200–220°C gradually increases as the hydration progresses. The exotherm at 372–413°C and 880–890°C are attributed to the inversion of $CaSO_4$ (III) to $CaSO_4$ (II) and slag divitrification, respectively. The C-S-H dehydration effect occurring at 120–140°C may have been masked by the ettringite peak. Based on DTA and IR results, it has been concluded that the increased strengths with curing are due to activated hydration of the gypsum anhydrite-slag system.

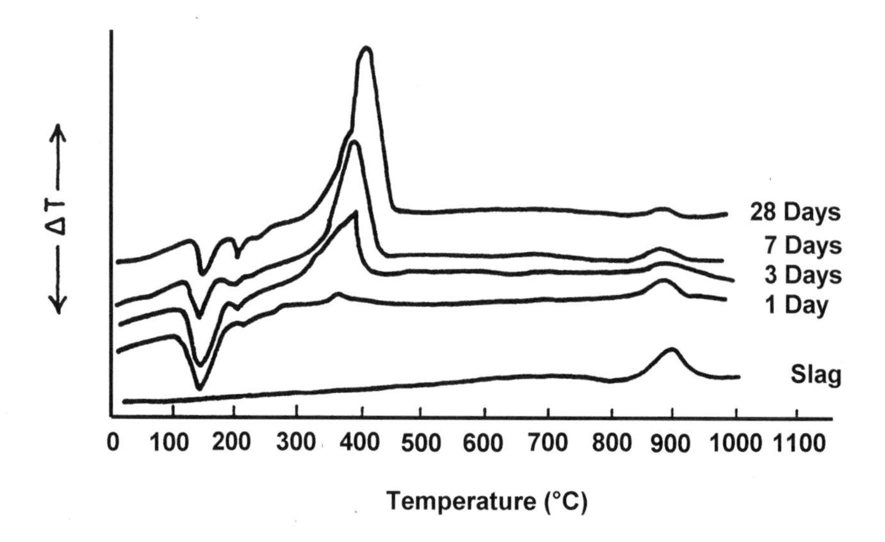

Figure 20. Differential thermograms of slag anhydrite mixtures hydrated with activator mixtures of lime and calcium sulfate or ferrous sulfate.

The role of activators in slag mixtures may also be studied by conduction calorimetry. The effect of 1% NaF, NaCl, NaBr, and NaI on calorimetric curves are shown in Fig. 21.[50]

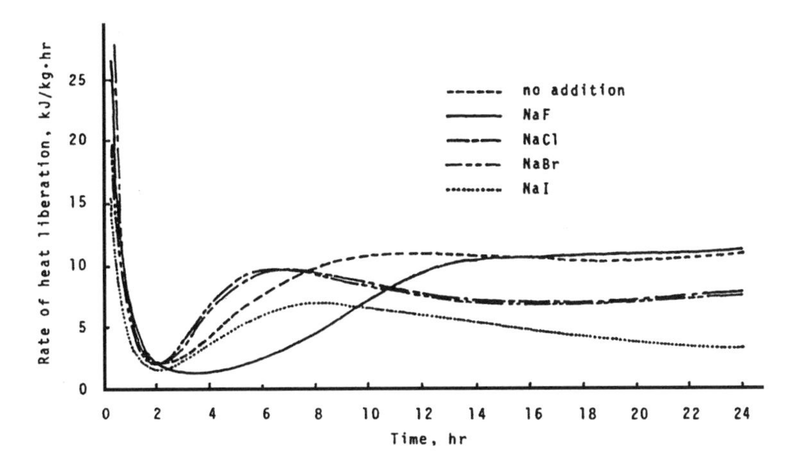

Figure 21. Conduction calorimetric curves of slag activated by some sodium salts.

The rate of reaction is increased by the addition of NaCl and NaBr in the 2–8 hour period. By the addition of these compounds, an increase in strength results at all ages up to 91 days.

5.0 RICE HUSK ASH

Rice husks, also called rice hulls, are the shells produced by the dehusking operation of paddy rice. Each ton of paddy produces about 400 pounds of husk which on combustion produces about 90 pounds of rice husk ash (RHA). Amorphous silica may be obtained by maintaining the combustion temperature of RHA below 500°C under oxidizing conditions for prolonged periods or up to 680°C, provided the total time is less than a minute.[52]

The predominant constituent of RHA is SiO_2, which exceeds 90% of the total ash. The RHA may have a specific surface of 40 m^2/g and is highly pozzolanic. Rice husk ash concrete is fabricated by utilizing high dosages of superplasticizers. The compressive strengths of RHA concrete are generally higher than the concrete containing no ash. Typically the 3 day strength of mortars with 0, 30, 50, and 70% RHA is 22.7, 32.3, 26.5, and 24.3 MPa respectively, and the corresponding values at 28 days are 43, 59.5, 58.3, and 43.3 MPa.[53]

Thermal analysis techniques have been applied to determine the decomposition of RHA and also to investigate the hydration characteristics of RHA-cement pastes exposed to different conditions. The reported results of DTA, DSC, and TG of RHA-cement pastes show variations. The thermal curves of RHA from different sources are not comparable because of variation in their physico-chemical characteristics, the conditions under which the husk is heated, and other factors.[4][54]–[58] Thermograms generally show endothermal and exothermal peaks. Those ashes formed at lower temperatures show an exothermal effect for the oxidation of the unburned carbon. Endothermal effects at about 100°C denote the expulsion of water from the adsorbed surface. The oxidation reactions correspond to the loss of weight in TG. Some DSC results have been obtained for the ash obtained at 1200°C.[56] The peaks in DSC (Fig. 22) were interpreted from XRD studies. An exothermic peak at 135°C is attributed to the transformation of the trydimite phase (T_α to T_β) and four endothermal effects at 190, 220, 235, and 250°C represent the transformation of the low form of disordered cristabolite. The endothermic peak at 250°C is caused by a transformation of the well-crystallized form.

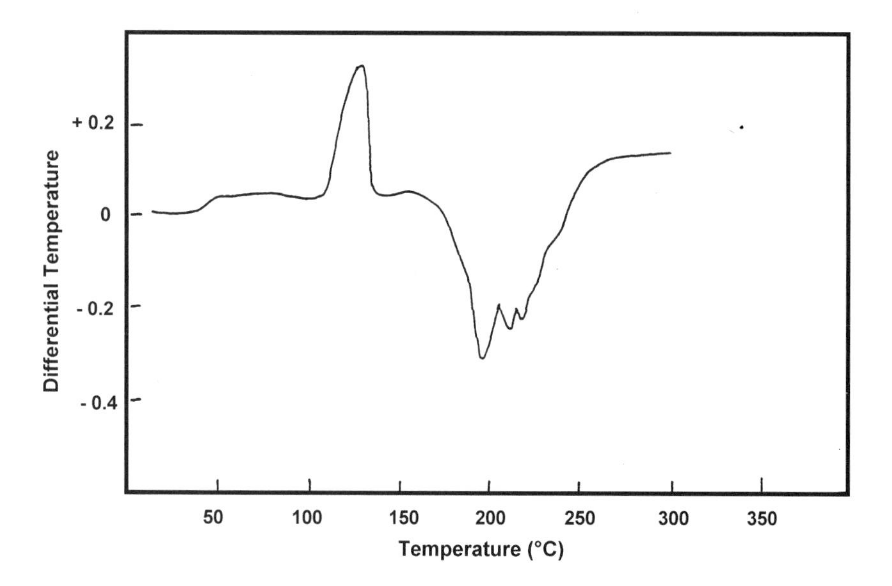

Figure 22. DSC curves of the rice husk ash fired at 1200°C for three hours.

Because RHA has a high surface area, it acts as a good pozzolan. A mixture of CH and RHA promotes the formation of a C-S-H product of composition $Ca_{1.5}$•$SiO_{3.5}$•XH_2O.[59] The TG/DTA curves for the system RHA-CH mixtures are shown in Fig. 23. The evaporation of water is denoted by an endotherm at about 140°C. Thereafter, there is a gradual loss of water and at 827°C crystallization to wollastonite is indicated by an exothermic peak.

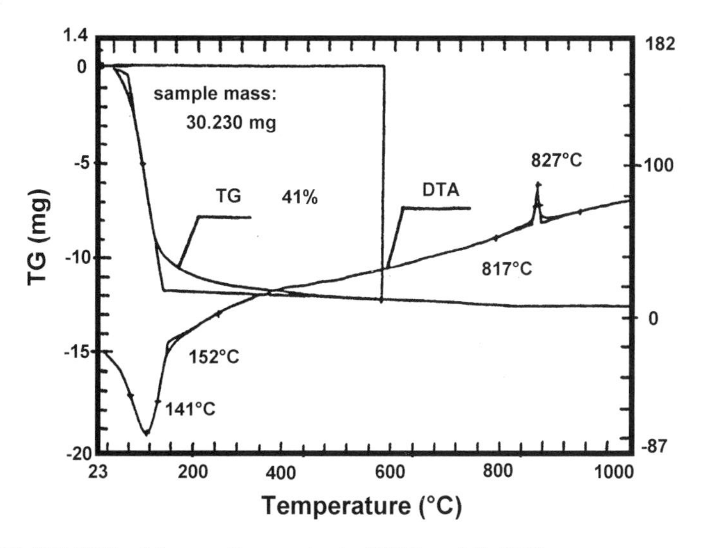

Figure 23. TG-DTA of the reaction product of RHA and $Ca(OH)_2$.

The rate of heat development in RHA cement pastes containing from 5 to 20% RHA has been followed by Hwang and Wu.[58] The shapes of the curves for all the mixes including the curve for the reference cement are similar (Fig. 24). The first peak that appears within a few minutes is lower in all samples containing RHA. The second peak occurring at about 10 hours is lower in samples with RHA-cement pastes. The lower heat effect is caused by the reaction of K^+ and $Ca(OH)_2$ with RHA. The dilution effect also plays a part.

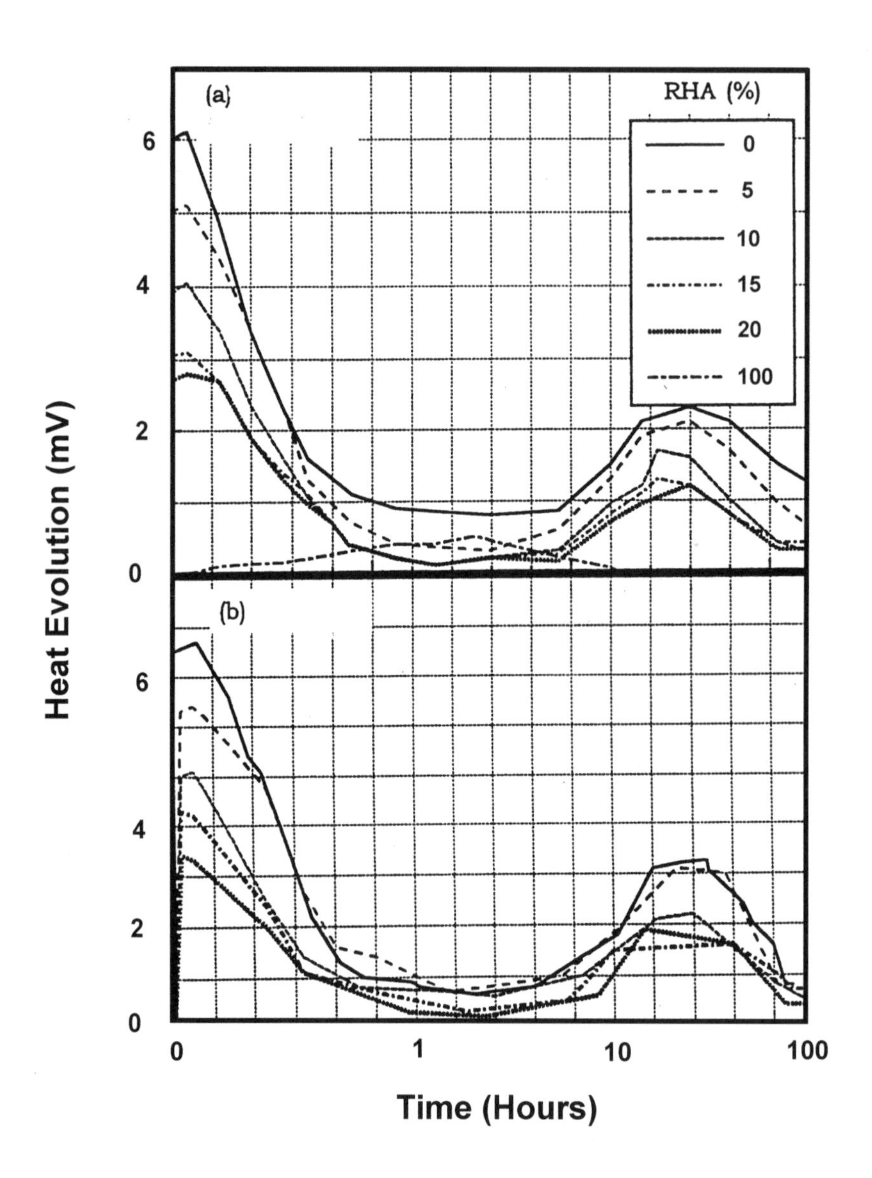

Figure 24. Heat evolution in cement paste containing different amounts of RHA. *(a)* w/c ratio = 0.35; *(b)* w/c ratio = 0.47.

6.0 METAKAOLINITE

Soils and clays, in general, when calcined give off adsorbed, interlayer, and hydrated types of water. These effects produce endothermal peaks or loss of weight in DTA and TG, respectively. The endothermal peaks are followed by exothermal peaks that are caused by re-crystallization. Although many types of clay minerals such as montmorillonite, illite, and some shales show these effects, they are not suitable as pozzolans in concrete. *Metakaolin*, formed by heating kaolinite, seems to be the most suitable additive material for cement. Heating of kaolinite involves removal of adsorbed water at about 100°C and dehydroxylation at above 600°C, followed by the formation of metakaolinite, an almost amorphous product. The sequence of reactions is as follows:[60]

$$Al_2Si_2O_5(OH)_4 \text{ (Kaolinite) at 600°C}$$

$$\text{forms } Al_2Si_2O_7 \text{ (Metakaolinite)} + 2H_2O$$

At higher temperatures, metakaolinite transforms to crystalline compounds as follows:

- Metakaolin at 925°C forms silicon spinel
- Silicon spinel at 1100°C produces pseudo-mullite
- Pseudo-mullite at 1440°C forms mullite + cristobalite

Metakaolin (MK) is not cementitious by itself, but, having a high surface area of about 20 m^2/g, it reacts with calcium, sodium and potassium hydroxides, and gypsum/cement. The hydrated lime + MK reaction yields gehelinite hydrate (C_2ASH_8) and C-S-H. The C/S ratios of the calcium silicate hydrate vary between 0.8 and 1.5. At higher hydration temperatures, gehelinite is the stable phase. The hydration reaction at 50°C produces a hydrogarnet phase of approximate composition $C_3AS_{0.3}H_{5.3}$. When MK is mixed with portland cement and hydrated, the lime formed as a hydration product of cement is consumed by MK and the C-S-H product is formed. At higher MK contents, gehlenite may form. The C/S ratio of the product may vary between 1.0 (at 50% MK) and 1.6 (at 20% MK).[61] The addition of MK to portland cement results in the increased strength of the product. For example, at 10% MK addition, the compressive strengths at 1, 3, and 28 days are 25, 33, and 40 MPa respectively, compared to the corresponding values of 21, 25.5, and 36.4 MPa for the reference samples.[62]

Thermo-analytical techniques are applied widely to monitor the reactions occurring in the raw clay materials and also to follow the pozzolanic effects of metakaolinite when combined with lime or cement.

Not much work has been carried out on studies related to the effect of MK on the hydration of individual cement compounds. The effect of 30–90% MK addition to C_3S and C_3A hydration has been examined.[61] At a curing period of 28 days, the DTA curves indicate that as the C_3S:MK ratio decreases from 2.4 to 0.6, the peak due to CH (C) is decreased (Fig. 25). The main products were found to be C-S-H (A) and stratlingite (B)(C_2ASH_8). At 90 days, C_3AH_6 is also formed. These results show that CH formed during the hydration is consumed by MK. It is also evident from the figure that the amount of C-S-H formed is larger in the C_3S:K mixtures compared to that in pure C_3S. The accelerating effect of MK on C_3S hydration is also confirmed by conduction calorimetric studies. The hydration of C_3A is not influenced by MK.[61]

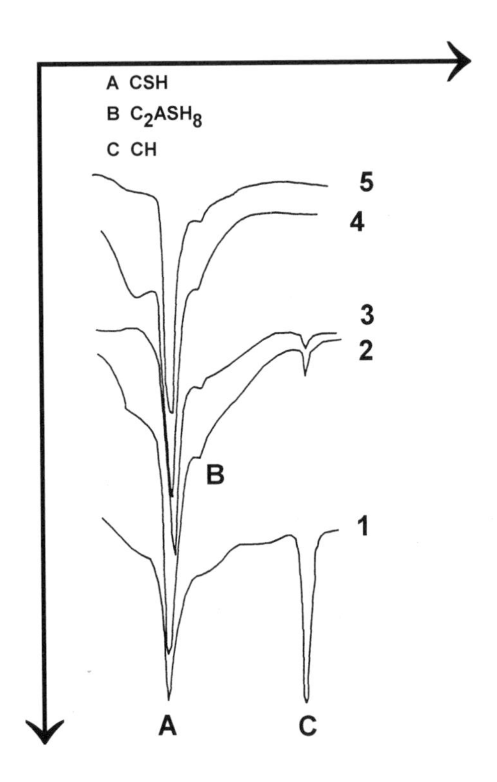

Figure 25. DTA of C_3S pastes containing different proportions of metakaolinite. 1. C_3S; 2. C_3S:MK = 2.4; 3. C_3S:MK = 1.4; 4. C_3S:MK = 0.9; 5. C_3S:MK = 0.6.

The addition of MK influences the hydration of cement significantly. The DTA curves of cement pastes treated with various amounts of MK are presented in Fig. 26.[63] The first peak at 135°C is due mainly to the dehydration of C-S-H (I). The endothermal effect at about 175°C is caused by the decomposition of C_4AH_{13} and that at 480°C is due to the decomposition of CH. Two other endotherms at 740° and 765°C are ascribed to the decomposition of the amorphous and crystalline forms of calcium carbonate respectively. The peak effects of C-S-H (I) and C_4AH_{13} increase as the amount of MK increases up to a 30% addition signifying an acceleration effect. The decrease in the peak area of CH is attributed to the reaction between MK and CH liberated by the hydration of the cement.

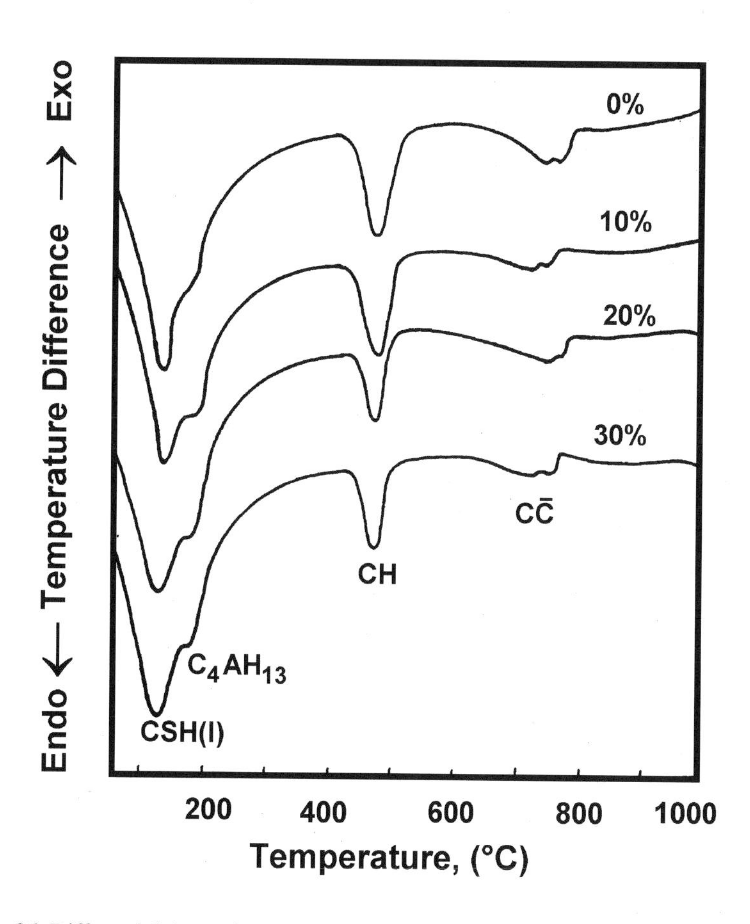

Figure 26. Differential thermal curves of cement pastes containing different amounts of MK.

The reactivity of MK depends on the temperature conditions under which it is obtained, its surface area, and types of other materials contained in it.[64][65] Ambroise, et al.,[66] examined the pozzolanic activity of four clays of different compositions, by reacting them with lime. The consumption of lime was followed by DTA (Table 8). The most reactive mixtures were Nos. 1 and 2, containing 92% kaolinite, and the peak areas corresponding to lime in these samples are 6 and 0 respectively. Samples 3 and 4, which contained illite, quartz, carbonate, and only 19–29% kaolinite, reacted only to a limited extent with lime, as evident from large areas for the lime peak. The mortars with mixes 1 and 2 show higher compressive strengths than the others.

Table 8. Dependence of Clay Composition on Reactivity and Strength

Clay Composition	DTA Peak Area (mm^2)	Compressive Strength (MPa)
1. Kaolinite (92.5%)	6.0	25.9
2. Kaolinite (92%)	0.0	24.1
3. Kaolinite (19%), Illite (37.5%), Quartz (29%), Carbonate (8%)	500.0	10.7
4. Kaolinite (29%), Illite (30%), Quartz (18%), Carbonate (11%)	630.0	12.0

Metakaolin reacts with lime to yield calcium silicate hydrate.[60] Metakaolin may also be activated by other materials such as alkali metal hydroxides, water glass, etc. Activation leads to a polycondensation product with cementing properties. The type of MK, composition, temperature at which it is produced, surface area, etc., determine the strength development characteristics of the product.[67][68] In conduction calorimetry, an exothermic peak results by the reaction of MK and the activator. A strong asymmetric peak in calorimetry is associated with an amorphous inorganic polymer formation which is responsible for strength development. A shoulder also appears corresponding to a zeolite crystallization process.[67] In DTA, an endothermal peak appearing at 165–175°C is attributed to the process of dehydroxylation of the amorphous and crystalline zeolite.[68]

At ambient temperatures, MK is known to react with Ca, Na and K hydroxides, and hydrated cement. Attempts have been made to increase the rate of strength development by autoclaving mixtures of cement-quartz or lime-quartz.[69][70] DTA of the system cement-quartz treated with MK are shown in Fig. 27.[71] Aluminum substituted C-S-H is identified by a peak at 865–880°C. The presence of hydrogarnet is indicated by a small exotherm at 940°C. Unreacted quartz can be recognized by the endotherm at 573°C. The gradual decrease of the endothermal effect of C-S-H at 170°C and also that at 865–880°C indicates formation of decreased amounts of C-S-H and that more aluminum is substituted in the C-S-H phase. According to the work of Klimesch and Ray,[70] a better delineation of the thermal peaks is possible in the autoclaved systems using the second derivatives of thermal curves.

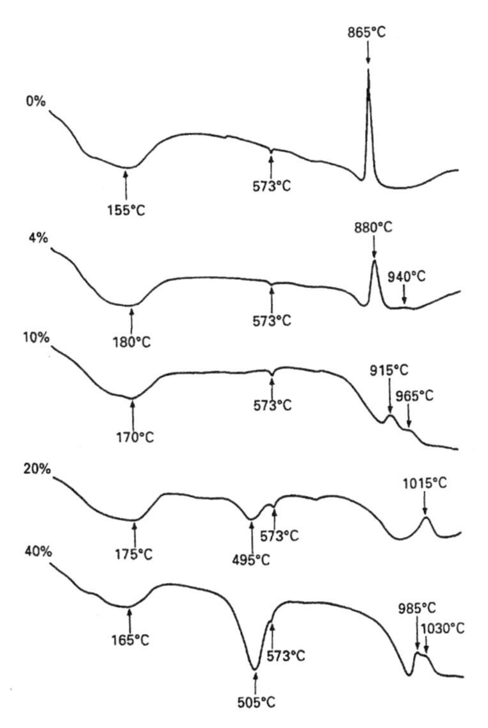

Figure 27. Thermograms of cement-quartz (50:50) mixture autoclaved with different amounts of MK.

7.0 NATURAL POZZOLANS

A *pozzolan* may either be a material of natural origin or an artificial preparation. The natural pozzolans are generally of volcanic origin or are derived from sedimentary rocks. The well-known natural pozzolans are volcanic glass, tuffs, and siliceous materials such as diatomite and rocks. Natural pozzolans are used in combination with lime or portland cement, and the resultant product possesses good sulfate resistance and produces low heats. Some of them are known to reduce the alkali-aggregate expansion reaction. These pozzolans may serve as partial replacements of cement. The reactivity of the pozzolan is determined by its chemical and mineralogical composition, morphology, the amount of glassy phase, and fineness. Most natural pozzolans have a high $SiO_2 + Al_2O_3$ content and a glassy or amorphous structure, with the exception of the zeolite. The products formed in the pozzolan-cement paste mixtures are similar to those formed in pure cement pastes. The products that are produced by the reaction of a natural pozzolan with lime include C-S-H (calcium silicate hydrate), $C_3A{\bullet}CaCO_3{\bullet}12H_2O$ (calcium carboaluminate), $C_3A{\bullet}3CaSO_4{\bullet}32H_2O$ (ettringite), C_4AH_x (hexagonal calcium aluminate hydrate), C_2ASH_8 (hydrated gehelinite), and $C_3A{\bullet}CaSO_4{\bullet}12H_2O$. The types and the amounts of products depend on the chemical constituents in the pozzolan and the curing conditions. In general, the addition of pozzolan promotes the hydration of cement. Also, iron and aluminum can be incorporated in the C-S-H structure. Adequate strengths, improved durability, and low permeability can only be obtained in pozzolan concrete after a longer curing period, compared to that required for normal portland cement concrete.

Studies have been conducted on the thermal behavior of natural pozzolans. For example, the TG analysis indicates that the weight loss between 60° and 400°C in zeolites amounts to 80–90% of the total amount. This is due mainly to the loss of capillary, inner layer, and zeolitic water. A very small weight loss occurs between 600° and 700°C. The loss at higher temperatures is due to the dehydration of the zeolitic water. Zeolites may be classified by the decomposition reactions that they undergo when subjected to thermal analysis. The stability range of harmotone, phillipsite, pualingite, etc., is between ambient temperature and 250°C. Gismondine, yugawaralite, stilbite, etc., have the stability range between 250° and 400°C. Laumonite, faujasite, natrolite, and others have the stability range between 400° and 600°C, and the values for analcime, erionite, offretite, etc., exceed 600°C. The TG/DTA/TMA curves of various pozzolans such as kaolin, tuff, basalt, diatomite, and activated kaolin have been published.[35]

Conduction calorimetry and DTA have been applied to the study of the effect of natural pozzolans on the hydration of cement phases.[1][20][72] Ogawa, et al.,[72] examined the role of five pozzolans on the hydration of the tricalcium silicate phase using TG/conduction calorimetric techniques (Fig. 28). The samples V (white clay), F (volcanic ash), and R (volcanic ash) are natural pozzolans whereas T is a fly ash and S is a slag. All were hydrated at a w/c ratio of 0.5 and C_3S:pozzolan ratio of 6:4. The amount of $Ca(OH)_2$ formed was estimated by DSC/TG. All the pozzolans modify the calorimetric curves to a great extent. The natural pozzolans have an accelerating effect on hydration as evidenced by the intense peaks which appear somewhat later with respect to the reference. The induction period does not vary. The accelerating effect is explained by the surface of the pozzolan acting as a precipitation site and also by its ability to adsorb Ca^{+2} ions. Lowering of the calcium ion in the solution phase leads to the acceleration effect. The amount of CH determined by thermal methods does not follow the same trend as the heat evolution data. This can be explained by the reaction and adsorption of CH by the pozzolans, especially those having high surface areas and reactivity. Conduction calorimetry of C_3S containing montmorillonite and magnesium silicate clays indicates that the peaks in the clay-containing samples are of larger intensity than that of C_3S and also appear earlier. The induction period is reduced from 6 hours to 2 hours in pozzolan-containing samples. This confirms that these clays accelerate the hydration of C_3S.

The products of the lime-pozzolan reaction include calcium silicate hydrate, calcium aluminate hydrate of the form C_4AH_x, in which H varies from 9 to 13, hydrated gehelinite (C_2ASH_8), calcium carboaluminate ($C_3A{\bullet}CaCO_3{\bullet}H_{12}$), high sulfoaluminate ettringite ($C_3A{\bullet}3CaSO_4{\bullet}32H_2O$), and low sulfoaluminate ($C_3A{\bullet}CaSO_4{\bullet}12H_2O$). All the above compounds yield characteristic thermal effects in DSC/DTA/DTG. In general, all the products will not be present at any one time. For example, calcium aluminate mono-sulfate hydrate exists at early stages of hydration and is converted to ettringite subsequently. Only pozzolans with high alumina contents form gehelinite, and they also reduce the amount of C-S-H. Jambor,[35][38][73] studied extensively by thermal methods, the system lime-natural pozzolan-water containing various types of pozzolanic materials (Fig. 29). The first ten mixtures (A–J) contained natural pozzolans, the 11th sample (K) was activated kaolin, the 12th to 15th referred to fly ashes (L–O), the 16th to 18th to siline gels (P–R), the 19th was silica gel (S), and the 20th to 22nd (T–V) were chemical glasses. All these mixtures were cured for 400 days at 20°C. There are differences in the intensity of various peaks and some of them do not exhibit the high temperature exotherm. The endothermal effect at 140–170°C is caused by the dehydration of the bound water from the C-S-H (I) phase. The endothermal peak at 560°C is due to the decomposition of the $Ca(OH)_2$. Most mixes have calcium hydroxide in them. There is also an endothermal effect at 800–900°C representing the decomposition of $CaCO_3$. The high temperature exothermic peaks above 800°C indicate the development of wollastonite or β-C_2S.

In the study of the lime-natural pozzolan systems, lime-rich and lime-poor calcium silicate hydrates have been differentiated by DTA.[74] The endothermal peak at 90°C was attributed to the lime-rich C-S-H phase, and that at 125°C, to the lime-poor C-S-H phase. An exothermic peak that appears in this system at about 850°C is caused by the crystallization of wollastonite.

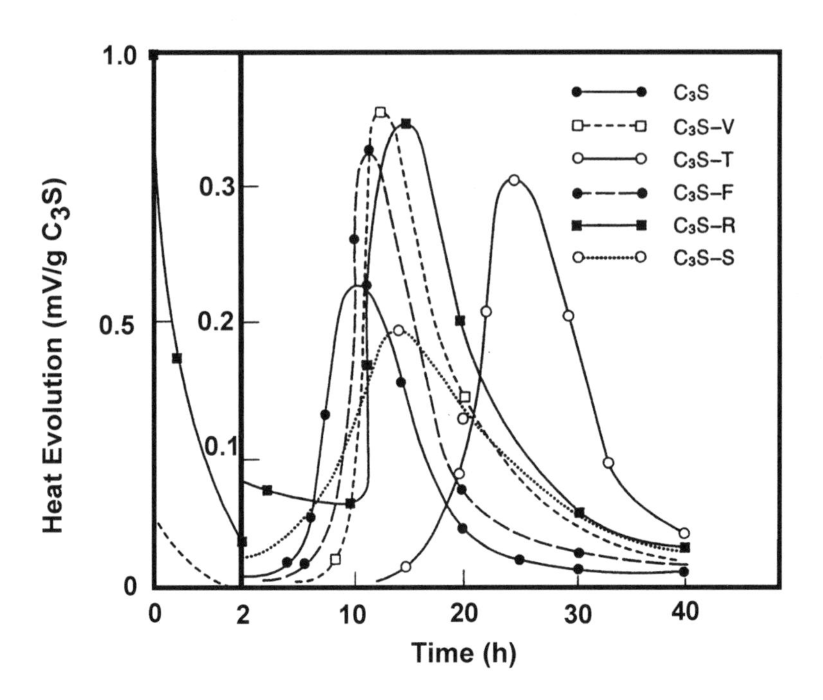

Figure 28. Conduction calorimetric curves of tricalcium silicate treated with natural pozzolans.

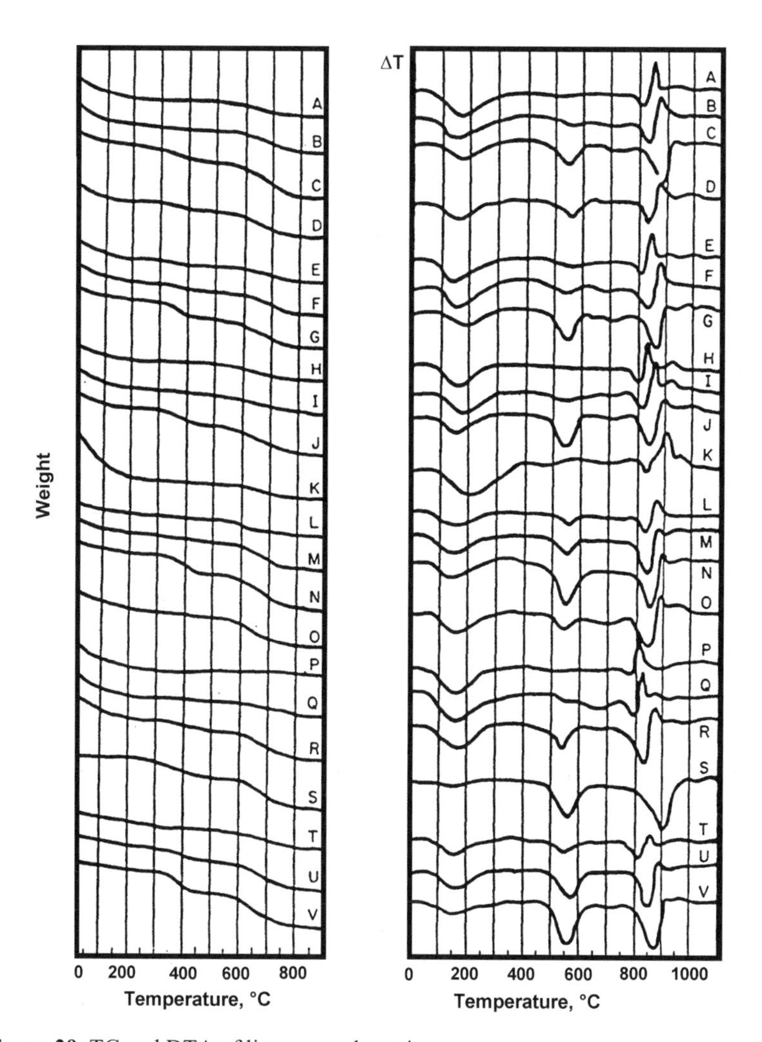

Figure 29. TG and DTA of lime-pozzolan mixtures.

Conduction calorimetric investigations have been carried out on portland cement in which natural pozzolans have been incorporated.[75] It has been found that in the pozzolan-containing pastes, the dormant period is decreased and the normalized height of the second main peak is greater than that given by the reference cement. This indicates that the pozzolan accelerates the hydration of alite. It is possible that the acceleration effect is due to a combination of factors, viz., higher specific surface area of the pozzolan, its chemical composition, the physical state of the surface, and the rate of release of the alkalis.

The conduction calorimetric curve of cement with 45% diatomite shows that the pozzolan modifies considerably the heat generated in the samples (Fig. 30).[2][76] Not only does the main exothermic peak appear earlier, but it is also of greater intensity than that exhibited by the reference cement. This indicates that the tricalcium silicate phase of the cement is accelerated by diatomite.

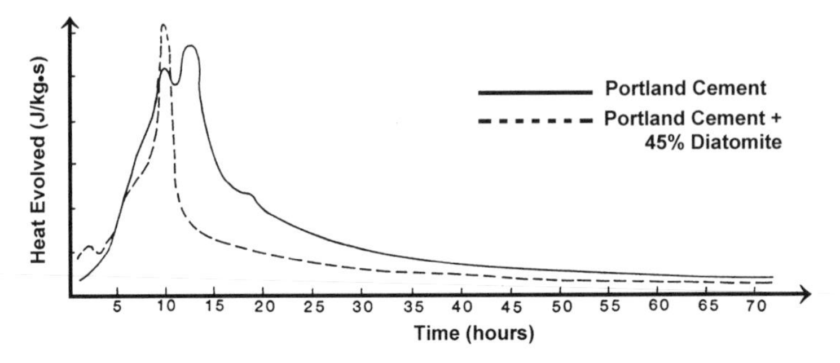

Figure 30. Heat generation in the portland cement-diatomite mixture.

Thermal analytical techniques have been applied to investigate the causes leading to the deterioration of concrete subjected to various environmental factors. A mortar nearly two thousand years old obtained from El Tajan near Mexico City was analyzed by TG and electron microscopy.[77] TG analysis was done on samples taken from different areas and depths. The loss of water below 100°Cwas caused by the adsorbed water from the volcanic tuff, while the endothermal effect at >700°C corresponded to the carbonated lime and carbonated silicates and aluminates derived from the pozzolan. The extent of the reaction of lime with pozzolan was computed to be 6.91%.

8.0 RELATIVE EFFECTS OF POZZOLANS AND THEIR MIXTURES

The role of individual supplementary cementing materials and additives has been described in the previous sections. The effect of these materials, viz., silica fume, slags, rice husk ash, fly ash, metakaolinite, or natural pozzolans on the properties of concrete such as workability, setting, rate of hydration, strength development, heat, durability, etc., is known to vary significantly. Attempts have, therefore, been made to study the properties of cement paste and concrete into which are blended two or more additives in order to derive the optimum properties and benefits. These have implications in terms of availability of the materials and economy. This section describes the effect of supplementary cementing-material mixtures, and also the comparative effects of different supplementary cementing materials on cement hydration. Kasai, et al.,[78] applied DTA/TG techniques to characterize zeolites and other supplementary materials. In zeolites, peaks at 80–400°C represented the loss of zeolitic water. An exothermic peak at 600°C in fly ash signified combustion of carbon, and that at 800–900°C corresponded to the phase transition of the glassy phase. Silica fume exhibited a small peak at 500°C for the crystallization of a cristobalite from the amorphous phase.

Not much data is available on the relative effects of pozzolans on the individual cement phases. The DTA curves of β-C_2S treated with 5% fly ash, silica fume, or slag hydrated for 1 to 90 days are compared with respect to the untreated sample in Fig. 31.[79] The reference C_2S exhibits endothermal effects at 100–150, 160–300, 350–450, 500–600, 700–800, and 800–900°C. The endothermal effects in the range 100–150°C are attributed to the loss of free water and the dehydration of interlayer water from the C-S-H phase. The decomposition of calcium carbonate occurs with the peak at 700°C. The sample, hydrated for one day, shows the peaks for the presence of C-S-H, $Ca(OH)_2$, and $CaCO_3$ which increase in intensity with the curing period. It is also evident that $Ca(OH)_2$ is converted to $CaCO_3$ at 90 days. In the presence of slag, C_2S shows several endothermal effects including an exothermal peak that can be ascribed to the oxidation of sulfide in the slag. The presence of smaller amounts of $Ca(OH)_2$ and larger amounts of C-S-H indicate that $Ca(OH)_2$ is consumed by the slag. In samples containing silica fume, an exothermal peak at 870°C is indicative of the formation of wollastonite from C-S-H (I). This peak appears even at 1 day of hydration. The rapid consumption of calcium hydroxide is followed by a corresponding enhancement of the exothermal effect. The fly ash-cement mix also indicates the formation of the C-S-H (I) phase and a peak at about 230°C due to C_4AH_{13} or gehelinite hydrate. In general, the amount of hydration products in the β-C_2S treated with pozzolans is higher than that in the untreated sample. Conduction calorimetry shows that the total heat evolved in cement blends with 30% sand, fly ash, or silica fume is higher than that produced by the reference cement paste (Fig. 32).[22] These results indicate that fly ash, silica fume, and ground sand accelerate the hydration of cement, silica fume being the most effective admixture.

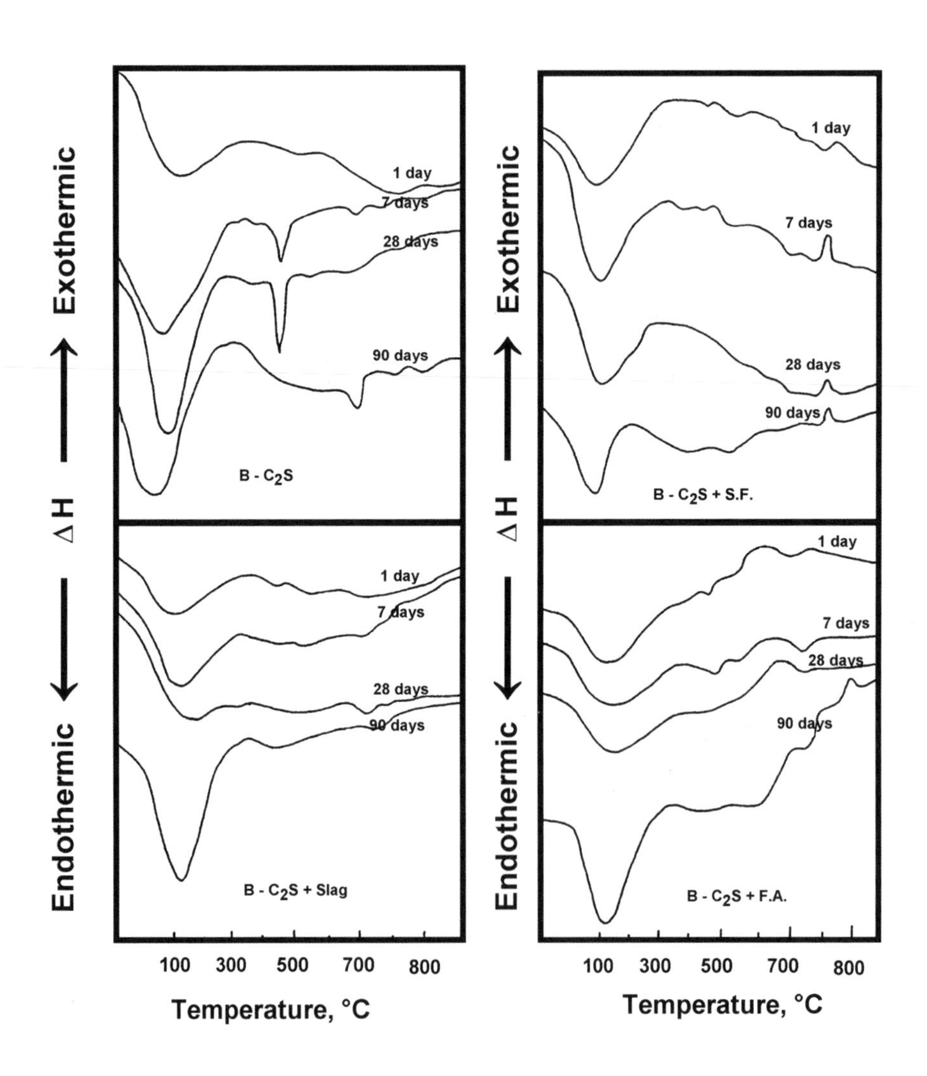

Figure 31. Thermograms of β-C_2S treated with silica fume, slag, and fly ash.

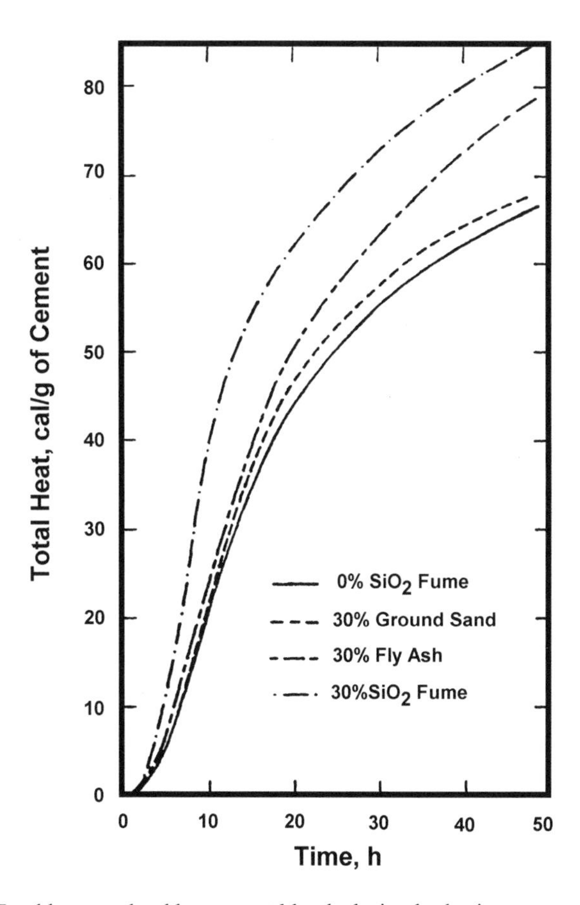

Figure 32. Total heat evolved by cement blends during hydration.

The addition of supplementary materials such as fly ash and silica fume affects the induction period in cement hydration. The relative effects have been illustrated by conduction calorimetric curves.[80]

The relative effects of 10% of fly ash, limestone, or sludge on the hydration of portland cement have been investigated by DTA/DTG/TG and XRD. Based on $Ca(OH)_2$ estimation, it has been concluded that lime sludge and limestone enhance hydration of cement compared to that by slag or fly ash.[81]

A detailed study on the influence of binary and ternary blends of portland cement mixed with silica fume, fly ash, and slag has been presented by Uchikawa and Okamura.[4] Applying conduction calorimetry, it was found that in blends with fly ash and slag, the hydration of the interstitial phase is accelerated, and that of alite at stages I and II is slightly delayed. In blends with slag and silica fume, the hydration of the interstitial phase and that of alite was accelerated with the reduction in the induction period. In fly ash-silica fume blends, the hydration of the interstitial phase decreased, and the induction period for alite hydration was lengthened.

Fly ash is available in large quantities and is more economical to use than silica fume. The addition of fly ash to cement causes an increase in the setting time and a reduction in strength. Improvement in hydration rate could be obtained by the addition of silica fume. By blending 5–10% silica fume with 15–30% fly ash, the rate of hydration as well as the strength is improved, as confirmed by conduction calorimetry and DTA investigations.[82] The heat of hydration of mortars containing a mixture of slag (30–90%) and silica fume (10–30%) indicates that the values decreased with the silica fume content. The results would not be the same if they were based on the cement content.[83]

The relative activities of pozzolans can be followed by estimating the lime content at different times in the cement-pozzolan-water systems. The DTG/DTA/TG techniques are well suited to estimate the amount of lime in these systems. Chatterjee, et al.,[84] compared the amounts of lime in the systems containing fly ash, silica fume, and three natural pozzolans (Fig. 33). The pozzolan:cement ratio was kept at 1:4 by weight. In the control and that containing Segni and fly ash, the amount of calcium hydroxide increases with time. The other mixes indicate a dip in CH content at 7 days. Segni and fly ash have lower lime-fixing ability. The dip in the curve may indicate the pozzolan has been completely consumed by this time. Any further increase in CH is due to continued hydration of the cement, or the rate of reaction between the pozzolan and CH has been reduced without a corresponding drop in the hydration rate of portland cement. Only low amounts of lime are recorded in mixtures with silica fume, diatomaceous earth, and Sacrafano indicating the consumption of lime by these pozzolans.

DTA has been used successfully to identify the products formed in mixtures of calcined gypsum, slag or fly ash, and cement (with or without hydrated lime) that were cured at 55, 70, and 85°C for different periods.[85] The compounds C-S-H, calcium trisulfate aluminate hydrate, calcium monosulfate aluminate hydrate, gypsum, $Ca(OH)_2$, and tricalcium aluminate hydrate were identified by their characteristic peaks at about 80 ± 5°, 115 ± 15°, 200°, 160 ± 20°, 470 ± 10°, and 300°C respectively.

A high early strength calcium sulfoaluminate cement can be formed from the raw materials slag, clay, fly ash, silica, etc.[86] Hydration of the system $C_4A_3\bar{S}$-$C\bar{S}$-C_4AF was carried out by DTA to investigate the rate of conversion of ettringite (high calcium sulfoaluminate phase) to the monosulfoaluminate phase.

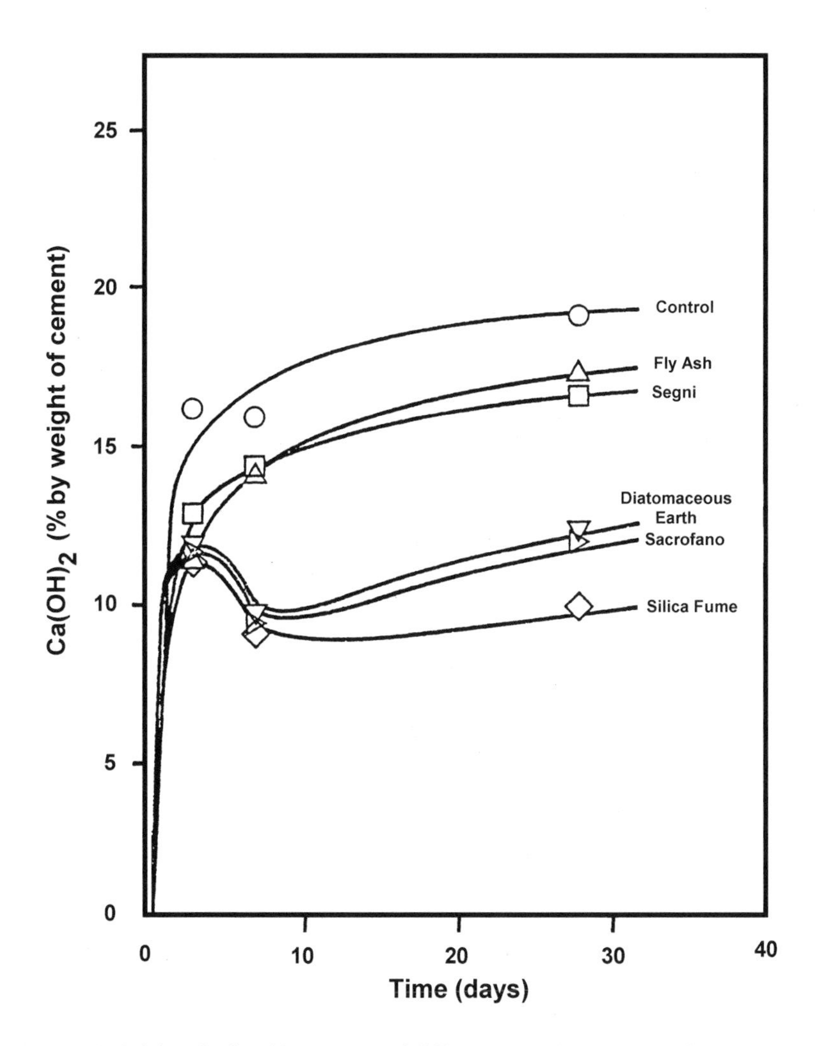

Figure 33. Calcium hydroxide contents of different pozzolan-cement mixtures.

An autogenous shrinkage occurs in cement-slag and cement-silica fume mixtures. In order to compare the rate of these shrinkages and also the mechanisms involved, the heat effects were determined. It was concluded that the shrinkage was due to the free water being consumed by the C-S-H phase, and increased shrinkage in slag-cement pastes was attributed to the formation of larger amounts of C-S-H.[87]

Thermal methods have been applied to explain the causes leading to the deterioration of various types of concrete. Normal concrete subjected to calcium chloride deteriorates, and this can be reduced or prevented by using supplementary cementing materials such as fly ash, slag, and silica fume. Torii, et al.,[88] working with mortar mixes, found that with 10% silica fume and 50% slag a good resistance to chloride was obtained, but with 30% fly ash slight deterioration occurred. DSC-TG curves of mortars were able to identify ettringite, Friedel's salt, CH and a complex chloroaluminate salt in deteriorated specimens. In blends containing lower amounts of lime, the deterioration due to chloride could be controlled.

Another form of deterioration in concrete involves the chemical reaction between the products of cement hydration and carbon dioxide. This reaction decreases the pH of the solution, and this may lead to the corrosion of the embedded steel. Carbonation shrinkage is another phenomenon that has been documented. Maslehuddin, et al.,[89] examined by DTA/TG, the products formed in mortar specimens exposed to CO_2 at 55–75°C for 54 weeks. Some were contaminated with chloride and sulfate ions. The amount of $Ca(OH)_2$ and $CaCO_3$ formed in several cement mixes containing fly ash, silica fume, and slag was determined (Table 9). In general, the amount of $Ca(OH)_2$ is lower in samples exposed to CO_2, and the amount of $CaCO_3$ in samples exposed to CO_2 is higher. There is also an accelerated carbonation in contaminated specimens.

9.0 MISCELLANEOUS ADDITIVES

Many types of industrial by-products and chemicals have potential for use in concrete. They are too numerous to mention. Experiments have been carried out to assess the possibility of using them in practice. The thermal techniques have been applied to monitor the hydration of mixes containing these substances. These techniques also provide valuable data on the durability aspects. In this section, a few of those that are not covered in other sections are described.

Table 9. Amounts of Calcium Hydroxide and Calcium Carbonate Formed in Mortar Blends Exposed to Carbon Dioxide

Material	Contaminant	Calcium Hydroxide (%)		Calcium Carbonate (%)	
		Unexposed	Exposed	Unexposed	Exposed
Cement	None	5.48	1.83	4.68	10.64
Cement	Cl + SO_4	6.10	1.70	1.70	16.70
Cement + Fly Ash	None	3.13	3.71	2.30	5.20
Cement + Fly Ash	Cl + SO_4	1.98	4.48	4.48	10.44
Cement + Silica Fume	Cl + SO_4	—	7.10	—	13.60
Cement + Slag	Cl + SO_4	—	5.30	—	13.20

Admixtures in concrete may remain in a free state, at the surface, or chemically combine with the hydrating constituents of cement. The air-entraining action of the air-entraining agent involves adsorption at the air/water and solid/water interfaces. An anion air-entraining agent, such as vinsol resin, reacts with $Ca(OH)_2$ to form a precipitate of calcium salt. Conduction calorimetry of cement treated with a sodium oleate-based air-entraining agent indicates that the C_3S is peak unaffected, but the peaks due to C_3A hydration are accelerated.[90][91] The ettringite and monosulfate reactions are retarded, possibly caused by an impermeable layer of Ca oleate aluminate hydrate salt. In the presence of neutralized wood resins, high dosages lead to the retardation of C_3S hydration while the C_3A hydration is accelerated.

Unsaturated fatty acids, such as oleates and others used as water-proofers, affect the hydration of cement. The tricalcium silicate peak in the calorimetry is unaffected, but the ettringite and monosulfate reactions are affected substantially.[91]

A composite concrete material containing polymers and cement has superior properties to those of conventional concrete. Several types of polymer concretes have been fabricated and examined for their physico-chemical and durability characteristics. Some work has also been carried out to investigate the influence of polymers on the hydration of C_3S.[92] In Fig. 34, DTA curves of neat C_3S pastes (*a*) are compared with the paste mixed with polyvinyl sulfonic acid (PESA)(*b*).[92] The sample ESA(*c*) was prepared by using vinyl sulfonic acid monomer with an initiator potassium persulfate and sodium bisulfite. In the figure, the period of curing is shown in the circles. The endotherm in the range 100–300°C (not shown) is mainly attributable to the loss of water from the C-S-H phase, and that at 450–530°C, to the decomposition of calcium hydroxide. The endotherm for the composite material is weaker than the corresponding peak for the neat paste, and it also appears at a lower temperature. The decrease in the endotherm is taken as evidence that there is a reaction between the sulfonate groups of the polymer and the calcium ions or calcium hydroxide released during the hydration of the silicate. It is also possible that the sulfonate groups, being hygroscopic, absorb water thus diminishing water available for the hydration of the silicate. Similar conclusions were drawn from the TG results.

Cook, et al., and others have examined several polymer-cement systems by DTA.[6][93] The polymers were mixed with monomers and polymerized either by irradiation or thermal catalysis. The mixes were hydrated for 14 days before examination by DTA (Fig. 35). The intensity of the calcium hydroxide peak varies, depending on the polymer. Polystyrene does not have much effect, but acrylonitrile, methyl methacrylate (MMA), vinyl acetate, and polyester styrene reduce the intensity of the $Ca(OH)_2$ peak, suggesting the occurrence of an interaction between these polymers and calcium hydroxide. Caution should be exercised in the interpretation of the thermal curves in the polymer-cement systems. The decrease in intensities could be due to the dilution effect, masking by the exothermal peaks, and interactions during heating.

Polymer-impregnated concrete is obtained by impregnating concrete with a monomer and then polymerizing it by various means. Such concrete, in addition to exhibiting a three-fold increase in compressive strength, manifests better durability than normal concrete under different exposure conditions. Several investigators have reported that an interaction occurs between the polymer and $Ca(OH)_2$ of the cement paste. Ramachandran and Sereda studied the DTA of polymethyl methacrylate-impregnated cement pastes obtained at w/c ratios of 0.3 and 0.7.[94] Thermograms of cement and that impregnated with polymethyl methacrylate are shown in Fig. 36. In cement paste *A,* the endothermal effects at 215–225°C and 490–505°C are due to the dehydration of water from the

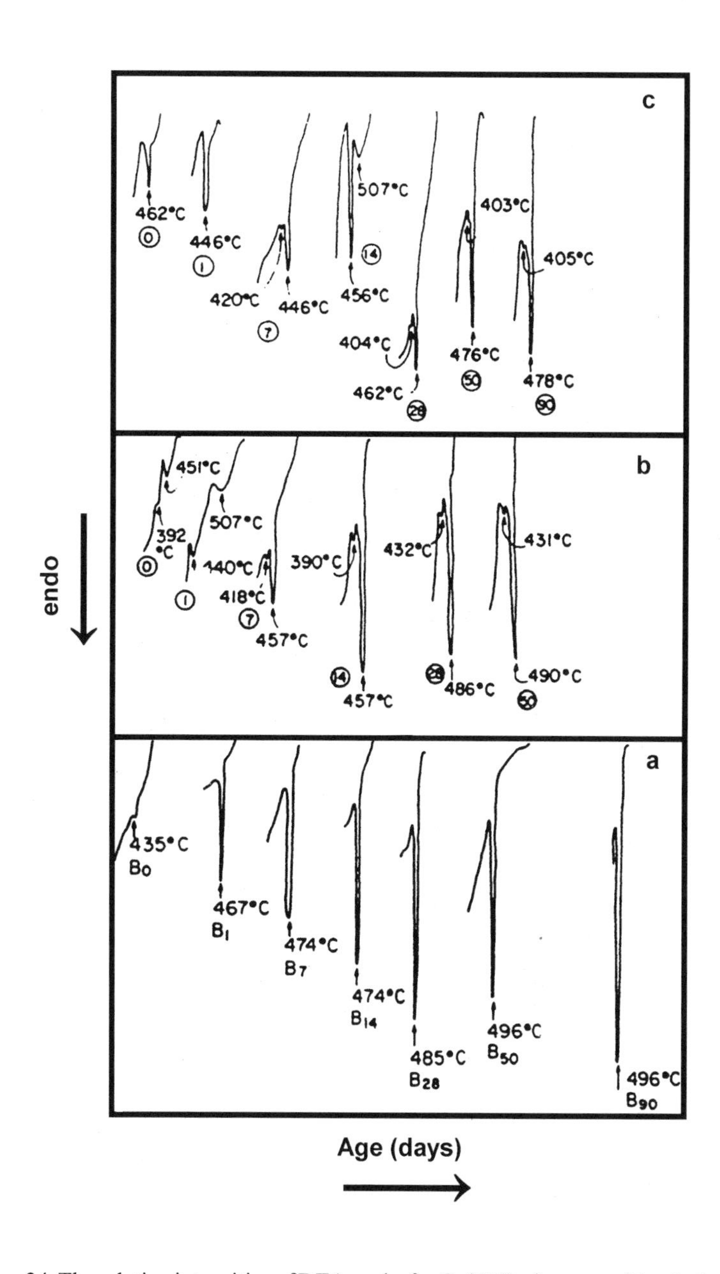

Figure 34. The relative intensities of DTA peaks for $Ca(OH)_2$ decomposition in the system containing tricalcium silicate and vinyl sulfonic acid polymer.

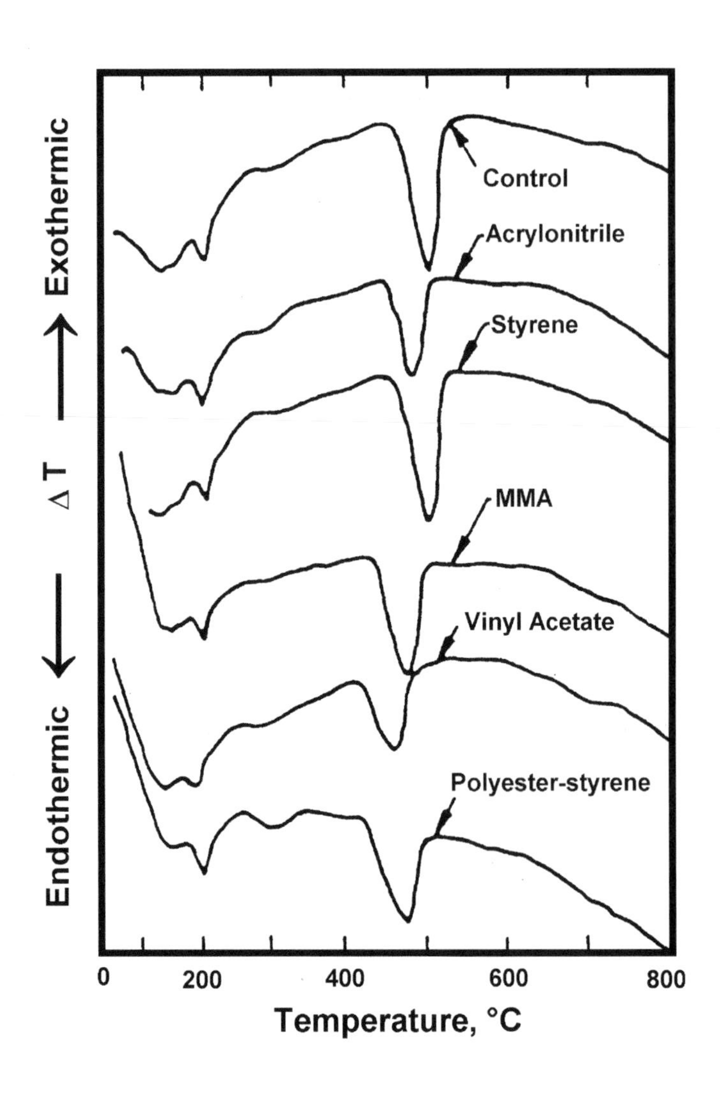

Figure 35. Thermograms of the premix cement-polymer systems.

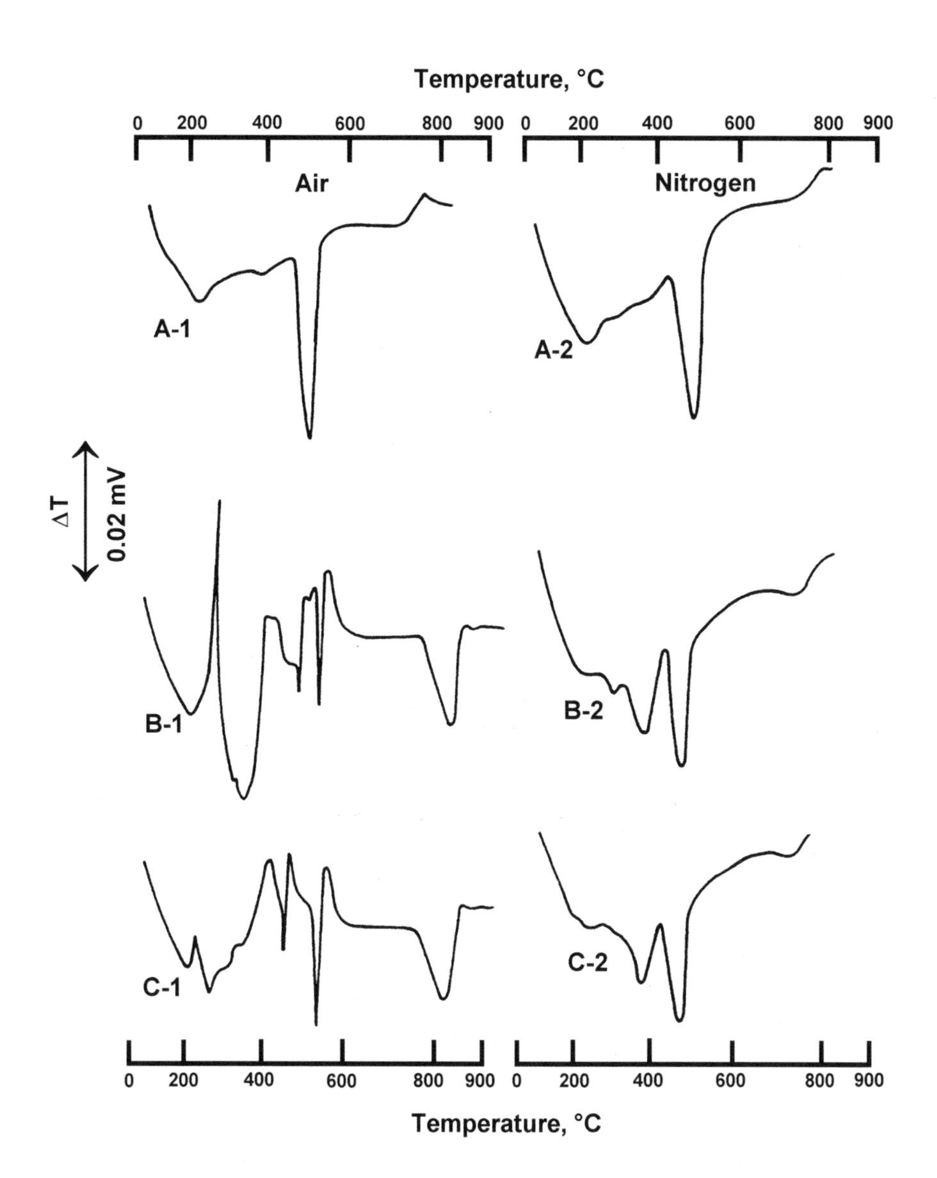

Figure 36. Thermograms of polymethyl methacrylate-impregnated cement paste.

C-S-H phase/sulfoaluminate hydrate and $Ca(OH)_2$ respectively. The polymer impregnated paste *B* shows several endothermal and exothermal effects. The exothermal peaks also occur in pure polymers, hence this effect may be ascribed to the oxidation of the polymer. There is also an indication that the $Ca(OH)_2$ peak in the impregnated samples is decreased in intensity. This decrease in intensity could be due to the masking effect of the exothermal peak occurring at about the same temperature and a dilution effect. Curve *C-1* refers to a mechanical mix of the polymer and cement paste. Both *B-1* and *C-1* are similar in terms of exo- and endothermal effects, except the sharp endothermal effects at 230–270°C and 250–350°C are not apparent in sample *C-1*. This may be due to the masking effect of the endotherm of portland cement paste. Other experimental evidence suggests that the emergence of new thermal effects in the polymer-impregnated cement paste are not due to $Ca(OH)_2$-polymer interaction during impregnation. It is rather due to complex interaction that occurs between hydrated cement decomposition products and PMMA during heating in the DTA apparatus. In samples exposed to N_2 only, endotherms result.

The effect of polymers on the fly ash concrete has also been studied by thermal methods. The amount of lime formed in mixes containing cement, fly ash, and a non-ionic polymer (Versicol) and cured for 1 to 90 days has been determined by thermal analysis.[95] At all curing times, the amount of lime formed in the cement-fly ash-polymer mixtures was higher, especially at a w/c ratio of 0.25. It was concluded that the polymer interacts preferentially with the silica and alumina of the fly ash rather than with lime produced in the reaction.

Stabilized laterite soils have been applied in highways, dams, and foundations. Laterites consist of quartz, hematite, gibbsite, etc. It has been found that heating laterite to about 800°C and curing with cement produces a strong body with good durability characteristics. By replacing cement with 20% laterite, the 28 day strength reaches a value of 80% of that of the reference.[96] DTA studies indicate that in laterite-cement mixes the amount of lime produced is decreased due to the pozzolanic reaction, and the amount of the C-S-H phase is increased. Other phases, such as C_2ASH_8 and C_4AH_{13}, were also identified.

Red mud is a by-product derived from the aluminum industry and consists mainly of Si, Al, Fe, Ca, and compounds kaolinite and gibbsite. Pera and Momtazi[97] evaluated the pozzolanic activity of red mud by heating it at different temperatures in the range 600° to 800°C. The cement was mixed with 30–50% red mud and cured for different times in water. The products were subjected to DTA and XRD studies. Lime was completely consumed in a mixture with a 30% red mud replacement. The hydrated products consisted of C-S-H, hydrated aluminum silicates and hydrogarnets. The blended cement showed better durability to acids and sea water compared with the untreated cement.

Pressurized fluidized bed combustion (in a coal-fired thermal plant) offers efficiency and reduction in environmental loads. The coal ash produced in this process is a consequence of firing a mixture of pulverized limestone and coal. In this process, the mixes are burned at a lower temperature than that carried out in the conventional power plant which produces fly ash. There is a possibility of utilizing coal ash as an admixture in concrete. Thermal analysis of the hydration products obtained at different w/s ratios from coal ash-cement pastes indicated substantially lower amounts of lime than that estimated in pure cement pastes.[98] This indicates that lime in the mixes is consumed by the coal ash through a pozzolanic action. In another study, the residue of fluidized combustion of bituminous and sub-bituminous coal was tested for its hydraulic properties.[99] The residue had a carbon content of 0.1 to 10% and compounds such as iron oxide, calcite, feldspar, anhydrite, and quartz. DTA/TG was successfully used for determining the residual carbon content, $CaCO_3$, and anydrite.[15][99]

REFERENCES

1. Hewlett, P. C., (ed.), *Lea's Chemistry of Cement and Concrete,* p. 1053, John Wiley & Sons, New York (1998)
2. Swamy, R. N., (ed.), *Cement Replacement Materials,* p. 259, Surrey University Press, Glasgow (1986)
3. Taylor, H. F. W., *Cement Chemistry,* p. 459, Thomas Telford, London (1997)
4. Sarkar, S. L., and Ghosh, S. N., (eds.), *Mineral Admixtures in Cement and Concrete,* p. 565, ABI Books Priv. Ltd., New Delhi, India (1993)
5. Muller, C., Hardtl, R., and Shielbl, P., High Performance Concrete with Fly Ash, *3rd Int. Symp. on Advances in Concr. Technol.,* (V. M. Malhotra, ed.), SP-171:173–200 (1997)
6. Ghosh, S. N. (ed.), *Advances in Cement Technology,* p. 804, Pergamon Press, New York (1983)

7. Sturrup, V. R., Hooton, R. G., and Calendenning, T. G., Durability of Fly Ash Concrete, *Proc. CANMET/ACI Int. Conf. on the Use of Fly Ash, Silica Fume, Slag and Other By-Products in Concrete,* (V. M. Malhotra, ed.), SP-79:71–86 (1983)
8. Bouzoubaa, N., Zhang, M. H., Malhotra, V. M., and Golden, D. M, Blended Fly Ash Cements—A Review, *6th CANMET/ACI Int. Conf. on Fly Ash, Silica Fume, Slag and Natural Pozzolans in Concr.,* supplementary papers, pp. 717–749 (1998)
9. Giergiczny, Z., and Werynska, A., Influence of Fineness of Fly Ashes on Their Hydraulic Reactivity, *3rd CANMET/ACI Int. Conf. on Fly Ash, Slag and Natural Pozzolans in Concr.,* (V. M. Malhotra, ed.), SP-114:97–115 (1989)
10. Malhotra, V. M., and Ramezanianpour, A. A., *Fly Ash in Concrete,* p. 307, CANMET, Canada (1994)
11. Feldman, R. F., Carette, G. G., and Malhotra, V. M., Studies on Mechanism of Development of Physical and Mechanical Properties of High Volume Fly Ash Cement Pastes, *Cement and Concr. Composites,* 12:245–251 (1990)
12. Fan, Y., Yin, S., Wen, Z., and Zhong, J., Activation of Fly Ash and its Effect on Cement Properties, *Cement and Concr. Res.,* 29:467–472 (1999)
13. Sarkar, S. L., Kumar, A., Das, D. K., and Banerjee, G., Utilization of Fly Ash in the Development of a Cost Effective Cementitious Product, *5th CANMET/ACI Int. Conf. on Fly Ash, Silica Fume, Slag and Natural Pozzolans in Concr.,* Vol. 1, SP-153:439–457 (1995)
14. Malhotra, V. M., (ed.), *6th CANMET/ACI/JCI Conf. on Fly Ash, Silica Fume, Slag and Natural Pozzolans in Concr.*, p. 603 (1998)
15. Brandstetr, J., and Havlica, J., Properties and Some Possibilities of Utilization of Solid Residue of Fluidized Bed Combustion of Coal and Lignite, *6th CANMET/ACI/JCI Conf. on Fly Ash, Silica Fume, Slag and Natural Pozzolans in Concr.*, pp. 133–160 (1998)
16. Paya, J., Monzo, J., Borrachero, M. V., Parris, E., and Amahjour, F., Thermo-gravimetric Methods for Determining Carbon Content in Fly Ashes, *Cement and Concr. Res.,* 28:675–686 (1998)
17. Nocun-Wczelik, W., The Hydration of Tricalcium Silicate with Active Silica, *4th CANMET/ACI Int. Conf. on Fly Ash, Silica Fume, Slag and Natural Pozzolans in Concr., Istanbul, Turkey,* supplementary papers, pp. 485–492 (1992)
18. Traetteberg, A., Silica Fume as Pozzolanic Materials, *Il Cemento,* 75:369–375 (1978)

19. Malhotra, V. M., Ramachandran, V. S., Feldman, R. F., and Aitcin, P. C., *Condensed Silica Fume in Concrete.,* p. 221, CRC Press, Boca Raton, Florida (1987)

20. Beedle, S., Groves, G. W., and Rodger, S. A., The Effect of Fine Pozzolanic and Other Particles on the Hydration of C_3S, *Advances in Cement Res.,* 2:3–8 (1989)

21. Wu, Z. Q., and Young, J. F., The Hydration of Tricalcium Silicate in the Presence of Colloidal Silica, *J. Mater. Sci.,* 19:3477–3486 (1984)

22. Chengyi, H., and Feldman, R. F., Hydration Reactions in Portland Cement-Silica Fume Blends, *Cement and Concr. Res.,* 15:585–592 (1985)

23. Domone, P. L., and Tank, S. B., Use of Condensed Silica Fume in Portland Cement Grouts, *CANMET/ACI Int. Conf. on Fly Ash, Silica Fume, Slag and Natural Pozzolans in Concrete* (V. M. Malhotra, ed.), Vol. 2, SP-91:1231–1260 (1986)

24. Meland, I, Influence of Condensed Silica Fume and Fly Ash on the Heat Evolution in Cement Pastes, *Proc. CANMET/ACI 1st Int. Conf. on Fly Ash, Silica Fume, Slag and Other By-Products in Concr.*, Vol. 2, SP-79:665–676 (1983)

25. Sellevold, E. J., Bager, D. H., Klitgaad, J. K., and Knudsen, T., Silica Fume- Cement Pastes:Hydration and Pore Structure, Report BML 82:610, pp. 19–50, The Norwegian Inst. Of Technol., Trondheim, Norway (1982)

26. Khayat, K. H., and Aitcin, P. C., *Silica Fume- A Unique Supplementary Cementitious Material,* (S. L. Sarkar and S. N. Ghosh, eds.), pp. 226–265, ABI Books Priv. Ltd., New Delhi, India (1993)

27. Meng, B., Wiens, U., and Schiessl, P., *6th CANMET/ACI Int. Conf. on Fly Ash, Silica Fume, Slag and Natural Pozzolans in Concr.,* Vol. 1, SP-178:109–128 (1998)

28. Lilkov, V., Dimitrova, E., and Petrov, O. E., Hydration Process of Cement Containing Fly Ash and Silica Fume: The First 24 Hours, *Cement Concr. Res.*, 27:577–588 (1997)

29. Weber, S., and Reinhardt, H. W., Improved Durability of High Strength Concrete Due to Autogenous Curing, *Int. Conf. Durability of Concr.,* (V. M. Malhotra, ed), SP-170:93–121 (1997)

30. Kawamura, M., Takemoto, K., and Hasaba, S., Effectiveness of Various Silica Fumes in Preventing Alkali-Aggregate Expansion, *Concr. Durability,* (J. M. Scanlon, ed.), Vol. 2, SP-100:1809–1819 (1987)

31. Zhang, M. H., and Gjorv, O. E., Effect of Silica Fume on Cement Hydration in Low Porosity Cement Pastes, *Cement Concr. Res.,* 21:800–808 (1991)

32. Atlassi, E. H., Non-Evaporable Water and Degree of Cement Hydration in Silica Fume-Cement Pastes, *CANMET/ACI Int. Conf. on Fly Ash, Silica Fume, Slag and Natural Pozzolans in Concr.,* (V. M. Malhotra, ed.), Vol. 2, SP-153:703–717 (1995)
33. Kovler, K., Setting and Hardening of Gypsum-Portland Cement-Silica Blends, Part 2: Early Strength, DTA, XRD and SEM Observations, *Cement Concr. Res.,* 28:523–531(1998)
34. Meland, I., Hydration of Blended Cements, *Nordic Concr Res.,* No. 2, pp. 183–196 (April 1983)
35. Ramachandran, V. S., *Applications of Differential Thermal Analysis in Cement Chemistry*, p. 308, Chemical Publ. Co., New York (1969)
36. Schwiete, H. E., Dolbar, F. C., and Schroder, F., Blast Furnace Slags and Slag Cements, *Proc. 5th Int. Symp. Chem. Cements, Tokyo, Japan*, Part IV, pp. 149–199 (1968)
37. Ramachandran, V. S, Feldman, R. F., and Beaudoin, J. J., *Concrete Science,* p. 427, Heyden and Son, London (1981)
38. Webb, T. L., and Kruger, J. E., Building Materials, in: *Differential Thermal Analysis* (R. C. Mackenzie, ed.), 2:181–205, Academic Press, London (1972)
39. Swamy, R. N., Sakai, M., and Nakamura, N., Role of Superplasticizers and Slag for Evaluating High Performance Concrete, *5th CANMET/ACI Int. Conf. of Superplasticizers and Other Chemical Admixtures in Concr.,* (V. M. Malhotra, ed.), SP-148:1–26 (1994)
40. Swamy, R. N., Design for Durability and Strength Through the Use of Fly Ash and Slag in Concrete, *3rd CANMET/ACI Int. Conf. on Advances in Concr. Technol.,* (V. M. Malhotra, ed.), SP 171:1–72 (1997)
41. Kokubu, K., Takahashi, S., and Anzai, H., Effect of Curing Temperature on the Hydration and Adiabatic Temperature Characteristics of Portland Cement – Blast Furnace Slag Concrete, *3rd CANMET/ACI Int. Conf. Fly Ash, Silica Fume, Slag and Natural Pozzolans in Concr.,* (V. M. Malhotra, ed.), SP-114:1361–1376 (1989)
42. Cook, D. J., Hinczak, I., and Cao, H. T., Heat of Hydration, Strength and Morphological Development in Blast Furnace Slag/Cement Blends, *Int. Workshop on Granulated Blast Furnace Slag Concr., Toronto*, pp. 67–78 (1987)
43. Tenoutasse, N., and Dieryck, V., Effect of Cement-Plasticizer Interactions on the Rheological Properties of Slag Cement Paste, *6th CANMET/ACI Int. Conf. on Fly Ash, Silica Fume, Slag and Natural Pozzolans in Concr., Bangkok, Thailand,* supplementary papers, pp. 293–306 (1998)

44. Duyang, D., Shao, J., Yang, B., He, J., and Xie, Y., Study on Steel-Iron Slag Clinker Systems, *4th CANMET/ACI Int. Conf. on Fly Ash, Silica Fume, Slag and Natural Pozzolans in Concr., Istanbul, Turkey*, supplementary papers, pp. 649–666 (1992)
45. Douglas, E., Mainwaring, P. R., and Hemmings, R. T., Pozzolanic Properties of Canadian Non-Ferrous Slags, *Int. Conf. on Fly Ash, Silica Fume, Slag and Natural Pozzolans in Concr.,* (V. M. Malhotra, ed.), Vol. 2, SP-91:1525–1550 (1986)
46. Schutter, G. D., Hydration and Temperature Development in Concrete Made with Blast-Furnace Slag Cement, *Cement Concr. Res.,* 29:143–149 (1999)
47. Chopra, S. K., and Kishan, L., See Ramachandran, V. S., *Applications of Differential Thermal Analysis in Cement Chemistry*, p. 202, Chemical Publ. Co., New York (1969)
48. Litvan, G. G., and Meyer, A., Carbonation of Granulated Blast Furnace Cement Concrete During 20 Years of Field Exposure, *Int. Conf. on Fly Ash, Silica Fume, Slag and Natural Pozzolans in Concr.,* (V. M. Malhotra, ed.), Vol. 2, SP-91:1445–1462 (1986)
49. Singh, M., and Garg, M., Activation of Gypsum Anhydrite-Slag Mixtures, *Cement Concr. Res.,* 25:332–338 (1995)
50. Tashiro, T., and Yoshimoto, T., Effect of Sodium Compounds on the Strength and Microstructural Development of Blast Furnace Slag Cement Mortars, *Proc. 3rd CANMET/ACI Conf. on Fly Ash, Silica Fume, Slag and Natural Pozzolans in Concr.* (V. M. Malhotra, ed.), Vol. 2, SP-114:1307–1323 (1989)
51. Wang, S. D., and Scrivens, K. L., Hydration Products of Alkali Activated Slags, *Cem. Concr. Res.,* 25:561–571 (1995)
52. Mehta, P. K., The Chemistry and Technology of Cements Made from Rice Husk Ash, *Proc. UNIDO/ESCAP/RCTT Workshop on Rice Husk Ash, Peshawar, Pakistan*, pp. 113–122 (1979)
53. Ramachandran, V. S., (ed.), *Concrete Admixtures Handbook,* p. 626, Noyes Publ., NJ, (1984)
54. Borthakur, P. C., Saikia, P. C., and Dutta, S. N., Physico-Chemical Characteristics from Paddy Husk; Its Reactivity and Probable Field of Application, *Indian Ceramics,* 25:25–29 (1980)
55. Shah, R. A., Khan, A. H., Chaudhry, M. A., and Qaiser, M. A., Utilization of Rice Husk Ash for the Production of Cement-Like Materials in Rural Areas, *Pakistan J. Sci. Ind. Res.,* 24:121–124 (1981)
56. Hanafi, S., El-Enein, A., Ibrahim, D. M., and El-Hemaly, S. A., Surface Properties of Silicas Produced by Thermal Treatment of Rice-Husk Ash, *Thermochimica Acta,* 37:137–143 (1980)

57. Olmez, H., and Heren, Z., Making a Pozzolanic Rice Husk Ash and Its Use in Concrete, *4th CANMET/ACI Conf. on Fly Ash, Silica Fume, Slag and Natural Pozzolans in Concr., Istanbul, Turkey,* supplementary papers, pp. 851–861 (1992)

58. Hwang, C. L., and Wu, D. S., Properties of Cement Paste Containing Rice Husk Ash, *3rd CANMET/ACI Int. Conf. on Fly Ash, Silica Fume, Slag and Natural Pozzolans in Concr.,* Vol. 1, SP-114:733–762 (1989)

59. Yu, Q., Sawayama, K., Sigita, S., Shoya, M., and Isojima, Y., The Reaction Between Rice Husk Ash and $Ca(OH)_2$ Solution and Nature of Its Product, *Cement Concr. Res.,* 29:37–43 (1999)

60. Pera, J., and Ambroise, J., Pozzolanic Properties of Metakaolin Obtained from Paper Sludge, *6th CANMET/ACI Int. Conf. on Fly Ash, Silica Fume, Slag and Natural Pozzolans in Concr.,* (V. M. Malhotra, ed.), SP-178:1007–1020 (1998)

61. Ambroise, J., Maximilien, S., and Pera, J., Properties of Metakaolin Blended Cements, *Advanced Cement Based Mater.,* 1:161–168 (1994)

62. Ramachandran, V. S., Malhotra, V. M., Jolicoeur, C., and Spiritos, N., *Superplasticizers: Properties and Applications in Concrete,* p. 404, Materials Technology Laboratory, CANMET, Government Services, Ottawa, Canada (1998)

63. Morsy, M. S., Abo El-Enein, S. A., and Hanna, G. B., Microstructure and Hydration Characteristics of Artificial Pozzolan Cement Pastes Containing Burnt Kaolinite Clay, *Cement Concr. Res.,* 27:1307–1312 (1997)

64. Curcio, F., DeAngelis, B. A., and Pagliolico, S., Metakaolin as Pozzolanic Microfiller for High Performance Mortars, *Cement Concr. Res.,* 28:803–809 (1998)

65. Oriol, M., and Pera, J., Pozzolanic Activity of Metakaolin Under Microwave Treatment, *Cement Concr. Res.,* 25:265–270 (1995)

66. Ambroise, J., Martin-Calle, S., and Pera, J., Pozzolanic Behavior of Thermally Activated Kaolin, *4th CANMET/ACI Int. Conf. on Fly Ash, Silica Fume, Slag and Natural Pozzolans in Concr.,* (V. M. Malhotra, ed.), SP-132:731–748 (1992)

67. Granizo, M. L., and Blanco, M. T., Alkaline Activation of Metakaolin, *J. Thermal Analysis,* 52:957–965 (1998)

68. Palamo, A., Blanco-Varela, M. T., Granizo, M. L., Puertas, F., Vazquez, T., and Grutzeck, M. W., Chemical Stability of Cementitious Materials Based on Metakaolinite, *Cement Concr. Res.,* 29:997–1004 (1999)

69. Klimesch, D. S., and Ray, A. S., Metakaolin Additions to Autoclaved Cement–Quartz Pastes:Evaluation of the Acid Insoluble Residue, *Advances in Cement Res.,* 9:157–165 (1997)

70. Klimesch, D. S., and Ray, A. S., Use of Second Derivative Differential Thermal Curve in the Evaluation of Cement-Quartz Pastes With Metakaolin Autoclaved at 180°C, *Thermochimica Acta,* 307:167–176 (1997)

71. Ray, A., Cantill, E., Stevens, M. G., and Aldridge, L., Use of DTA to Determine the Effect of Mineralizers on the Cement-Quartz Hydrothermal Reactions, Part 2, Clay Addition, *Thermochimica Acta,* 250:189–195 (1995)

72. Ogawa, K., Uchikawa, H., Takemoto, K., and Yasui, I., The Mechanism of Hydration in the System C_3S-Pozzolan, *Cement Concr. Res.,* 10:683–696 (1980)

73. Jambor, J., Relation Between Phase Composition, Overall Porosity and Strength of Hardened Lime-Pozzolan Pastes in the Presence of Silica Fume, *Mag. Concr. Res.,* 15:131–142 (1963)

74. Ubbriaco, P., and Tasselli, F., A Study of the Lime-Pozzolan Binder, *J. Thermal Analysis,* 52:1047–1054 (1998)

75. Takemoto, K., and Uchikawa, H., Hydration of Pozzolanic Cements, *Proc. 7th Int. Congr. on the Chemistry of Cements, Paris,* Vol. I, Theme IV-2, pp. 1–21 (1980)

76. Gryzymek, K., Roszczniałski, W., and Gustaw, K., Hydration of Cements with Pozzolanic Additions, *Proc. 7th Int. Congr. Chem. Cements, Paris,* Vol. III, Theme IV, pp. 66–71 (1980)

77. Cabrera, J. G., Rivera-Villarreal, R., and Sri-Ravindrarajah, R., Properties and Durability of Pre-Columbian Lightweight Concrete, *CANMET/ACI Int. Conf. Durability of Concr.,* (V. M. Malhotra, ed.), SP-170:1215–1230 (1997)

78. Kasai, Y., Tobinai, K., Asakura, E., and Feng, N., Comparative Study of Natural Zeolites and Other Admixtures in Terms of Characterization and Properties of Mortars, *4th CANMET/ACI Int. Conf. on Fly Ash, Silica Fume, Slag and Natural Pozzolans in Concr.,* (V. M. Malhotra, ed.), Vol. 1, SP-132:615–634 (1992)

79. Sharara, A. M., Didamony, H. El., Ebied, E., and El-Aleem, A., Hydration Characteristics of β-C_2S in the Presence of Some Pozzolanic Materials, *Cement Concr. Res.*, 24:966–974 (1994)

80. Dahl, P. A., and Meland, I., Influence of Different Pozzolan Types Upon the Effect of Plasticizing Admixture, *3rd CANMET/ACI Int. Conf. on Fly Ash, Silica Fume, Slag and Natural Pozzolans in Concr.,* (V. M. Malhotra, ed.), SP-114:689–711 (1989)

81. Sharma, R. L., and Pandey, S. P., Influence of Mineral Additives on the Hydration Characteristics of Ordinary Portland Cement, *Cement Concr. Res.,* 29:1525–1529 (1999)

82. Giergiczny, Z., The Properties of Cements Containing Fly Ash Together with Other Admixtures, *4th CANMET/ACI Int. Conf. on Fly Ash, Silica Fume, Slag and Natural Pozzolans in Concr.,* (V. M. Malhotra, ed.), Vol. 1, SP-132:439–456 (1992)

83. Shizawa, M, Joe, Y., Kato, H., and Morita, T., Study on Hydration Properties of Blended Cement for Ultra High Strength, *4th CANMET/ACI Int. Conf. on Fly Ash, Silica Fume, Slag and Natural Pozzolans in Concr., Istanbul, Turkey*, supplementray papers, pp. 987–1003 (1992)

84. Chatterjee, S., Collepardi, M., and Moriconi, G., Pozzolanic Property of Natural and Synthetic Pozzolans: A Comparative Study, *Proc. CANMET/ ACI 1st Int. Conf. on the Use of Fly Ash, Silica Fume, Slag and Other Mineral By-Products in Concr., Montebello, Canada,* SP-79:221–233 (1983)

85. Cioffi, R., Marroccoli, M., Santoro, L., and Valenti, G. L., DTA Study of the Hydration of Systems of Interest in the Field of Building Materials Manufacture, *J. Thermal Analysis,* 38:761–770 (1992)

86. Belz, G., Beretka, J., Marroccoli, M., Santoro, L., Sherman, N., and Velenti, G. L., Use of Fly Ash, Blast Furnace Slag and Chemical Gypsum for the Synthesis of Calcium Sulfoaluminate-Based Cements, *5th CANMET/ ACI Int. Conf. on Fly Ash, Silica Fume, Slag and Natural Pozzolans in Concr.,* (V. M. Malhotra, ed.), Vol. 1, SP-153:513–530 (1995)

87. Uchikawa, H., Hanehara, S., and Hirao, H., Influence of Structural and Thermal Changes and the Inner Part of Hardened Cement Paste on Autogenous Shrinkage, *CANMET/ACI Int. Conf. Durability of Concr.,* (V. M. Malhotra, ed.), SP-170:949–964 (1997)

88. Torii, K., Sasatani, T., and Kawamura, M., Effects of Fly Ash, Blast Furnace Slag and Silica Fume on Resistance of Mortar to Calcium Chloride Attack, *5th CANMET/ACI Int. Conf. on Fly Ash, Silica Fume, Slag and Natural Pozzolans in Concr.,* (V. M. Malhotra, ed.), Vol. 2, SP-153:2:931–949 (1995)

89. Maslehuddin, M., Shirokoff, J., and Siddiqui, M. A. B., Changes in the Phase Composition in OPC and Blended Cement Mortars Due to Carbonation, *Advances in Cement Res.,* 8:167–173 (1996)

90. Bruere, G. M., Air Entraining Action of Anionic Surfactants in Portland Cement, *J. App. Chem. Biotechnology.,* 21:61–64 (1971)

91. Rixom, M. R., *Chemical Admixtures for Concrete,* E. & F. N. Spon, London, p. 234 (1978)

92. Ben-Dor, L, Heitner-Wurguin, C., and Diab, H., The Effect of Ionic Polymers on the Hydration of C_3S, *Cement Concr. Res.,* 15:681–686 (1985)

93. Cook, D. J., Morgan, D. R., Sirivivatnan, V., and Chaplin, R. P., Differential Thermal Analysis of Premixed Polymer Cement Materials, *Cement Concr. Res.,* 6:757–764 (1976)

94. Ramachandran, V. S., and Sereda, P. J., Differential Thermal Studies of Polymethyl Methacrylate-Impregnated Cement Pastes, *Thermochimica Acta,* 5:443–450 (1973)

95. Bhatty, J. I., Dollimore, D., Gamlen, G. A., Mangabhai, R. J., and Olmez, H., Estimation of Calcium Hydroxide in OPC, OPC/PFA and OPC/PFA/ Polymer Modified Systems, *Thermochimica Acta,* 106:115–123 (1986)

96. Marwin, T., Pera, J., and Ambroise, J., The Action of Some Aggressive Solutions on Portland and Calcined Laterite Blended Cement Concrete, *4th CANMET/ACI Int. Conf. on Fly Ash, Silica Fume, Slag and Natural Pozzolans in Concr.,* (V. M. Malhotra, ed.), SP-132:763–779 (1992)

97. Pera, J., and Momtazi, A. S., Pozzolanic Activity of Calcined Mud., *CANMET/ACI Int. Conf. on Fly Ash, Silica Fume, Slag and Natural Pozzolans in Concr,* (V. M. Malhotra, ed.), Vol. 1, SP-132:749–761 (1992)

98. Fukudome, K., Shintani, T., Kita, T., and Sasaki, H., Utilization of Coal Ash Produced from Pressurized Fluidized Combustion Power Plant as a Concrete Mineral Admixture, *CANMET/ACI Int. Conf. on Fly Ash, Silica Fume, Slag and Natural Pozzolans in Concr.,* (V. M. Malhotra, ed.), SP-178:522–544 (1998)

99. Chandra, S., *Waste Materials Used in Concrete Manufacturing,* p. 651, Noyes Publ., NJ (1997)

9

Introduction to Non-Portland Cement Binders and Concrete

1.0 INTRODUCTION

Portland cement-based concretes are the most widely used in the construction industry. Non-portland cement-based systems, particularly those that have fast hardening and strength development characteristics, are well suited for special applications, many of these related to the repair of concrete structures.

The performance of many non-portland cement-based repair materials is dependent on the chemistry and physics of the system, the properties of the existing concrete, and the environment. Thermal analysis techniques are increasingly being used to investigate the physico-chemical behavior of non-portland cement binders exposed to aggressive environments.

This chapter provides a description of the hydration and engineering behavior of special cements including: magnesium oxychloride and oxysulfate cement; calcium aluminate cement; portland-calcium aluminate blended cement; phosphate cement; regulated set cement; and gypsum.

2.0 MAGNESIUM OXYCHLORIDE CEMENT

2.1 Description

The first oxychloride cement was reported by Sorel.[1] Magnesium oxychloride cement (MOC), referred to as Sorel cement can be formed by mixing fine particles of calcined magnesite with magnesium chloride solution (specific gravity 1.18–1.20). Many of its properties are superior to those of portland cement. Most are dependent on the chemical composition and the reactivity of calcined magnesite with magnesium chloride solution.[2] For example, coarse-grained magnesite compressed of grains coated with iron minerals results in reduced strength compared to magnesite consisting of primarily microcrystalline material.

MOC has excellent fire resistance, low thermal conductivity, good abrasion resistance, and high values of compressive and flexural strength. The cement has a high rate of early strength development, is lightweight, and resistant to attack by alkalis, organic solvents, common salts, and sultates. Its excellent binding capability enables the formation of suitable composites containing a variety of fillers and by-product materials including: wood flour, sawdust, asbestos fines, and stone chips.[3] MOC is also commonly used as a flooring material, for heat and sound insulation purposes, and for making artificial stone, dental cement, tiles, refractory bricks, and foam concrete walls.

The high water solubility of MOC has limited the extent of its use as a construction material for structural (exterior) applications. The excellent engineering characteristics, however, have encouraged researchers to seek methods to improve the water resistance of MOC.

2.2 Hydration Reactions

The reaction products in the system MgO-$MgCl_2$-H_2O are primarily $Mg_3(OH)_5Cl{\cdot}4H_2O$ (5 form), $Mg_2(OH)_3Cl{\cdot}4H_2O$ (3 form), and $Mg_{10}(OH)_{18}Cl{\cdot}5H_2O$.[4]–[6] The 5 and 3 form are present at room temperature. A molar ratio of MgO:$MgCl_2$:H_2O of 5:1:13 with some excess MgO is recommended to form essentially the 5 form and some $Mg(OH)_2$. The activity of MgO depends on the temperature of calcination (800–1000°C) and influences the speed of reaction and strength development. Variability of MgO due to the source of magnesite may require adjustments to the MgO:$MgCl_2{\cdot}H_2O$ ratio to achieve maximum strength.

Two compounds, $9Mg(OH)_2 \cdot MgCl_2 \cdot 5H_2O$ and $2Mg(OH)_2 \cdot MgCl_2 \cdot 4H_2O$ (9 and 2 form) are known to be present at temperatures higher than 100°C. Carbonation of MOC can occur on prolonged exposure to ambient. This generally results in the formation of magnesium chlorocarbonate, $Mg(OH)_2 \cdot MgCl_2 \cdot 2MgCO_3 \cdot 6H_2O$.

Small quantities of hydraulic aluminate minerals can cause the conversion of the reaction products in MOC paste.[7] Pastes with a molar ratio of $MgO/MgCl_2$ greater than 5 generally contain the 5 form as the main reaction product. The same paste containing CA or C_4AF additions results in formation of the 3 form. The conversion is attributed to the hydration products of the hydrated aluminate minerals. Small additions of the latter do not appear to influence strength of the hardened MOC paste. It is argued that the polycrystalline composite containing either the 5 form or the 3 form in combination with MgO or $Mg(OH)_2$ behaves similarly in response to mechanical loads. The addition of the inert aluminate minerals (e.g., C_2AS) does not activate the phase conversion effect.

2.3 Microstructure Development

The evolution of microstructure is described as follows.[8] A needle morphology is observed at about 2 hours after mixing the MOC paste. Short, stubby prismatic crystals with substantial intergrowth form a fine grained structure within 4 hours. Crystal morphology in the form of longer prisms with larger aspect ratio is more predominant in regions of high porosity. Monolithic structures prevail in regions of low porosity.

2.4 Strength Development

A study by Matkovic and Young[8] suggests that the low porosity regions in MOC paste are analogous to contact zones in C_3S pastes critical to strength development. The fracture process is consistent with that of crystalline material. Mechanical interlocking is not considered to be a major strength contributing factor.

A strength development curve is shown in Fig. 1. Strength (related to pulse velocity) changes little after one day. The first appearance of needle-like crystals corresponds to the onset of rigidity concommitant with rapid microstructural development during the next few hours. Strength development is a result of dense structures forming in limited space. Formation of needle-like crystals is not a requirement.

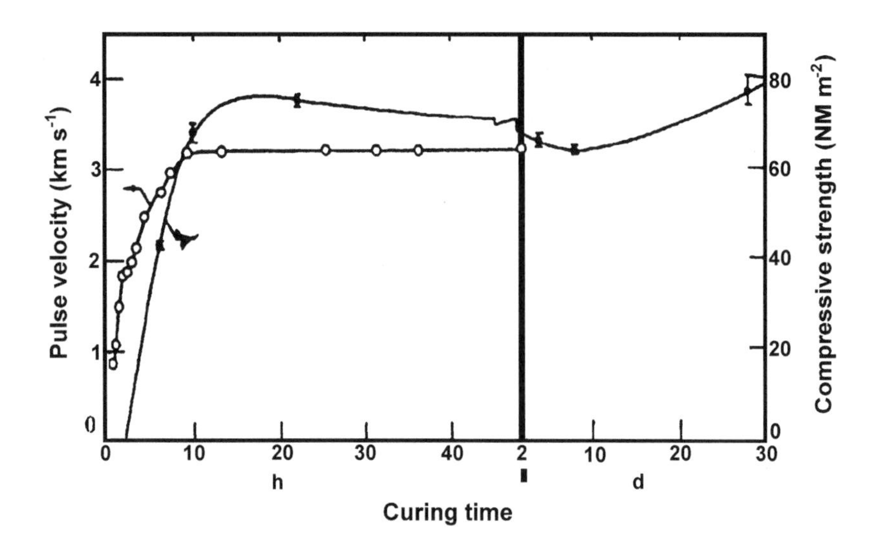

Figure 1. Development of rigidity and compressive strength in magnesium oxychloride cement.[8]

Beaudoin and Ramachandran compared the strengths (in terms of microhardness) of various cements using microhardness-porosity plots.[9] Strengths of various cement systems for different porosity values are indicated in Fig. 2.

The MOC paste has the highest strength values over the porosity range indicated. Normal portland cement paste has the next highest values. The magnesium oxysulfate paste has the lowest values. The strength values of MOC paste, obtained by extrapolating the strength-porosity curve to zero porosity, exceeds that of the other cement systems. The curves representing the latter converge to an approximately common strength value at zero porosity. This suggests that the solids comprising MOC paste have a higher intrinsic strength. The relative position of the curve for MOC paste also suggests that the bonding between solid particles is superior to that for the other cements over the porosity range investigated. The nature of these bonds was investigated employing compacted specimens prepared from ground MOC pastes. It is apparent that unlike portland cement the interparticle bonds formed during hydration cannot be remade by compaction.

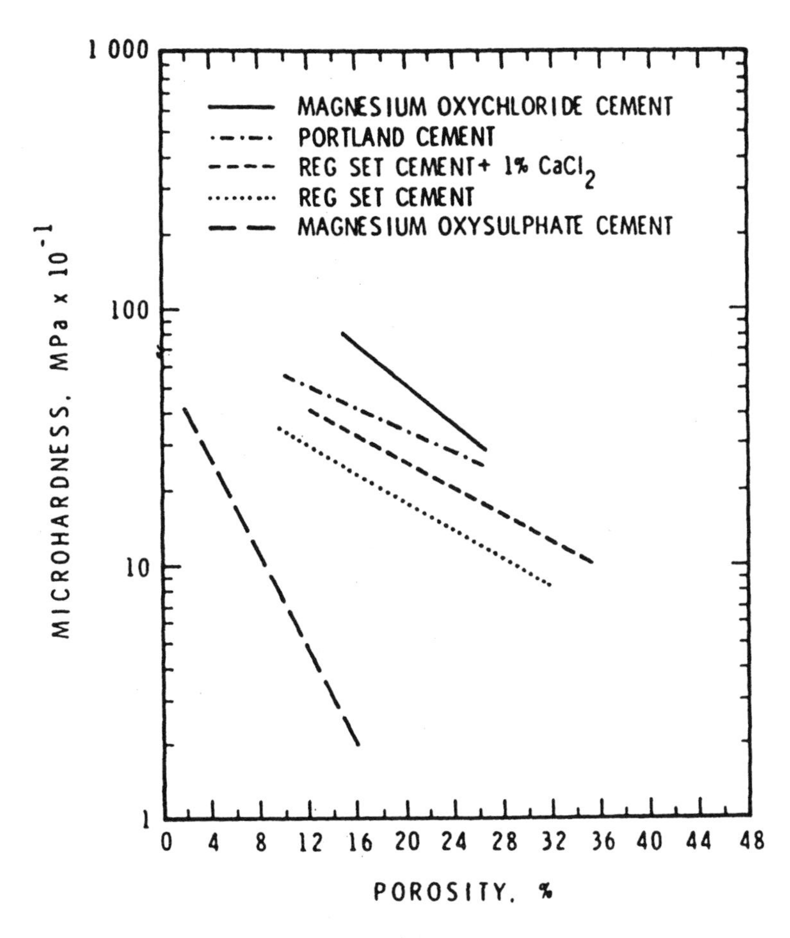

Figure 2. Microhardness vs porosity of several inorganic cementitious systems.[15]

The binding mechanisms in MOC pastes are not completely understood. The compositions of the gel formed and crystals of ternary oxychloride phases are uncertain. The relatively large contact areas between particles of MgO manifesting the inherent binding force are attributed to highly dispersive action occurring in the MOC system. Matkovic and Young[8] attribute the initial stiffening to the rapid growth of crystals into interstitial space between MgO grains and interlocking with the particles. The correspondence between strength and density is attributed to crystal intergrowth.

Strength development due to mechanical bonding of strong whisker crystals of hydrates on MgO crystals was a mechanism considered by Tooper and Cartz.[10] The bonding sites were augmented by interpenetration of crystals and solid-solid contact with other scroll-type crystal morphologies.

2.5 Resistance To Water

MOC paste undergoes significant strength retrogression (up to 50%) when continuously immersed in water.[11]–[13] This is a dissolution process and requires continuous contact with water for leaching of magnesium chloride to be significant. The water resistance of MOC can be improved by addition of ferrous phosphate and aluminum phosphate with magnesium oxide powder.[11] Coatings with bitumen emulsion have also been effective in minimizing strength retrogression. Sulphur impregnation of MOC paste improves water resistance significantly.[14] Loss in strength is reduced from 48.5% to 12.6% after complete immersion in water for seven days. Re-impregnation after exposure to water for 28 days increases strength and stabilizes the materials on further exposure to water. The more readily soluble material is removed through exposure of the initially impregnated material to water. Subsequent impregnation results in a body with more sulphur distributed in a matrix more resistant to water.

Additives which have been shown to be effective in lessening the weakening effects of water include the following:[15] fly ash, slag, baked clay, diatomite, silicate cement, coal gangue, baked diatomite, H_3PO_4, $Al(H_2PO_4)_3$, H_3BO_3, Na_2SO_4, Na_3PO_4, and K_3PO_4. The addition of copper powder in the amount of 10–12% has also been reported to increase strength and water resistance of MOC paste. This is attributed to formation of $3CuO{\bullet}CuCl_2{\bullet}3H_2O$.

3.0 MAGNESIUM OXYSULFATE CEMENT

3.1 Hydration

Magnesium oxysulfate cement pastes can be produced through reaction of MgO and an aqueous solution of $MgSO_4{\bullet}7H_2O$.[16] Alternatively magnesium chloride solutions can be added to calcium sulfates or calcium phosphate–sulfate mixtures. The presence of phosphates enhances the water resistance of this cement system.

Four magnesium oxysulfate complexes are formed in the MgO-$MgSO_4$-H_2O system in the temperature range, 30–120°C. These are:

$5Mg(OH)_2 \cdot MgSO_4 \cdot 2\text{–}3H_2O$

$3Mg(OH)_2 \cdot MgSO_4 \cdot 8H_2O$

$Mg(OH)_2 \cdot MgSO_4 \cdot 5H_2O$

$Mg(OH)_2 \cdot 2MgSO_4 \cdot 3H_2O$

The complex $3Mg(OH)_2 \cdot MgSO_4 \cdot 8H_2O$ is stable below 35°C. The complex $5Mg(OH)_2 \cdot MgSO_4 \cdot 3H_2O$ has also been detected below 35°C.

Magnesium oxysulfate cement systems have many similar characteristics to those of magnesium oxychloride cement systems. They are, however, less sensitive to elevated temperatures, hence more suitable to some industrial processes incompatible with the use of magnesium oxychloride cement.

3.2 Strength Development

Magnesium oxysulfate cement pastes are generally considered to be weaker than oxychloride pastes.[17] Beaudoin and Ramachandran examined the strength (related to microhardness)–porosity characteristics of paste-hydrated and compacted magnesium oxysulfate cement pastes.[18] The curves in Fig. 3 show the effect of particle size on the performance of compacted materials in addition to the behavior of the paste-hydrated materials. The pastes were prepared at different solution-solid ratios. A series of pastes made using a saturated solution of $MgSO_4 \cdot 7H_2O$ was included in the investigation. There is a linear semi-logarithmic relation for all the paste-hydrated materials. The compacted specimens show significantly greater strength than the paste-hydrated specimens at porosity values in the range 10–35%. The compaction process evidently increases the number of interparticle bonds possibly involving comminution processes. Larger values of microhardness for compacts made with finer starting material lend support to this argument. Increased strength at lower porosities (compacted samples) is attributed to an increase in the volume concentration of finer pores in the specimens and an increase in the number of contacts between particles.

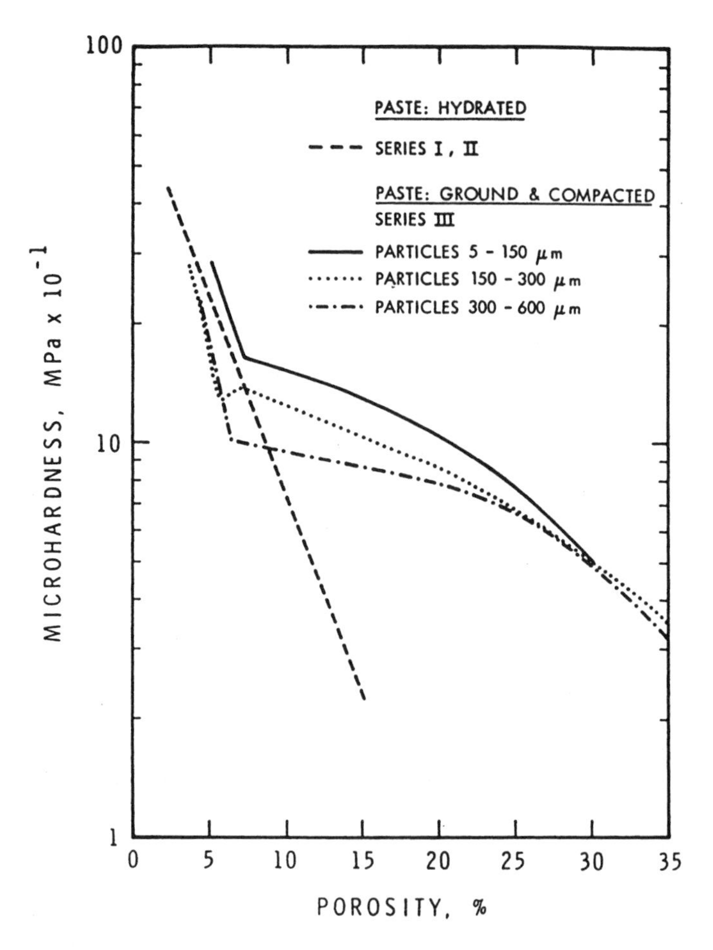

Figure 3. Microhardness vs porosity for magnesium oxysulfate cement.[9]

4.0 CALCIUM ALUMINATE CEMENTS

4.1 Description

Calcium aluminate cement (often referred to as high alumina cement) was patented in 1909.[19] Its development was a result of investigations directed at resolving the deleterious effects of sulfate solutions on portland cement concretes. Calcium aluminate cement (CAC) has several important characteristics including early strength gain, chemical resistance, and excellent refractory properties.[20]–[23]

The principal cementing compound in CAC is calcium monoaluminate (CA) present in amounts up to 60%. $C_{12}A_7$ and CA_2 are present; β-C_2S and C_2AS may also exist. Minor compounds include FeO, aluminoferrites, pleochroite and pervoskite. Glass content varies between 5 and 25% depending on the cooling conditions.

4.2 Hydration

Calcium monoaluminate reacts rapidly with water as follows:

Eq. (1) $$CA + H \rightarrow CAH_{10} + C_2AH_8 + AH_x$$

The relative amounts of reaction products are dependent on external factors. The reaction products are metastable and can enter into the following reactions:

Eq. (2) $$AH_x \text{ (alumina gel)} \rightarrow \gamma\text{-}AH_3 \text{ (gibbsite)} + (x\text{-}3)H$$

Eq. (3) $$3CAH_{10} \rightarrow C_3AH_6 + 2AH_3 + 18H$$

Eq. (4) $$3C_2AH_8 \rightarrow 2C_3AH_6 + AH_3 + 9H$$

The last two reactions showing the formation of hydrogarnet (C_3AH_6) are known as conversion reactions. Direct formation of hydrogarnet from CA can occur in the presence of water at elevated temperatures:

Eq. (5) $$3CA + 12H \rightarrow C_3AH_6 + 2AH_3$$

The iron analogue of the hydrogarnet phase can also form:

Eq. (6) $$3C_2F + xH \rightarrow 2C_3FH_6 + F + (x\text{-}12)H$$

Strätlingite formation results from the interaction of β-C_2S, alumina and water:

Eq. (7) $$\beta\text{-}C_2S + A + 8H \rightarrow C_2ASH_8$$

CAC of structural quality can be made even if complete conversion occurs provided the w/c ratio is sufficiently low. At low w/c ratios, the water released during the conversion reacts with the anhydrous grains not utilized in the initial hydration reaction, and fills the pores. A major factor preventing loss in strength is the grain size of the converted minerals, viz., C_2AH_8 and AH_3. At low w/c ratios, the low porosity is maintained by the packing of the small crystallites while at higher w/c ratios, the porosity is increased by the larger crystals.[24]

It is also suggested that the cubic hydrate (C_3AH_6) does not have a binding capacity. For example, the strengths of the aluminate hydrates are thought to be in the order: $CAH_{10} > C_2AH_8 > C_3AH_6$.

Work by Ramachandran and Feldman[25] has indicated that the conversion of C_3A to C_3AH_6 under certain conditions, in fact, enhances strength. Using high compaction methods, they obtained calcium monoaluminate discs for which the effective water/aluminate ratio for hydration was 0.15. They found that, compared with a microhardness value of 195 MPa for the unhydrated compacted CA sample, those hydrated at 20 or 80°C for 2 days developed hardness of 1074 and 1574 MPa respectively.[26] These results show that neither the high temperature nor the formation of the cubic phase is detrimental to strength development. At 80°C, the reaction proceeds very rapidly and accelerates the conversion of CA to C_3AH_6 and AH_3 phases. As the particles of CA in the compact lie very close to each other, direct bond formation between C_3AH_6 products is enhanced. At 20°C, however, direct bond formation due to C_3AH_6 may not be favored, as C_3AH_6 and AH_3 products are transported and recrystallized in the pores by the initial formation of hexagonal phases.

CAC may contain appreciable amounts of calcium aluminoferrite.[27] The contribution of this compound becomes important if CAC is made at low w/c ratios. Ramachandran and Beaudoin[28] investigated the physico-mechanical characteristics of C_4AF prepared at w/s ratios between 0.08 and 1.0 and hydrated for various periods at temperatures of 23 or 80°C or autoclaved at 216°C. In pastes hydrated at 23°C, with w/s ratios of 0.3–1.0, it was found that the lower the w/s ratio, the higher the microhardness values: 373 and 59 MPa, respectively, at a w/s ratio of 0.3 and 0.5. Significant increases in microhardness were observed in samples hydrated at an effective w/s ratio of 0.08. Unhydrated, pressed C_4AF had a value of 314 MPa; that hydrated at 23 and 80°C had values of 1128 and 1933 MPa respectively. A few samples prehydrated at 23 or 80°C were autoclaved at 216°C. The unhydrated pressed C_4AF sample and the two prehydrated at 23 and 80°C, having initial microhardness values of 314, 1128, and 1933 MPa, gave autoclave treatment values of 2717, 1825, and 2717 MPa, respectively. These results demonstrate that enhanced strengths occur by the direct formation of the cubic phase on the original sites of C_4AF. This results in a closely welded, continuous network with enhanced mechanical strength.

4.3 Strength Development

CAC concrete can develop up to 90% of its ultimate strength within the first 24 hours after mixing.[20] The CAC concrete develops substantially higher strength than ordinary portland cement concrete or concrete made with rapid hardening portland cement.

The high early strength characteristic facilitates emergency repair work and is particularly useful in the production of precast concrete units. The 28 day strengths decrease with temperature (above 25°C) and generally with water-cement ratio. Strength is also reduced if CAC cured at ambient temperature is suddenly exposed to a higher temperature environment.

The strength decreases cited are attributed to the conversion reaction. It is influenced by several factors such as temperature, water-cement ratio, stress, and the presence of alkalis. The precise mechanism responsible for the deterioration of strength is not completely understood.

4.4 Strength and the Conversion Reaction

The conversion of hexagonal hydrate (CAH_{10}) to the hydrogarnet or cubic phase (C_3AH_6) results in a solid volume decrease of about 50% whereas that of C_2AH_8 results in a decrease of about 65% of the original volume of the reactants. This is due to the higher specific gravity of the products compared with that of the reactants. The specific gravities of CAH_{10}, C_2AH_8, C_3AH_6, and AH_3 are 1.72–1.78, 1.95, 2.52, and 2.40, respectively. In the CAC pastes, the specific gravity of the unconverted material may be 2.11 and that of the converted material 2.64. In the hardened concrete, this conversion may be accompanied by the formation of pores resulting in lower strength. The strength loss has been explained by the decrease in the combined water in the converted product and also by the released water filling the small pores.

Lehman and Lecks[29] ascribe the higher strengths in CAC to CAH_{10} having a high surface area, and the conversion, producing hydrogarnet with a lower surface area, being responsible for the low strengths. It was found that, during the conversion reaction, the surface area is actually increased owing to the formation of alumina gel.[30] Mehta has affirmed that a poor bonding capacity of the hydrogarnet and its low surface area are important factors contributing to a loss of strength. It is also thought that the conversion of the flat plates, with the overlapping or interlocking system, to cubic morphology dislocates the intercrystalline bonds.

The fall in strength has also been attributed to the formation of macrocrystalline alumina hydrate from the microcrystalline alumina hydrate.

Although there is no unanimity on the details of the mechanism of strength deterioration in CAC, success has been achieved in explaining the strength loss in terms of the conversion of the hexagonal to the cubic phase. The loss in strength, as the temperature is increased, can be explained by the increase in the rate and amount of conversion. Similarly, the increase in the conversion rate at higher w/c ratios may explain the low strengths at these conditions.

4.5 Inhibition of C_3AH_6 Formation

The formation of C_3AH_6 (hydrogarnet) is often associated with volume instability in CAC concretes. It has been demonstrated that an additive comprised of a pozzolan and an inorganic sodium or potassium salt can be effective in eliminating the conversion reaction with minimal effect on early strength development.[3] The anion can be selected from the group consisting of sulfate, carbonate, nitrate, silicate, phosphate, chloride, and bromide. The pozzolan is selected from the group consisting of: zeolites, calcined zeolites, granulated blast-furnace slag, fly ash, silica fume, rice hull ash, and metakaolin. The resulting calcium aluminate cement composition is such that at temperatures between 25–40°C strätlingite is formed preferentially and formation of hydrogarnet is limited.

Strätlingite is preferentially formed from hexagonal calcium aluminate hydrates and anhydrous calcium aluminate cement at high temperature (e.g., 35–45°C). Conversion of those hydrates or anhydrates to hydrogarnet is prevented. Some anions such as sulfate, silicate, phosphate, carbonate, etc., also play an important role in preventing the conversion. This is a result of two primary reactions: (1) reaction of these anions with calcium ions to form insoluble calcium salts such as gypsum; the calcium concentration in the hydration system is significantly reduced and delays calcium dependent hydrogarnet formation; (2) substitution of sulfate in strätlingite-sulfate can be found in the strätlingite plates in calcium aluminate cement paste containing sodium sulfate and zeolite, indicating that sulfate plays a role in the crystallization of strätlingite. This confirms the postulation cited by F. M. Lea[20] that sulfate may be combined in strätlingite resulting in crystals having the formula $9CaO{\bullet}4SiO_2{\bullet}\ Al_2O_3{\bullet}7CaSO_4{\bullet}80H_2O$. The substitution of sulfate may reduce the crystallization energy of strätlingite plates and accelerate its formation. A large amount of preferentially formed strätlingite in a calcium aluminate cement system makes hydrogarnet formation unlikely. This process prevents strength reduction of the high alumina cement products due to the conversion.

4.6 Durability

The hydration of CAC does not result in the formation of $Ca(OH)_2$. This contributes to enhanced resistance in the presence of phenols, glycerols, sugars, etc. The formation of aluminum hydroxide gel during hydration provides resistance to dilute acids and sulfates. CAC has many industrial applications. It is used in flues, sewers, effluent-tanks, coal hoppers, ash sluices' flues, and in industrial plants, e.g., oil refineries, breweries, dairies, and tanneries.

CAC concretes are not resistant to alkali which readily attack the protective gels (possibly also iron-containing gels). The alkalis may also affect setting and strength development and accelerate the conversion reactions.

CAC concretes are resistant to solutions containing dissolved CO_2 and sea water. The formation of chloroaluminates is likely to affect the setting and hardening characteristics, and, therefore, the use of sea water as mix water should be avoided.

CAC concretes can also be used to encapsulate amphoteric metals such as Pb, Al, and Zn due to reduced reaction between the hydrates formed and these materials.

4.7 Chemical Admixtures

The addition of fully hydrated CAC to hydrating CAC systems can reduce setting times significantly. This behavior, also observed in portland cement systems, can be attributed to a nucleation mechanism.[21] Set acceleration can also occur with the addition of dilute solutions of sodium, potassium and lithium salts, or organic bases, e.g., triethanolamine.

Portland cement is an accelerator for CAC and vice versa. The effect of proportions of each component on setting time is illustrated in Fig. 4. Ternary blends of portland blast-furnace cement and CAC also have an acceleratory effect.

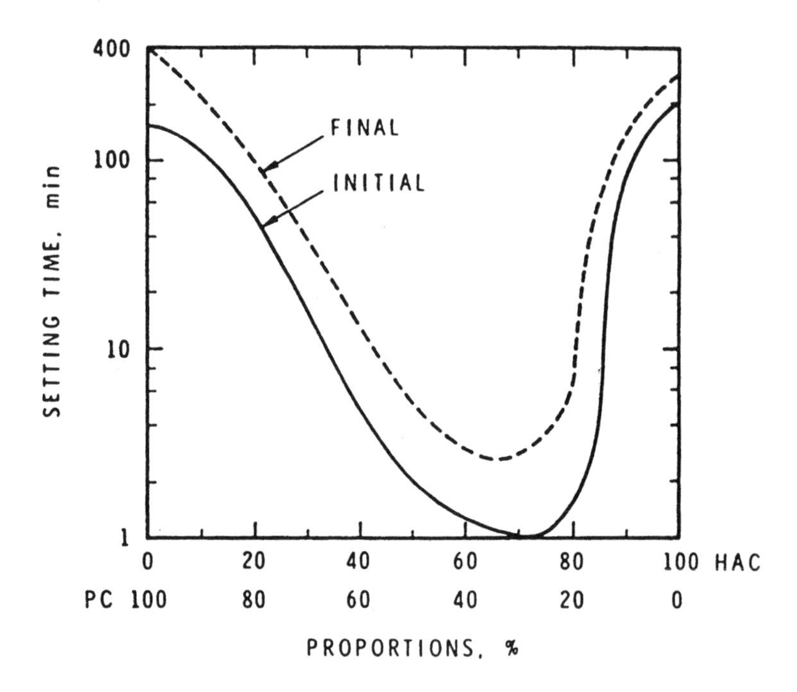

Figure 4. The setting times of mixtures of high alumina cement (HAC) and ordinary portland cement.[21]

The long-term strength and chemical resistance of CAC is affected by the addition of portland cement and chemical admixtures.[21] Calcium chloride retards CAC, the opposite of its effect on portland cement. Strength reduction is also a consequence of its use. Retarding action is provided by inorganic compounds such as phosphates, lead salts, and borax, and organic compounds that include cellulose products, starch, glycerine, sugars, flour, and cosein. Most sulfates in small amounts retard the set of CAC but concentrations between 0.5 and 1.0% produce quick setting.

Plasticizing agents, e.g., alkyl aryl sulfonates, secondary alkyl sulfates, or their sodium salts, are used to overcome the harshness of many CAC concrete mixes.[21] Workability can be improved with calcium lignosulfonate addition; larger doses, however, promote excessive retardation. Other additives that have a plasticizing effect on CAC concrete include: methyl cellulose, soya bean flour, bentonite, finely divided fly ash, and granulated slag. Alkalis present in the additives can also affect the physical properties of CAC.

The use of superplasticizers, e.g., polynaphthalene formaldehyde (PNS) and polymelamine formaldehyde (PMS), appear to improve flowability of CAC concrete; loss in workability appears to be too rapid. Normal range water reducers, e.g., hydrocarboxylic acids, function in a manner similar to their role in portland cement concrete.

Ma and Brown reported that sodium phosphate modified calcium aluminate cement (at addition levels of 10 and 20 mass%) can enhance strength.[33][34] The phosphates included $(NaPO_3)_r$, $(NaPO_3)_n \cdot Na_2O$, $Na_5P_3O_{10}$ and $(NaPO_3)_3$. They concluded that the phosphate-based additions do not promote the crystallization of calcium aluminate hydrates during the CAC hydration. Morphologically distinct hydration products can form. An amorphous C-A-P-H phase serves to interlink the other constituents forming between crystalline calcium aluminate hydrate grains. This phenomenon appears to enhance the structural integrity of the calcium aluminate cement and reduce the strength retrogression which occurs on conversion.

4.8 Refractory Applications

Strength of CAC concrete is generally superior to portland cement concrete at temperatures exceeding 1000°C.[21] The CAC systems are usually stable at temperatures up to 1300°C. CAC materials containing alumina or corundum are resistant to temperatures approaching 1600°C. CAC concrete when fired loses most of its combined water at about 700°C. Strength losses and porosity increases reach a maximum between 900 and 1100°C. Increased strength and new bond formation occur above about 900°C. Alumina gel passes through several crystal modifications to finally form α-Al_2O_3. CAH_{10} and other hexagonal phases form CaO and $C_{12}A_7$ between 600 and 1000°C.

CAC concrete containing fire brick aggregate produces anorthite (CAS_2) and gehlemite (C_2AS).[35] The results are similar for aluminosilicate aggregate. CA_2 and CA_6 form when white alumina cement is fired with fused alumina aggregate. Spiral (MA) and forsterite form when the aggregate used is dead burnt magnesite. Applications for CAC concrete include brick and tunnel kilns, kiln and chimney linings, kiln doors and coke ovens, and furnace foundations.

5.0 PORTLAND CEMENT–CALCIUM ALUMINATE CEMENT BLENDS

5.1 Introduction

The development of OPC-CAC based expansive cements can be traced back to the 1920s.[36]–[38] Fast set control was a problem that limited the actual application of these systems. The setting time varies with the proportion of the two cements. The optimum proportion resulting in minimum setting time is about 70% CAC and 30% OPC.[39] Methods for controlling rapid set involving the use of pre-hydrated CAC (H-CAC) have been recently proposed.[40] Satisfactory workability can be attained without loss of expansive and mechanical properties. Understanding of the hydration mechanisms and the microstructural development in these expansive cement systems is important to optimize potential application. This involves evaluation of factors affecting: (i) fast set behavior of the OPC-CAC system and (ii) the hydration mechanism of the OPC-H-CAC system.

5.2 Hydration

Electrochemical methods appear to have distinct advantages in the study of cement hydration. Methods involving potential measurement (including pH, zeta potential, and selected ion potential), conductivity measurement, and a.c. impedance measurement provide useful information related to both ion concentration of pore solution and microstructural change in hydrating cement paste. The early hydration and setting behavior of OPC-CAC and OPC-Hydrated-CAC paste systems can be determined using these techniques.

It appears that the impedance behavior of the OPC-CAC paste system is determined by the following microstructural events:[41]

i) Nucleation of ettringite accompanied by significant consumption of free water and fast set. This process is concomitant to the first rapid increase in impedance value at about 20 minutes (curves 2–5, Fig. 5).

ii) An "induction" period while hydration continues at a relatively slow rate. Accumulation of hydration products (mainly C-S-H, CAH_{10}, C_2AH_8) on the unhydrated particle surfaces are observed. The impedance value remains relatively constant.

iii) Rapid hydration of CAC leading to a very dense matrix. This stage corresponds to the second large increase of impedance value.

iv) Matrix densification and decrease of impedance after the second maximum.

The latter stage involves various reactions such as:

Eq. (8) $2CAH_{10} \rightarrow C_2AH_8 + AH_3 + 9H;$
$3C_2AH_8 \rightarrow 2C_3AH_6 + AH_3 + 9H$

Eq. (9) $2C_3A + C_6A\bar{S}_3H_{32} + 4H \rightarrow 3C_4A\bar{S}H_{12}$

Eq. (10) $C_2S + 2H \rightarrow C\text{-}S\text{-}H + CH$

Eq. (11) $C_3S + 3H \rightarrow C\text{-}S\text{-}H + 2CH$

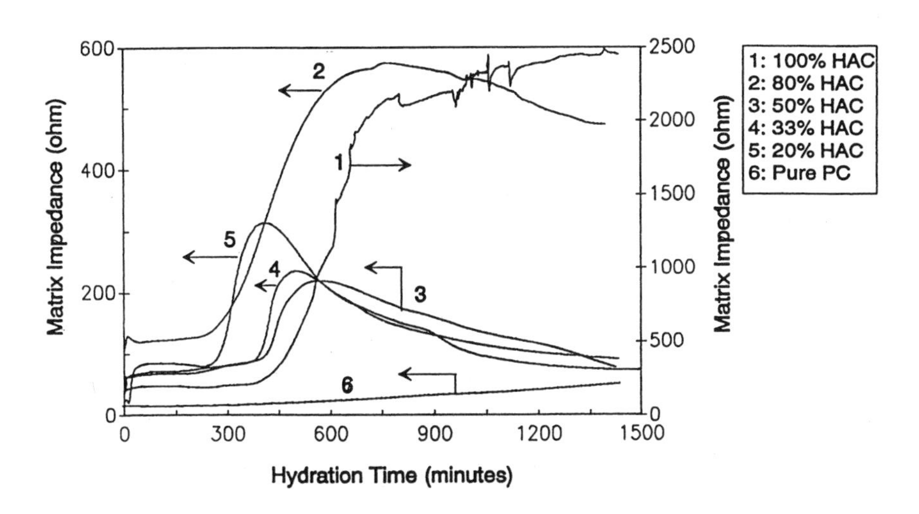

Figure 5. Plots of matrix impedance vs hydration time for OPC, HAC, and OPC-HAC pastes. The numbers 1–6 correspond to those pastes containing 100, 80, 50, 33, 20, and 0% HAC respectively.[41]

The hydrates CAH_{10} and C_2AH_8 (Eq. 8) have hexagonal crystal structures and are metastable with respect to conversion to the cubic hydrate C_3AH_6 and gibbsite (AH_3). The microstructural changes associated with a reduction in volume of solids result in an increase of porosity in the paste matrix. This process is unlikely the reason for the decrease of impedance after the maximum since OPC-CAC paste is still in a very early hydration stage. Moreover, there was no decrease of matrix impedance in the pure CAC paste (curve 1, Fig. 5). The second equation (Eq. 9) describes the process of conversion of ettringite to monosulfoaluminate accompanied by release of sulfate ions and free water. Free ions re-entering the liquid phase reduce the matrix impedance of the paste; however, this is not likely the main reason for the decrease of matrix impedance in OPC-CAC paste systems. At room temperature, ettringite, (solubility product, 1.1×10^{-40}), is more stable than monosulfoaluminate, (solubility product, 1.7×10^{-28}).[42] In addition, the amount of ettringite generated and its conversion rate are negligible in the period studied. The most probable explanation for the decrease of matrix impedance is the formation of CH due to the OPC hydration as indicated in the third equation (Eq. 10). CH is a source of Ca^{++} and OH^- ions for the pore solution and increases the conductivity of the liquid phase.

5.3 Setting Behavior and Ettringite Nucleation

The fast set behavior due to the formation of ettringite, controlling calcium, sulfate, and aluminate ions in the liquid phase is apparently the key in delaying the setting of OPC-HAC pastes.[41] Use of pre-hydrated HAC to control the aluminate ion concentration in the liquid phase provides satisfactory results. Impedance behavior of three OPC-CAC paste systems is illustrated in Fig. 6. The numbers 1–3 correspond to pastes containing 80, 50, and 20% of H-CAC, respectively. In contrast to OPC-HAC pastes in Fig. 5, there is no rapid increase of impedance value in all three pastes. The behavior is closer to that of OPC paste. The pre-formed layer of hydration products on the unhydrated CA particle surface is effective in delaying the setting time through control of the aluminate ion concentration in the pore solution. Dissolution of aluminate ions into the liquid phase is, therefore, controlled by a slow diffusion process.

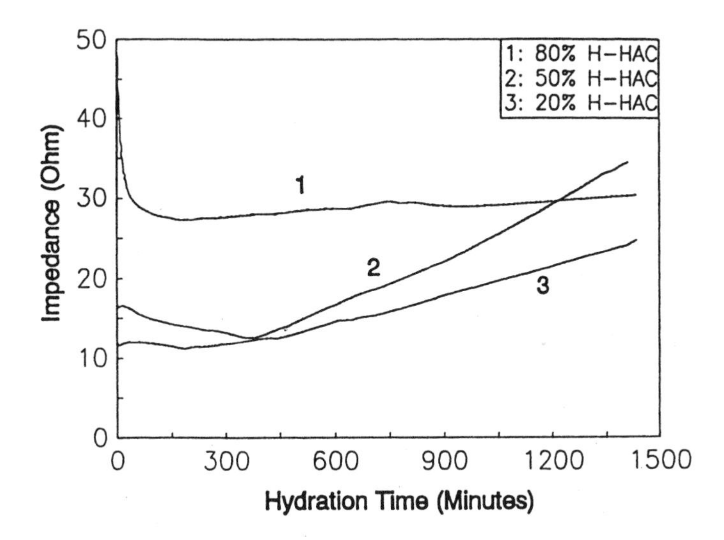

Figure 6. Plots of matrix impedance vs hydration time for OPC-H-HAC pastes. The numbers 1–3 correspond to those pastes containing 80, 50, and 20% of H-HAC respectively, w/c = 0.50.[41]

5.4 Early Strength Development

Compressive strength results for three pastes (OPC/CAC ratios: 92.5/7.5; 80/20; 20/80) prepared at w/c ratio = 0.40 are presented in Table 1.[43] The early strength development of the 92.5/7.5 and 20/80 pastes is very similar to that for pure OPC and CAC pastes. The 80/20 paste develops strength earlier than the other two pastes, and it appears to have unique strength development behavior.

Many factors affect the strength development of OPC/CAC systems. In the OPC/CAC binary cement pastes, ettringite can form in the first few minutes of the hydration mainly through the following reactions:

Eq. (12) $$CA + 3CSH_2 + 2C + 26H \Rightarrow C_6AS_3H_{32}$$

Eq. (13) $$C_3A + 3CSH_2 + 26H \Rightarrow C_6AS_3H_{32}$$

Table 1. Compressive Strength Values of the Three OPC/CAC Pastes

Hydration Time (h)	92.5/7.5 (MPa)	80/20 (MPa)	20/80 (MPa)
1	—	0.7	—
2	—	0.8	—
4	—	1.2	3.5
8	0.7	1.7	25.5
16	4.5	2.8	36.1
21	5.8	2.7	36.8
24	8.2	2.7	43.8
48	16.2	2.6	40.7
72	23.5	10.7	46.7

The early strength of the binary paste systems is attributed to the above reactions.[44] The OPC normally provides the sulfate and the CAC is the primary source of aluminates. The amount of ettringite formed depends on the available amount of sulfate and aluminates. In the 92.5/7.5 paste, the small amount of ettringite formed does not contribute significantly to strength of the paste (Table 1). It is apparent that the aluminates are the limiting reactants. An estimate of the CAC/CS ratio is 0.78 (assuming a 4% gypsum content in OPC and 40% aluminate content in CAC in the calculation) which is much smaller than the value of 1.43 suggested as required to develop early strength.[44]

In the 80/20 paste, the CAC/CS ratio is, however, about 2.50. Rapid formation of a large amount of needle-like ettringite crystals occurs. This contributes to the quick-set. The early strength is usually less than a few MPa (Table 1). It is much lower than that contributed by C_3S, C_2S, and CA hydration. It was also suggested by Bensted[45] that the fast formation of CAH_{10}/C_2AH_8 due to the CA/C_3A hydration may contribute to the early strength of the binary paste system. The work referred to in Table 1 does not seem to support Bensted's suggestion. Significant early strength development in the pure CAC paste or even in the 20/80 paste system does not occur in spite of rapid formation of CAH_{10}/C_2AH_8.

In the 20/80 paste, appreciable strength is developed at 4 hours hydration (Table 1). SEM examination at a half hour hydration failed to identify ettringite crystals. X-ray diffraction also failed to detect early ettringite formation. There is apparently not enough sulfate available and the hydration and strength development behavior of this system are similar to those for pure CAC paste.[46]

The strength development of the 80/20 paste is quite unique. SEM studies at 0.5 and 8 hours hydration provide evidence of a morphology change during this period—small needle-like crystals observed at 0.5 hours were transformed to larger plate-like crystals. It has been suggested that a slow conversion process of ettringite to monosulfoaluminate hydrate may take place during this period. The monosulfoaluminate hydrate is detectable by x-ray diffraction at 24 hours hydration.

The observation of a $Ca(OH)_2$ peak at 48 hours by x-ray diffraction analysis suggests that the slow strength development in the 80/20 paste may be due to a delay in the hydration of C_3S. The observation that C_3S remains unhydrated until the conversion is completed suggests an incompatibility between the OPC and CAC hydration products.

5.5 CAC-Based Expansive Cement Reactions

The use of calcium alumina cement (CAC) as an alumina-bearing component in expansive additives or cements has been reported by researchers since the 1920s.[47]–[49] The CAC-based expansive additive is generally a dry mixture comprising different types of particulates including calcium alumina cement, calcium sulfate hemihydrate or gypsum, and lime or hydrated lime. It can be simply obtained by mixing commercially available materials. Its application is, however, limited due to the problems of quick setting and unstable expansion. The use of hydrated calcium alumina cement (H-CAC) instead of high alumina cement in the composition of expansive cement was reported to be a potential solution to the quick-setting problem.[50] The expansion characteristics of expansive cement containing pre-hydrated alumina cement have also been reported.[51]–[53]

A recently developed chemically compounded-expansive additive has shown promise.[54] The compounded-expansive additive comprises mainly calcium aluminates, calcium sulfate, and calcium hydroxide. The primary difference between the C-HAC system and the compounded-expansive additive system is that the latter contains all the expansive components in the compounded material. Particulate products are formed during the process of compounding.

Particulates in the form of different calcium aluminate hydrates, e.g., CAH_{10}, C_2AH_8, and C_3AH_6, or pre-hydrated CAC/CH (H-CAC/CH) have varying effects on ettringite formation and expansion of expansive cement pastes.

Mixtures containing different calcium aluminate hydrates can be prepared by hydrating calcium aluminate cement (CAC) at different temperatures (Table 2). Either CAH_{10}, C_2AH_8, or C_3AH_6 is the major component for each different mixture. The major component of the H-CAC/CH mixture also is C_3AH_6. Addition of CH to CAC appeared to promote C_3AH_6 formation. The use of calcium aluminate mixtures prepared from CAC hydration rather than pure synthetic calcium aluminates is based on practical considerations.

Table 2. Preparation Conditions of Calcium Aluminate Hydrate Mixtures

Name of Mixtures	Curing Temperature (°C)	Main Mineral Component	Moisture State
M-CAH_{10}	0	CAH_{10}	Paste
M-C_2AH_8	23	C_2AH_8	Paste
M-C_3AH_6*	38	C_3AH_6	Dry Powder
H-CAC/CH	23	C_3AH_6	Paste

*Cured at 38°C for 2 months followed by 24 hours drying at 85°C.

The effect of different calcium aluminate hydrates on free expansion is shown in Fig. 7. It is apparent that the expansion of expansive cement pastes containing the CAH_{10} or C_2AH_8 mixture started at very early ages (about 1 hour), and developed mainly in the first 3 hours. Their 3-hour expansion and ultimate expansion values are 0.58 and 0.65% respectively. The expansion of specimens containing additives based on anhydrous CAC and H-CAC/CH mixture developed after 3 hours. The former had the largest ultimate expansion (around 0.8%) at about 10 days; the latter a much lower expansion than the cement pastes containing expansive additives based on CAH_{10} or C_2AH_8 mixtures. Its ultimate expansion was about 0.45%. The cement paste containing the expansive additive based on the C_3AH_6 mixture showed very delayed and lower expansion compared to the other samples.

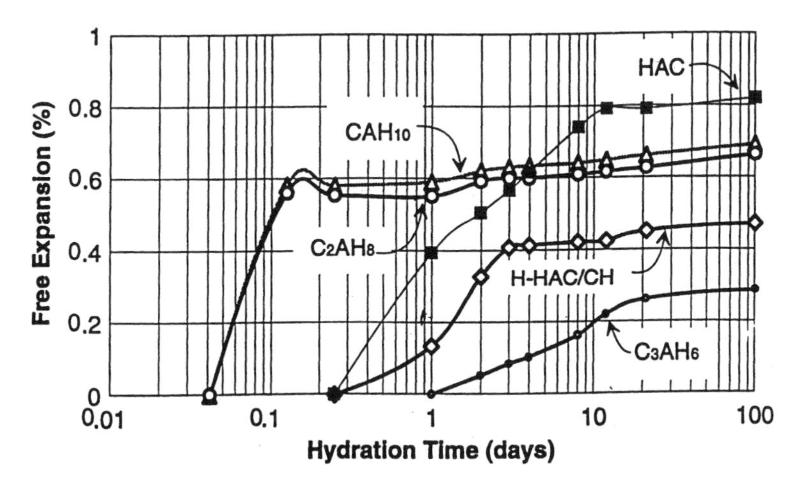

Figure 7. Effect of expansive additives based on different calcium aluminate hydrates or HAC on free expansion of cement paste.[54]

The different calcium aluminate hydrates exhibit very different behavior with respect to ettringite formation and expansion. The reaction rate of CAH_{10} and C_2AH_8 to form ettringite appears to be much faster than the other forms of calcium aluminates described above. The use of CAH_{10} or C_2AH_8 alone as the Al-bearing component in the formulation of expansive cement may result in little expansion in concrete members. This is due to the timing effect postulated by Aroni, et al.[55] The magnitude of expansion is relatively small when the ettringite crystals grow too quickly in the early stage, as the structure of cement paste is not well formed. Ettringite formation from the reaction of C_3AH_6 is apparently delayed and reduced. This makes the Al-bearing material inefficient. Delayed ettringite formation also causes stability problems. The expansion period of such expansive cement can last up to 28 days. It is not desirable for the application of shrinkage-compensating cement concrete. A good CAC-based expansive additive may have an optimum composition comprising anhydrous CAC, CAH_{10} and/or C_2AH_8 present in the surface layer of CAC particles. This surface layer prevents quick setting. It will react with sulfate to form ettringite and adjust the hydration rate of the inner anhydrous CAC. This optimized structure of expansive additive particles provides expansive cement with controllable expansion characteristics. The compounded-expansive additive approach appears to be an example of this material-design philosophy.[55]

5.6 Chemical Admixtures

The hydration process of OPC/CAC mixtures is affected by the presence of admixtures. Phosphonate compounds have attracted attention because of their super-retarding capability with respect to ordinary portland cement and calcium aluminate cement hydration.[56] Phosphonate compounds, aminotri (methylene-phosphonic acid) (ATMP), 1-hydroxyethylidene-1, 1-diphosphonic acid (HEDP), and diethylene-triaminepenta (methylene-phosphonic acid) (DTPMP), appear to be much more efficient retarders than many others used in concrete practice. Low dosages (0.05%) of phosphonate admixtures generally appear to have little effect on the hydration reactions in OPC/CAC binary cement systems associated with quick-setting phenomena, e.g., ettringite formation and simultaneous hydration of CAC and OPC. Higher dosages of phosphonate admixtures (0.2%) substantially reduce the rate of ettringite formation and simultaneous hydration of CAC in OPC/CAC binary cement systems. OPC hydration is significantly retarded. It is apparent that the phosphonate compounds interfere with the gypsum-cement reactions especially in the pastes containing 0.2% ATMP and HEDP resulting in portland cement hydration (mainly C_3S and C_2S phases) retardation in an OPC/CAC blended system.

The retarding action of phosphonates on CAC paste hydration alone can be explained as follows.[57] Phosphonates may adsorb or form complexes incorporating cations such as Ca and Al on the surface of unhydrated cement particles and stabilize the very early hydration product (CAH_{10} or C_2AH_8) retarding further hydration. Poisoning nucleation may also occur. The efficiency of retardation does not appear to relate to the chelating capability of the phosphonates but rather to the spatial effects in the complex. This may be due to the small cations (Al^{+3} in the form of AH_6^{-3} or AH_3) and large chelating molecules which render the larger chelating molecule less effective. The efficiency of retardation is in the order ATMP and HEDP > DTPMP in CAC hydration.

Lithium salts cause rapid setting and hardening of CAC and promote very early strength development.[58] It has also been shown that the delay of OPC hydration in OPC/CAC mixtures can be overcome by using lithium-salt-based chemical admixtures. Table 3 lists the compressive strength of the binary cement pastes containing 0.8% lithium salts prepared at a water-cement ratio of 0.40. An improvement of strength is obtained in the pastes containing LiOH at < 3 days. However, all the samples reach about the same strength at 7 days hydration. The paste containing Li_2CO_3 had highest strength.

Table 3. Compressive Strength of 80% OPC-20% CAC Pastes Containing Lithium Salts

Hydration Time (days)	OPC/CAC Paste (MPa) w/c = 0.4	Paste 0.8% LiOH (MPa) w/c = 0.4	Paste 0.8% LiBr (MPa) w/c = 0.4	Paste Sat. Li_2CO_3 (MPa) w/c = 0.4
1	2.06	2.51	—	2.30
2	2.57	17.22	2.36	2.19
3	16.70	21.79	5.24	15.50
7	31.12	27.72	31.28	33.01

6.0 PHOSPHATE CEMENT SYSTEMS

6.1 Description

Phosphate cement systems, because of their quick setting property, are utilized in many civil engineering repair applications.[59] Applications of phosphate cement include: dental castings, phosphate-bonded refractory bricks, mortars, ramming mixes, highway patching, cement pipe, sprayable foamed insulation, flame-resistant coatings, and patching of preformed concrete products. A discussion of the physical, chemical, and mechanical properties of selected phosphate cement systems follows, with particular emphasis on those systems used as construction materials.

Different methods can be used to produce phosphate cements.[60]–[62] These include:

1) Reaction of metal oxides with phosphoric acid.

2) Reaction of acid phosphates with weakly basic or amphoteric oxides.

3) Reaction of siliceous materials with phosphoric acid.

4) Reaction of metal oxides with ammonium phosphates, magnesium acid phosphates, aluminum acid phosphates, and other metal phosphates.

Phosphate binders contain hydroxy-groups and undergo the following condensation-polymerization reaction on dehydration.

Eq. (14)
$$\mathrm{HO-\overset{\overset{\displaystyle O}{\|}}{\underset{\underset{|}{\displaystyle O}}{P}}-OH + HO-\overset{\overset{\displaystyle O}{\|}}{\underset{\underset{|}{\displaystyle O}}{P}}-OH \rightarrow HO-\overset{\overset{\displaystyle O}{\|}}{\underset{\underset{|}{\displaystyle O}}{P}}-O-\overset{\overset{\displaystyle O}{\|}}{\underset{\underset{|}{\displaystyle O}}{P}}-OH + H_2O}$$

The function of the phosphoric acid is to provide the phosphate linkages, to act as a dispensing medium to promote cohesion and to serve as a wetting liquid to increase adhesion.

There is, with acid phosphate reaction cements, a relationship with bond strength and the rate of reaction. The rate of reaction of oxides with phosphoric acid depends on the basicity of the oxide. The reaction between aluminum oxide and phosphoric acid is slow enough to permit gradual development of the phosphate bonds. The zinc oxide reaction is so rapid that the structure develops no cohesion.

A few specific examples of phosphate cement systems are listed below.

1) Mixes comprising MgO, dolomite, and ammonium phosphate solutions set quickly at ambient temperatures and are useful as repair materials.

2) A hard solid is formed when phosphoric acid reacts with a metal oxide (e.g., Al_2O_3) at 20–200°C. Acid phosphates of the type $Al(H_2PO_4)_3$ are formed initially; subsequent heating produces the $AlPO_4$ bonding phase.

3) Aluminum acid phosphate [$Al(H_2PO_4)_3$], reacts with MgO to form crystalline $MgHPO_4{\bullet}3H_2O$ as well as $AlH_3(PO_4)_2$ gel. If $P/(Al + Mg) > 1$, problems may result owing to the formation of water soluble $Mg(H_2PO_4)_2$.

4) A sodium hexametaphosphate ($Na_6P_6O_{18}$) solution prepared at 25°C is mixed with magnesite powder and used as a mortar for bonding magnesite brick. This cement gives high strength mortars when used with fireclay aggregate. The cement is cured for 24 hr at 120°C.

5) Aluminum chlorophosphate hydrate decomposes on heating to form $AlPO_4$. It can be used with the addition of MgO in formulations designed to set at ambient temperatures.

7.0 MAGNESIA PHOSPHATE CEMENT BINDERS

These cements are based on the chemical reaction of solid basic magnesia powder and an aqueous acidic ammonium phosphate solution.[63][64] The acid-base reactions that occur on mixing lead to the formation of insoluble magnesium ammonium phosphate hydrates, the primary binding material in the hardened product. Commercial products generally contain inert fillers and a set retarder to facilitate control of the exothermic setting process.

The principal reaction product in the system based on magnesia and ammonium dihydrogen orthophosphate is magnesium ammonium phosphate hexahydrate (struvite):

Eq. (15) $$MgO + NH_4H_2PO_4 + 5H_2O \rightarrow MgNH_4PO_4{\bullet}6H$$

Hall, et al., report that $MgNH_4PO_4{\bullet}H_2O$ (dittmorite) and $Mg(NH_4)_2(HPO_4)_2{\bullet}4H_2O$ (schertelite) can form when the water content is too low to ensure complete reaction of $NH_4H_2PO_4$.[63] Abdelrazig[64] reports the presence of schertelite as a primary intermediate reaction product:

Eq. (16) $$MgO + 2NH_4H_2PO_4 + 3H_2O \rightarrow Mg(NH_4)_2\,(HPO_4)_2{\bullet}4H_2O$$

Eq. (17) $$Mg(NH_4)_2\,(HPO_4)_2{\bullet}4H_2O + MgO + 7H_2O \rightarrow MgNH_4PO_4{\bullet}6H_2O$$

The formation of magnesium ammonium phosphate monohydrate, $MgNH_4PO_4{\bullet}H_2O$ at low water contents was reported by Popovics, et al.[65] The use of excess water in magnesia-phosphate cement systems has been found to lead to increased porosity and strength reduction.

7.1 Mechanical Properties

Compressive strength, flexural strength, and the elastic modulus of magnesia-phosphate cement products are affected by composition and water content. The compressive strength, at different times, for seven phosphate-cement concretes is plotted in Fig. 8.[66]

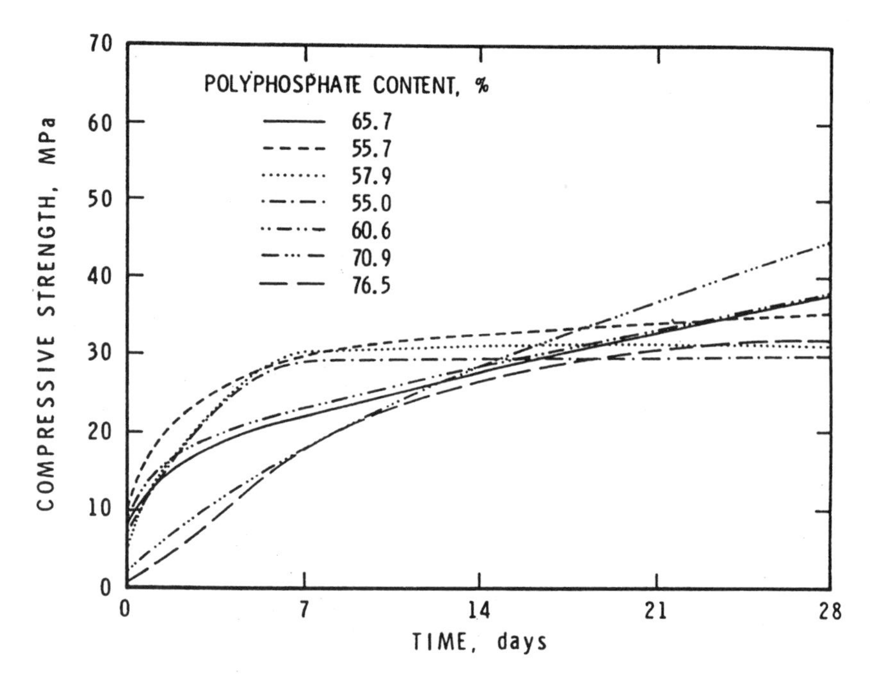

Figure 8. Compressive strength vs time for seven phosphate cement concretes made with ammonium phosphate solution.[66]

The concrete mixes contained 22.5 parts solution to 100 parts aggregate; the aggregate contained 80% sand and 20% dead burnt magnesia, and the phosphate solutions all contained approximately 10% ammoniacal nitrogen and 34% P_2O_5. The pH ranged from 5.8 to 6.3, and the specific gravity was approximately 1.39. The polyphosphate contents of the solutions varied from 55 to 76.5%. Concretes made with phosphate solutions containing < 58% polyphosphate have greater strengths below 14 days than concretes made with phosphate solutions containing > 58% polyphosphate. Also, the rate of strength development from 2 to 3 days is greater for the concretes containing < 58% polyphosphate. At 28 days, the compressive strengths of those concretes made with phosphate solutions containing > 58% polyphosphate are generally higher.

The compressive strength and setting time (Gillmore needle ASTM C266-74) for several phosphate-cement concretes containing different amounts of polyphosphate in ammonium phosphate solutions are plotted in Fig. 9. At 2 hours, 24 hours, and 7 days the compressive strength increases with a decrease in the polyphosphate content. At 28 days, there is a small increase in the compressive strength at higher polyphosphate contents. For polyphosphate contents < 58%, the compressive strength exceeds 8 MPa after 2 hours at ambient temperature and, at 7 days, at least 75% of the 28 day strength was achieved. The setting time increases as the polyphosphate content increases.

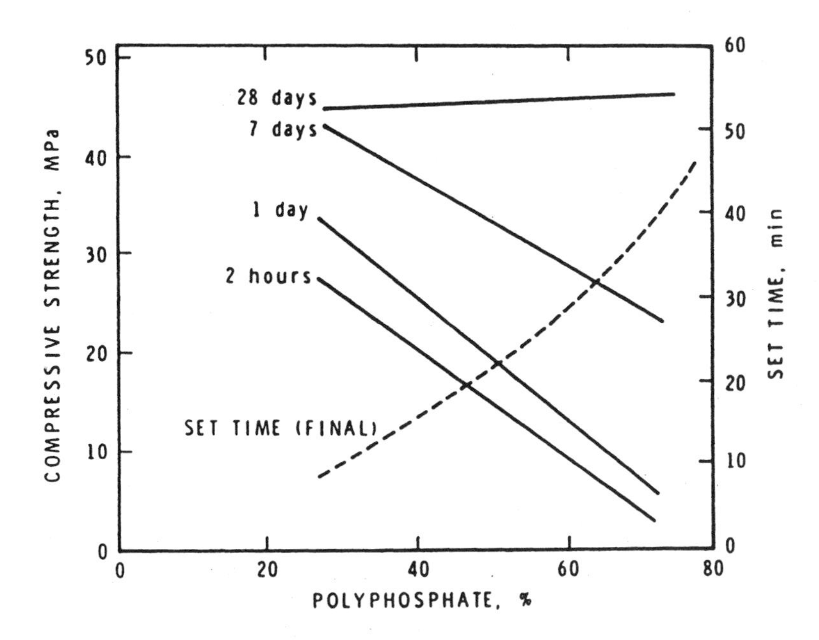

Figure 9. Compressive strength vs polyphosphate content of the ammonium phosphate solution used for several phosphate cement concretes.[66]

The effect of water content (of mortars containing silica sand) on elastic modulus, flexural strength, and compressive strength is evident in Figs. 10 and 11.[63] The net effect is a decrease in modulus with water content at each test age. Decreases of 80% in flexural strength and compressive strength occurred at a water content of 12% at 7 days. It was concluded that the crystal structure and morphology of the binder was highly sensitive to the water content. It was confirmed that an increase in the quantity of ammonium phosphate hexahydrate, $MgNH_4PO_4{\bullet}6H_2O$, occurred with an increase in water content from 5 to 8%. Reaction products with a more clearly defined crystalline morphology were present when the water content exceeded 8%. The result was a greater amount of unreacted water that produced on evaporation a finer pore structure and a significant degradation of the mechanical properties. These characteristics (at higher water content) were also found by Abdelrazig, et al., who observed that at lower water contents struvite morphology was needle-like while schertelite had a platey morphology.[67] The morphology at low water contents was considered a contributing factor for maximizing strength.

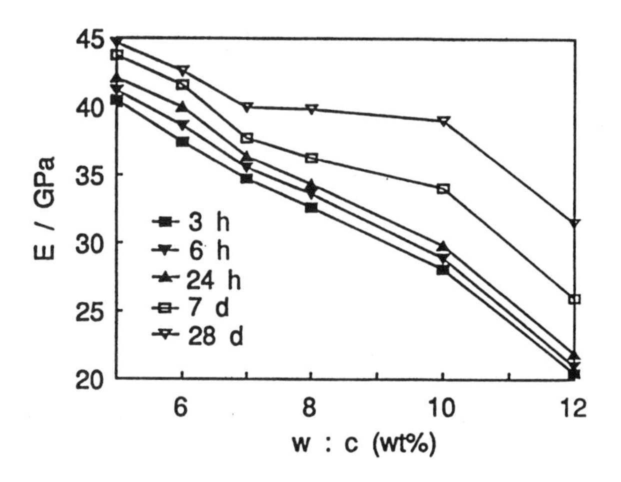

Figure 10. Young's modulus as a function of water content, measured at various aging times.[63]

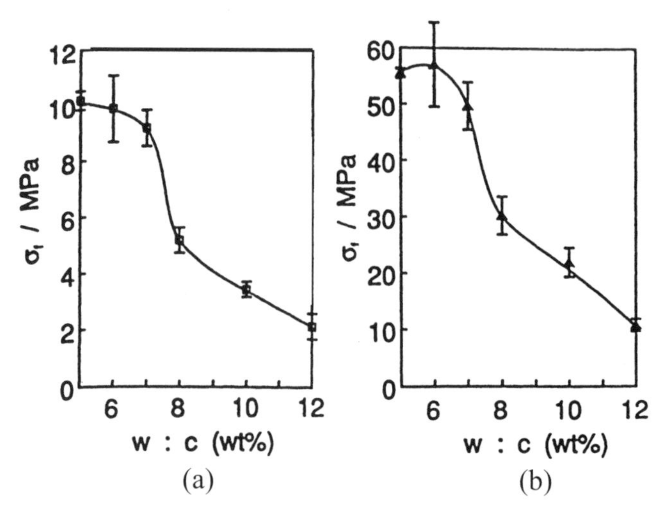

Figure 11. Flexural and compressive strength as a function of water content measured after 7 days aging.[63]

7.2 Additives

Additives such as sodium tetraborate decahydrate (borax), and ammonium pentaborate can be used to control the setting time and rate of strength development of phosphate cement concrete made with ammonium phosphate solutions.[68] The early strength is reduced with the addition of 1% borax and the setting time is increased from 9.5 minutes to 15 minutes. The effects of various additives on the setting times of the phosphate cement are given in Table 4.[68] The compressive strengths at 28 days for phosphate cement incorporating some additives are given in Table 5.[68]

Table 4. Effect of Additives on Setting Times of Phosphate Cement[a]

	Setting time (min)					
	Concentration (%)					
Additive	0	0.5	1.0	1.5	2.0	3.0
Borax	9.5		15.0		23.0	30.0
Ammonium pentaborate	7.5	14.5		22.0		
Boric acid		7.5				12.5
Trimethylborate	7.5				13.5	

[a] Typical mix contains 1335 parts sand, 150 parts fly ash, 495 parts MgO, 198 parts ammonium phosphate and 150 parts water.

Table 5. Effect of Additives on 28 day Compressive Strength

	28 day Compressive Strength (MPa)					
	Concentration (%)					
Additive	0	0.5	1.0	1.5	2.0	3.0
Borax	55.5		60.0		50.0	45.1
Ammonium pentaborate	65.0	69.3		63.5		
Boric acid	40.4				52.3	

Factors affecting the compressive strength include aggregate type and curing environment. Sand aggregate, in lieu of dolomite, significantly reduces early strength (approximately 25% at 7 days); the 28 days strengths are unaffected.[69] Moist curing results in a significant decrease in strength (25% after 24 hours).

7.3 Calcium Phosphate-Based Materials

Calcium phosphate-based systems have wide applications in biomedical areas. Brown has outlined the similarities between the hydration of calcium silicates and calcium phosphates.[70] The hydration products in both systems have high surface areas, variable composition, and poor crystallinity. Pozzolanic reactions and Hadley-like grains form in both systems. The primary cement-water reactions for C_3S and tetracalcium phosphate are as follows:

Eq. (18) $$3CaO{\bullet}SiO_2 + {\sim}\, 5.3\, H_2O \rightarrow {\sim}\, 1.7\, CaO{\bullet}SiO_2{\bullet}4H_2O + {\sim}\, 1.3\, Ca(OH)_2$$

Eq. (19) $$3[4CaO{\bullet}P_2O_5] + 3H_2O \rightarrow Ca_{10}(PO_4)_6(OH)_2 + 2Ca(OH)_2$$

The reaction products in the phosphate system are stoichiometric hydroxyapetite (HAp) and calcium hydroxide (CH).

The above reactions are specific to one composition of C-S-H or HAp. It is known that C-S-H can exist over a range of compositions (c/s extending from 0.83 to 2.0). The Ca/P ratio varies from ~ 1.40 to 1.84.

In the pozzolanic reaction analog, tetracalcium phosphate is the source of $Ca(OH)_2$. A variety of acidic calcium phosphates can be considered as the SiO_2 analog. The analogy is extended by the following reaction:

Eq. (20) $$6CaHPO_4\,(2H_2O) + 4Ca(OH)_2 \rightarrow Ca_{10}(PO_4)_6(OH)_2 + (nH_2O)$$

where $n = 5$ or 18.

The hydrolysis reactions of tetracalcium phosphate are coupled and solid CH actually does not form. Calcium deficient HAp then forms through the following reaction:

Eq. (21) $$6CaHPO_4(2H_2O + 3Ca_4(PO_4)_2O \rightarrow 2Ca_9(HPO_4)(PO_4)_5(OH) + (nH_2O)$$

where $n = 1$ or 13.

7.4 Lime Silico-Phosphate Cement

Quick setting silico-phosphate cements have been considered for use as pavement patching material. These are cement compositions which have a sufficiently high and rapidly developed early strength to support normal highway loads within one hour after application.

The fully developed compressive strengths are equivalent to those obtained with conventional portland cement. The silico-phosphate cement should be compatible with conventional portland cements. This is achieved by reacting wollastonite powder with a buffered phosphoric acid to form a lime silico-phosphate cement having a high early set strength. A typical phosphate solution consists of the following: 918 ml H_2O, 1611 ml 85% H_3PO_4, 510 g $Zn_3(PO_4)$, 456 g $AlPO_4$, and 21 g $Mg_3(PO_4)_2$.[69][71]

Concrete nominally contains 1 part phosphate solution, 1 part magnesia and 4 parts dolomite. Setting usually occurs in less than 30 minutes and up to 50 MPa compressive strength is developed within 4 hours. Factors affecting strength development are the particle size of the wollastonite, the P_2O_5 content of the liquid, and the liquid/powder ratio.

X-ray diffraction analysis of this cement system gives only those peaks attributable to wollastonite. It is suggested that an acid-crystal surface interaction occurs, reducing the apparent wollastonite surface area and freeing the bonding constituents for matrix development.[71] Microprobe analysis shows that the matrix binding wollastonite grains contains phosphorus, calcium, silicon, and some aluminum and zinc.

Typical strength development data of air cured silico-phosphate cement indicate that this system develops about 30% of its 28 day compressive strength in 2 hours. The tensile bond strength, compared with conventional concrete substrates, is very poor.[69]

A minor amount of blast-furnace fines can be added to the wollastonite-sand mixture to decrease the set time of the cement. A corresponding reduction of compressive strength also results. The use of 20% blast-furnace fines has resulted in satisfactory properties.

8.0 REGULATED-SET CEMENT

8.1 Description

The current state of disrepair of the North American concrete infrastructure has had a significant impact on strategic planning related to maintenance, repair, and service-life prediction of the concrete structures. The relatively slow setting properties of portland cement and portland cement concretes are inadequate for many repair applications. The development of a modified portland cement composition capable of developing a high early set strength upon hydration was an attempt to address this need. The high early strength development is due to the rapid formation of ettringite. This cement is known as regulated-set cement.

Regulated-set cement (often referred to as Reg Set or Jet Set) is a modified portland cement which contains as one ingredient a substantial amount of a ternary compound, essentially a calcium halo-aluminate, having the chemical formula $11CaO{\bullet}7Al_2O_3{\bullet}CaX_2$ wherein X is a halogen, i.e., fluorine, chlorine, bromine, or iodine.[72] The preferred ternary calcium halo-aluminate to be incorporated into the cement is calcium fluoroaluminate having the formula $11CaO{\bullet}7Al_2O_3{\bullet}CaF_2$. The modified cement contains between 1 and 30 % by weight of the calcium halo-aluminate.

Reg Set can be made by mixing finely divided $11CaO{\bullet}7Al_2O_3{\bullet}CaF_2$ with portland cement or directly by the production of clinker containing the fluoroaluminate compound, by using fluoride as one of the raw materials in the manufacture of portland cement clinker. Chemical analysis of Reg Set cement used by various investigators is given in Table 6. Reg Set cements contain more SO_3 than portland cements because the fluoride complex requires additional amounts of SO_3 for the formation of ettringite.

8.2 Paste and Mortar Hydration

Uchikawa and Tsukiyama have followed the hydration of Jet Set cement paste (w/c = 0.40) at 20°C.[73] They utilized two Jet Set cements — one containing 2% calcium carbonate and citric acid (0.2%), the other 2.5% calcium sulfate hemihydrate. In the early stages of hydration, the cement using calcium sulfate hemihydrate as a set-regulator displayed a higher rate of hydration than that regulated by citric acid; at a later period this behavior was reversed. The cement regulated with citric acid then had a higher degree of hydration, lower porosity, a larger percentage of small pores, and a larger

Table 6. Chemical Analysis of Various Reg Set Cements[15]

SiO_2	Al_2O_3	Fe_2O_3	CaO	MgO	SO_3	K_2O	Na_2O	F	Cl
13.48	12.37	2.31	57.86	1.47	6.80	1.00	0.64	0.78	0.06
14.71	8.99	2.37	58.52	1.52	6.57	1.20	0.90	0.74	
14.25	10.07	1.71	58.57	0.97	10.86	0.40	0.45	1.40	
17.10	10.40	0.90	60.90	1.00	6.00	0.16	0.64		
15.00	10.20	2.10	58.10	1.90	6.20	0.98	0.44		
[a]13.70	10.80	1.70	58.60	0.70	11.00	0.40	0.60	0.90	
[a]13.90	11.00	1.70	58.90	0.80	11.00	0.40	0.60	1.00	

[a] These cements contained about 60% alite, 1–2% belite, 20% halo-aluminate, and 4–5% ferrite.

amount of ettringite. The estimated rate of hydration of alite in jet cement is about twice that in portland cement. Hardened Jet Set cement paste has a much denser microstructure than portland cement paste.

Observation of the hydration process indicated the following. In the first few minutes of the dissolution of free lime, the hydration of anhydrite and hemihydrate and the formation of calcium aluminate hydrate and monosulfate hydrate occur. Ettringite is formed within one hour, monosulfate hydrate within 2–6 hours and C-S-H gel within 1–16 hours, with the maximum heat of hydration of C_3S at about 10 hours. Measurements of the non-evaporable water indicate that the amount of combined water is 60–80% of the theoretically determined total amount of combined water at complete hydration. (The total amount of combined water was estimated to be 36%.)[73] The amount of ettringite in the paste is estimated to be about 18–25% for the period one hour to seven days. The monosulfate content increases from about 10% at six hours to about 15–25% in one day. The alite in Reg Set cement paste is approximately 65–70% hydrated in one day and 80–95% hydrated in seven days. It is suggested that fluoride is tied up as $Al(OH)_2F$. The possibility of fluoride-substituted ettringite and the formation of halo-aluminate hydrates of the form $C_3A \cdot CaX_2 \cdot nH_2O$ is conceivable.

Uchikawa and Uchida have studied the hydration of the haloaluminates, extensively.[74][75] Several kinds of calcium aluminate hydrates form immediately after mixing. The hydrates C_2AH_8, C_4AH_{13}, and cubic C_3AH_6 form within one hour. The hexagonal calcium aluminate hydrate converts to cubic C_3AH_6 after one day. The reaction is 60% complete after one hour. Addition of calcium hydroxide accelerates the formation of calcium aluminate hydrates and the conversion of the hexagonal hydrate into C_3AH_6. The degree of hydration, however, is lower than for the controls.

Addition of calcium sulfate anhydrate has little effect in the intial stage of hydration. Calcium aluminate monosulfate hydrate, cubic hydrate, and ettringite form and coexist for a long period. Hexagonal hydrate is depleted with the formation of monosulfate hydrate and ettringite.

The addition of calcium sulfate dihydrate at a low SO_3/Al_2O_3 ratio (0.5) results in the formation of monosulfoaluminate and cubic C_3AH_6. At higher ratios (1.0), rectangular ettringite crystals and hexagonal plates of monosulfate hydrate coexist. As the quantity of ettringite increases and that of gypsum decreases, the amount of monosulfate hydrate increases.

The addition of calcium sulfate hemihydrate (the solubility is greater than that of the anhydrite and dihydrate) is conducive to the co-existence of ettringite and gypsum. The ettringite and gypsum crystals are larger than those produced in the systems initially containing anhydrite and dihydrate. The degree of hydration is about 60% at six hours. This is maintained for an extended period due to formation of a dense barrier layer of ettringite surrounding unreacted $C_{11}A_7 \cdot CaF_2$. At low SO_3/Al_2O_3 ratios the barrier layer is disrupted by the partial transition from ettringite to monosulfate hydrate several hours after mixing. This resulted in further formation of monosulfate hydrate and cubic C_3AH_6.

In the ternary system, $C_{11}A_7 \cdot CaF_2$ – $CaSO_4 \cdot 2H_2O$ – $Ca(OH)_2$ acceleration occurs, as a result of the $Ca(OH)_2$ addition. Large amounts of $Ca(OH)_2$ cause very rapid formation of ettringite. They coat the unhydrated particles and significantly inhibit the hydration process. In general, an increase in SO_3 concentration in the liquid phase decreases the handling time of regulated set cement concrete. The greater the solubility and rate of dissolution of calcium sulfate, the greater is the effect on handling time in the presence of a suitable amount of calcium hydroxide. Excess addition of calcium hydroxide retards the hydration of $C_{11}A_7 \cdot CaF_2$ and interferes with the hardening process.

Admixtures can be used to control the rate of hardening in regulated set cement products.[75] The hydration of the system $C_{11}A_7 \cdot CaF_2 - C_3S - CaSO_4$ in the presence of various additives has been reported. The concentration of calcium ions in the liquid phase controls the rate of hydration of C_3S. Retardation of aluminate hydration is caused by excess hemihydrate addition and C_3S hydration is accelerated. Citric acid retards the hydration of both $C_{11}A_7.CaF_2$ and C_3S. Sodium sulfate suppresses the hydration of $C_{11}A_7 \cdot CaF_2$, but accelerates the hydration of the C_3S. Sodium carbonate retards hydration of $C_{11}A_7 \cdot CaF_2$, but, in the presence of sodium sulfate, acts as an accelerator.

The addition of carboxylic acid lowers the solubility of the calcium ion in the liquid phase and severely retards the hydration of $C_{11}A_7 \cdot CaF_2$. Hydration proceeds gradually if calcium ion is supplied from calcium hydroxide and carboxylic acid is sufficiently depleted. Surface active agents, e.g., β-naphthalene sulfonic acid, influence the morphology of the hydrates, but do not appear to significantly affect the hydration process.

9.0 MECHANICAL PROPERTIES AND DURABILITY OF JET SET-BASED CEMENT SYSTEMS

9.1 Strength, Microhardness, and Modulus of Elasticity

Early strength development of Jet Set mortars is generally obtained (up to 15 MPa) within six hours. The strength-time curves for mortars (Fig. 12) indicate slow strength gain between six hours and one day followed by a second period of rapid strength development.

The formation of ettringite, and to some extent monosulfate hydrate, mainly contribute to strength development within the first 12 hours. Calcium silicate hydrate formation contributes to the strength after 12 hours. It has been suggested that the low rate of strength gain from 6 to 24 hours is due to the conversion of ettringite to monosulfate, which has about 50% of the theoretical strength of ettringite.[73] Alite in Reg Set cement hydrates much faster than monoclinic alite in portland cement.

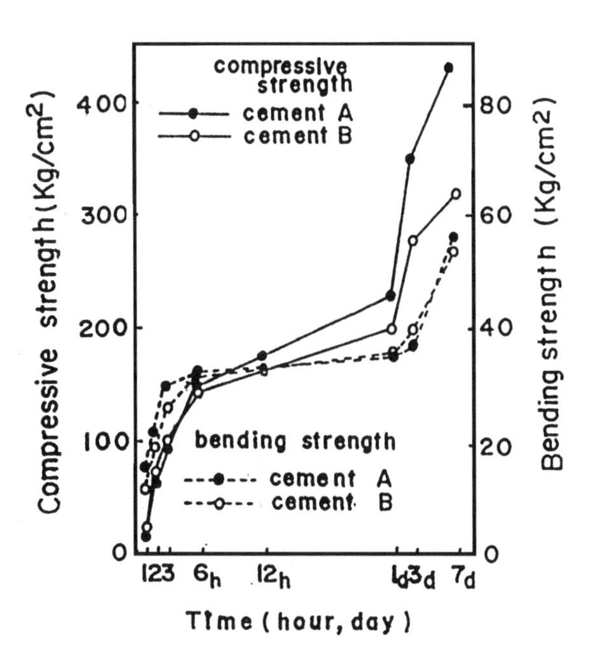

Figure 12. Compressive and bending strength of jet cement mortar (w/c = 0.65, sand/cement = 2, at 20°C).[73]

The curing of Reg Set in air results in low strengths because the large amounts of heat created in the reaction result in the drying out of the material, which slows down the hydration reaction.

The compressive strengths for Reg Set cement during the first 24 hours are higher than those for ultra-rapid hardening portland cement. The strength characteristics of the cements are similar beyond 24 hours.

Reg Set cement mortar with a citric acid retarder has greater strength than Reg Set cement regulated with calcium sulfate hemihydrate. The increased strength in Reg Set cement paste or mortar made with the addition of citric acid is usually a result of an increased degree of hydration, increased amount of ettringite, increased volume concentration of small pores, and lower total porosity.

The strength development of a typical Reg Set concrete (w/c = 0.60, water cured at 20°C) indicates that the strengths up to two years are superior to those of portland cement concrete.[76] The strength at two years for concrete with 20% Reg Set cement replaced with fly ash is slightly higher than the strength of concrete containing no fly ash. Such fly ash concretes are expected to exhibit low early strengths because of the lesser amounts of the binding material.

Calcium lignosulfonate or calcium chloride can be used as alternatives to calcium sulfate hemihydrate in order to control the setting time of Reg Set cement. Calcium lignosulfonate (2.2 ml/kg) increases the compressive strength of Reg Set cement paste by approximately 70% at 14 days.[77] Also, a 1% addition of calcium chloride increases the compressive strength by about 40% at 14 days. At ages greater than 14 days, free chloride may form from the aluminate phases and accelerate silicate hydration. Both calcium lignosulfonate and calcium chloride increase the time of setting from two minutes up to 30 minutes depending on the dosage. The morphology may also be affected because the N_2 surface areas of Reg Set cement paste at w/c = 0.60 are 12, 14, 20, and 21 m^2/g for 0, 1, 2, and 5% $CaCl_2$ respectively.

Microhardness has been correlated with compressive strength for several cementitious systems.[78] Figure 13 is a plot of microhardness versus porosity for the following cement systems: hydrated Reg Set cement paste with 0, 1, 2, and 5% $CaCl_2$ and hydrated portland cement paste.[79] The microhardness of Reg Set cement paste is significantly increased with the addition of 1% $CaCl_2$, and increased further with additions of 2 and 5% $CaCl_2$, for the porosity range studied. The increase in microhardness is larger at higher porosities. The microhardness of hydrated portland cement paste is significantly higher than that of Reg Set cement paste. The curve for portland cement paste lies between the curves for Reg Set cement paste containing 2 and 5% $CaCl_2$. In general, pore size analysis of Reg Set cement paste indicates that the concentration of small pores is greater for those pastes hydrated in the presence of $CaCl_2$.

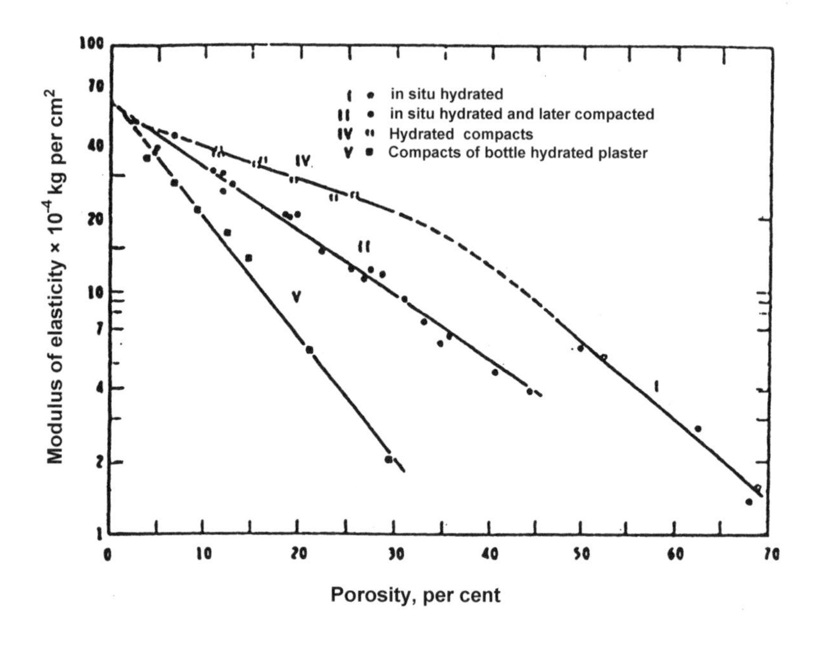

Figure 13. Modulus of elasticity vs porosity for various gypsum systems.[84]

Modulus of elasticity versus porosity curves for Reg Set cement paste hydrated in the presence of $CaCl_2$ are all higher than the curve for Reg Set cement paste with 0% $CaCl_2$. All curves for Reg Set cement paste are below the curve for hydrated portland cement paste.

Whilst the microhardness values of hydrated portland cement paste lie between those for Reg Set cement paste with 2 and 5% $CaCl_2$, the modulus of elasticity values for hydrated portland cement paste are higher than for any of the Reg Set cement pastes. It is suggested that the processes of microstructural deformation occurring in microhardness and modulus of elasticity measurements are influenced in different ways by bond formation during the hydration of Reg Set cement in the presence of $CaCl_2$.

9.2 Durability

The sulfate resistance of Reg Set cement concrete to magnesium and sodium sulfate solutions is generally poor. However, it appears that Reg Set containing larger amounts of fluoroaluminate promotes better resistance to sulfate attack.[76] A 20% replacement of Reg Set cement by fly ash provides a substantial improvement in sulfate resistance; this is, at least partly, due to the consumption of CH by fly ash.

The curing of Reg Set cement concrete at elevated temperatures (up to 38°C) appears to have no adverse effects; it even results in increased strengths. Concrete made with Reg Set cement having greater than normal amounts of $C_{11}A_7{\bullet}CaF_2$ and anhydrite appears to be more frost resistant, giving comparable frost resistance to an air-entrained portland cement concrete.

Concrete with a high early strength development is desirable for cold weather concreting. Reg Set cement concrete can be used for this application. Strength determinations for Reg Set concrete cured for one hour at 20°C and subsequently at –10°C for 28 days give similar strengths to those for Reg Set concrete cured continuously at 20°C for the same period.[80]

9.3 Gypsum

Calcium sulfate in the form of gypsum ($CaSO_4{\bullet}2H_2O$) is added in small quantities to the other constituents of cement during its manufacture to control setting. The calcium sulfate can also be the hemihydrate ($CaSO_4{\bullet}1.2H_2O$); anhydrite (anhydrous calcium sulfate, $CaSO_4$) or a combination. The use of blends of gypsum (dehydrated to the more soluble hemihydrate) and natural anhydrite can be effective in controlling stiffening or false-setting problems.

Pure gypsum ($CaSO_4{\bullet}2H_2O$) contains 79.1% of calcium sulfate and 20.9% of water or 32.5–32.6% calcium oxide, 46.5–46.6% of sulfur trioxide, plus the water. Gypsum crystallizes in the monoclinic system. Well-formed crystals are colorless, transparent, and flattened parallel to the plane of symmetry. Commercial gypsum is massive consisting of an aggregate of crystals which have interfered with the growth of one another.

Gypsum when heated at temperatures between 128° and about 163°C loses approximately 75% of its hydrate water and hemihydrate forms. This is mixed with some gypsum and "soluble anhydrite." This material increases its volume on rehydration and hardening. Heating of gypsum or hemihydrate to about 185–200°C yields soluble anhydrite. At 400–500°C, gypsum is dehydrated to an insoluble anhydrite of the composition $CaSO_4$.

The exact mechanism by which gypsum addition to portland cement controls setting behavior is not known although much work has been reported in this field.[81] It is clear, however, that interactions between C_3A and gypsum are integral to the stiffening and hardening processes.

Feldman and Ramachandran investigated the hydration character of C_3A-$CaSO_4$•$2H_2O$ systems containing up to 20% $CaSO_4$•$2H_2O$.[82] They made extensive use of thermal analysis and coupled the hydration study with length change measurements using compacted solid bodies with an equivalent water/solid ratio of 0.1. It was concluded that the formation of ettringite had no direct effect upon reaction rate, but that the reactivity of C_3A was reduced by the sorption of SO_4^{2-} ions on active sites of its surface. A second mechanism retarding the normal hydration of C_3A is the reduced rate of conversion of the hexagonal hydroaluminates to the cubic hexahydrate probably by sorption of SO_4^{2-} ions. The retardation of the hydration of C_3A at a particular temperature is affected by a balance between the following: the concentration of SO_4^{2-} ions on and in proximity to the surface of the C_3A; the rate of reaction of SO_4^{2-} ions with the hexagonal hydroaluminates; and the thickness of the hexagonal hydroaluminate layer around the C_4A grain. The large volume of sulfoaluminate product may contribute to a general decrease in porosity and retard the reaction of C_3A in this way. The disruptive expansions that occur when the sulfoaluminate is formed suggest that it cannot form an impermeable layer.

The engineering properties of cement minerals including gypsum have been determined using the compact technique.[83] The use of *compacts* (porous bodies formed by pressure compaction of powdered material) as structural models of cement systems has been extensively studied.[84] The co-linearity of mechanical property—porosity relationships for in-situ hydrated and compacted cement paste powders—provides strong evidence for the absence of "chemical" bonds between particulates as it is unlikely that these would not be broken (if present) during compaction. The fact that the values for modulus of elasticity of compacts fit so closely to that of the paste supports the idea that the system has none or very few chemical bonds. This concept of the nature of inter-particle bonds in cement paste allows for the possibility of one particle breaking its bond with one neighboring particle and remaking a similar bond with another neighbor. The result would be permanent deformation, but no net loss of strength when the system is subjected to sustained load. These bonds may approach the character of the "chemical" bonds when the porosity approaches zero. The extrapolated regression lines (modulus of elasticity versus porosity) of the gypsum systems in Fig. 13 intersect at zero porosity. This suggests that the extrapolated values have a definite meaning. Differences in the relationships for the various gypsum preparations can be attributed to some inherent differences in their structure such as the particle size and shape, the effective area of particle contact, or the degree of crystal inter-growth and interlocking. The in-situ hydrated (System I) and hydrated compacts (System IV) should be represented by a single regression line. The deviation noted suggests that the relation does not hold over such a wide range of porosity (11 to 70%). It appears that during the formation of System II the primary structure of System I is progressively destroyed and replaced by a new structure corresponding to that of System V. This is verified by the fact that the regression lines of Systems II and V intersect at approximately zero porosity.

The modulus of elasticity values of hydrated gypsum compacts are similar in magnitude to those for cement paste systems (10 to 30% porosity). This information supports the view that gypsum itself is mechanically compatible with other constituents in hydrated portland cement paste.

REFERENCES

1. Sorel, S., Procedure for the Formation of a Very Solid Cement by the Action of a Chloride on the Oxide of Zinc, *C. R. Hebd. Seances Acad. Sci.*, 41:784–785 (1955)
2. Kacker, K. P., Rai, M., and Ramachandran, V. S., Suitability of Almora Magnesite for Making Magnesium Oxychloride Cement, *Chem. Age of India*, 20:506–510 (1969)
3. Mathur, R., Chanorawat, M. P. S., and Nagpal, K. C., X-ray Diffraction Studies of the Setting Characteristics of Magnesium Oxychloride Cement, *Res. and Ind.*, 29:195–201 (1984)
4. Demediuk, T., Cole, W. F., and Hueber, H. V., Studies on Magnesium and Calcium Oxychlorides, *Aust. J. Chem.*, 8:215–233 (1955)
5. Cole, W. F., and Demediuk, T., X-ray, Thermal and Dehydration Studies on Magnesium Oxychlorides, *Aust. J. Chem.*, 8:234–251 (1955)
6. Matkovic, B., Popovic, S., Rogic, V., Zunic, T., and Young, J. F., Reaction Products in Magnesium Oxychloride Cement Pastes, System MgO-$MgCl_2$-H_2O, *J. Am. Ceram. Soc.*, 60:504–507 (1977)

7. Dehua, D., and Chuanmei, Z., The Effect of Aluminate Minerals on the Phases in Magnesium Oxychloride Cement, *Cement Concr. Res.*, 26:1203–1211 (1996)
8. Matkovic, B., and Young, J. F., Microstructure of Magnesium Oxychloride Cements, *Nat. Phys. Sci.*, 246:79–80 (1973)
9. Beaudoin, J. J., and Ramachandran, V. S., Strength Development in Magnesium Oxychloride and Other Cements, *Cement Concr. Res.*, 5:617–630 (1975)
10. Tooper, B., and Cartz, L., Structure and Formation of Magnesium Oxychloride Sorel Cements, *Nature*, 211:64–66 (1966)
11. Srivastova, R. S., and Rai, M., Water-Proofing of Magnesium Oxychloride Cement and Cement Products, *Res. and Ind.*, 28:203–206 (1983)
12. Lu, H. P., Wang, P. L., and Jiang, N. X., Design of Additives for Water-Resistant Magnesium Oxychloride Cement Using Pattern Recognition, *Mater. Sci. Lett.*, 20:217–223 (1994)
13. Lu, H., and Wang, P., Design of Additives for Water Resistant Magnesium Oxychloride Cement, *J. Inorg. Mater.*, 8:341–346 (1993)
14. Beaudoin, J. J., Ramachandran, V. S., and Feldman, R. F., Impregnation of Magnesium Oxychloride Cement with Sulfur, *Am. Ceram. Soc. Bull.*, 56:424–427 (1977)
15. Ramachandran, V. S., Feldman, R. F., and Beaudoin, J. J., *Concrete Science*, p. 421, Heyden & Sons Ltd. (1981)
16. Beaudoin, J. J., and Ramachandran, V. S., Strength Development in Magnesium Oxysulfate Cement, *Cement Concr. Res.*, 8:103–112 (1978)
17. Herrera, M. E., A Comparison of Critical Properties of Magnesium Oxysulfate and Magnesium Oxychloride Cements as Used in Sprayed Fire Resistive Coatings, ASTM STP, 826:94–101 (1983)
18. Beaudoin, J. J., and Feldman, R. F., The Flow of Helium into the Microspaces of Magnesium Oxysulfate Cement Paste, *Cement Concr. Res.*, 7:585–596 (1977)
19. Bied, J., (J & A Pavin de Lafarge), British Patent No. 8193 (1909)
20. Scrivener, K. L., and Capmas, A., *Calcium Aluminate Cements in Lea's Chemistry of Cement and Concrete*, 4th Ed., (P. C. Hewlett, ed., Ch. 13, pp. 709–778, Arnold, London (1998)
21. Robson, T. D., *High Alumina Cements and Concretes*, p. 263, Wiley, New York (1962)
22. French, P. J., Montgomery, G. J., and Robson, T. D., High Strength Concrete Within the Hour, *Concr.*, 5:253–257 (1971)
23. Mangabhai, R. J. (ed.), *Proc. Int. Symp. Calcium Aluminate Cements,* p. 380, E. & F. N. Spon, London (1990)

24. Midgley, H. G., and Pettifer, K., Electron Optical Study of Hydrated High Al_2O_3 Cement Pastes, *Trans. Br. Ceram. Soc.*, 71:55–60 (1972)
25. Ramachandran, V. S., and Feldman, R. F., Significance of Low Water-Solid Ratio and Temperature on the Physico-Mechanical Characteristics of Hydrates of Tricalcium Aluminate, *J. Appl. Chem. Biotech.*, 23:625–633 (1972)
26. Ramachandran, V. S., and Feldman, R. F., Hydration Characteristics of Monocalcium Aluminate at a Low Water-Solid Ratio, *Cem. Concr. Res.*, 3:729–750 (1973)
27. George, G. M., Aluminous Cements—A Review of Recent Literature, *Proc. 7th Int. Congr. Chem. Cement*, 1:V-1/3–V-1/26 (1980)
28. Ramachandran, V. S., and Beaudoin, J. J., Significance of Water-Solid Ratio and Temperature on the Physico-Mechanical Characteristics of Hydrating $4CaO{\bullet}Al_2O_3{\bullet}Fe_2O_3$, *J. Mater. Sci.*, 11:1893–1910 (1976)
29. Lehman, H., and Lecks, K. J., *Tonin, Ztg.*, 87:29–36 (1963)
30. Sereda, P. J., Feldman, R. F., and Ramachandran, V. S., Structure Formation and Development in Hardened Cement Pastes, *7th Int. Congr. Cement Chem., Paris,* I:V1-1/3–V1-1/43 (1980)
31. Wells, L. S., and Carson, E. T., Hydration of Aluminous Cements and Its Relation to the Phase Equilibria in the System Lime-Alumina: Water, *J. Res. Natl. Bus. Stand.*, 57:335–353 (1956)
32. Fu, Y., Ding, J., and Beaudoin, J. J., Conversion Preventing Additive for High Alumina Cement Products, p. 13, U.S. Patent 5,624,489 (1997)
33. Ma, W., Brown, P. W., Hydration of Sodium Phosphate-Modified High Alumina Cement, *J. Mater. Res.*, 9:1291–1297 (1994)
34. Ma, W., and Brown, P., Mechanical Behavior and Microstructural Development in Phosphate Modified High Alumina Cement, *Cement Concr. Res.*, 22:1192–200 (1992)
35. Neville, A. M., and Brooks, J. J., *Concrete Technology,* p. 438, Longman Scientific and Technical, J. Wiley & Sons, New York (1987)
36. Lafuma, H., Les Aluminates de Calcium et la Chimie des Ciments, *Le Ciment*, 30:174 (1925)
37. Thomas, W. N., Roy, J., Ciments Fondu and Mixtures of Cement Fondu and Portland Cement, *Inst. Brit. Architect*, 31(20):670–675 (1924)
38. Touche, M., Note sur l'emploi du ciment, Fondu sur la Ligne de Nice a Com, *Le Ciment*, 31:240 (1926)
39. Lea, F. M., and Desch, C. H., *The Chemistry of Cement and Concrete*, Edward Arnold Pub. Ltd., London, Ch. XVI, p. 426 (1956)

40. Fu, Y., Xie, P., Gu, P., Beaudoin, J. J., Characteristics of Shrinkage Compensating Expansive Cement Containing a Pre-hydrated High Alumina Cement-based Expansive Additive, *Cement Concr. Res.*, 24:267–276 (1994)

41. Gu, P., Fu, Y., Ping, X., and Beaudoin, J. J., A Study of the Hydration and Setting Behavior of OPC-HAC Paste, *Cem. Concr. Res.*, 24:682–694 (1994)

42. Brantervik, K., and Niklasson, G. A., Circuit Models for Cement Based Materials Obtained from Impedance Spectroscopy, *Cement Concr. Res.*, 21:496–508 (1991)

43. Gu, P., Beaudoin, J. J., Quinn, E., Myers, R., Early Strength Development and Hydration of Ordinary Portland Cement/Calcium Aluminate Cement Pastes, *Adv. Cem. Based Mater.*, 6:53–56 (1997)

44. Robson, T. D., The Characteristics and Applications of Mixtures of Portland and High-Alumina Cements, *Chem. and Ind.*, 1:2–7 (1952)

45. Bensted, J., A Discussion of a Study of the Hydration and Setting Behavior of OPC-HAC Pastes, *Cement Concr. Res.*, 25:221–222 (1995)

46. Gu, P., and Beaudoin, J. J., A Conduction Calorimetric Study of Early Hydration of OPC/HAC Pastes, *J. Mater. Sci.*, 32:3875–3881 (1997)

47. Mikhailov, V. V., Stressing Cement and the Mechanism of Self-stressing Concrete Regulation, *Proc. 4th Int. Symp. on Chem. Cement,* 2:927–955 (1960)

48. Monford, G. E., Properties of Expansive Cement Made with Portland Cement, Gypsum and Calcium Aluminate Cement, *J. PCA Res. Dev. Lab*, 2:2 (1964)

49. Budnikov, P. P., and Krovchenko, I. V., Expansive Cements, *Proc. 5th Int. Symp. on Cement Chem., Tokyo*, IV:319–329 (1968)

50. Fu, Y., Expansive Concrete for Use in Drilled Shafts, M.S. Thesis, p. 125, Dept. of Civil Eng., Univ. Toronto (July 1992)

51. Fu, Y., Xie, P., Gu, P., and Beaudoin, J. J., Characteristics of Shrinkage Compensating Expansive Cement Containing a Pre-Hydrated High Alumina Cement-Based Expansive Additive, *Cement Concr. Res.*, 24:267–276 (1994)

52. Fu, Y., Sheikh, S. A., and Hooton, R. D., Microstructure of Highly Expansive Cement Paste, *ACI Mater. J.*, 91:46 (1994)

53. Sheikh, S. A., Fu, Y., and O'Neill, M. W., Expansive Cement Concrete for Drilled Shafts, *ACI Mater. J.*, 91:237–245 (1994)

54. Fu, Y., Ding, J., and Beaudoin, J. J., Effect of Different Calcium Alumina Hydrates on Ettringite Formation and Expansion of High Alumina Cement-Based Expansive Cement Pastes, *Cement Concr. Res.*, 26:417–426 (1996)

55. Aroni, S., Polwka, M., and Bresler, B., *Structure and Materials Research Report No. 66-67*, Dept. of Civil Eng., Univ. California-Berkeley (1966)

56. Gu, P., Beaudoin, J. J., and Ramachandran, V. S., Portland Cement/High Alumina Cement Pastes Containing Phosphonate Compounds, *Indian J. Eng. Mater. Sci.*, 3:63–69 (1996)
57. Gu, P., Ramachandran, V. S., and Beaudoin, J. J., Study of Early Hydration of High Alumina Cement Containing Phosphonic Acid by Impedance Spectroscopy, *J. Mater. Sci. Lett.*, 14:503–505 (1995)
58. Gu, P., and Beaudoin, J. J., Effect of Lithium Salts on Portland Cement/ High Alumina Cement Paste Hydration, *J. Mater. Sci. Lett.*, 14:1207–1209 (1995)
59. Ramachandran, V. S., Feldman, R. F., and Beaudoin, J. J., Special Cementitious Systems, *Concrete Science*, 9:309–339, Heyden and Sons Ltd., UK (1981)
60. Cassidy, J. E., Phosphate Bonding Then and Now, *Am. Ceram. Soc. Bull.*, 56:640–643 (1977)
61. Cartz, L., Other Inorganic Cements, *Cements Res., Prog. Am. Ceram. Soc.*, pp. 159–171 (1975)
62. Kingery, W. D., Fundamental Study of Phosphate Bonding in Refractories: 1, Literature Review, *J. Am. Ceram. Soc.*, 33:239 (1950)
63 Hall, D. A., Steven, R., and El-Jazairi, B., Effect of Water Content on the Structure and Mechanical Properties of Magnesia-Phosphate Cement Mortar, *J. Am. Ceram. Soc.*, 81:1550–1556 (1998)
64. Abdelrzig, B. E. I., Sharp, J. H., and El-Jazairi, B., The Chemical Composition of Mortars Made from Magnesia-Phosphate Cement, *Cement Concr. Res.*, 18:415–425 (1988)
65. Popovics, S., Rajendran, N., and Penko, M., Rapid Hardening Cements for Repair of Concrete, *ACI Mater. J.*, 84:64–73 (1987)
66. Limes, R. W., and Russell, R. O., Process for Preparing Fast-Setting Aggregate Compositions and Products of Low Porosity Produced Therewith, U.S. Patent 3,879,209 (April 22, 1975)
67. Abdelrazig, B. E. I., Sharp, J. H., and El-Jazairi, B., The Microstructure and Mechanical Properties of Mortars made from Magnesia-Phosphate Cement, *Cem. Concr. Res.*, 19:247–258 (1989)
68. Stierli, R. F., Gaides, J. M., and Tarver, C. C., Magnesia Cement Mixture Based on Magnesium Oxide and an Ammonium Phosphate Capable of Reacting with the Mixture, German Patent 2,551,140,26 (May 1976)
69. Pike, R. G., and Baker, W. M., Concrete Patching Materials, Federal Highway Admin. Report, FHWA-RD-74-55 (April 1974)
70. Brown, P., Hydration Behavior of Calcium Phosphates is Analagous to Hydration Behavior of Calcium Silicates, *Cement Concr. Res.*, 29:1167–1171 (1999)

71. Semler, C. E., Quick Setting Wollastonite Phosphate Cement, *Am. Ceram. Soc. Bull.*, 55:983–985 (1976)
72. Greeming, N. R., Copeland, L. E., Verbeck, G. J., Modified Portland Cement and Process, U.S. Patent 3,628,973 (1971)
73. Uchikawa, H., and Tsukiyama, K, The Hydration of Jet Cement at 20°C, *Cement Concr. Res.*, 3:263–277 (1973)
74. Uchikawa, H., and Uchida, S., The Hydration of $11CaO•7Al_2O_3•CaF_2$ at 20°C, *Cement Concr. Res.*, 2:681–695 (1972)
75 Uchikawa, H., and Uchida, S., The Influence of Additives Upon the Hydration of the Mixture of $11CaO•7Al_2O_3•CaF_2$, $3CaO•SiO_2$ and $CaSO_4$ at 20°C, *Cement Concr. Res.*, 3:607–624 (1973)
76. Osborne, G. J., Sulphate Resistance and Long-Term Strength Properties of Regulated-Set Cements, *Mag. Concr. Res.*, 29:213–224 (1977)
77. Berger, R., private communication.
78. Beaudoin, J. J., and Feldman, R. F., A Study of the Mechanical Properties of Autoclave Calcium Silica Systems, *Cement Concr. Res.*, 5:103–118 (1975)
79. Ramachandran, V. S., Feldman, R. F., and Beaudoin, J. J., *Concrete Science*, p. 427, Heyden & Sons (1981)
80. Hoff, G. C., Houston, B. J., and Sayles, F. H., Use of Regulated-Set Cement in Cold Weather Environments, U.S. Army Engineers Waterway Experiment Sta. Concr. Lab., misc. paper C-75-5, p. 19 (1975)
81. Kovler, K., Setting and Hardening of Gypsum-Portland Cements-Silica Fume Blends, Part 2: Early Strength, DTA, XRD and SEM Observations, *Cement Concr. Res.*, 28:523–531 (1998)
82. Feldman, R. F., and Ramachandran, V. S., The Influence of $CaSO_4•2H_2O$ Upon the Hydration Character of $3CaO•Al_2O_3$, *Mag. Concr. Res.*, 18:185–196 (1966)
83. Soroka, I., and Sereda, P. J., The Structure of Cement Stone and the Use of Compacts as Structural Models, *Proc. 5th Int. Symp. Chem. Cement, Tokyo*, III:67–72, (1968)
84. Soroka, I., and Sereda, P. J., Interrelation of Hardness, Modulus of Elasticity and Porosity in Various Gypsum Systems, *J. Am. Ceram. Soc.*, 51:337–340 (1968)

10

Non-Portland Rapid Setting Cements

1.0 INTRODUCTION

Cements other than portland cement, either alone or in combination with portland cement, are often more suitable for specific applications. Requirements of construction or environmental conditions can place constraints on conventional practice. Rapid setting, high early strength, and durability specifications often dictate the choice of an alternate cement system. The application of thermal analysis to selected rapid setting cements is presented in this chapter. These include the following systems: calcium aluminate cements (high alumina cement, Jet Set or regulated-set cement) and related pure mineral systems; magnesium oxychloride and oxysulfate cement; and magnesia-phosphate cements, precursors to the formation of hydroxyapatite. Cement systems containing supplementary materials and other additions are discussed in Ch. 8 of this book.

2.0 CALCIUM ALUMINATE CEMENTS

2.1 Basic Reactions

Calcium aluminate cements (CAC) have a wide range of alumina content (38 to 90%). The chemistry and principle cement-water reactions for high alumina cement—a widely used non-portland cement—are described in detail in Ch. 9. The primary binding phase is calcium monoaluminate ($CaAl_2O_4$ or CA). Refractory cements contain higher alumina contents (70 to 90%).

In CAC, the CA reacts with water to form a series of calcium aluminate hydrates. These include CAH_{10}, C_2AH_8, C_3AH_6, and AH_3 (an amorphous phase). The metastable hydrates, CAH_{10} and C_2AH_8, convert to C_3AH_6. The following scheme summarizes the conversion reactions.

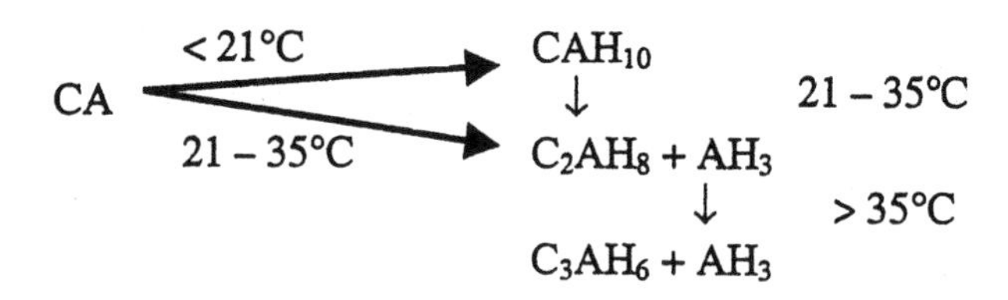

The conversion that occurs as a consequence of the transformation of the hexagonal phases, CAH_{10} or C_2AH_8, into the cubic phase, C_3AH_6, is known to be accompanied by a loss of strength of the hardened alumina cement.

The conversion reactions themselves are described according to the following equations:

Eq. (1) $$3CAH_{10} \rightarrow C_3AH_6 + 2AH_3 + 18H$$

Eq. (2) $$3C_2AH_8 \rightarrow 2C_3AH_6 + AH_3 + 9H$$

The conversion of CAH_{10} to C_3AH_6 results in a volume decrease to about 50% whereas that of C_2AH_8 to the cubic phase results in a decrease of about 65% of the original volume of the reactants. It is apparent that methods to identify and determine the amounts of the aluminate hydrates in CAC concretes are useful for a meaningful diagnosis of potential problems.

2.2 Thermal Analysis of Hydrated Calcium Aluminate Cements

Differential thermal analysis (DTA), differential scanning calorimetry (DSC), and differential thermogravimetric analysis (DTG) have been used to provide estimates of the degree of conversion in CAC concretes.

An idealized thermogram for hydrated calcium aluminate cement (after Bushnell-Watson and Sharp)[1] is presented in Fig. 1. Peak temperature ranges for the CAC hydrates were observed as follows: gel (60–130°C); CAH_{10} (100–160°C); C_2AH_8 (140–200°C); AH_3 (260–330°C); C_3AH_6 (290–350°C). For samples of concrete or mortar, either or both of the major doublets (between 60 and 170°C and 260 and 330°C) may appear as single peaks and the minor peaks (170–150°C) may not be observed. The sand present yields a small, sharp, reversible exotherm at 573°C due to the polymorphic transformation of α-quartz to β-quartz. It is important to recognize that peak temperatures depend on several variables including sample mass, heating rate, degree of compaction, gas composition and flow rate, and crucible geometry. Figure 2 is a DTA thermogram of CAC paste (Secar 71, about 70% Al_2O_3) hydrated at 10°C for 24 hours. It contains overlapping C-A-H gel and CAH_{10} peaks. The height of the major XRD peak for each hydrate is also indicated by the bar segments at the bottom of Fig. 2.

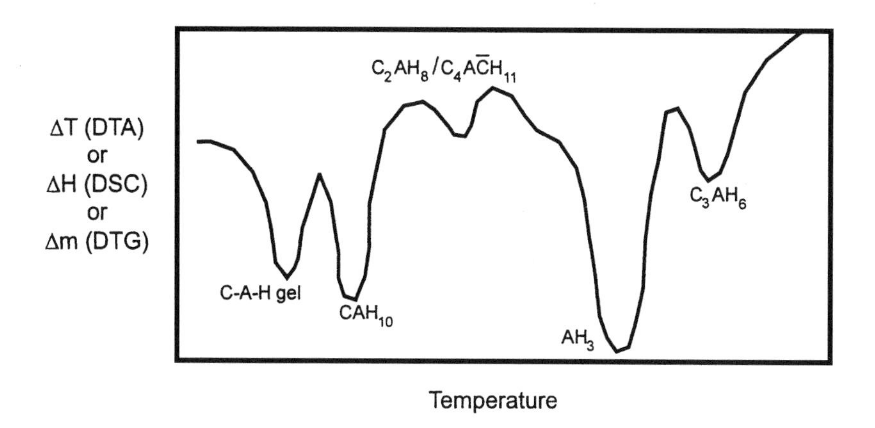

Figure 1. An idealized thermal analysis curve for hydrated calcium aluminate cement.[1]

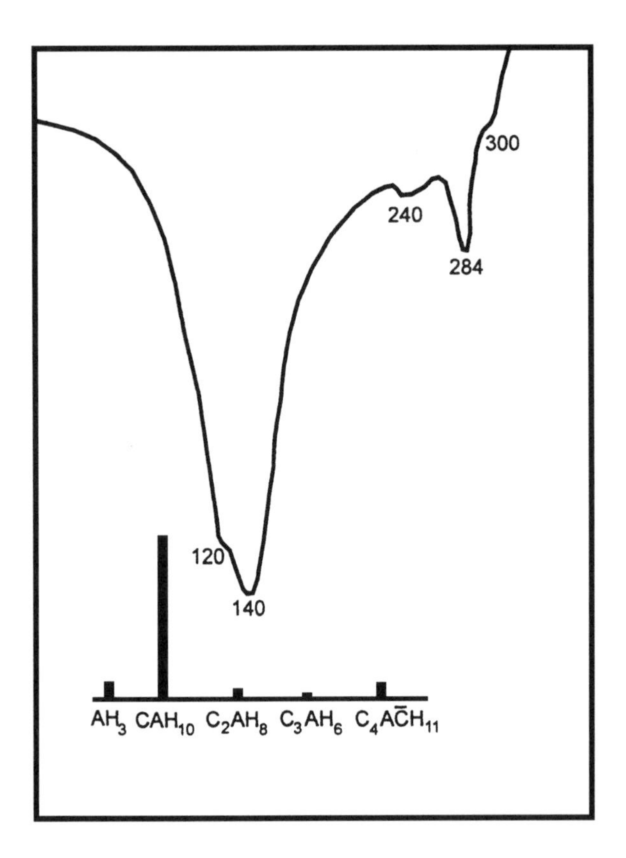

Figure 2. DTA and XRD data for a Secar 71 paste hydrated at 10°C for 24 hours.[1]

Carboaluminate phases (e.g., $C_4A\bar{C}H_{11}$) can form when hydration takes place in normal atmospheres. The presence of this phase is indicated by a DTA peak in the region 180–200°C, which can overlap with the peak due to the presence of C_2AH_8. A shoulder may be present for the phase present in the lesser amount.

Figure 3 is a plot of various DTA thermograms (Secar, Alcoa CA, Secar + $C\bar{C}$) depicting the presence of C_2AH_8 and $C_4A\bar{C}H_{11}$. Thermograms for pastes to which $CaCO_3$ had been added are included in Fig. 3. Supplementing Fig. 3 are bar graphs indicating the relative amounts of the phases determined by XRD.

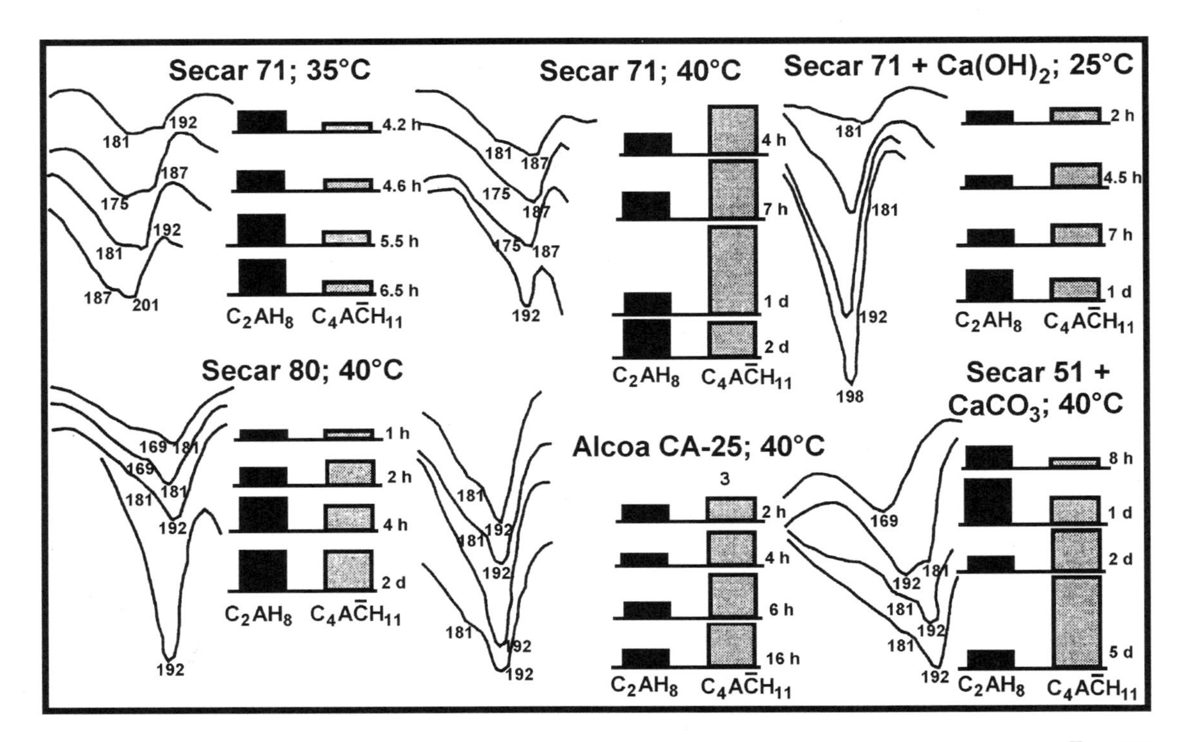

Figure 3. Various DTA traces (with complementary XRD data) showing the presence of C_2AH_8 and $C_4A\bar{C}H_{11}$.[1]

Addition of metakaolinite or zeolites to CAC systems can favor the formation of strätlingite (C_2ASH_8).[2] This phase has a DTA peak in the temperature range 201–203°C. Metakaolinite-calcium hydroxide mixtures (1:1) prepared at a water:solid = 0.75 and hydrated at 40°C produce a significant amount of C_2ASH_8 and C_4AH_{13}, and large quantities of $C_4A\bar{C}H_{11}$ and C_3AH_6 (detected by XRD).

Figure 4 is a DTA thermogram of the hydrated mixture indicating a large peak at 120°C and a sharp peak at 192°C. The low temperature peak is attributed to alumina gel and C-S-H. The higher temperature peak represents all four aluminate phases referred to previously. This underscores the necessity (in this case) for a complementary technique such as XRD to separate out and identify the individual aluminate phases.

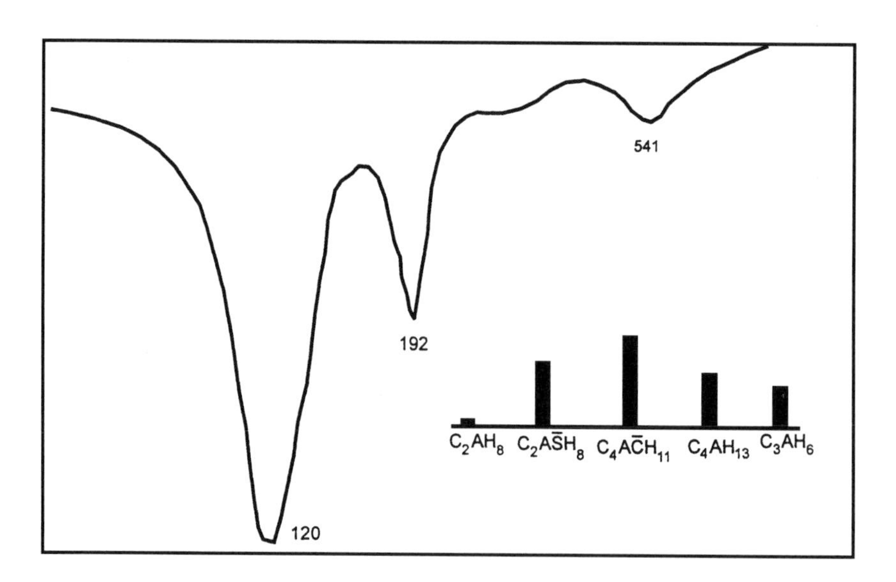

Figure 4. DTA and XRD data obtained from the metakaolin-calcium hydroxide mixture hydrated at 40°C.[1]

The presence of AH_3 can be detected by DTA showing a strong endothermic peak at about 270 to 300°C and (when large amounts form) an additional endothermal peak at temperatures up to 240°C.

Triple DTA peaks have been observed in the region 270–340°C with Secar 71 hydrated at 40°C and 50°C (and other Secar systems). Thermograms are shown in Fig. 5. The peak at the lower temperature (290–300°C) can be attributed to AH_3, and that at 311–325°C to C_3AH_6. An extra phase (attributed to the third peak) was not detected by XRD suggesting that this phase was amorphous or poorly crystalline. There is doubt, however, with this explanation given that the additional peak occurs at a higher temperature than that observed for crystalline AH_3, contrary to expectation.

Typical DSC, TG, and DTG curves for high alumina cement (HAC) pastes are provided by Brown and Cassel[3] in Figs. 6 and 7. They illustrate that the techniques are complementary, as discussed previously.

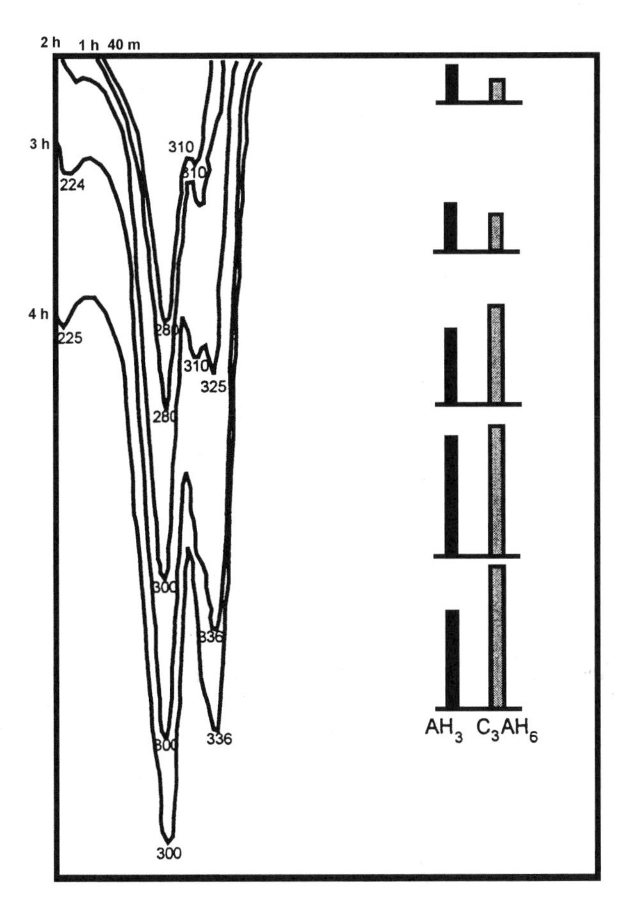

Figure 5. DTA traces of Secar 71 hydrated at 50°C showing a triple peak. XRD data is also included.[1]

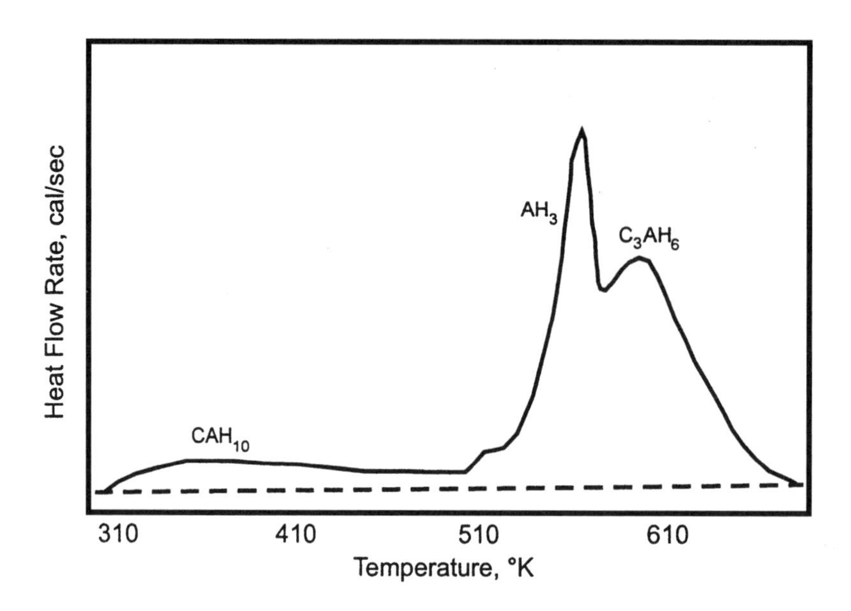

Figure 6. Typical DSC curve for high alumina cement.[3]

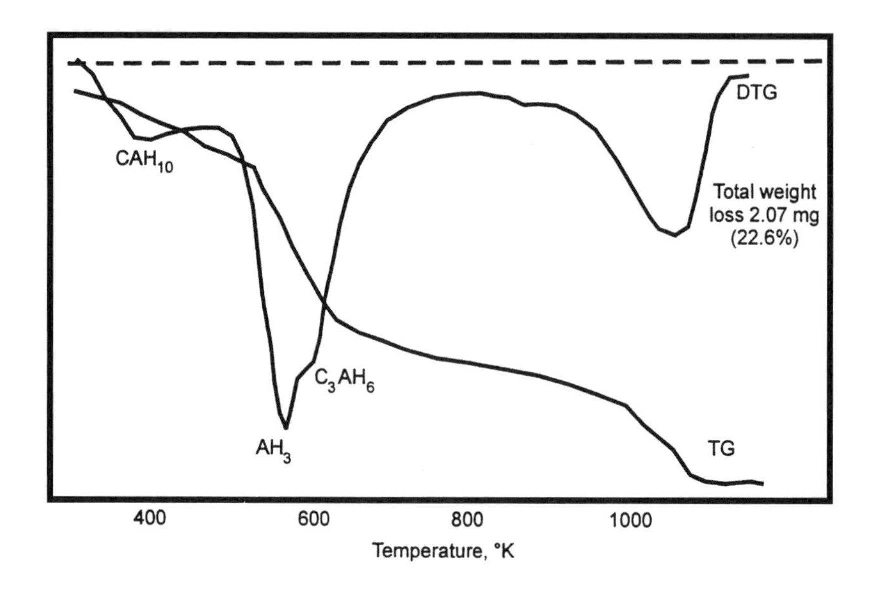

Figure 7. Typical thermogravimetric curves for high alumina cement.[3]

Estimates of the degree of conversion of high alumina cement (HAC) concrete can be made from thermal analysis. Midgley[4] provides an expression for the degree of conversion based on the simplified equation: $3CAH_{10} \rightarrow C_3AH_6 + 2AH_3 + 18H$. It is:

$$\text{Eq. (3)} \qquad \text{Degree of conversion}, D_c = \frac{100}{1 + K^1 \dfrac{\text{amount of } CAH_{10}}{\text{amount of } AH_3}}$$

The relationship between peak area and the quantities of the minerals present in mixtures was determined from calibration mixtures of the pure synthetic mineral and corundum.[5] More reproducible results for the ratio of CAH_{10} to AH_3 were obtained using peak heights instead of peak areas. It was reported that DTA, DSC, and DTG all gave accurate analyses for an HAC sample with about 65% conversion with a standard deviation not exceeding 1.4%. An illustration of high conversion and low conversion is provided by the DTA thermograms in Fig. 8. Midgley developed an expression for the normalized peak temperature or temperature that a fixed amount of the mineral in a mixture would give

$$\text{Eq. (4)} \qquad T_{peak}(°C) = k + b \ln \text{peak area}$$

This enables detection of variations in polymorphism and polytypism of the alumina hydrates, e.g., those hydrates formed on carbonation of HAC.[6]

The original definition of conversion was based on estimates of the amounts of C_3AH_6 and CAH_{10}; this was subject to significant error due to the carbonation of C_3AH_6.[7] Further, Eq. (3) for degree of conversion, D_c, described above ($K^1 = 1$) is modified to account for instrument effects by an empirical constant, $K^1 \neq 1$, determined by the use of calibration standards. Bushnell-Watson and Sharp note that K^1 itself may vary with the degree of conversion.[1]

The formation of $3CaO•Al_2O_3•CaCO_3•12H_2O$ (a calcium carboaluminate hydrate) is also described by Midgley as a reaction product of the HAC-calcium carbonate reaction. A typical DTA curve for hydrated HAC-$CaCO_3$ paste is given in Fig. 9. The method used for determining amounts of minerals present involved determination of calibration factors (peak height in mm per mg of mineral present) and is described by Midgley.[4] Estimates of calcium carboaluminate present (in HAC-$CaCO_3$

pastes containing 30% $CaCO_3$ by mass of solids prepared at 2.7 water/solids ratio) after curing for different times at different temperatures are plotted in Fig. 10. At 50° and 20°C, there was no evidence of decomposition of the calcium carboaluminate hydrate leading to the conclusion that calcium carboaluminate hydrate decomposes at temperatures in excess of 60°C. At this temperature, there is a continuous increase in the carboaluminate content with time. No evidence was found that calcium carbonate addition increases strength. Rather, when cured at 50°C the results show a loss in strength with curing time.

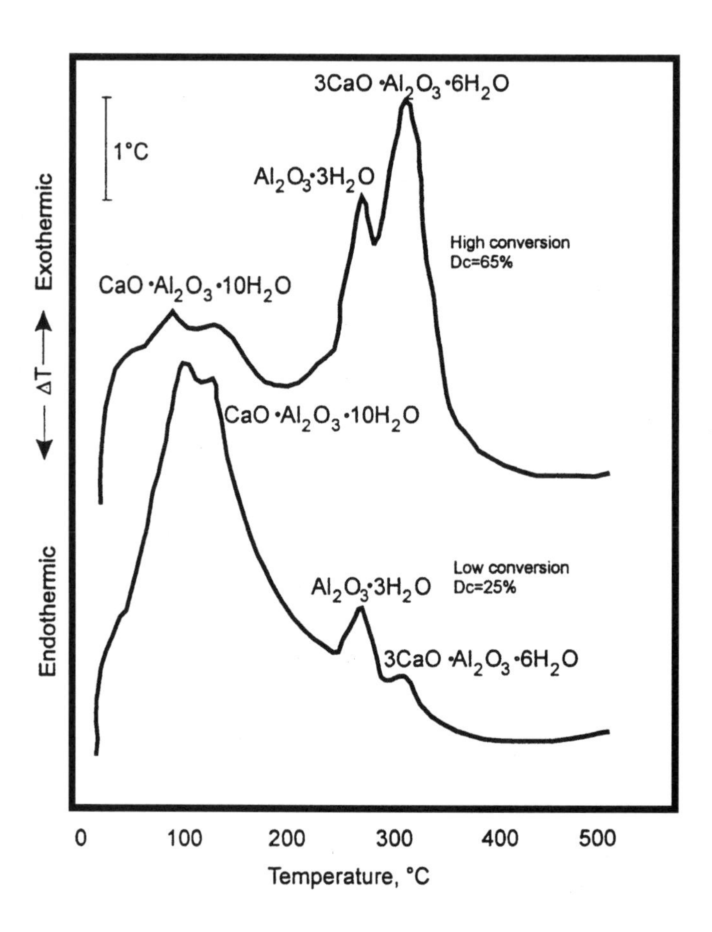

Figure 8. DTA curves for high alumina cements of different degrees of conversion. Heating rate: 20°C/min.[4]

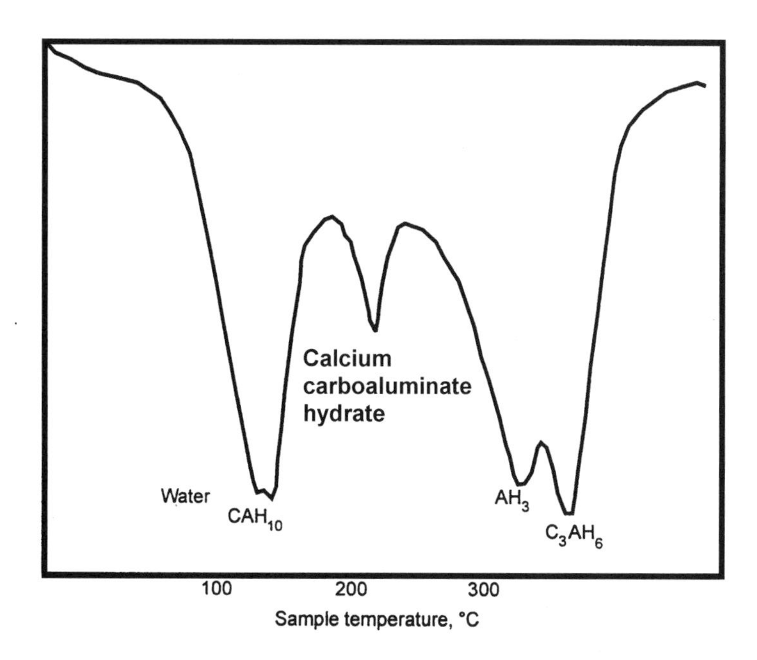

Figure 9. Typical DTA curve for hydrated HAC-$CaCO_3$ paste.[7b]

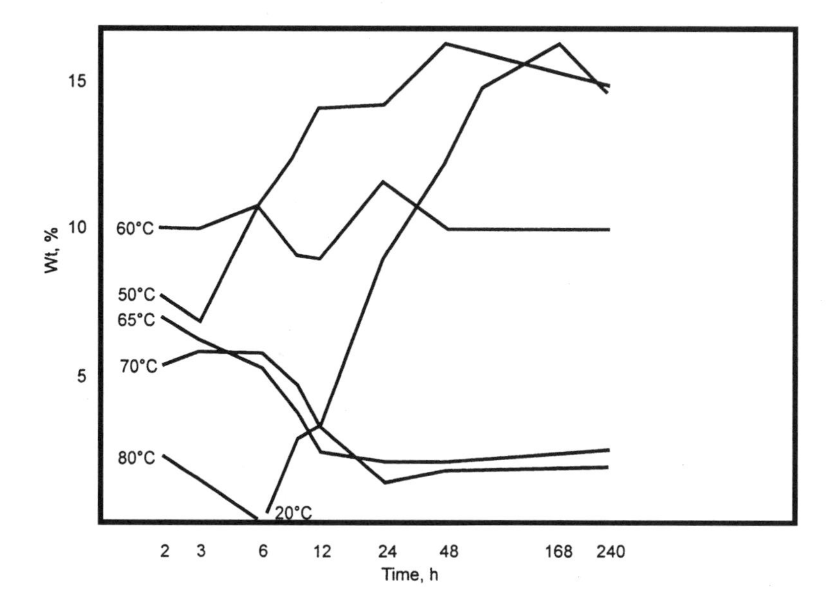

Figure 10. Estimates of calcium carboaluminate hydrate found after curing for different times at different temperatures.[7b]

Monocalcium aluminate (CA) is the principal binding mineral in HAC. Ramachandran and Feldman have studied the hydration of CA at low water/solid ratio using DSC methods.[8] Samples were in the form of compacted powders formed at 415 MPa. Disks 31.75 mm in diameter by 1.27 mm thick were formed yielding an effective water:aluminate ratio of 0.15. The product at 80°C shows a much higher strength than that hydrated at 20°C. The main initial hydration products are $2CaO{\bullet}Al_2O_3{\bullet}8H_2O$ and alumina gel. The data indicate that it is possible to obtain durable high alumina cement products by using a low water/cement ratio and hydrating at higher temperatures. The formation of the cubic C_3AH_6 is initiated within an hour in the sample hydrated at 80°C. A C_3AH_6-C_3AH_6 bond is favored under these conditions. Thermograms (DSC) of CA hydrated to different periods at 20°C and 80°C are shown in Figs. 11 and 12.

At 20°C, an endothermal valley with a peak at about 100°C appears along with a small endothermal doublet (in the range 175° to 225°C) at 10 hours (Fig. 11). At 1 day, a sharp endothermal peak appears at about 175°C and increases in intensity as hydration progresses. Additionally, a small endothermic peak at about 260°C appears at 1 day and continues to grow in intensity. Also, an endothermal peak appears at about 300°C at 5 days and increases in intensity up to 60 days. The large endotherm around 100°C may be attributed to the removal of water from alumina gel. The endotherm at 125°C (not apparent in some curves) may be ascribed to the presence of C_2AH_8. The endotherm appearing at about 175°C is due to the presence of CAH_{10}. The dual peaks occurring in the temperature range 200° to 325°C represent dehydration reactions involving gibbsite and C_3AH_6.

The thermal behavior of CA hydrated at 80°C is significantly different from that hydrated at 20°C (Fig. 12). Large endothermal effects appear after 30 minutes at about 100, 145, 210, and 280°C. The first effect is caused by alumina gel and is practically absent after 2 days of hydration. The peak at about 150°C, present in all samples up to 1 day, is attributable to the presence of C_2AH_8. The endothermal effect at 280°C is due to C_3AH_6 and gibbsite; it increases in intensity with hydration. The endotherm at 225°C which emerges at 2 days may be due to dehydration of CAH_{10} and possibly gibbsite. The resolution of the large endothermal effect into two effects at about 300°C and 340°C at 5 days confirms the formation of gibbsite and C_3AH_6. The endothermal peak at about 500°C represents the typical stepwise dehydration effect of C_3AH_6. The degree and rate of conversion of the hexagonal phases and alumina gel to the cubic and gibbsite phases, respectively, are also enhanced at a higher temperature. It is also suggested that conversion to C_3AH_6 and gibbsite phases occurs

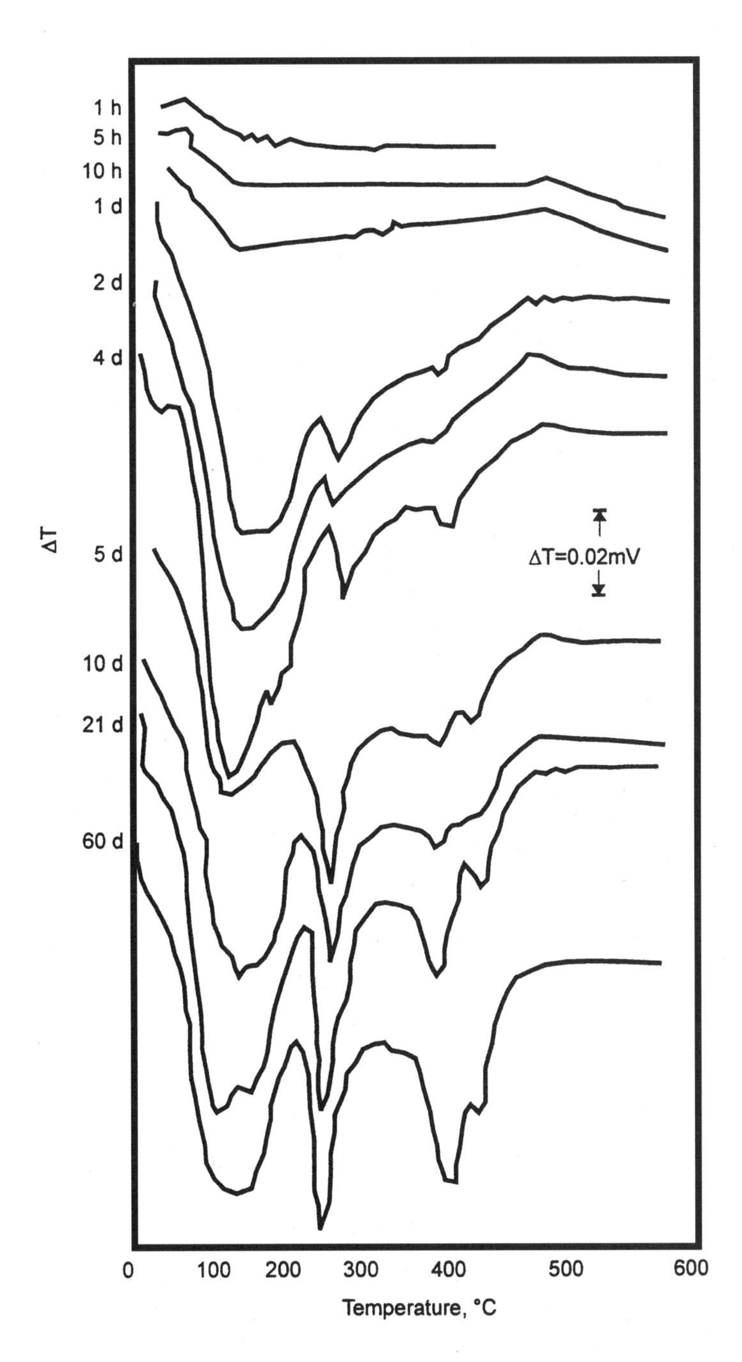

Figure 11. Thermograms of CA hydrated for different periods at 20°C at a water/solid ratio of 0.15.[8]

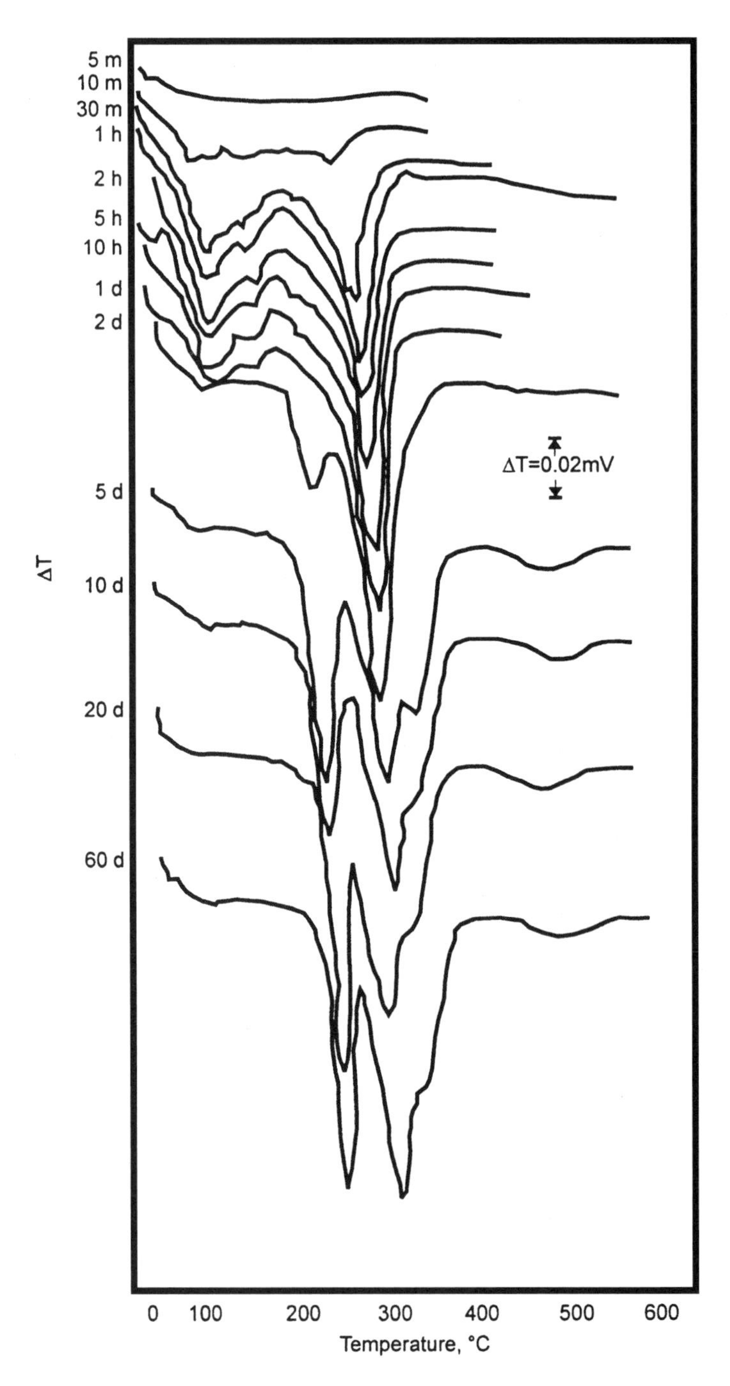

Figure 12. Thermograms of CA hydrated at 80°C to different periods at a water/solid ratio of 0.15.[8]

directly on the surface of CA particles. The rate of formation and the conversion effect are dependent on both the temperature and the water/solid ratio. A higher water/solid ratio (at 20°C), e.g., w/s = 0.50, results in faster conversion and crystallization reactions compared with CA hydrated at 20°C and at a lower water/solid ratio (Fig. 13).

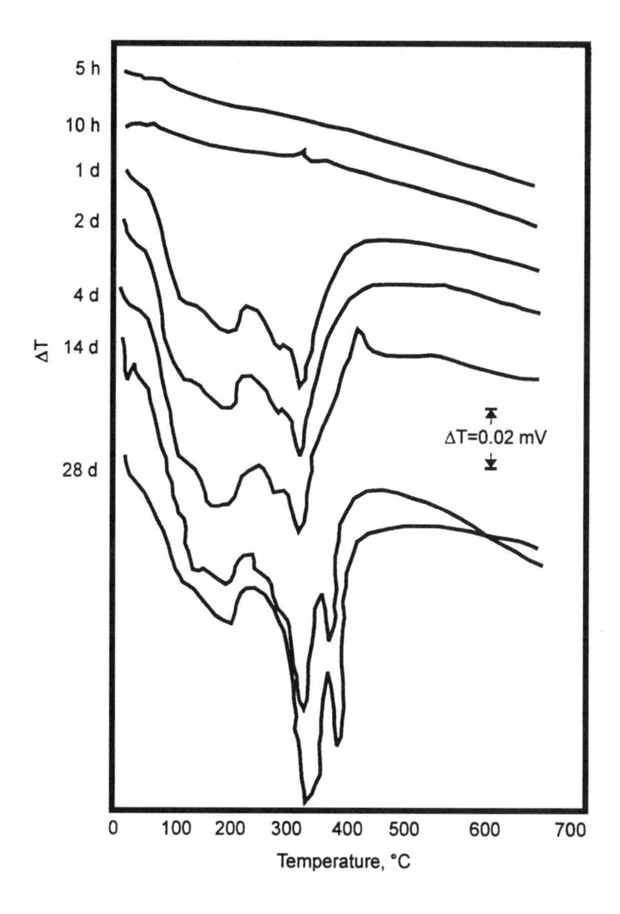

Figure 13. Thermograms of CA hydrated at 20°C at a water/solid ratio of 0.5.

Comparison of DSC results (CA) with those of conduction calorimetry strengthens interpretation of the hydration behavior (Fig. 14). In conduction calorimetry, the maximum rate of hydration (at 25°C) is found to proceed earlier and the total heat development is also of greater magnitude at a higher water/solid ratio in the first few days. The total heat developed at 25°C (paste, w/c = 0.50) is also more than that registered for the compact at 72°C. The total amounts of heat evolved for compacts hydrated at 72°C (first 30 min), 25°C (2 days), and 25°C (w/s = 0.50, 1 day) are, respectively, 56.0, 39.0, and 72.6 cal/g. It is apparent that at 80°C a direct bond formation of C_3AH_6 may occur on the surface of CA even within 30 minutes due to rapid conversion of the hexagonal phase. Strength development occurring at 80°C in the first few days does not change during further hydration over several months. This is contrary to general opinion and is contingent upon a low initial water/solids ratio.

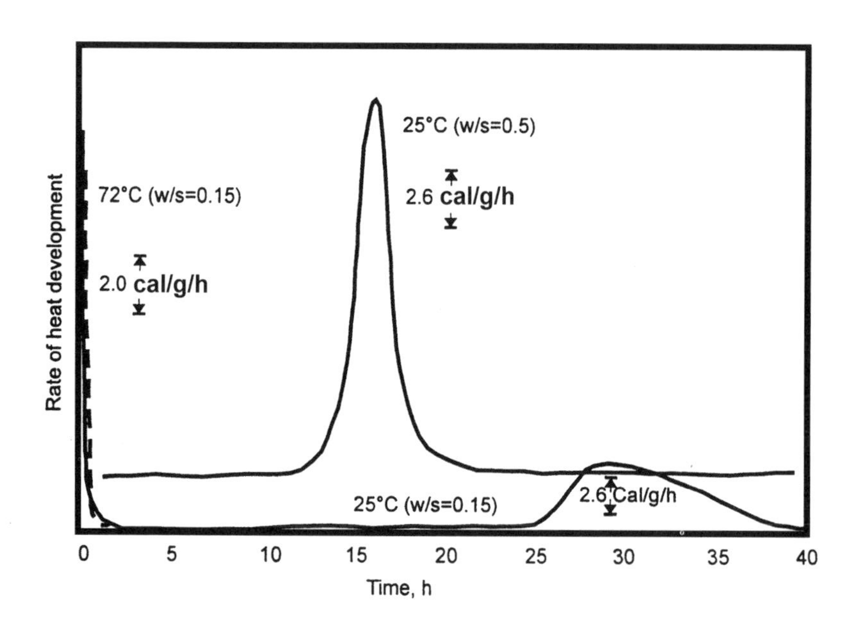

Figure 14. Conduction calorimetric curves of CA hydrated at 25°C or 72°C.

Strength of high alumina cement concrete has been estimated from exponential expressions relating strength and porosity, e.g., $S = S_o e^{-bp}$ where S is the strength, p is the porosity and S_o and b are constants. Talaber, et al., have shown that thermal analysis (DTG, TG, and DTA) can be used as input for calculation of a porosity change due to both transformation (conversion to C_3AH_6) and carbonation.[9] The thermal method involves analysis of both the original HAC concrete and the same concrete that has been autoclaved. Hydrothermal treatment completes the reactions of the active cement components. The parameters which affect strength, e.g., degrees of hydration (D_h), transformation (D_t), and carbonation (D_c) can be estimated from TG curves.

The transformation (conversion) and carbonation processes are considered to be diffusion controlled first order reactions with rate constants $k_c = 0.48$ and $k_{carb} = 0.007$. One unit of conversion results in a 0.57 fold increase in porosity and a corresponding unit of carbonation, 0.18 times decrease in porosity. Therefore, the change in porosity (ΔP) due to these processes can be expressed as follows:

Eq. (5) $$\Delta P = 0.57\,(1 - D_t)\, e^{-0.46t} - 0.18\,(1 - D_c)\, e^{-0.007t}$$

where t is the time in years and D_t and D_c are as previously defined.

The hydration of CA, CA_2, and $C_{12}A_7$ after different days of hydration was reported by Das, et al.[10] They used DSC methods in their analysis. The CA calorimetry peaks were similar to those described previously (Fig. 15a).

Endothermic peaks at 175° and 300°C were attributed to the dehydration of CAH_{10} and AH_3 gel, respectively. The peak observed at 275°C (1 day) may be due to the formation of a complex C_2AH_8 phase. C_2AH_8 formed after 28 days as indicated by the presence of a peak at 230°C. The peak at 320°C denotes the dehydration of C_3AH_6 (significant at 7 days of hydration). XRD results (dehydration of AH_3 gel and gibbsite transformation occur at about the same temperature) reveal the formation of crystalline AH_3 and the C_3AH_6 phases after 28 days hydration. Dehydration of AH_3 and C_3AH_6 occur at very close temperatures, and the DSC peaks overlap (particularly with a slow scanning rate, e.g., 10°C min^{-1}). These two dehydration peaks are well separated in DSC at high scanning rates (Fig. 16). The activation energy, E_a for the dehydration of AH_3 and C_3AH_6 have been calculated using Kissinger's relation.[11]

Eq. (6) $$\frac{V_m}{T_m^2} = \exp(-E_a / RT_m)$$

where E_a = activation energy, V = scanning rate, T_m = peak transition temperature, and R = gas constant. Estimates of E_a for the dehydration of AH_3 and C_3AH_6 are 107.16 kJ mol^{-1} and 35.58 kJ mol^{-1}, respectively.

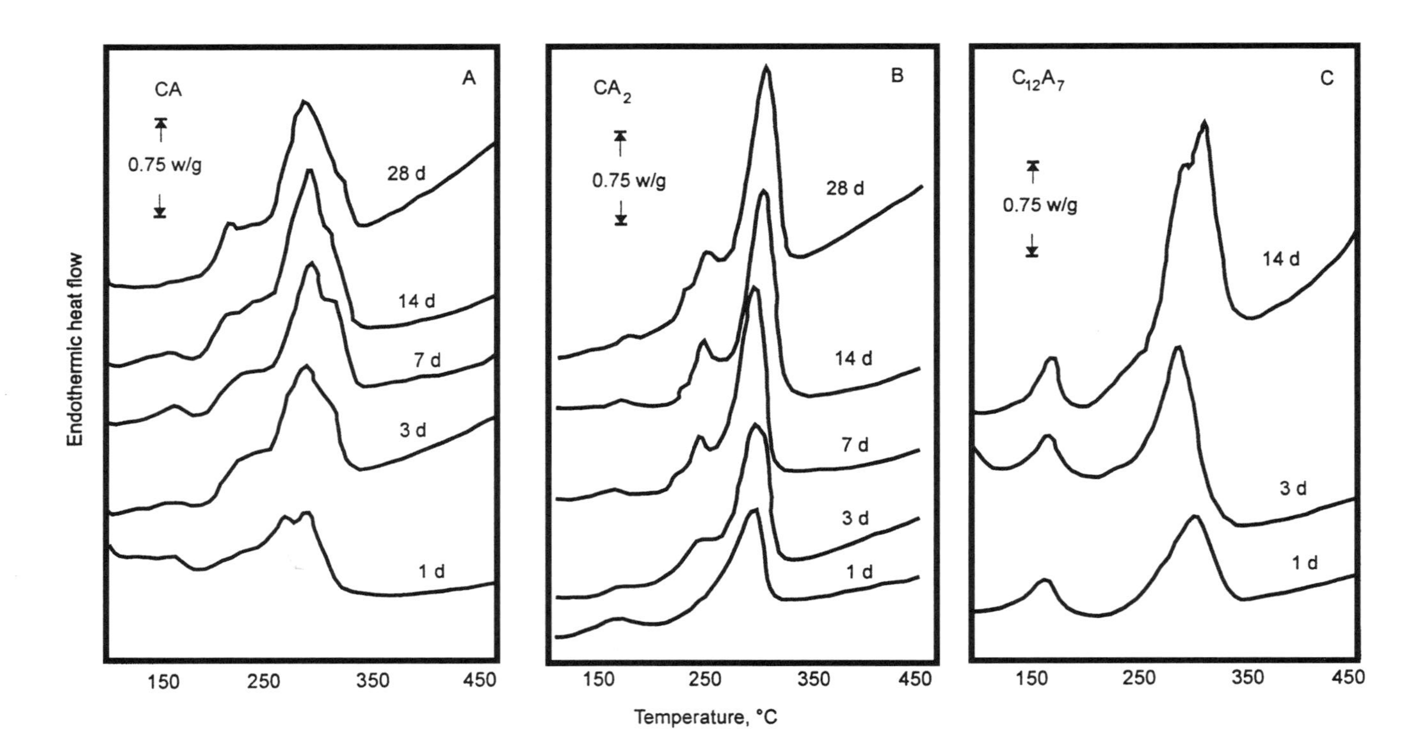

Figure 15. (*a*) Calorimeter curves of hydrated CA at different hydration periods. (*b*) Calorimeter curves of hydrated CA_2. (*c*) Calorimeter curves of hydrated $C_{12}A_7$.[10]

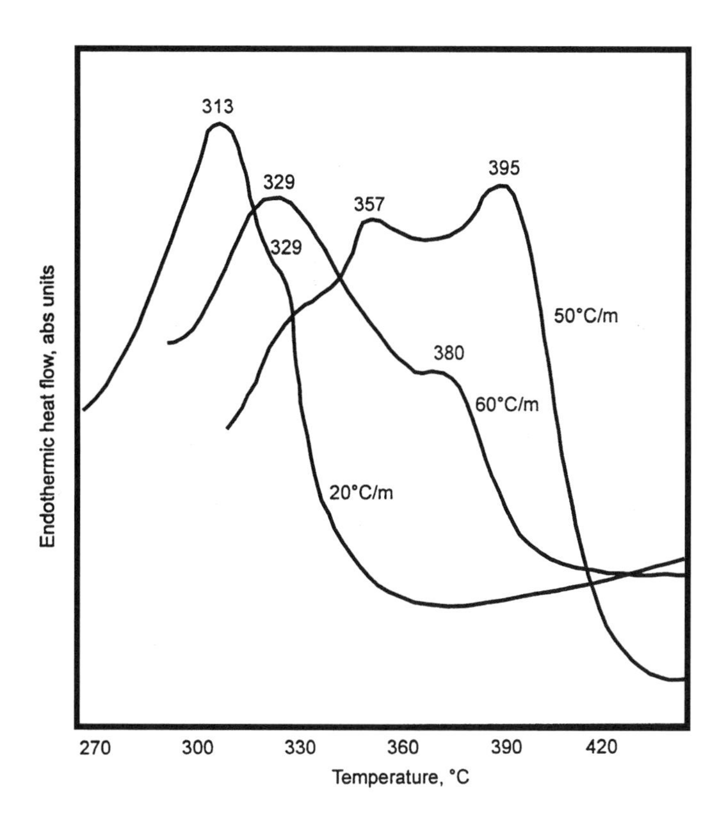

Figure 16. Calorimeter output of hydrated CA at different scanning rates after 28 days hydration.[10]

The DSC curves for hydration of CA_2 are presented in Fig. 15b. Primarily the CAH_{10} and AH_3 phases appear after 1 day of hydration (165° and 295°C). C_2AH_8 appears after 3 days of hydration. The strength of hydrated CA_2 increases up to 14 days (maximum period measured). This behavior suggests that use of mixtures of CA and CA_2 may be of practical benefit. In this work,[11] there was no indication of the formation of C_3AH_6.

DSC curves for hydrated $C_{12}A_7$ are shown in Fig. 15c. The broadening of the dehydration peak at 310°C indicates that C_3AH_6 together with AH_3 phases are formed within 1 day of hydration. The peak at 170°C corresponds to the dehydration of CAH_{10}. A significant amount of the C_3AH_6 phase is formed after 14 days of hydration. There is also a significant decrease in strength after 1 day, corresponding to the formation of the C_3AH_6 phase.

Figure 16 represents the differential scanning calorimeter output of hydrated CA at different scanning rates (at 28 days hydration). The dehydration peaks of AH_3 and the C_3AH_6 phases are well separated, the heat energy being much higher for AH_3.

Satava and Veprek (using DSC methods) studied the CA + $C\bar{S}H_2$ and the $C_{12}A_7 + C\bar{S}H_2$ systems.[12] In both systems (the former containing up to 30% $C\bar{S}H_2$ and the latter 3%), ettringite is primarily formed at temperatures below 110°C as a result of the reaction with the saturated $C\bar{S}H_2$ solution. Monosulfate is formed only above 110°C. The formation of a film which retards the hydration of $C_{12}A_7$ and CA is less pronounced than that with $C_3A + C\bar{S}H_2$.

3.0 JET SET (REGULATED-SET) CEMENT

Jet Set or regulated-set cement is a modified portland cement composition capable of developing a high early set strength upon hydration.[13] The cement consists essentially of portland cement and from about 1 to 30% by mass of a calcium halo-aluminate having the formula $11CaO{\bullet}7Al_2O_3{\bullet}CaX_2$ in which X is a halogen. The system's most studied compound appears to be $11CaO{\bullet}7Al_2O_3{\bullet}CaF_2$.

Uchikawa and co-workers[14]–[16] have reported extensively on the use of thermal analysis to monitor the hydration process of Jet Set cement and combinations of its mineral constituents.

3.1 Hydration of $11CaO{\bullet}7Al_2O_3{\bullet}CaF_2$

DTA and conduction calorimetry were used to follow the hydration of pastes (water/solid ratio = 0.60).[14] In Fig. 17, DTA thermograms of the reference calcium fluoroaluminate (top left) with calcium sulfate (top right), gypsum (bottom left), and hemihydrate (bottom right) are shown. The curves indicate a gradual increase in height of the endothermal peak of C_3AH_6 at 320°C, and a decrease in height of the broad peak at 120°C indicative of the hexagonal calcium aluminate hydrate. In the presence of anhydrite after 10 hours, the endothermal peak of the monosulfate hydrate appeared at 200°C. The peaks of ettringite at 140°C and 270°C also appeared after 1 day.

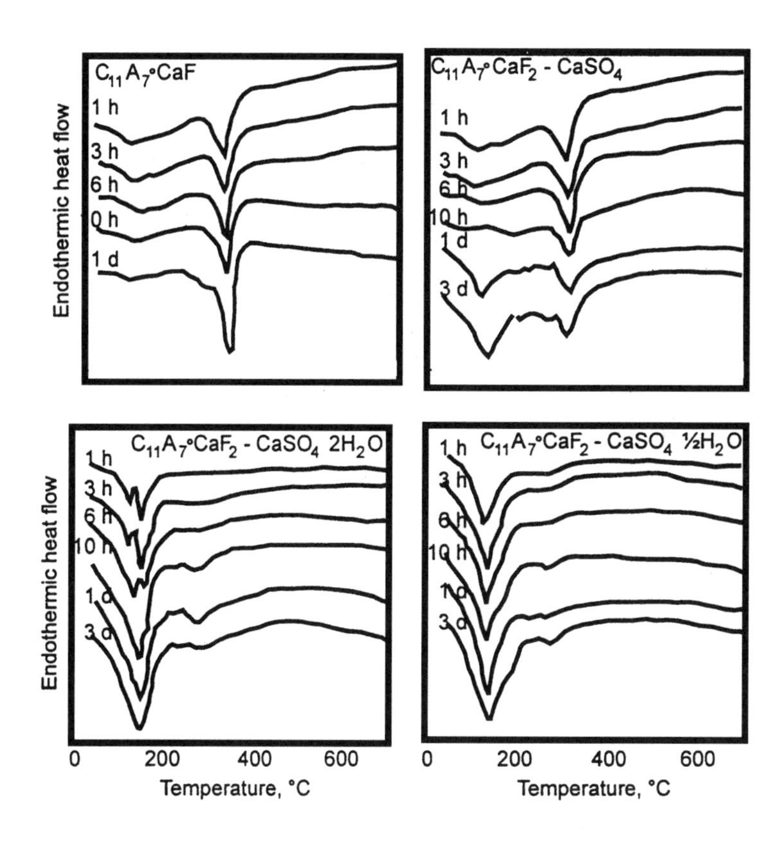

Figure 17. DTA curves of $C_{11}A_7$•CaF_2 paste with and without calcium sulfates. (S–/A = 1.0, w/s = 0.60, 20°C).[14]

The influence of a gypsum addition is also illustrated in Fig. 17. The increase in the quantity of ettringite, the decrease of gypsum, and the appearance of monosulfate hydrate are indicated by the relevant peaks at 120°C (270°C), 160°C, and 200°C respectively. The presence of hemihydrate resulted in the formation of both ettringite and gypsum (confirmed by XRD). The peaks in the system containing hemihydrate in Fig. 17 (120–140°C, 270°C) support the XRD results.

Conduction calorimetry curves are presented in Fig. 18 for $C_{11}A_7{\bullet}CaF_2$ containing $Ca(OH)_2$, $CaSO_4$, $CaSO_4{\bullet}2H_2O$, and $CaSO_4{\bullet}½H_2O$ respectively.

The heat of evolution curve in the initial stage of hydration (of $C_{11}A_7{\bullet}CaF_2$ alone) typically consists of three peaks. The first large peak appears immediately and is due to vigorous hydration of $C_{11}A_7{\bullet}CaF_2$ forming several kinds of calcium aluminate hydrate. The second and third peaks usually appear within one hour and one day respectively after mixing with water. The second peak corresponds to the "overlapped" formation of hexagonal calcium aluminate hydrate, e.g., C_2AH_8, C_4AH_{13}, and cubic C_3AH_6 as they are generally observed in the hydration of C_3A, $C_{12}A_7$, and CA. The third peak is due to the conversion of hexagonal calcium aluminate hydrate to cubic C_3AH_6 and the continued hydration of the remaining $C_{11}A_7{\bullet}CaF_2$.

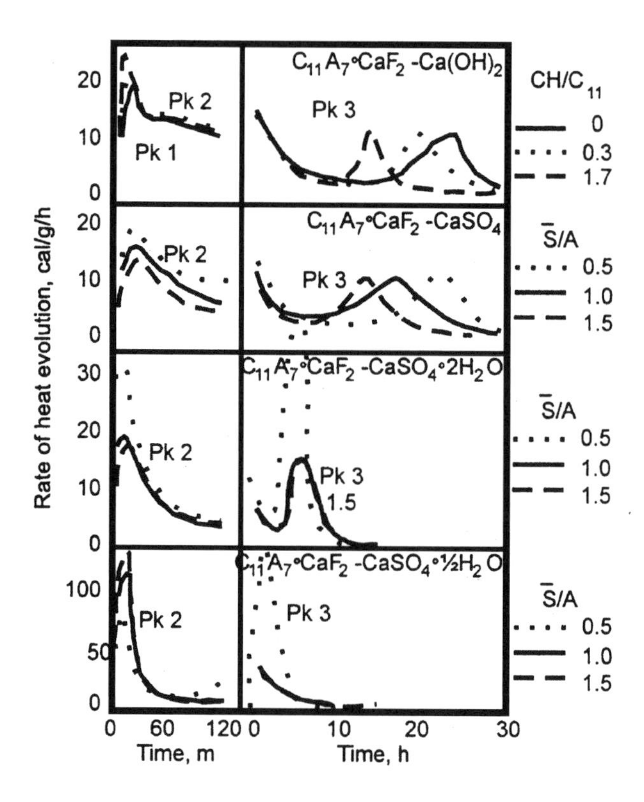

Figure 18. Heat evolution curves of $C_{11}A_7{\bullet}CaF_2$ pastes with $Ca(OH)_2$ and with calcium sulfates (w/s = 0.60, 20°C).[14]

The addition of CH (Fig. 18) shifts the second and third peak to an earlier stage. The formation of the calcium aluminate hydrates are accelerated, as is the conversion of the hexagonal hydrate into C_3AH_6.

Anhydrite addition (the rate of dissolution is slower than that of other calcium sulfates) shifts the second peak slightly toward a later stage (Fig. 18). The third peak shifts to an earlier stage as the SO_3/Al_2O_3 ratio increases. The third peak corresponds to the formation of monosulfoaluminate and partly to the continued hydration of $C_{11}A_7{\bullet}CaF_2$ forming ettringite on fresh surfaces of $C_{11}A_7{\bullet}CaF_2$. The conversion of calcium aluminate hydrate provides a source of water for the hydration.

In the case of the gypsum addition (Fig. 18), the heat evolution curves are significantly different. The second peak appears at about 15 minutes and the third peak at about six hours after mixing with water. The third peak shifts to a later stage as the amount of gypsum increases. The second peak is due to the formation of ettringite (XRD evidence supports).

Hydration of the remaining $C_{11}A_7{\bullet}CaF_2$ is accelerated when the third peak appears. Monosulfoaluminate and C_3AH_6 form during this period at a low SO_3/Al_2O_3 ratio (0.5). At higher SO_3/Al_2O_3 ratios, the third peak becomes smaller and the monosulfoaluminate phase appears later. At 7 days, ettringite and the monosulfate hydrate coexist.

The rate of heat evolution for the hemihydrate addition is large, and the peak positions correspond to very early times (Fig. 18). This is a result of the relatively high solubility and rate of dissolution of calcium sulfate hemihydrate. In the case of the high SO_3/Al_2O_3 ratio (1.5), the second peak corresponding to the formation of ettringite is higher; the third peak is not observed. XRD results confirm the co-existence of ettringite and gypsum.

The handling time of regulated set cement concrete depends on several factors. The addition of carboxylic acid, for example, lowers the solubility of the Ca^{++} ion in the liquid phase, and effectively retards hydration of $C_{11}A_7{\bullet}CaF_2$. The thermal analysis results show that increasing the SO_3 concentration in the liquid phase retards the handling time of regulated set cement concrete. The solubility and rate of dissolution of calcium sulfate, therefore, affects the regulation of the handling time. Excess addition of calcium hydroxide retards the hydration of $C_{11}A_7{\bullet}CaF_2$ and interferes with the hardening process.

The hydration of mixtures of $C_{11}A_7{\bullet}CaF_2$, C_3S, and $CaSO_4$ with various additives has been investigated.[15] The molar ratio of $C_{11}A_7{\bullet}CaF_2$ to C_3S was 1.0 to 15.4. The SO_3/Al_2O_3 molar ratio was 1.0. Hydration occurred at 20°C with a water/solid ratio = 0.60. The additives included calcium sulfate hemihydrate, sodium sulfate, sodium carbonate, and citric acid. The rate of hydration of C_3S increased dramatically as the rate for $C_{11}A_7{\bullet}CaF_2$ decreased. An excess of hemihydrate results in the excessive retardation of $C_{11}A_7{\bullet}CaF_2$ and the acceleration of C_3S hydration. Citric acid retards the hydration of both $C_{11}A_7{\bullet}CaF_2$ and C_3S. Sodium sulfate retards the hydration of $C_{11}A_7{\bullet}CaF_2$ and accelerates the hydration of C_3S. Sodium carbonate retards the hydration of $C_{11}A_7{\bullet}CaF_2$, but in combination with sodium sulfate, the opposite effect occurs. Calcium carbonate and superplasticizer additions have little effect on the hydration process.

Conduction calorimetry curves for the above systems are shown in Fig. 19 (a–i). Figures 19a and b are for the C_3S and the C_3S-sulfate systems. For systems containing $C_{11}A_7{\bullet}CaF_2$, there are up to five characteristic peaks observed as depicted in Fig. 19c. The first (immediate) is due to the rapid hydration of $C_{11}A_7{\bullet}CaF_2$ and the formation of various calcium aluminate compounds. The second peak (1 to 5 hrs) is due to the formation of hexagonal and cubic C-A-H. The third and fourth peaks are assigned to the formation of monosulfoaluminate hydrate and ettringite. The active hydration of C_3S is retarded in the presence of anhydrite (peak 5), and it occurs at 50 hours.

In the quaternary system containing both hemihydrate and anhydrite, the first peak is due to the dissolution of C_3S, the hydration of the hemihydrate, and the formation of ettringite immediately after mixing with water. The second peak is attributed to the formation of C-A-H and specifically C_3AH_6. The third and fourth peaks are due to the formation of monosulfate hydrate and newly formed ettringite. The hydration of C_3S begins after 40 hours (peak at 70 hrs). For the curve for the system containing the largest amount of hemihydrate, the second peak corresponds to the hydration of C_3S.

The effect of citric acid addition is depicted in Fig. 19e. The first and second peaks, corresponding to the dissolution of mineral compounds and the formation of C-A-H, were less intense and shifted to a later stage. The third peak, due to monosulfoaluminate and ettringite, was shifted to an earlier stage proportionally to the amount of citric acid addition. The fourth peak, corresponding to active C_3S hydration, was shifted to a later stage. The shift was dependent on the amount of citric acid added.

The effect of Na_2SO_4 addition is illustrated in Figs. 19f, g, and h. Heat evolution, due to the formation of ettringite, was observed immediately in the samples containing hemihydrate. In the absence of hemihydrate, no heat evolution peak was observed immediately after mixing. There is no significant difference in the heat evolution curve (after the first peak) for hydration in the presence of Na_2SO_4 except that hemihydrate

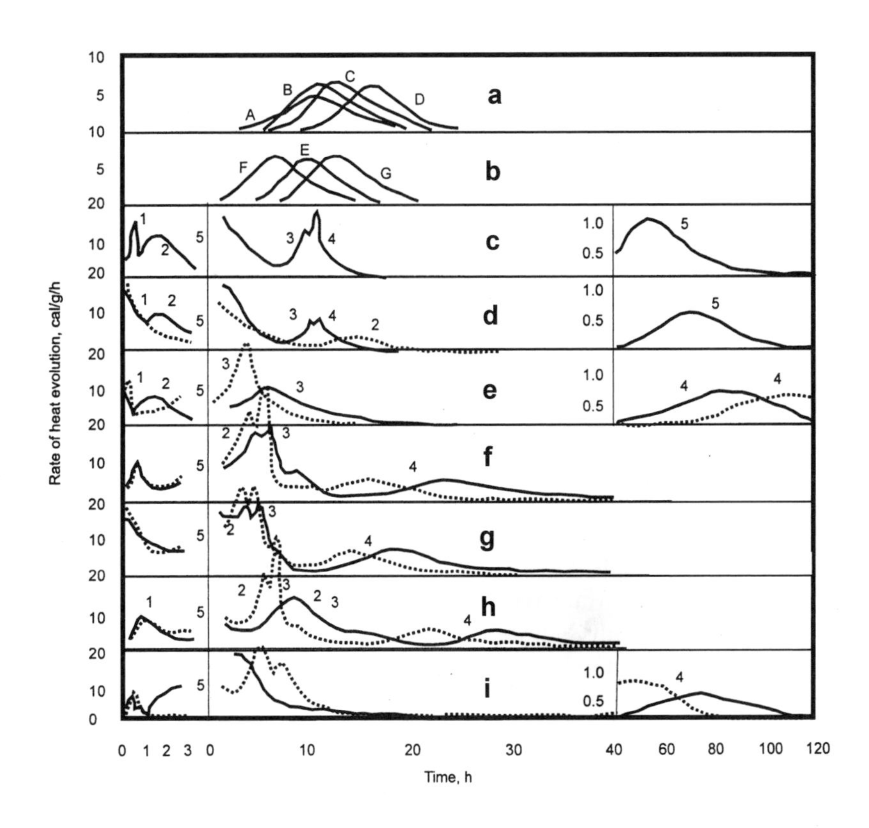

Figure 19. Heat evolution curves for C_3S paste and mixtures containing $C\overline{S}$, $C\overline{S}H_{1/2}$, $C_{11}A_7{\bullet}CaF_2$ and various additives.[15]

(*a*) **A:** C_3S; **B:** C_3S-$C\overline{S}$-$C\overline{S}H_{1/2}$ (H/A: 0.35); **C:** C_3S-$C\overline{S}$; **D:** C_3S-$C\overline{S}$-$C\overline{S}H_{1/2}$ (H/A = 0.14).

(*b*) **E:** C_3S-$C\overline{S}$-$N\overline{S}$ ($N\overline{S}$ = 0.06); **F:** C_3S-$C\overline{S}$-$N\overline{S}$ ($N/\overline{S}$ = 0.13).

(*c*) $C_3\overline{S}$-$C_{11}A_7{\bullet}CAF_2$-$C\overline{S}$.

(*d*) —— C_3S-$C_{11}A_7{\bullet}CAF_2$-$C\overline{S}H_{1/2}$ (H/A = 0.14); ······ (H/A = 0.34).

(*e*) —— C_3S-$C_{11}A_7{\bullet}CaF_2$-$C\overline{S}$ (0.05% citric acid); ······ (0.1% citric acid).

(*f*) —— C_3S-$C_{11}A_7{\bullet}CAF_2$-$C\overline{S}$-$N\overline{S}$ ($N/\overline{S}$ = 0.06); ······ ($N/\overline{S}$ = 0.13).

(*g*) —— C_3S-$C_{11}A_7{\bullet}CaF_2$-$C\overline{S}$-$C\overline{S}H_{1/2}$-$N\overline{S}$ (H/A = 0.14, $N/\overline{S}$ = 0.06); ······ (H/A = 0.14, $N\overline{S}$ = 0.13).

(*h*) —— C_3S-$C_{11}A_7{\bullet}CaF_2$-$C\overline{S}$-$N\overline{S}$-0.05% citric acid ($N/\overline{S}$ = 0.06); ······ ($N/\overline{S}$ = 0.13).

(*i*) —— C_3S-$C_{11}A_7{\bullet}CaF_2$-$N\overline{C}$ ($N/\overline{S}$ = 0.06); ······ C_3S-$C_{11}A_7{\bullet}CaF_2$-$C\overline{S}$-$N\overline{S}$-$N\overline{C}$ ($N/\overline{S}$ = 0.13).

accelerates and citric acid retards the time of appearance of the corresponding peaks. The second and third peaks shifted to earlier times in proportion to the amount of sodium sulfate added. The formation of C_3AH_6 was suppressed by the addition of sodium sulfate. The hydration of C_3S (peak at 14 to 35 hrs) in this series is retarded compared to pure C_3S, but it is accelerated compared to the case without Na_2SO_4. The acceleration is dependent on the added quantity of Na_2SO_4.

Uchikawa and Tsukiyama investigated the hydration of two Jet Set cements at 20°C.[16] The cements contained about 60% alite, 20% calcium fluoroaluminate, 4.5% ferrite phase, and 0.8–2.0% belite. One (A) contained about 2% calcium carbonate, a citric acid based retarder (0–2%) and 1% sodium sulfate. The other (B) contained 1% sodium sulfate and 2.5% hemihydrate. Both cements had surface areas of about 5000 cm^2/g. Pastes were prepared at water/cement ratio = 0.40. DTA curves for these pastes (hydration period ranging from 1 hour to 7 days) are presented in Fig. 20. The endothermic peaks at about 130°C and 280°C correspond to the dehydration of ettringite. A small peak at about 190°C corresponds to the dehydration of monosulfate hydrate. The results suggest that the amount of ettringite increases with hydration and subsequently decreases due to the conversion to monosulfate hydrate. The maximum amount of ettringite is formed in the samples at about 6 hours. The endothermic peak for $Ca(OH)_2$ (1 day), suggests the hydration of alite has progressed significantly.

Calorimetry curves for the two Jet Set cements described above are shown in Fig. 21. There are four main peaks in the heat evolution curves. The first peak appears immediately and is due to the following: dissolution of free lime, hydration of anhydrite and hemihydrate, and the formation of C-A-H and monosulfate hydrate. The second peak is attributed to the formation of ettringite, the third to the formation of monosulfate hydrate, and the fourth peak to the formation of C-S-H. The overlap of the second and third peaks (cement B) and the larger third peak are attributed to active conversion of ettringite to monosulfate hydrate. The broader fourth peak (cement B) occurred later indicating a less active formation of C-S-H gel than for cement A.

The differences between the two cements can be summarized as follows. The initial rate of hydration (2 hrs) in the presence of hemihydrate is greater than for citric acid. The reverse is the case at a later stage of hydration. Mechanical strength using citric acid is greater than for a hemihydrate addition. This corresponds to the following: a higher degree of hydration corresponds to a larger amount of ettringite, a smaller average pore size, and lower porosity.

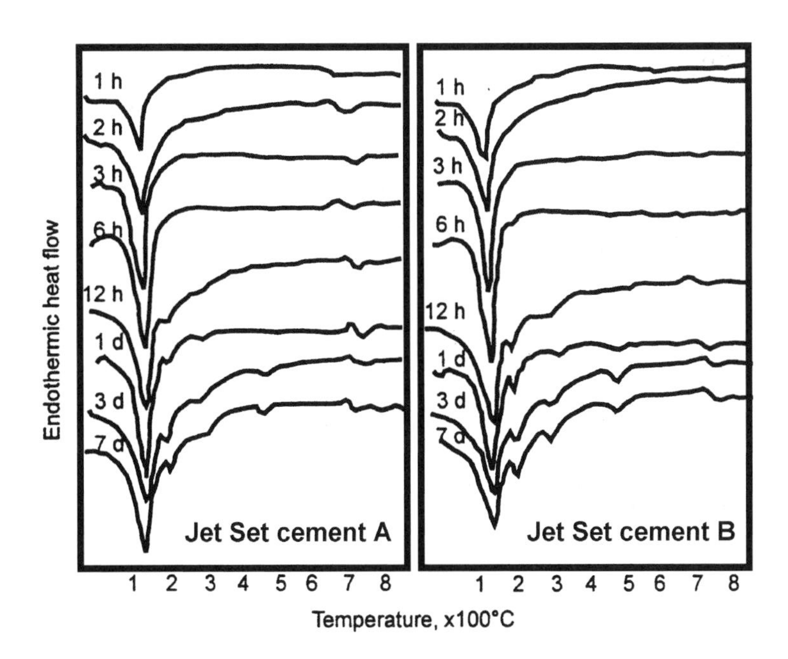

Figure 20. Differential thermal analysis curves of hardened Jet Set cement pastes.[16]

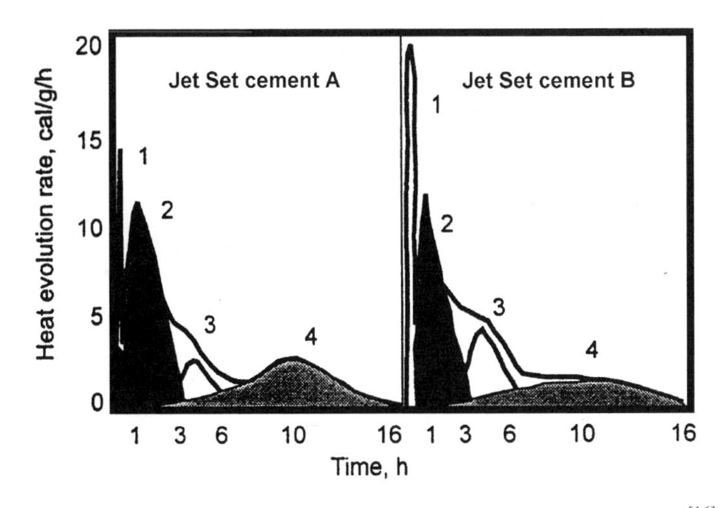

Figure 21. Heat evolution characteristics of hardened Jet Set cement pastes.[16]

4.0 MAGNESIUM OXYCHLORIDE AND MAGNESIUM OXYSULFATE CEMENT SYSTEMS

Magnesium oxychloride cement, also known as Sorel cement, is formed by mixing finely divided magnesium oxide with an aqueous solution of magnesium chloride.[17] It possesses many properties superior to those of portland cement. These include fire resistance, low thermal conductivity, higher abrasion resistance, and compressive and flexural strengths. Its excellent binding capability permits the use of many organic and inorganic aggregates which would be unsuitable for making portland cement concrete.

Four types of oxychloride complexes[18] are known to form in the $MgO–MgCl_2$-H_2O system. They are $5Mg(OH)_2•MgCl_2•8H_2O$ (5-form), $3Mg(OH)_2•MgCl_2•8H_2O$ (3-form), $2Mg(OH)_2•MgCl_2•5H_2O$ (2-form), and $9Mg(OH)_2•MgCl_2•6H_2O$ (9-form). The 2 and 9-forms are formed in solutions at temperatures above 100%. The formation of magnesium chlorocarbonate, $Mg(OH)_2•MgCl_2•2MgCO_3•6H_2O$, is detected upon continuous exposure of the oxychloride to air.

The stabilization of hardened oxychloride cement products against attack by water has been studied, as it is known that these products lose strength on prolonged exposure to water.

Magnesium oxysulfate cements can be produced by adding magnesium chloride solutions to calcium sulfates or calcium phosphate-sulfate mixtures.[19] The cements formed can be regarded as variants of Sorel cements. The following phases have been identified depending upon the temperature and pressure conditions: $3Mg(OH)_2•MgSO_4•8H_2O$; $3Mg(OH)_2•MgSO_4•3H_2O$; $Mg(OH)_2•MgSO_4•5H_2O$; $Mg(OH)_2•2MgSO_4•3H_2O$; $2Mg(OH)_2•3MgSO_4•5H_2O$; $MgSO_4•H_2SO_4•3H_2O$; $3Mg(OH)_2•MgSO_4•4H_2O$. DTA and DTG curves of the 3, 5, 2, and 9-forms are presented in Fig. 22. The numbers on the curves refer to the number of water molecules per mole remaining. The peaks and their corresponding temperatures are given in Table 1.

There is a close correspondence between the DTA and the DTG curves. The former occurs at a temperature of about 45°C higher than the latter. The corresponding heating rates were 10°C/minute and 0.4°C/minute.

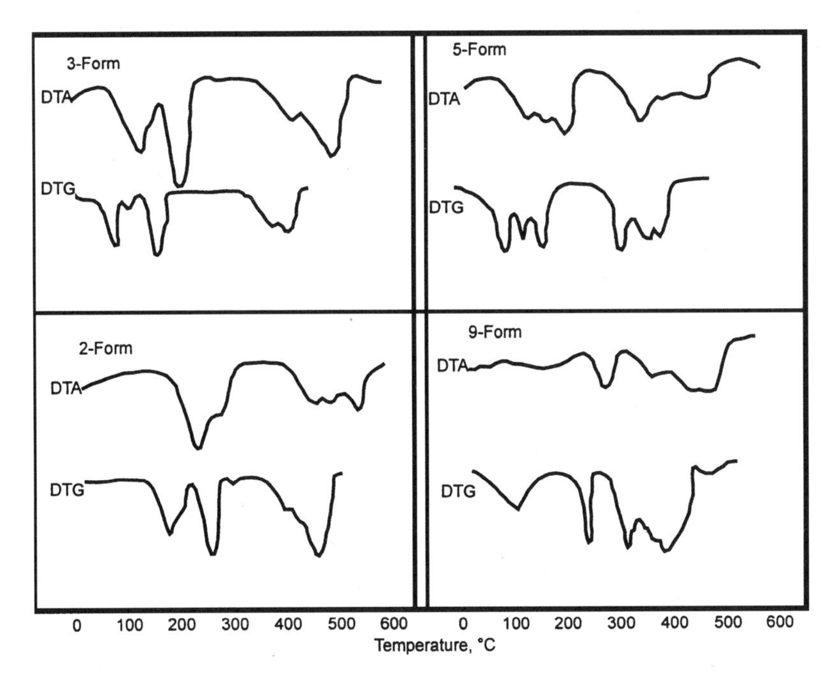

Figure 22. Differential thermal analysis curves (DTA) and differential weight loss curves (DTG) of magnesium oxychlorides. The composition of the phases identified are indicated above the curves:[18]

5-Form, $5Mg(OH)_2{\bullet}MgCl_2{\bullet}7.93H_2O$;

3-Form, $3Mg(OH)_2{\bullet}MgCl_2{\bullet}7.68H_2O$;

2-Form, $2Mg(OH)_2{\bullet}MgCl_2{\bullet}4.14H_2O$;

9-Form, $9Mg(OH)_2{\bullet}MgCl_2{\bullet}5.68H_2O$.

Table 1. Peak Temperatures for Magnesium Oxychloride Phases

Phase	DTA (°C)	DTG (°C)
3-form	145, 170, 220, (378), 430, 505	96, 123, 177, (348), 398, 425
5-form	157, 184, 218, 364, 411, 478	100, 135, 173, 328, (380), 398
2-form	245, 284, 462, 486, 546	188, 268, 418, 469
9-form	173, 292, 368, 452, 481	162, 242, 316, (366), 388
*Figures in parentheses refer to minima or maxima.		

All forms of oxychloride, upon heat treatment, dehydrate to anhydrous phases having the crystal structure of $Mg(OH)_2$ in which Cl ions replace (OH) ions to an extent determined by the composition of the original form. The anhydrous phases are represented by an extended plateau in the DTG and DTA curves. Beyond this plateau, the anhydrous phases decompose, and since they all have similar structure, the decomposition is similar.

There are up to three endothermic reactions which vary considerably in position and intensity from form to form. Throughout the decomposition process, hydrogen chloride is liberated, and the MgO content increases after each endothermic peak. It is suggested that decomposition involves loss of water and chlorine (endothermic) followed by exothermic reactions between the by-products.

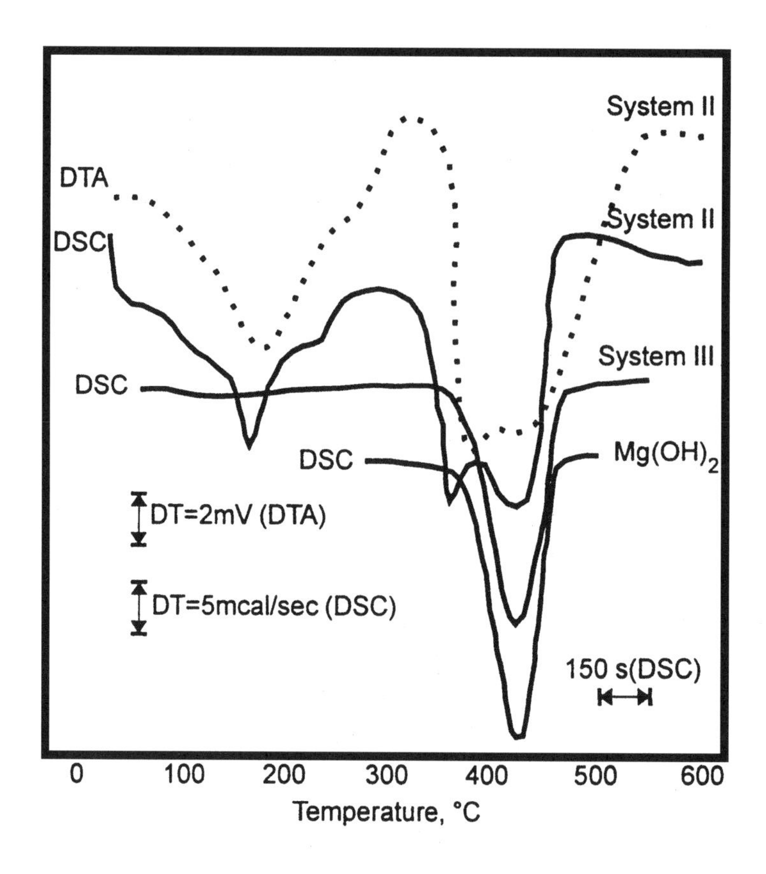

Figure 23. Differential scanning calorimetry traces of compacted magnesium oxychloride paste untreated (System II) and immersed in water at 85°C for 5 hours (System III).[17]

Compacts of hydrated magnesium oxychloride paste (designated System III) were also studied.[17] The effect of immersion in water at 85% for 5 hours on the compacts of paste (designated System III) hydrated magnesium oxychloride cement (chloride solution – solid ratio = 0.59) are illustrated in Fig. 23.[17] The endothermal dip in the DSC curves for System III at 425°C was due only to $Mg(OH)_2$. It is apparent the oxychloride complex became unstable in hot water.

Magnesite is used as a source material for making magnesium oxychloride cement.[20] Variations in crystallinity and composition of magnesite can affect the quality of oxychloride-based products including their mechanical strength. Significant strengths are obtained with cryptocrystalline magnesite with low iron and calcium content. The presence of forsterite (Mg_2SiO_4) is not desirable and was not detected in the two samples that gave the best results. Formation of dicalcium and tricalcium silicate can occur if the CaO/SiO_2 ratio is greater than 1.87. This would result in good strength as all the MgO is available to form oxychloride and additional hydraulic reactions of the calcium silicates can occur. There was no evidence for the presence of these silicates in the work cited.

The two superior magnesites are designated 1 and 3 in Fig. 24. Figure 25 contains thermograms of the magnesium oxychlorides produced with these magnesites. Figure 24 indicates that the content of the magnesium carbonate (peak at 700°C) is similar for all samples (verified by chemical analyses). Samples 2, 4, 5, and 6 show higher endothermal effects due to $FeCO_3$ and $CaCO_3$ at 500°C and 800–925°C respectively, and the presence of these compounds is not conducive to the development of strengths. Further, it was demonstrated that an oxychloride cement giving a higher endothermal peak area at 400°C gave higher flexural and compressive strengths.[21] This peak represents the primary strength-contributing component, $3Mg(OH)_2{\bullet}MgCl_2{\bullet}8H_2O$ in the set cement. It was also shown that larger dehydration losses between 50–250°C would mean greater uncombined magnesium chloride in the set cement and lower strength.

The thermograms in Fig. 25 (obtained after 28 days reaction) show that the oxides of samples 1 and 3 have definitely resulted in the formation of larger amounts of $3Mg(OH)_2{\bullet}MgCl_2{\bullet}8H_2O$ than those of the others as is evident by the large endothermal peaks at 400°C. The oxychloride cements prepared from samples 2, 4, 5, and 6 show more uncombined $MgCl_2$ than that prepared from samples 1 and 3 as exhibited by the endothermal peak between 50–250°C.

Figure 24. Thermograms of almora magnesite.[20]

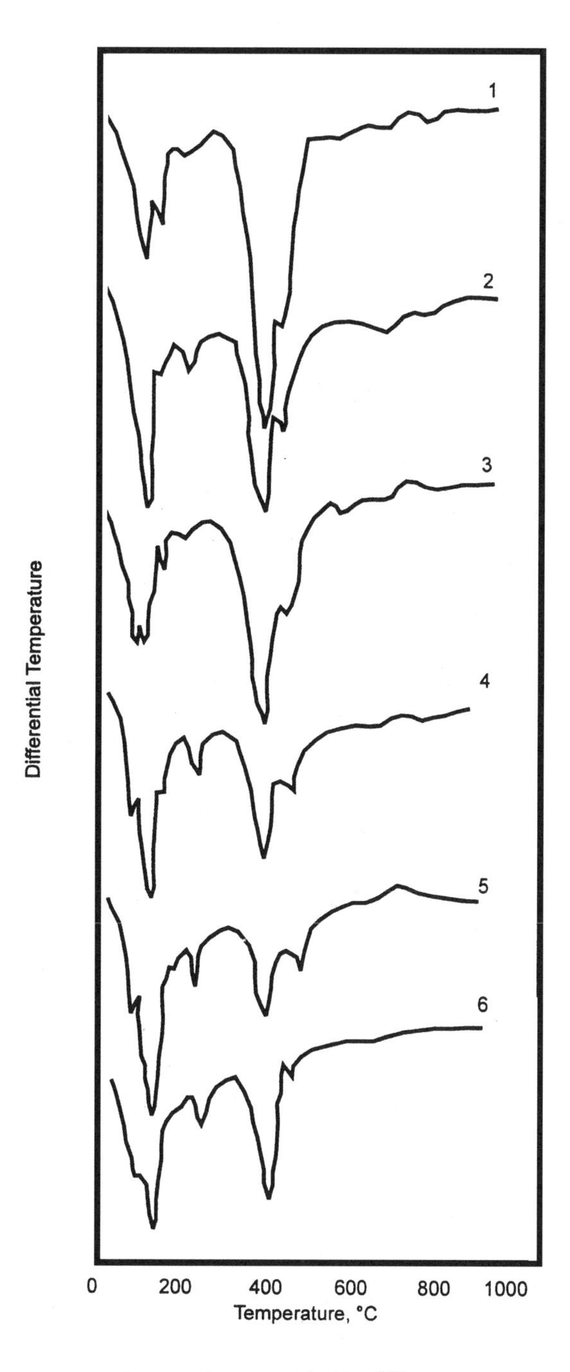

Figure 25. Thermograms of magnesium oxychlorides.[20]

It is, therefore, apparent that differential thermal analysis (in addition to chemical analysis) could be very useful in determining the suitability of magnesites for making oxychloride cement.

Typical thermograms for magnesium oxysulfate cement pastes (Series I, w/s = 0.59–1.43) prepared with $MgSO_4 \cdot 7H_2O$ solution having a specific gravity (s.g.) of 1.18 and similar pastes (Series II) prepared with $MgSO_4 \cdot 7H_2O$ solution (s.g. = 1.303) are presented in Fig. 26. There are two low temperature endothermal peaks at 105°C and 155°C. There is also a large endothermal peak at approximately 450°C. The two low temperature peaks have been attributed to the dehydration of $3Mg(OH)_2 \cdot MgSO_4 \cdot 8H_2O$.[22] The large peak at 475°C is due to the decomposition of $Mg(OH)_2$. There is less $Mg(OH)_2$ present in Series II samples. Also, the ratio of the low temperature peak heights to the peak height at 450°C is greater for Series II samples than for Series I samples. The areas of the two low temperature peaks are greater for Series II samples indicating more magnesium oxysulfate complexes have formed.

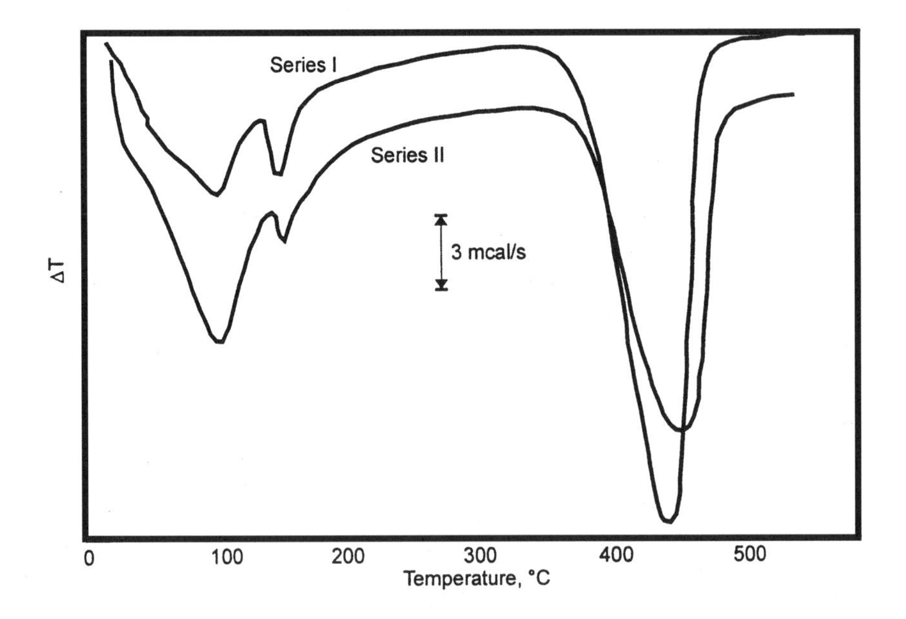

Figure 26. Thermograms of magnesium oxysulfates.[19]

5.0 ZINC OXYCHLORIDE CEMENT

The reaction of zinc oxide with zinc chloride in water gives an extremely hard zinc oxychloride cement product which is not attacked by acids or boiling water.[23] The main hydrated products are two zinc oxychloride hydrates, $4ZnO{\bullet}ZnCl_2{\bullet}5H_2O$ and $ZnO{\bullet}ZnCl_2{\bullet}2H_2O$. The former product is very stable and insoluble, but is associated with very poor workability while the latter is not stable in water and can cause excessive solubility and leaching. Zinc oxychloride would, therefore, not appear to have sufficiently reliable properties to justify practical application. TG and DTG curves for the 4:1:5 and 1:1:2 phases are plotted in Fig. 27. The 1:1:2 phase undergoes a two-step dissociation beginning at about 230°C. The weight loss corresponding to half the constituent water occurred above the melting point of anhydrous $ZnCl_2$, i.e., 275°C. Weight loss ceased at a level slightly above that corresponding to a complete loss of $ZnCl_2$ and H_2O. The 4:1:5 phase dissociates at 160°C to the 1:1:2 phase and ZnO, and at higher temperatures undergoes the same dissociation as the 1:1:2 sample.

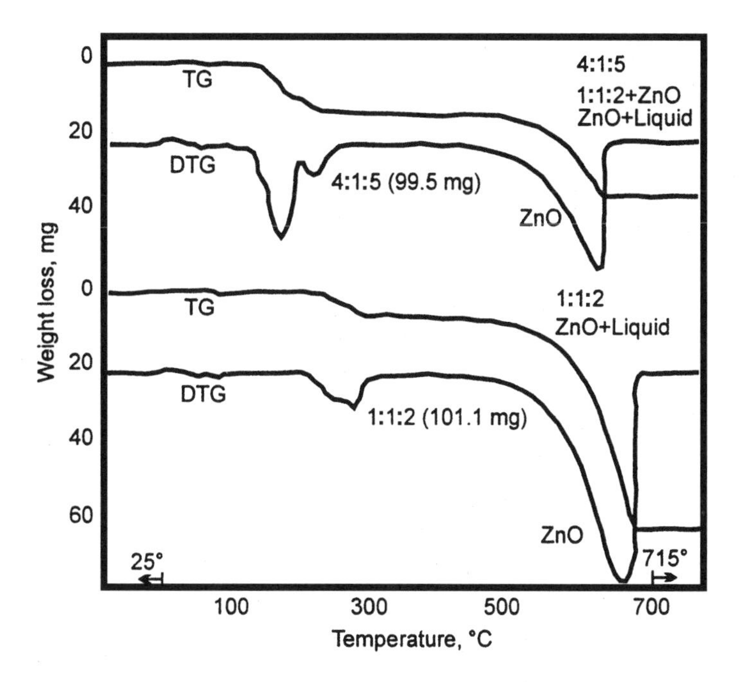

Figure 27. TG and DTG curves for 4:1:5 and 1:1:2 phases in zinc oxychloride cements.[23]

6.0 MAGNESIA-PHOSPHATE CEMENTS

There are several commercial rapid-setting cements used for the repair of airport runways and other concrete infrastructure where rapid setting and strength gain are mandated. These include magnesia-phosphate cements.

Several basic oxides will react with phosphoric acid or acid phosphates at ordinary temperature forming cohesive masses, setting and giving high compressive strengths. Magnesia-phosphate cements are quick setting cements based on magnesium ammonium phosphate. These cements consist primarily of magnesia, ammonium phosphate, and sodium tripolyphosphate. Magnesium ammonium phosphate hexahydrate and magnesium phosphate tetrahydrate have been identified as hydration products.[24] The initial hydration product, $Mg(NH_4)_2\text{-}H_2(PO)_4 \cdot 4H_2O$ (tetrahydrate), converts to $MgNH_4\text{-}PO_4 \cdot 6H_2O$ (hexahydrate). In the presence of colloidal silica, more tetrahydrate forms. There is some uncertainty about the reaction sequence, but it appears the hexahydrate formation is responsible for strength.[25]

Abdelrazig and co-workers studied the hydration (at 22°C) of MgO (75 g) and monoammonium phosphate (56 g) a composition ratio not dissimilar to commercial phosphate cements.[26] Mortars (Systems 1 and 2) and pastes (System 4) were investigated. The mortars (containing quartz sand) were prepared at water/solid ratios of 0.62 and 0.125; the pastes had water/solid ratios of 0.125. Thermograms (DTA) of mortar and paste are shown in Fig. 28a. The hydration time is one week.

A small endotherm at 51°C is due to the dehydration of $NaNH_4HPO_4 \cdot 4H_2O$. The dehydration of the hexahydrate into the monohydrate ($MgNH_4PO_4 \cdot H_2O$) is shown by the well-developed endotherm at 114°C. The small endotherms at 198°C and 232°C are attributed to the unreacted monoammonium phosphate and the dehydration of the monohydrate respectively. The sharp endotherm at 575°C is due to the α - β polymorphic transformation of quartz. The exotherms at 614°C and 812°C are due to the crystallization of $Mg_2P_2O_7$ and the formation of $Mg_3(PO_4)_2$ respectively.

The compressive strength of mortar continues to develop for several weeks. If colloidal particles of hydrate form around nuclei of hexahydrate, it may account for continuous strength development.

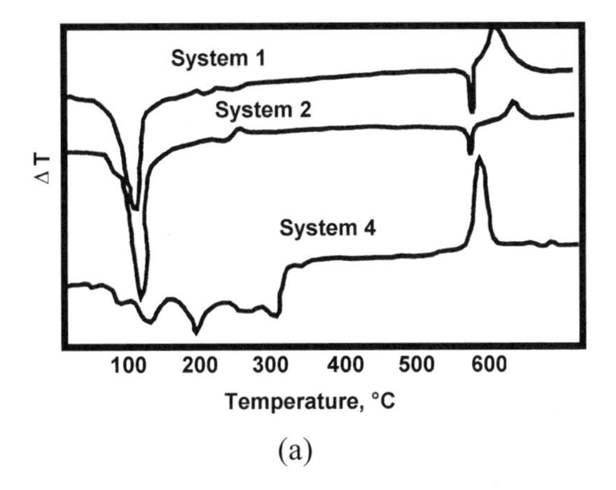

(a)

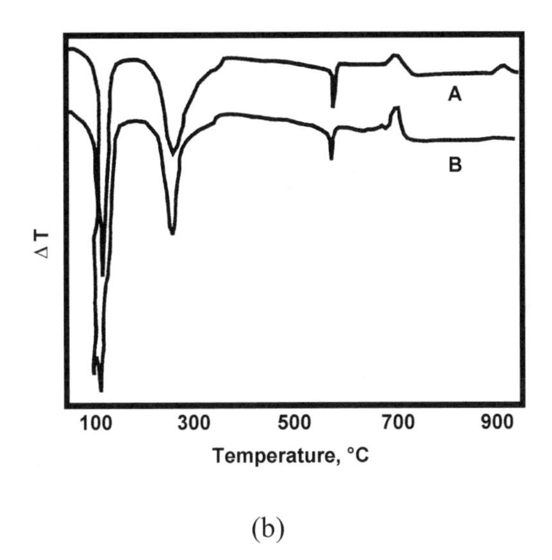

(b)

Figure 28. (*a*) DTA curves of magnesia-phosphate cement systems.[26] (*b*) DTA curves of phosphate cement mortars hydrated for *A*, 5 minutes and *B*, 672 hours.

In another study by Abdelrazig and co-workers,[27] thermal analysis of mortars made from magnesia-phosphate cement was reported. The magnesia was mixed with $NH_4H_2PO_4$ (ADP) and $Na_5P_3O_{10}$ (STPP). The water/solid ratio including silica sand was 1:8 or 1:16. The principal hydrate formed was struvite, $NH_4MgPO_4 \cdot 6H_2O$. Schertalite, $(NH_4)_2$

$Mg(HPO_4)_2 \cdot 4H_2O$, also forms initially. Minor amounts of dittmarite, $NH_4MgPO_4 \cdot H_2O$, and/or stercorite, $NaNH_4PO_4 \cdot 4H_2O$, were also detected. DTA indicated the presence of quartz (endotherm at 573°C, Fig. 28b). A double endotherm at approximately 95°C and 110°C is due to the presence of both struvite and schertalite. The appearance of an apparent double endotherm is due to the superposition of an exotherm at about 100°C superimposed on a large endotherm due to dehydration. The endotherm is due to the formation of $NH_4MgPO_4 \cdot H_2O$. Dittmarite is responsible for the endotherm at 248°C. The formation of $Mg_2P_2O_7$ is responsible for the exotherm at 705°C and $Mg_3(PO_4)_2$ for the one at 923°C.

In mortars originally formed with both ADP and STPP, there is a small endotherm around 70°C. At early times, e.g., 5 minutes, the superposition of exothermic and endothermic effects (at about 107°C) results in a double endotherm indicating the presence of schertalite. A small exothermic peak above 800°C may be due to the formation of $Mg(PO_4)_2$ coincident with an endothermal effect due to the melting of $Na_4P_2O_7$ formed from STPP.

In mortars prepared at the low water/solid ratio (1:16), a small endotherm at around 51°C is attributed to stercorite.

Phosphate additions on the hydration of portland cement were investigated by Ma and Brown.[28] There was no apparent literature on the subject prior to their investigation. Strength development in the CaO-SiO_2-P_2O_5-H_2O system may be due to the formation of C-S-H and C-S-P-H gels. The contribution of C-S-H may, however, be a prime factor.

Ma and Brown reported on the mechanical properties obtained with the addition of sodium and calcium phosphates. The portland cement and phosphate were mixed with deionized water and then pressed at 28 MPa. The water/solid ratio varied from 0.176–0.250 and the phosphate/cement ratio varied from 0.10–0.42.

Calorimetric curves for portland cement (OPC) and samples containing $(NaPO_3)_n$ and $(NaPO_3)_n \cdot Na_2O$ are shown in Fig. 29 for the first four hours of hydration. Phosphate addition increased the rates of hydration and more heat was evolved at the beginning of hydration. The $(NaPO_3)_n \cdot Na_2O$ had a greater effect than $(NaPO_3)_n$. At times greater than 4 hours (Fig. 30), OPC samples exhibit a second peak and exceed the total heat developed by the phosphate-modified cements at about 30 hours. Generally, the phosphate-modified cement samples had higher flexural strengths [except $(NaPO_3)_n \cdot Na_2O$]. In some cases, calcium phosphate-modified cements also give superior strength results.

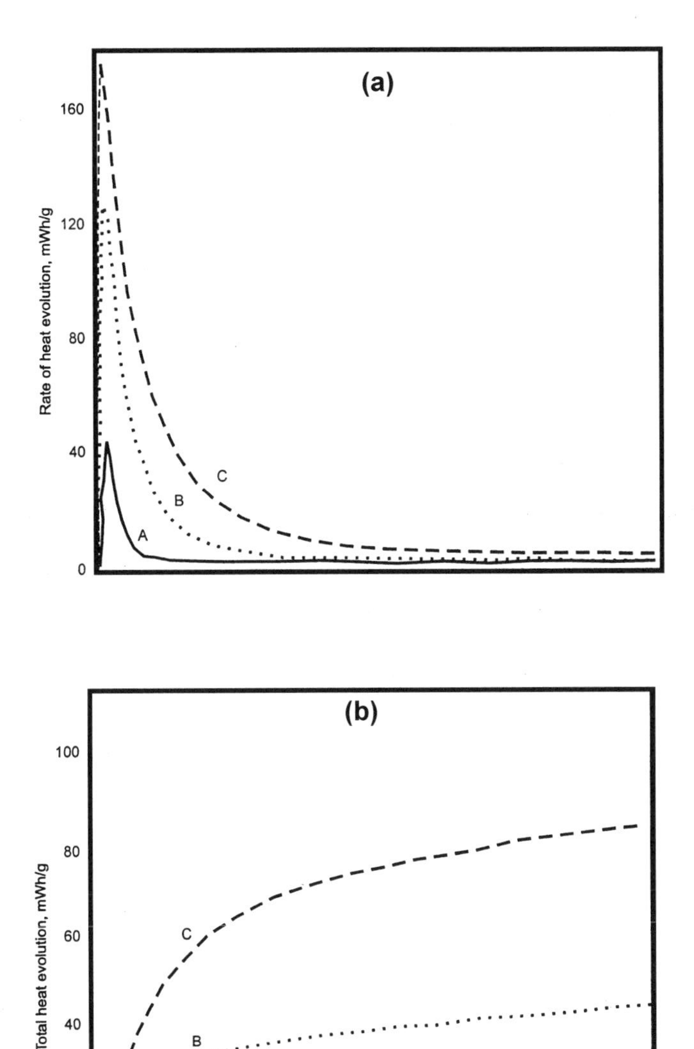

Figure 29. Calorimetric curves showing (*a*) rates of heat evolution and (*b*) total heat evolved as a function of time for portland cement (OPC) and portland cement modified by $(NaPO_3)_n$ and $(NaPO_3)_n \cdot Na_2O$ during the first four hours of hydration. *A:* OPC; *B:* OPC modified by 10% $(NaPO_3)_n$; *C:* OPC modified by 10% $(NaPO_3)_n \cdot Na_2O$.[27]

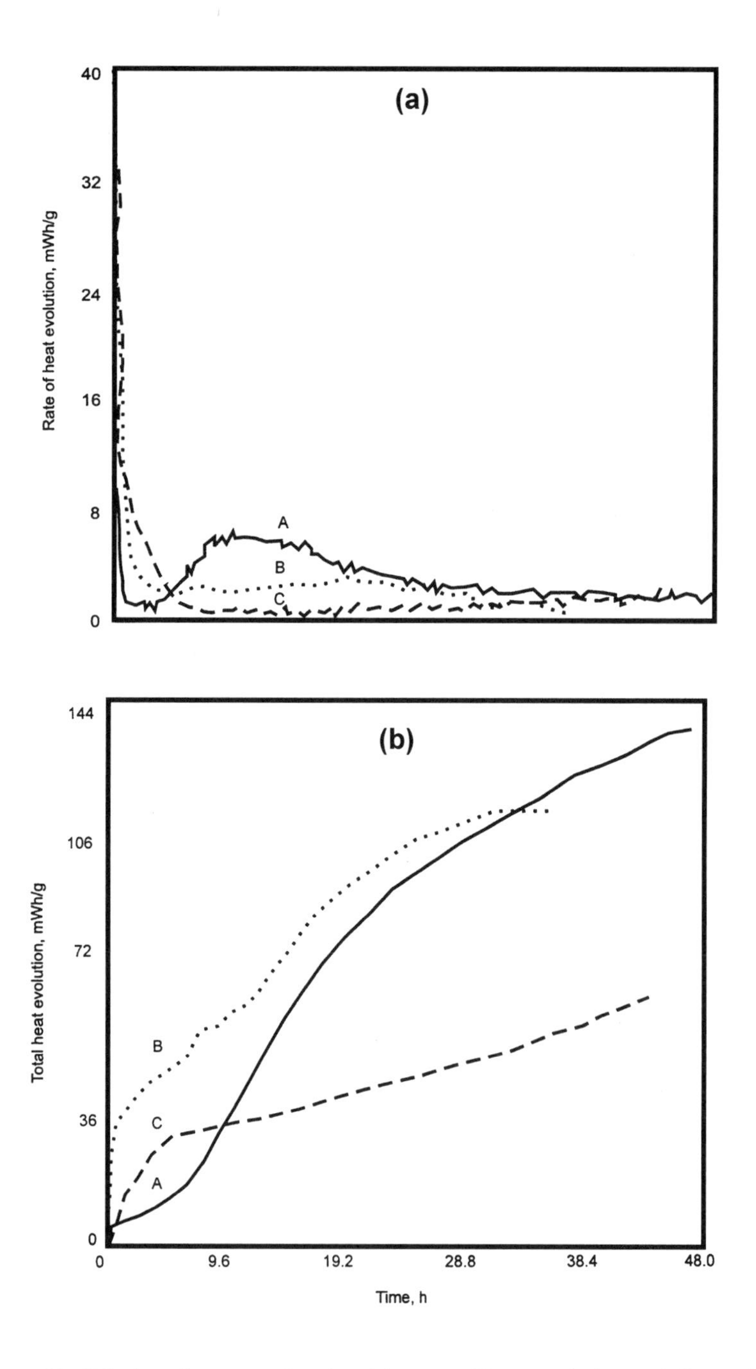

Figure 30. Calorimetric curves showing (*a*) rates of heat evolution and (*b*) total heat evolved as a function of time for portland cement (OPC) and portland cement modified by $Na_5P_3O_{10}$ and $(NaPO_3)_3$ during the first 48 hours of hydration. *A:* OPC; *B:* OPC modified by 10% $Na_5P_3O_{10}$; *C:* OPC modified by 10% $(NaPO_3)_3$.[27]

The heat evolved (over 24 hrs) for calcium pyrophosphate-modified cements is shown in the calorimetric curves in Fig. 31. Phosphate additions of 10, 25, and 40% were used. The modified cements have only a minor effect on the rate of heat evolution. Progressive additions of $2CaO{\bullet}P_2O_5$ reduce the total heat evolved per unit mass of sample.

The hydration products formed in the systems described above were x-ray amorphous. Hydrothermal treatment of the calcium phosphate-modified OPC resulted in the formation of crystalline hydroxyapatite.

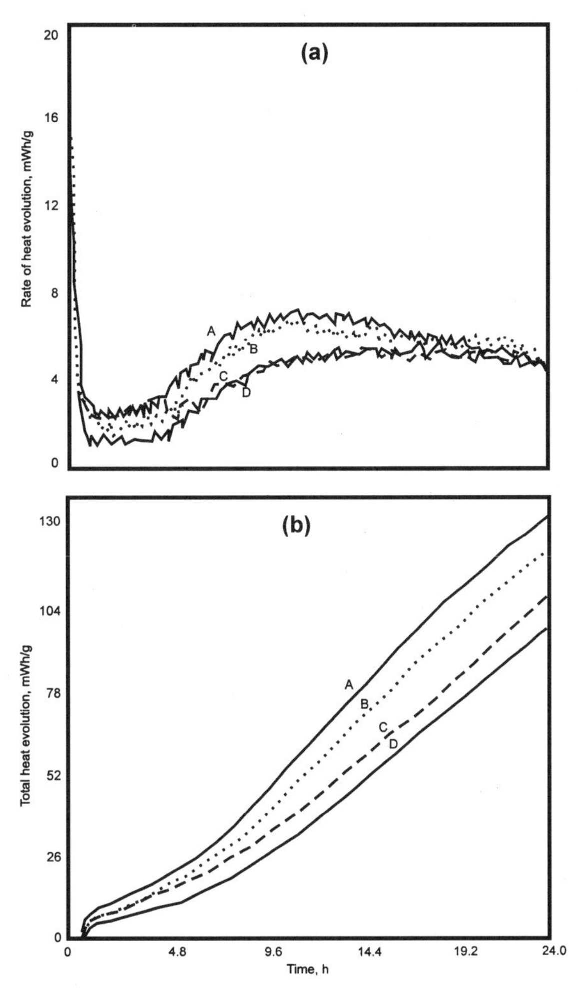

Figure 31. Calorimetric curves showing (*a*) rates of heat evolution and (*b*) total heat evolved as a function of time for OPC and OPC modified by $2CaO{\bullet}P_2O_5$ during the first 24 hours of hydration. *A:* OPC; *B:* 10% $2CaO{\bullet}P_2O_5$; *C:* 25% $CaO{\bullet}P_2O_5$; *D:* 40% $2CaO{\bullet}P_2O_5$.[27]

7.0 HYDROXYAPATITE

Hydroxyapatites have been studied extensively due to their similarity in composition to bone. Ten Huisen and Brown have described the formation of hydroxyapatite by low temperature cementitious reactions between various calcium phosphate precursors such as tetracalcium phosphate, $Ca_4(PO_4)_2O$, and brushite ($CaHPO_4{\bullet}2H_2O$).[29] The reaction involving the formation of stoichiometric hydroxyapatite (Ca/P = 1.67) (after Brown and Chow[30]) may be represented by the following reaction:

Eq. (7) $$2CaHPO_4 + 2Ca_4(PO_4)_2O \rightarrow Ca_{10}(PO_4)_6(OH)_2$$

The formation of calcium-deficient hydroxyapatite (Ca/P = 1.50) can be represented as follows:

Eq. (8) $$3CaHPO_4 + 1.5Ca_4(PO_4)_2O \rightarrow Ca_9(HPO_4)(PO_4)_5OH + \tfrac{1}{2}H_2O$$

Isothermal calorimetry results (37.4°C) are shown in Fig. 32. A two-step reaction mechanism was proposed. The first peak is associated with consumption of all the brushite and some tetracalcium phosphate in the formation of noncrystalline calcium phosphate and nanocrystalline hydroxyapatite. The second peak is associated with the consumption of the remaining tetracalcium phosphate and the calcium phosphate intermediate.

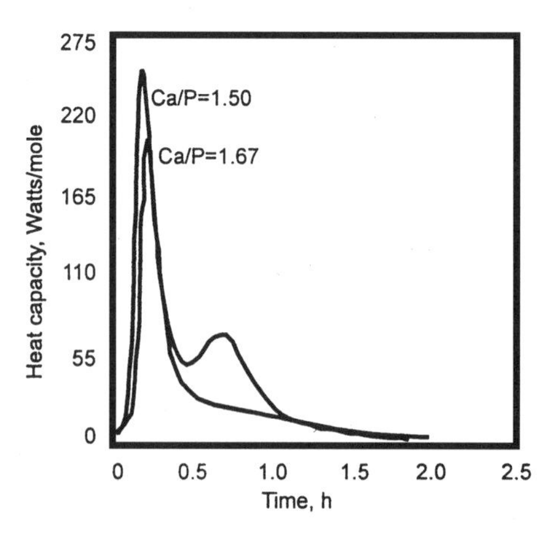

Figure 32. Heat evolution (watts per mol of hydroxyapatite formed) vs time of hydroxyapatite formation at 37.4°C from $CaHPO_4{\bullet}2H_2O$ and $Ca_4(PO_4)_2O$.[28]

Martin and Brown investigated the hydration of tetracalcium phosphate (C_4P) due to the fact that it is the only calcium phosphate more basic than hydroxyapatite.[31] Its hydration was considered to be conceptually similar to that of the silicates. The reaction was considered to be represented by the following equation:

Eq. (9) $$3Ca_4(PO_4)_2O + 3H_2O = Ca_{10}(PO_4)_6(OH)_2 + 2Ca(OH)_2$$

Heat evolution curves for the reaction at 25, 38, 55, and 70°C are presented in Fig. 33. Heat evolution is undetectable after 5 hours at all temperatures.

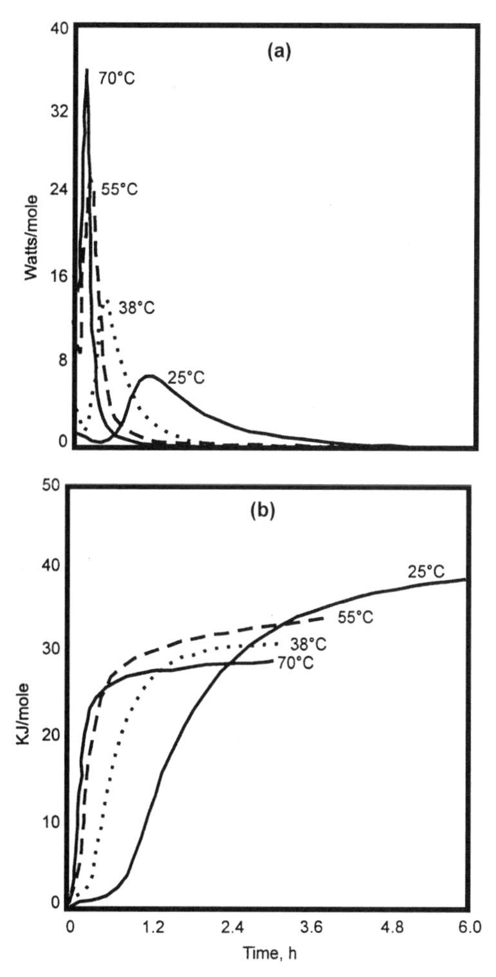

Figure 33. Hydration of $Ca_4(PO_4)_2$ at 25, 38, 55, and 70°C: (*a*) calorimetric rate curves; (*b*) calorimetric rate of reaction curves.[30]

Two hydration peaks are observed (similar to C_3S hydration). An initial peak is observed on mixing; peak height is directly related to the temperature of hydration. An induction period (0.5 hr) was observed at 0.5 hour at 25°C. The minimum rates of heat evolution (at temperatures > 25°C) increase with temperature. A slower rate of heat evolution, but a larger total amount of heat released was observed at 25°C. This behavior was attributed to the retrograde solubility of hydroxyapatite along with incongruent dissolution of the acidic constituent.

REFERENCES

1. Bushnell-Watson, S. M., and Sharp, J. H., The Application of Thermal Analysis to the Hydration and Conversion Reactions of Calcium Aluminate Cements, *Materiales de Construccion,* 42:15–32 (1992)
2. Majumdar, A. J., Edmonds, R. N., and Singh, B., Hydration of Secar 71 Aluminous Cement in the Presence of Granulated Blast Furnace Slag, *Cement Concr. Res.,* 20:7–14 (1990)
3. Brown, R. A., and Cassel, B., High Alumina Cements: Background and Application of Thermal Analysis Methods, *Am. Lab.,* 9:45–56 (1977)
4. Midgley, H. G., The Use of Thermal Analysis Methods in Assessing the Quality of High Alumina Cement Concrete, *J. Thermal Analysis,* 13:515–524 (1978)
5. Midgley, H. G., The Mineralogy of Set High Alumina Cement, Trans., *Brit. Ceram. Soc.,* 66:161–168 (1967)
6. Midgley, H. G., The Mineralogical Examination of Set Portland Cement, *4th Int. Symp. Chem. Cement,* Paper IV-82, 1:479–490, WA (1960)
7. (a) Midgley, H. G., and Midgley, A., The Conversion of High Alumina Cement, *Mag. Concr. Res.,* 27:59–77 (1975); (b) Midgley, H., Measurement of High Alumina Cement-Calcium Carbonate Reactions Using DTA, *Clay Minerals,* 19:857–864 (1984)
8. Ramachandran, V. S., and Feldman, R. F., Hydration Characteristics of Monocalcium Aluminate at a Low Water/Solid Ratio, *Cement Concr. Res.,* 3:729–750 (1973)
9. Talaber, J., Revay, M., and Wagner, Z., Thermal Analytical Measurements of Hungarian High-Alumina Cement Concretes, *2nd European Symp. on Thermal Analysis,* pp. 498–500, Univ. Aberdeen, Heyden (1981)
10. Das, S. K., Mitra, A., and Das Poddar, P. K., Thermal Analysis of Hydrated Calcium Aluminates, *J. Thermal Analysis,* 47:765–744 (1996)

11. Kissinger, H. E., Reaction Kinetics in Differential Thermal Analysis, *Anal. Chem.,* 29:1702–1796 (1957)
12. Satava, V., and Veprek, O., Investigations of Hydrothermal Reactions by Means of Differential Thermal Analysis II: Reactions of Calcium Aluminates and Gypsum, *Zement-Kalk-Gips,* 10:424–426 (1975)
13. Greening, N. R., Copeland, L. E., Verbeck, G. J., Modified Portland Cement and Process, p. 6, U.S. Patent 3,628,973 (1971)
14. Uchikawa, H., and Uchida, S., The Hydration of $11CaO{\bullet}7Al_2O_3{\bullet}CaF_2$ at 20°C, *Cement Concr. Res.,* 2:681–695 (1972)
15. Uchikawa, H., and Uchida, S., The Influence of Additives upon the Hydration of Mixtures of $11CaO{\bullet}7Al_2O_3{\bullet}CaF_2$, $3CaO{\bullet}SiO_2$ and $CaSO_4$ at 20°C, *Cement Concr. Res.,* 3:607–624 (1973).
16. Uchikawa, H., and Tsukiyama, K., The Hydration of Jet Cement at 20°C, *Cement Concr. Res.*, 3:263–277 (1973)
17. Beaudoin, J. J., and Ramachandran, V. S., Strength Development in Magnesium Oxychloride and Other Cements, *Cement Concr. Res.*, 5:617–630 (1975)
18. Cole, W. F., and Demediuk, T., X-ray, Thermal and Dehydration Studies on Magnesium Oxychlorides, *Australian J. Chem.*, 8:234–251 (1955)
19. Beaudoin, J. J., and Ramachandran, V. S., Strength Development in Magnesium Oxysulfate Cement, *Cem. Concr. Res.*, 8:103–112 (1978)
20. Kacker, K. P., Rai, M., and Ramachandran, V. S., Suitability of Almora Magnesite for Making Magnesium Oxychloride Cement, *Chem. Age of India*, 20:506–510 (1969)
21. Ramachandran, V. S., Kacker, K. P., and Rai, M., Chloromagnesial Cement Prepared from Calcined Dolomite, *Zh. Prikl. Khim*, 40:1687–1695 (1967)
22. Newman, E. S., Preparation and Heat of Formation of a Magnesium Oxysulfate, *Phys. and Chem. A, J. Res., Natl. Bur. Stand.*, 68A:645–650 (1964)
23. Sorrell, C. A., Suggested Chemistry of Zinc Oxychloride Cements, *J. Am. Ceram. Soc.,* 60:217–220 (1977)
24. Suguma, T., and Kukacka, L. E., Characteristics of Magnesium Phosphate Cements Derived from Ammonium Polyphosphate Solutions, *Cement Concr. Res.*, 13:499–506 (1983)
25. Neiman, R., and Sarina, A. C., Setting and Thermal Reactions of Phosphate Investments, *J. Dental Res.*, 59:1478–1485 (1980)
26. Abdelrazig, B., Sharp, J. H., Siddy, P. A., and El-Jazairi, B., Chemical Reactions in Magnesia-Phosphate Cement, *Proc. Brit. Ceram. Soc.*, No. 35, pp. 141–154 (1984)

27. Abdelrazig, B., Sharp, J. H., and El-Jazairi, B., The Chemical Composition of Mortars Made from Magnesia-Phosphate Cement, *Cement Concr. Res.*, 18:415–425 (1988)
28. Ma, W., and Brown, P. W., Effect of Phosphate Additions on the Hydration of Portland Cement, *Adv. Cement Res.*, 6:1–12 (1994)
29. Ten Huisen, K. S., and Brown, P. W., Formation of Variable Composition Hydroxyapatite, *Proc. 5th World Biomaterials Congr.*, *Toronto*, p. 366 (1996)
30. Brown, W. E., and Chow, L. C., A New Calcium Phosphate Water-Setting Cement, Cements Research Progress, 1987, (P. W. Brown, ed.), *Am. Ceram. Soc.,* Westerville, OH (1988)
31. Martin, R. I., and Brown, P. W., Hydration of Tetracalcium Phosphate, *Adv. Cement Res.*, 5:119–125 (1993)

11

Gypsum and Gypsum Products

1.0 INTRODUCTION

The use of gypsum to control setting in Portland cement accounts for considerable quantities of the use of this material.[1] Control of the reaction rate of tricalcium aluminate (C_3A), the constituent of cement that reacts most rapidly with water, is most commonly achieved through the addition of gypsum (sometimes hemihydrate is also used) to commercial portland cement.[2] This material is normally added to the cement clinker (in amounts of approximately 2 to 3%) before grinding. The cement manufacturers usually specify a sulfur trioxide content of about 36%. Excess sulfate in the form of hemihydrate can cause flash set in portland cement.

Calcined gypsum (e.g., plaster of Paris formed by grinding and heating gypsum to about 150–190°C) has many uses especially in the construction industry. Gypsum wallboard is widely used in the North American housing industry. At temperatures greater than 190°C, the soluble anhydrite forms followed by the formation of an insoluble anhydrite. Dehydration of the gypsum often results in the formation of a mixture of hemihydrate and anhydrite. All plasters eventually revert to gypsum on setting, the rate of transformation being dependent on the conditions of calcination.

Gypsum has useful fire-resistant properties due to its water of crystallization (20.9 %). Gypsum plaster is widely used as an insulating material for protecting columns and beams of wooden materials from the high temperatures that develop during a fire.

It is apparent that rapid methods of analysis for gypsum, hemihydrate, and anhydrite are of significant practical interest. Thermal methods of analysis appear to be particularly suited for this task.[3]

This chapter describes various aspects of thermal studies conducted on $CaSO_4 \cdot 2H_2O$ and α and β forms of $CaSO_4 \cdot ½H_2O$, using differential thermal analysis, differential scanning calorimetry, and thermogravimetry. The effect of environmental conditions on the quantitative deterioration of the various calcium sulfate compounds is also examined. The development of more recent techniques such as controlled reaction thermal analysis is also presented.[4]

2.0 DIFFERENTIAL THERMAL ANALYSIS (DTA) AND DIFFERENTIAL SCANNING CALORIMETRY (DSC)

Typical DTA curves of $CaSO_4 \cdot 2H_2O$ in a N_2 atmosphere are presented in Figs. 1 and 2 at cell pressures of 760, 590, 380, 150, and 197 torr.[5] The two endotherms in Fig. 1 occurring at 150 and 190°C are typical of numerous thermograms appearing in the literature, but with less overlap possibly due to sample size effects. A small endothermic dent observed by some investigators, immediately after the second endotherm, is not indicated in Fig. 1. An exothermic peak observed at 375°C is characteristic of hemihydrate.

Three transformations ascribed to the peak temperatures are described by the following three reactions occurring at 760 torr:

Eq. (1) $$CaSO_4 \cdot 2H_2O\ (150°C) \leftrightarrow \beta\text{-}CaSO_4 \cdot ½H_2O + {}^3\!/_2H_2O$$

Eq. (2) $$\beta\text{-}CaSO_4 \cdot ½H_2O\ (197°C) \leftrightarrow \gamma\text{-}CaSO_4 + ½H_2O$$

Eq. (3) $$\gamma\text{-}CaSO_4\ (375°C) \rightarrow \beta\text{-}CaSO_4$$

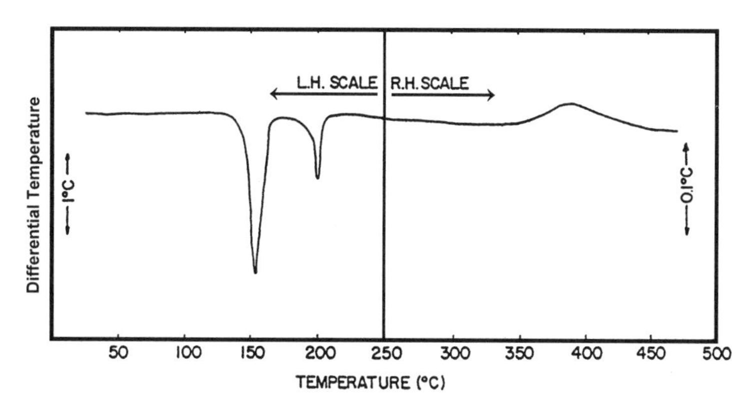

Figure 1. DTA curve of $CaSO_4{\bullet}2H_2O$ in a N_2 atmosphere at 760 torr.[5]

Reactions (1) and (2) are dehydration processes. The exothermic reaction (3) represents a lattice modification from the hexagonal to orthorhombic form. Decreases in peak temperatures (150 to 123°C and 197°C to an undetectable position) occurred when the gas pressure in the cell was reduced to 1 torr (Fig. 2). The peak for the exothermic effect was not sensitive to changes in gas pressure. The DTA curve for β-$CaSO_4{\bullet}\frac{1}{2}H_2O$ has a sharp endothermic peak at 195°C and a small exothermic peak at 375°C (Fig. 3). The position and shape of the small exothermic peak were not significantly affected by changes in atmospheric pressures within the DTA cell. The endothermic peak has the same pressure-temperature behavior previously noted for the second endothermic curve of gypsum (a shift occurs from 195°C to 132°C for 760 torr and 1 torr respectively).

The position of the endothermic effect also has a similar pressure-temperature behavior. The endothermic peak shifts from 198°C to 132°C at pressures of 760 torr (A) and 1 torr (B) (Fig. 4). The exothermic peak shifts from 217°C to 163°C.

The differential thermogram of an equal molar mixture of α- and β-$CaSO_4{\bullet}\frac{1}{2}H_2O$ (Fig. 5) includes the small exothermic effect of the respective components. This indicates that this method can be used to identify the two forms in the presence of each other. The DSC results (for gypsum) reported by Clifton[5] (heating rate 5°C/min) indicate the presence of a single endothermic effect (145°C). Thermal equilibrium was likely not obtained (with respect to the sample reference cells) after the first dehydration step was complete and before the second had commenced. Curves for α-, β-$CaSO_4{\bullet}\frac{1}{2}H_2O$ were essentially identical even with hermetically sealed crucibles.

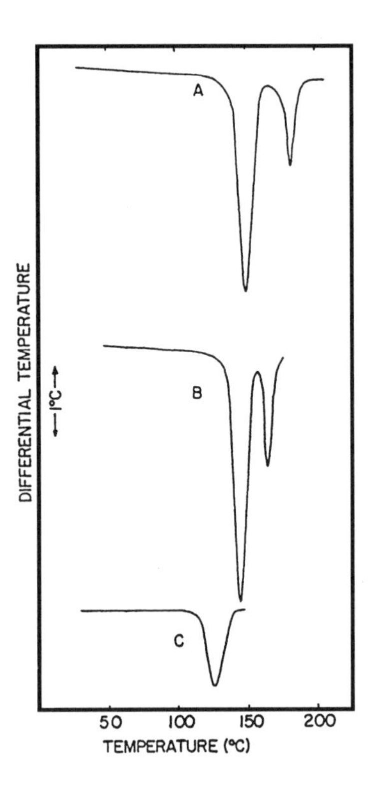

Figure 2. Differential thermograms of $CaSO_4 \cdot 2H_2O$ at reduced pressures in the DTA cell.[5]

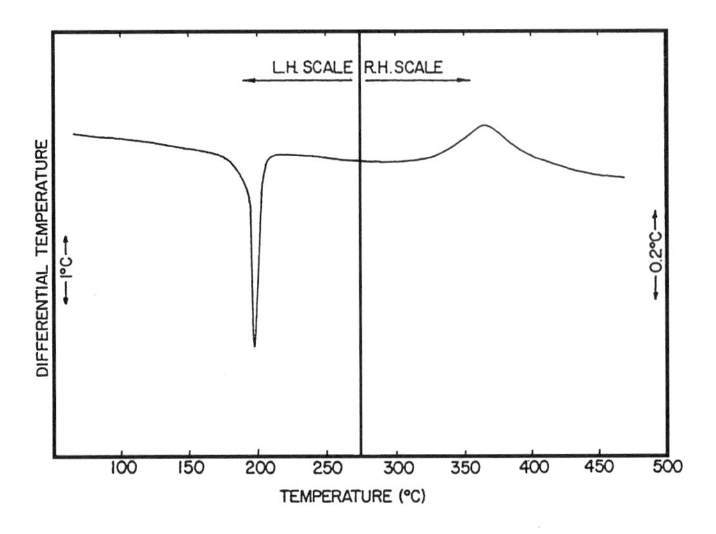

Figure 3. DTA curve of β-$CaSO_4 \cdot \frac{1}{2}H_2O$ in a N_2 atmosphere at 760 torr.[5]

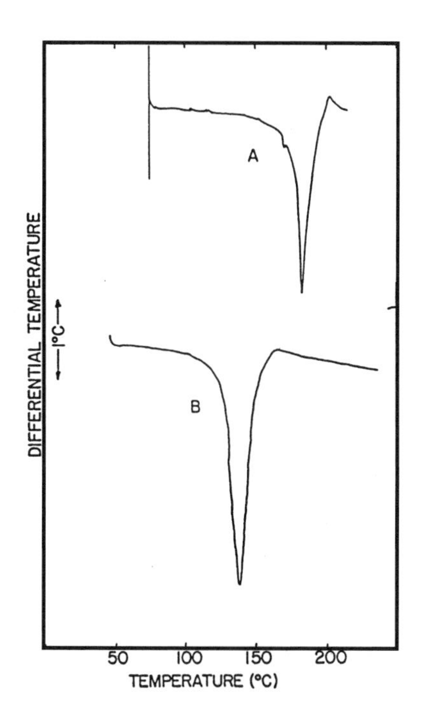

Figure 4. Differential thermograms of α-$CaSO_4 \cdot ½H_2O$ with reduced pressures in the DTA cell.[5]

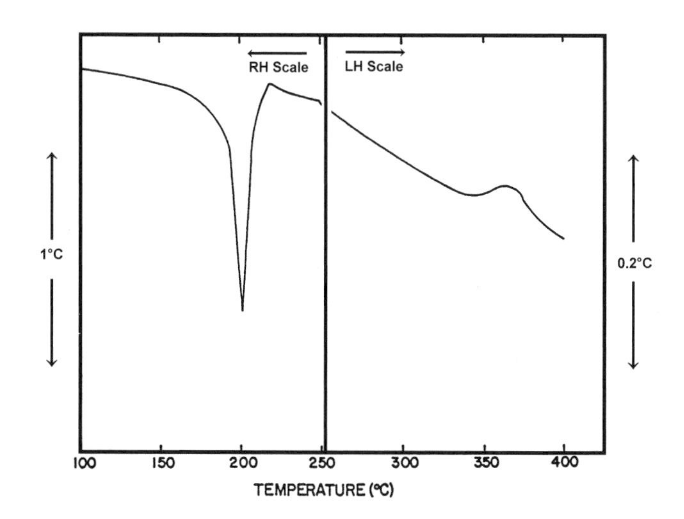

Figure 5. DTA curve of an equal molar mixture of α- and β-$CaSO_4 \cdot ½H_2O$ at 760 torr.[5]

3.0 THERMOGRAVIMETRIC ANALYSIS (TG)

There is no well-defined inflection point in the mass loss-temperature curve for $CaSO_4 \cdot 2H_2O$ (Fig. 6) although the change in slope is compatible with a two-stage hydration process. The small changes in slope observed between the two segments of the mass-loss curve indicate that there is little difference in the dissociative mechanisms of the dehydration reaction. Interrupting the TGA cycle (heating rate, 3°C/min) at a mass-loss equivalent of 1½ moles of H_2O produced a mixture of about 5% $CaSO_4 \cdot 2H_2O$ and γ-$CaSO_4$ and 95% β-$CaSO_4 \cdot ½H_2O$ indicating the difficulty of obtaining a precise end point for the conversion to hemihydrate. The TGA curves for α- and β-$CaSO_4 \cdot ½H_2O$ were essentially identical. Dehydration (as evidenced by a small mass change) was initiated at room temperature and continued at a relatively small rate up to about 100°C. This was attributed to the loss of loosely held zeolitic water prior to the removal of lattice water, corresponding to the more pronounced slope of the mass-loss–temperature curve at higher temperatures. The endothermic effects of $CaSO_4 \cdot 2H_2O$ and α- and β-$CaSO_4 \cdot 2H_2O$ (at 760 torr) were found to occur at higher temperatures (DTA) than the corresponding effects associated with the mass-loss process (TG). The apparent discrepancies are probably due to both differences in the heating rates and physical differences in the TG and DTA cells (e.g., differences in their respective heat capacities).

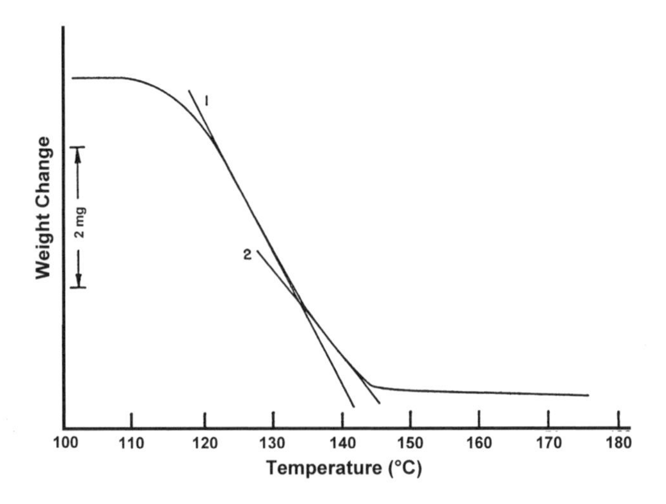

Figure 6. TG curve of $CaSO_4 \cdot 2H_2O$. Line 1 is drawn tangent to the curve corresponding to the dehydration of $CaSO_4 \cdot 2H_2O$ to β-$CaSO_4 \cdot ½H_2O$. Line 2 indicates the dehydration of β-$CaSO_4 \cdot ½H_2O$ or γ-$CaSO_4$.[5]

4.0 DEHYDRATION OF GYPSUM

Calcining gypsum at temperatures below 100°C for various periods (1 hour to 22 days) gives rise to a limiting value of mass loss around that caused by the evolution of 1.5 molecules of water, irrespective of the time of heating (15 days at 70°C, 2 days at 80°C, and 12 hours at 90°C).[6] DTA curves for isothermally heated gypsum are presented in Figs. 7 and 8. At relatively low temperatures (up to 90°C), dehydration of gypsum does not proceed beyond the hemihydrate state for any heating period (Fig. 7). Above 90°C, a continuous loss of mass with time is prolonged, resulting in the formation of γ-anhydrite (soluble) as a final product (Fig. 8). Trace amounts of residual water are reached towards the end of the dehydration process. The residual water is driven off completely below the conversion temperature of the soluble (γ-form) to the insoluble anhydrite (β-form) probably because of the comparatively long periods of heating used in these experiments. In other work, the residual water is considered attached to the γ-$CaSO_4$ and cannot be expelled below the temperature of conversion of γ to β-$CaSO_4$.[7]

Additional study on the effect of prolonged heating was reported by Khalil, et al.,[8] who noted that hemihydrate forms between 100 and 220°C when gypsum is heated at a constant rate (10°C/min). However, complete formation was confirmed to occur below 100°C with long heating times. The disappearance of the hemihydrate and the formation of the soluble anhydrite (γ-$CaSO_4$) occurs with additional heating at 250°C for a few minutes or between 100 and 130°C for a few hours. On further heating (around 360°C), insoluble anhydrite (β-$CaSO_4$) appears and transforms to the α-form at 1230°C (Fig. 9).

The presence of $CaSO_4$ persists up to 1300°C where it begins to partially decompose at 1350°C. It was reconfirmed that the heating at 120°C and 130°C causes a marked decrease in the hemihydrate content, and the simultaneous increase of the γ-anhydrite is suggested from the notable change in the magnitude of the endotherm. Heating at 110°C, 120°C, and 130°C for more than 5 hours causes the complete disappearance of gypsum (the product being mainly γ-anhydrite as the heating time is prolonged). It is noted that re-heating of previously heated gypsum samples results in markedly reduced DTA peaks compared to those for previously unheated gypsum. This could be related to disorder in the lattice of previously calcined gypsum samples.[9] A transformation from the monoclinic to the orthorhombic form could also be responsible for the phenomenon of lower dehydration of the previously calcined samples.

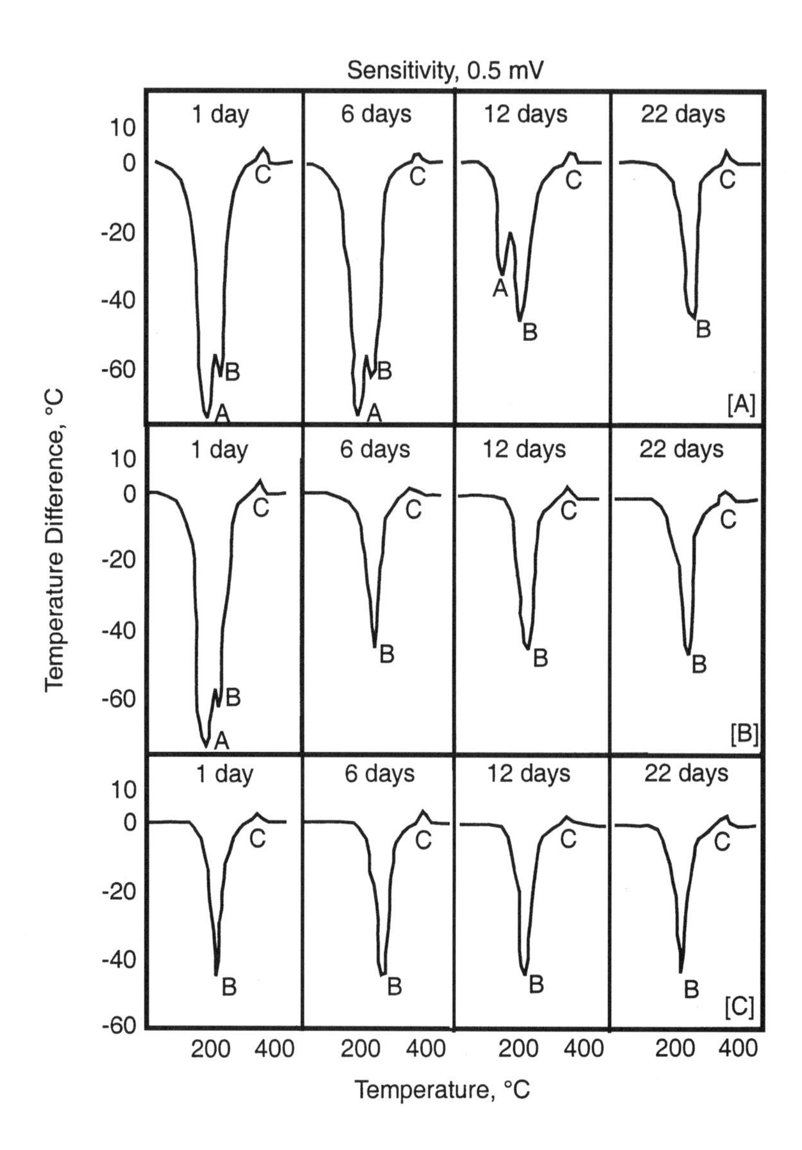

Figure 7. Differential thermal analysis curves of calcined gypsum. (*a*) calcined at 70°C; (*b*) calcined at 80°C; (*c*) calcined at 90°C.[6]

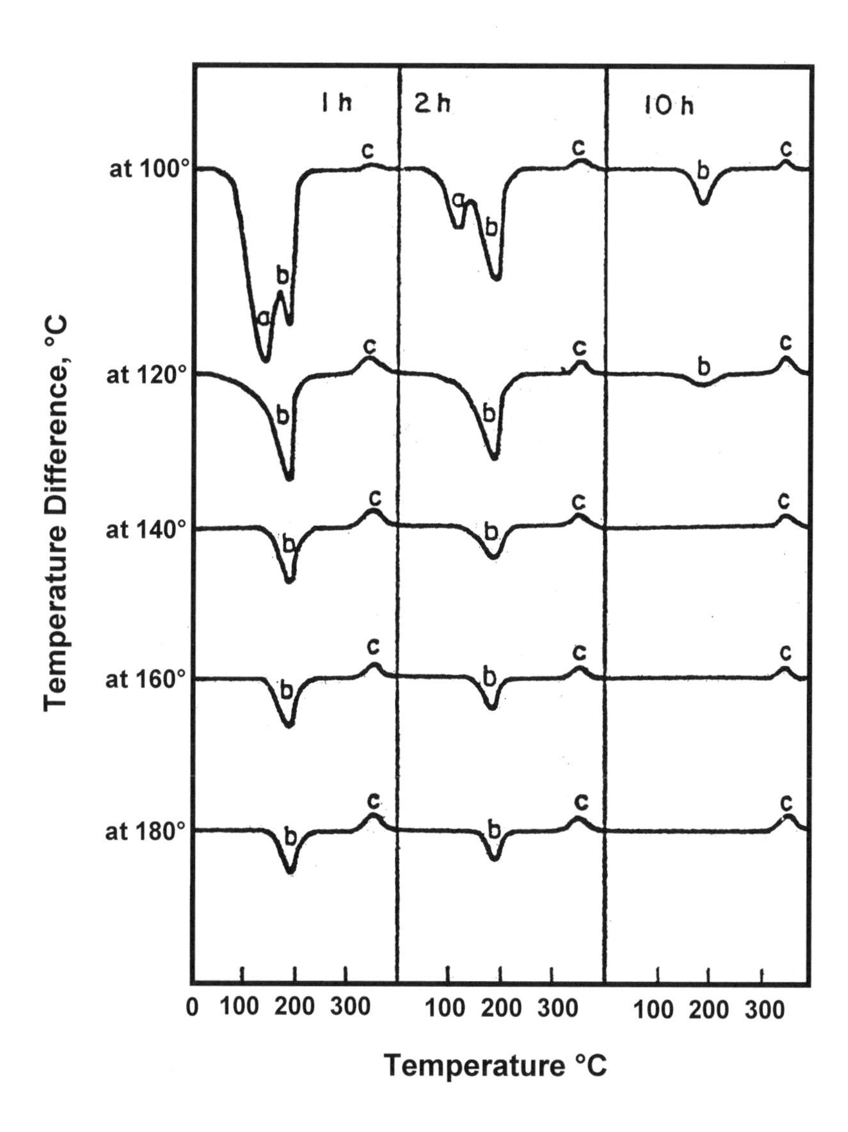

Figure 8. Differential thermograms of gypsum calcined at various temperatures for 1, 2, and 10 hours.[6]

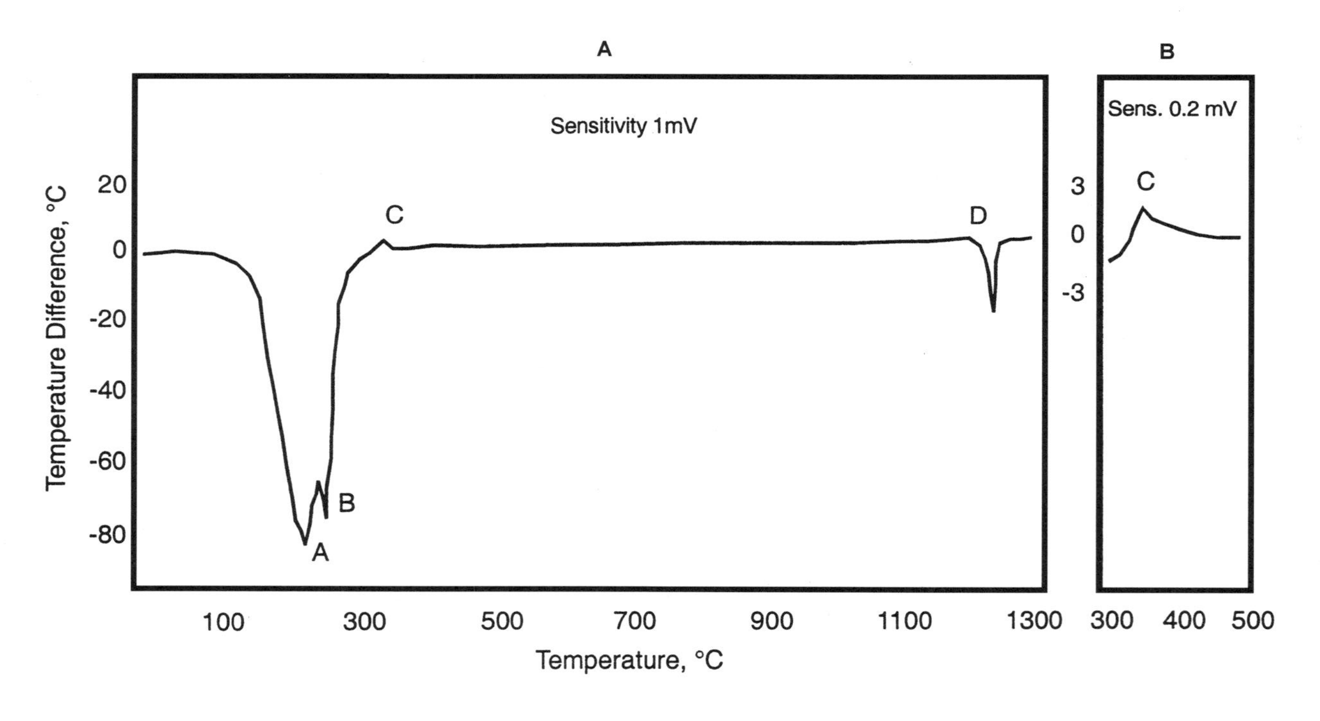

Figure 9. The differential thermal curve for gypsum.[8]

Khalil and Gad's results[8] indicate that heating gypsum continuously at a constant rate of 10°C/min results in a loss of one and a half molecules of its water of crystallization or about 15.3%. This leads to hemihydrate formation between 180 and 220°C. The γ-anhydrite formed between 220 and 300°C with a total loss of nearly all the combined water of crystallization amounting to about 20.4%. Complete hemihydrate formation occurs when gypsum is heated at relatively lower temperatures for longer periods. Heating at temperatures ranging between 100 and 130°C causes the loss of nearly all of the two molecules of water forming the γ-anhydrite with varying amounts of hemihydrate. Longer periods at these temperatures favor the formation of γ-anhydrite. Heating at higher temperatures causes the conversion of γ-anhydrite to the β-form at 360°C. The latter changes to the α- modification at about 1230°C. Further heating up to 1300°C shows insignificant loss in mass other than that due to the loss of the water of crystallization.

5.0 SIMULTANEOUS TG-DTG-DTA

Isa and Oruno describe a method that enables identification of intermediates (in the gypsum dehydration process) more easily.[10] The method involves the use of simultaneous TG-DTG-DTA under various sealed atmospheres corresponding to three systems—open completely, sealed, and quasi-sealed. Endothermic DTA peaks appear earlier (129 and 133°C) than the point of decreasing TG. This technique, resembling the quasi-isothermal and isobaric thermogravimetry (Q-TG), is superior to the latter in that it needs less of the sample.

Typical results for the open system are shown in Fig. 10. The first endothermic peak from the DTA, due to dehydration, begins at 102°C. The TG and DTG curves begin at 107°C and 102°C respectively (for the simultaneous technique) indicating the coincidence of the thermal and thermogravimetric behaviors of hydration. This temperature is, however, dependent on the sample amount, heating rate, the sensitivity of the TG, DTG, and DTA, and the signal to noise ratio. The DTA exhibits two endotherms (at 122 and 130°C), and there are considered to be two reaction steps.

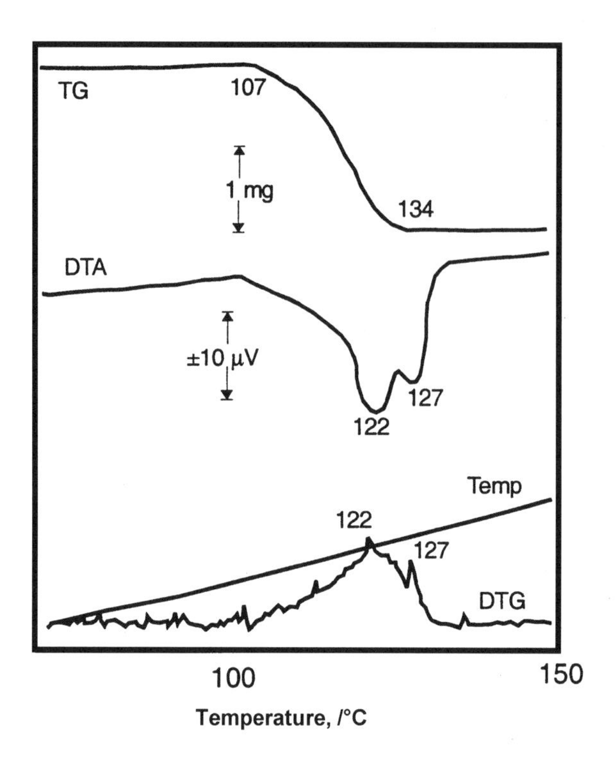

Figure 10. Simultaneous TG-DTG-DTA of gypsum in an open system sample mass is 0.067 m mol. Heating rate, 1°C/min.[10]

Results for the completely sealed system are given in Fig. 11. The first step of the TG appears to be a rapid decreasing reaction which corresponds to the loss of 3/2 moles of H_2O and the formation of $CaSO_4 \cdot \frac{1}{2}H_2O$. The DTA results indicate endothermic effects at 129, 133, 162, and 219°C, respectively. The two low temperature peaks (129°, 133°C) occur without an apparent decrease in TG. These peaks are designated O_A and O_B (in Fig. 11). The rapid descent of the first step of the TG is considered to be due to the desorption of water owing to the hydration at the O_A and O_B steps. Desorption continues gradually after these steps, and the equilibrium vapor pressure increases gradually (to about 6 atm at 159°C). The sample pan explodes above the critical temperature. The reaction scheme for the TG descent is considered as follows:

Eq. (4) $CaSO_4 \cdot 2H_2O \rightarrow CaSO_4 \cdot \frac{1}{2}H_2O + \frac{3}{2}H_2O$
Hemihydrate (15.7% loss)

Eq. (5) $CaSO_4 \cdot \frac{1}{2}H_2O \rightarrow CaSO_4 + \frac{1}{2}H_2O$
Anhydrite (5.2% loss)

The sealed pan is opened (above the critical pressure) under the vapor pressure of water, and the vapor is allowed to escape from the sealed pan through the leak. The unreacted water of the reaction (Eq. 4) begins to be liberated rapidly. The rapid descent for the first decomposition step is modified in such a way that the response rate may be slower than the reaction rate or the equilibrium is disturbed (with both rates fast), and the unreacted substance is changed under a superheated condition. The first reaction step was considered to be dehydration of structural water, and the second reaction step to be dehydration of mobile water.

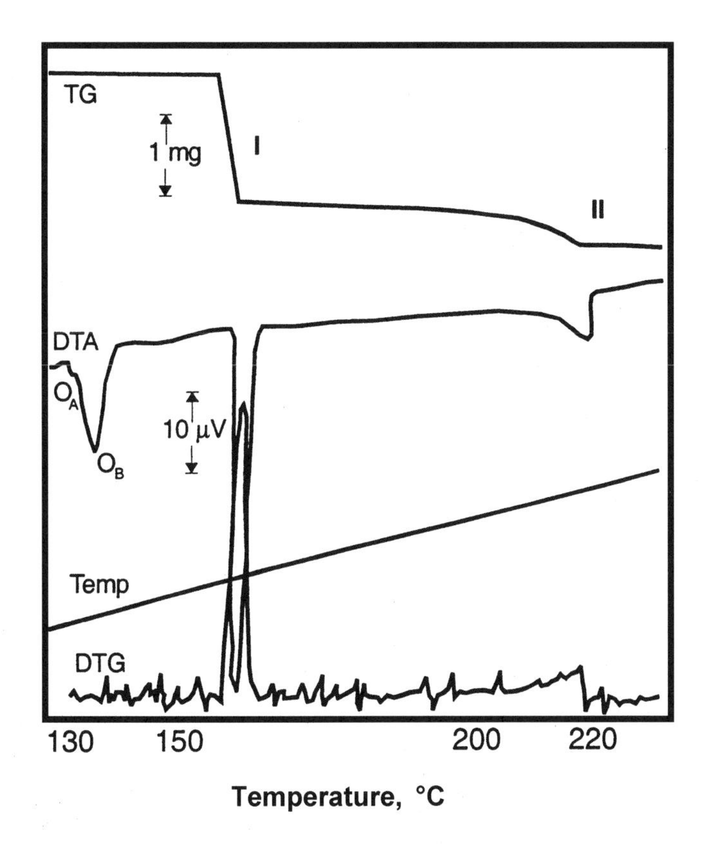

Figure 11. Simultaneous TG-DTG-TGA of gypsum in a completely sealed system sample mass is 0.067 millimole. Heating rate, 1°C/min.[10]

The results of the quasi-sealed method are illustrated in Fig. 12. A quasi-sealed condition was achieved by inserting a thin tungsten wire (12–250 μm in diameter) between the lid and sample cell. The two steps are easily separated using the quasi-sealed conditions. This is clearly seen for the curves obtained with 30 and 18.7 μm tungsten wires. The onset temperatures for the reaction are increased to 125 and 127°C. In the curve of the quasi-sealed system, dehydration first occurs after some minor pressure of water vapor has been reacted. The change in slope of the TG curve is then easily recognizable. It is concluded that two separate steps of $CaSO_4•2H_2O$ can be readily determined by using a quasi-sealed condition obtained by the application of 12–30 μm tungsten wires in the sealing mechanism of the apparatus.

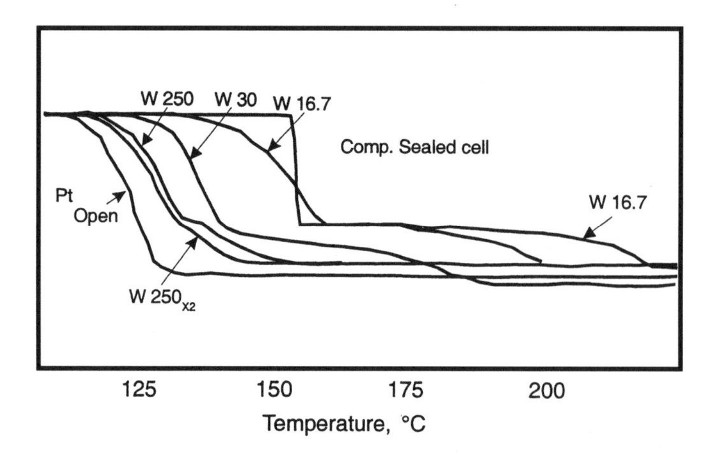

Figure 12. TG results of gypsum in a quasi-sealed, open and completely sealed system.[10]

6.0 CONVERSION REACTIONS

6.1 Dihydrate to β-Anhydrite

The influence of the partial pressure of water vapor during calcium sulfate dihydrate decomposition was explored by Lehmann and Rieke.[11] This work provided confirmation of the direct conversion of the dihydrate to *β*-anhydrite. This is dependent on sample holder and particle packing. It has been shown that a two-step mechanism (even under vacuum) for the decomposition of dihydrate may be operative if the fine-powdered material is placed in a crucible with a high length to diameter ratio.[12] The one-step conversion of dihydrate to *β*-anhydrite requires the quick removal of the water of crystallization. It is possible to conduct a one-step conversion in a DTA-apparatus without forced water vapor removal if the sample is distilled in a monolayer of grains (Fig. 13). The outward diffusion of the liberated water vapor takes place without any difficulties, and with a heating rate of 5°C/min, one peak only results. The presence of a pinhole in the crucible lid increases the potential pressure sufficiently to give two separate peaks, one for the dihydrate to hemihydrate conversion and the other for the hemihydrate to anhydrite reactions. Two peaks also result when the sample bed thickness increases for the open lid condition. This is consistent with an increase in water vapor pressure in the deepest part of the bed. The shape of the peak for the dehydration of the hemihydrate is marked more distinctly. It is clear that it is meaningless to look for correlation between dihydrate habit and peak shapes and lengths if the partial pressure is not completely identical. It is apparent that the DTA curves are not in contradiction when the result is a one-step decomposition of dihydrate to soluble anhydrite at low partial pressures, given that experimental and material parameters influence the partial pressure during dehydration. These experiments cannot prove, however, if the lattice of the hemihydrate is passed over and a direct conversion of dihydrate to soluble anhydrite is possible.

DTA curves for the liberation of non-stoichiometric water of *β*-hemihydrate are presented in Fig. 14. The curve (*p.a.*) refers to the sample of particle size < 20 μm. The influence of the temperature for the dihydrate dehydration and the time for rehydration on the slope of the peak doublet are apparent. The doublet reflects mass loss from both the non-stoichiometric lattice water and the sorbed water associated with the high specific surface area. This suggests that the micro-porosity of the *β*-hemihydrate pseudo-morph is responsible for the appearance of the peak doublet. Examination of Fig. 14 shows a small effect of the temperature of dehydration on the shape of the peak doublet. The first peak can be attributed to the liberation of micropore water, which does not appear to depend on the dehydration temperature. The absorption of lattice water is a slow process as reflected in the much longer time of rehydration required for the growth of the second peak. It is also apparent that the slow water absorption of the lattice depends on the grain size [compare curves (*c*) and (*f*) in Fig. 14].

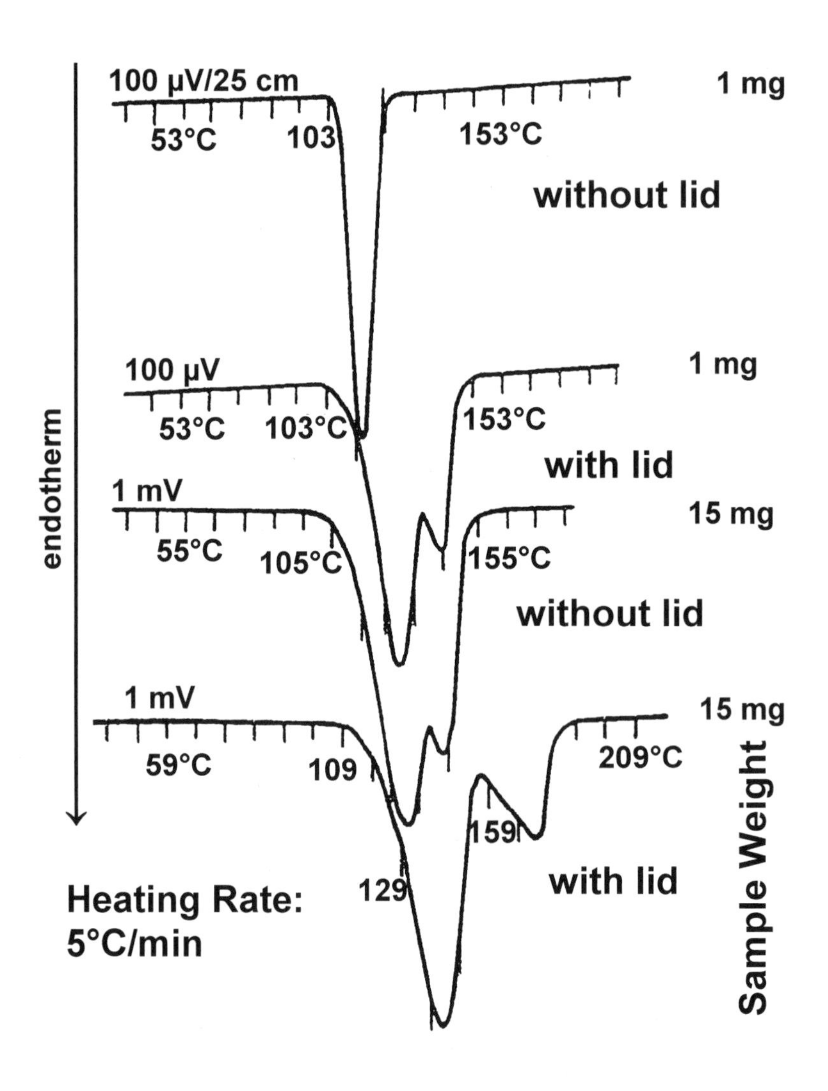

Figure 13. Influence of experimental parameters on the DTA curves of granular gypsum (<63 µm).[11]

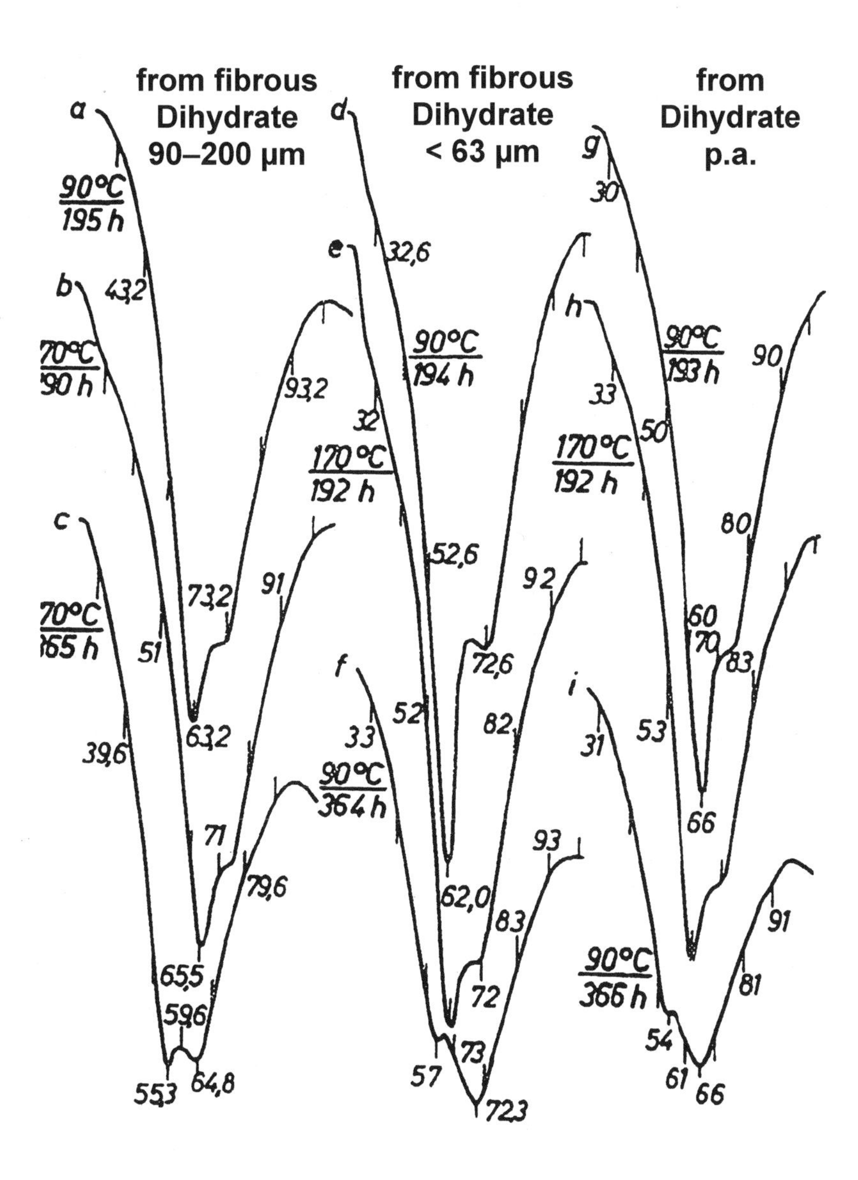

Figure 14. DTA showing the influence of the temperature of dihydrate dehydration and the time of rehydration on the shape of the peak doublet (10 mg, 5°C/min, crucible with lid).[11]

The influence of the experimental parameters on the shape of the peak doublet in the DTA curves of the liberation of the non-stoichiometric water of β-hemihydrate is indicated in Fig. 15. The influence of the partial pressure of the water vapor or the speed of liberation of the non-stoichiometric water can be discerned from the curves. Curve (1) was obtained without a crucible lid. The water can be driven off much faster and a low partial pressure of water vapor can be obtained. A higher partial pressure results in a displacement of the peak to higher temperature and produces improved resolution of the second peak. The areas of curves (2) and (3) are approximately the same if the decreased heating rate and reduced chart speed are taken into account. The sensitivity of the measurements was held constant. Curve (3) illustrates that a lower heating rate results in a lower partial pressure of water vapor, lower peak temperatures, and a decrease in the intensity of the second peak.

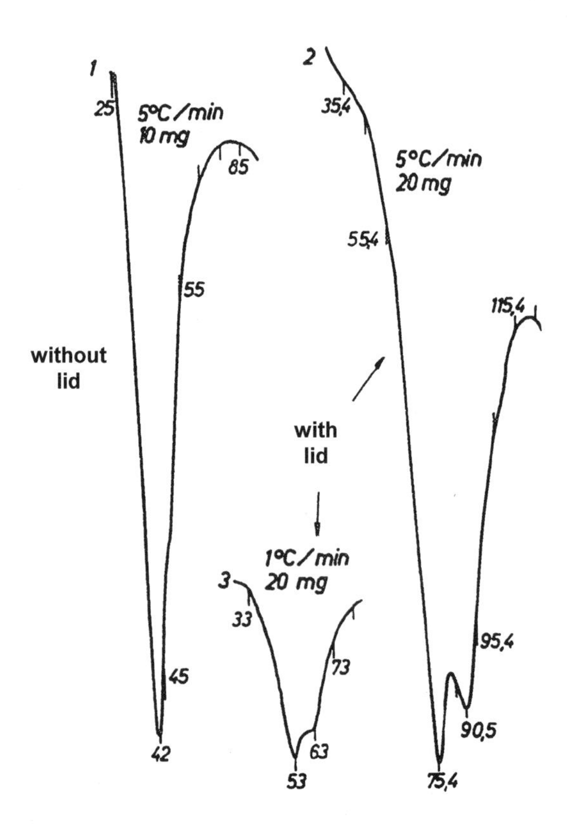

Figure 15. The influence of experimental parameters on the shape of the peak doublet exhibited by the DTA curves of β-hemihydrate.[11]

It should be noted that other workers, e.g., Holdridge and Walker, indicate that for prolonged storage, e.g., 18 months, the first peak is larger than the second one.[13] At shorter times of storage, the second peak was of larger intensity. Comparison of the data is rendered difficult unless the experiments are carried out under identical conditions.

6.2 Conversion of Soluble to Insoluble Anhydrite

The soluble form of anhydrite (the descriptor soluble or insoluble will be used as the soluble form has previously been referred to as the γ or β form and sometimes to the III form) has a hexagonal-trapezohedric lattice, a lower specific gravity (2.587) than the insoluble II form (2.985), and is permeated by interstitial cavities. The insoluble form has a rhombic pyramidal lattice (cone packed). The $CaSO_4$ chains are deposited in the interstitial cavities during the conversion. As previously discussed, a small partial pressure of water vapor during the DTA measurements of the α-hemihydrate influences the exothermic lamination peaks (Figs. 16 and 17). Crystals of various shapes and sizes dehydrate at different rates and form insoluble anhydrite lamellae at different temperatures. The curve in Fig. 16 shows that the exothermic peak disappears at a low rate of heating (2°C/min) in open atmosphere. The concept of "concealing" the exotherm describes the possibility that partial conversion of dehydrated crystals has occurred while larger crystals are still dehydrating, masking the overall effect. The curve in Fig. 17 demonstrates that low heating rates are not the only factor governing the shape and appearance of the exothermic peak. The sample in Fig. 17 was heated at a rate of 1°C/min with the use of a pinhole in the crucible lid. The use of a lid and a higher sample mass resulted in a higher partial pressure of the water vapor and consequently, a large and sharp α-peak. It is apparent that contradictions in the literature may arise from differences in material parameters and experimental details.

7.0 CONTROLLED TRANSFORMATION RATE THERMAL ANALYSIS (CRTA)

Conventional thermal analysis requires that the temperature of the sample follows some predetermined program as a function of time. *Controlled transformation rate thermal analysis* (CRTA), referred to as the

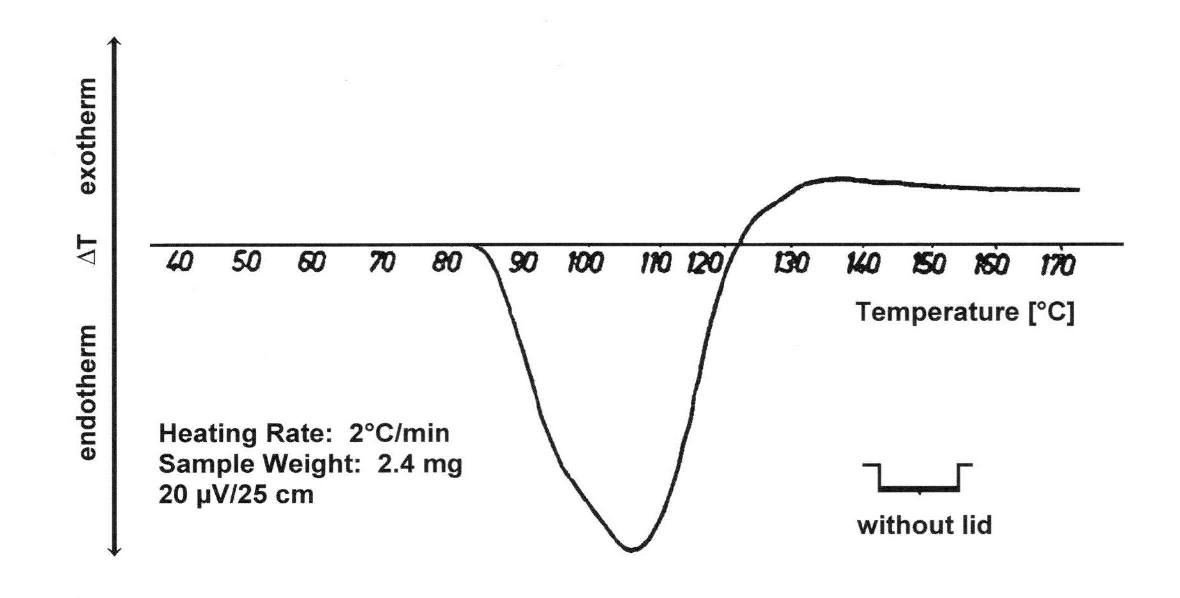

Figure 16. DTA curve of α-hemihydrate (2.4 mg, 2°C/min).[11]

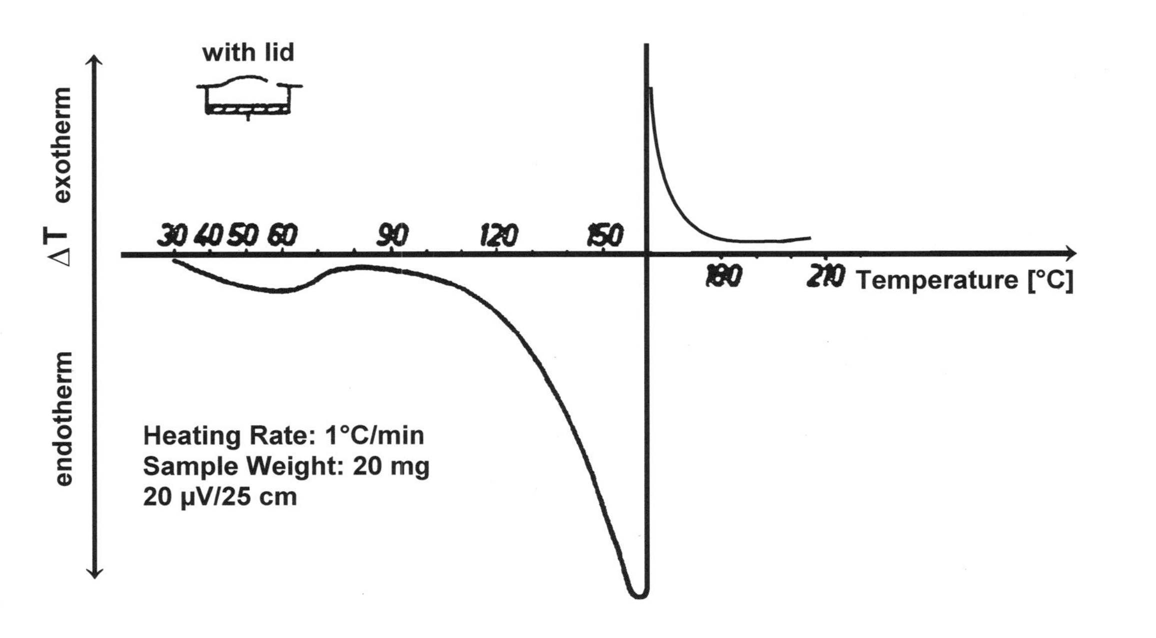

Figure 17. DTA curve of α-hemihydrate (20 mg, 1°C/min).[11]

quasi-isothermal and quasi-isobaric method, involves the use of some parameter of the sample that follows a programmed time function (achieved by adjusting the sample temperature).[4][14] Both techniques maintain a constant reaction rate and control the pressure of the evolved species in the reaction environment. The International Confederation on Thermal Analysis and Calorimetry has approved the term CRTA for techniques that monitor the temperature versus the time profile needed to maintain a chosen, fixed rate of change of a property of a sample in a specific atmosphere. CRTA represents a new approach for thermal analysis that offers significant advantages over conventional methods. In particular, it gives improved sensitivity and resolution of the thermal analysis curve and provides better kinetic data.

Anii and Fujii developed a controlled-rate thermogravimetric system (CDRC) which utilizes constant decomposition rate control without using a vacuum.[14] The basic principle involves monitoring the sample temperature from a derivative TG signal that is directly related to the decomposition rate of the sample. In the CDRC technique, the rate of decrease in mass is monitored, and the rate of temperature increase is regulated such that the rate of mass change remains constant. The heating rate in CDRC is controlled in such a way that the absolute value of the rate of the sample temperature decrease is expressed as a monotonic function of the decrease in sample mass. The heating rate is, therefore, controlled so as to decrease promptly as the mass loss occurs.

Badens and co-workers reported studies on the dehydration of gypsum carried out by heating under a constant pressure of water vapor using CRTA.[4] No intermediate formation of calcium sulfate hemihydrate occurred under a water vapor pressure of 1 and 500 Pa (at the slow reaction rates used) where as the transformation of gypsum into calcium sulfate hemihydrate was observed at 900 Pa. This transformation occurred without any lattice transformation when micron-sized needle-shaped crystals of gypsum were used, although a partial lattice transformation was observed when the starting sample was a centimeter-sized single crystal of gypsum.

A schematic of a set-up used for CRTA is shown in Fig. 18. The apparatus permits a thermal pathway to be characterized at each point, not only by the sample temperature, but also by the rate of reaction, and more importantly by the gaseous environment. Control of the residual pressure above the sample at different values permits an understanding of this parameter.

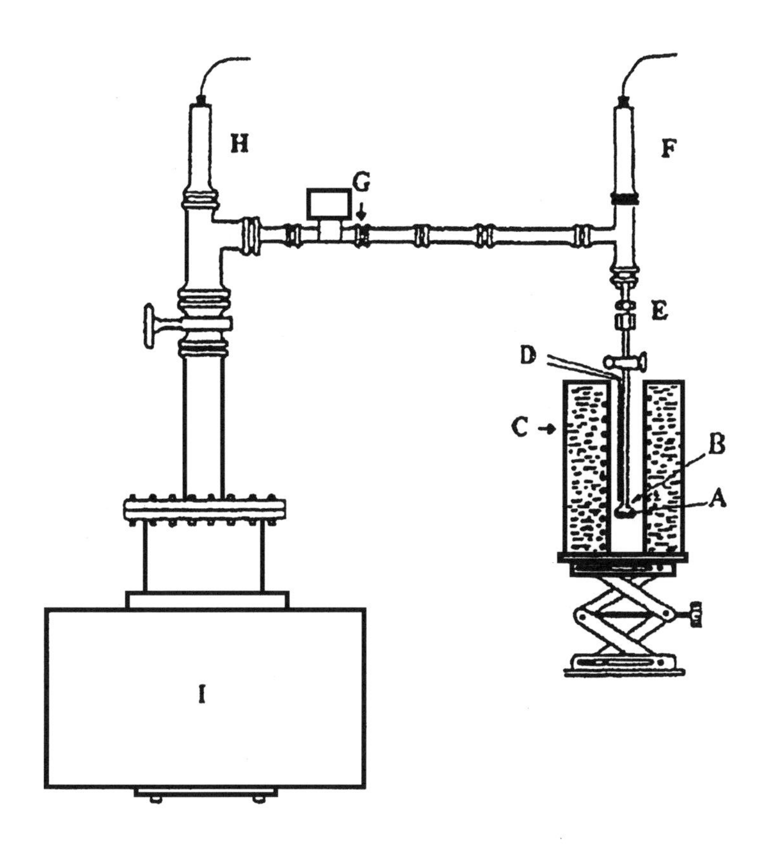

Figure 18. Schematic of apparatus used for controlled transformation rate thermal analysis. *(A)* sample; *(B)* glass or fused-silica cell; *(C)* furnace; *(D)* thermocouple; *(F)* and *(H)*, Pirani gauges; *(G)*, diaphragm; *(I)*, pumping system.[4]

Dehydration curves for gypsum at 1 Pa and 900 Pa are presented in Figs. 19 and 20. The curve obtained at 1 Pa shows only one dehydration step. The XRD analysis for the final product indicates the presence of β-$CaSO_4$. This agrees with the mass analysis. The mass loss between points *(A)* and *(B)* was about 20.9% comparing favorably with the theoretical mass loss of water of recrystallization for the transformation of gypsum to anhydrite. A step at 840°K (567°C) for the gypsum powder is observed at 1 Pa. This corresponds to the removal of the $CaCO_3$ impurity (about 2.5%).

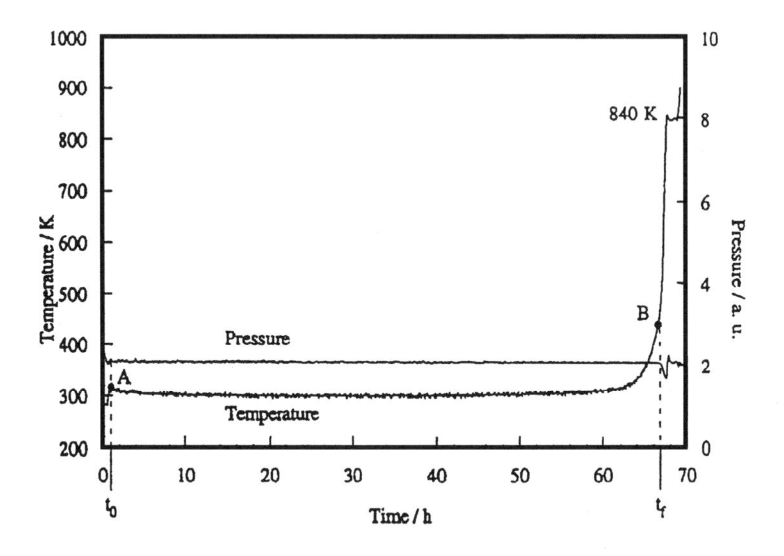

Figure 19. Dehydration of a powder of gypsum needles (100 mg; P = 1 Pa; reaction rate, 0.015 h^{-1}).[4]

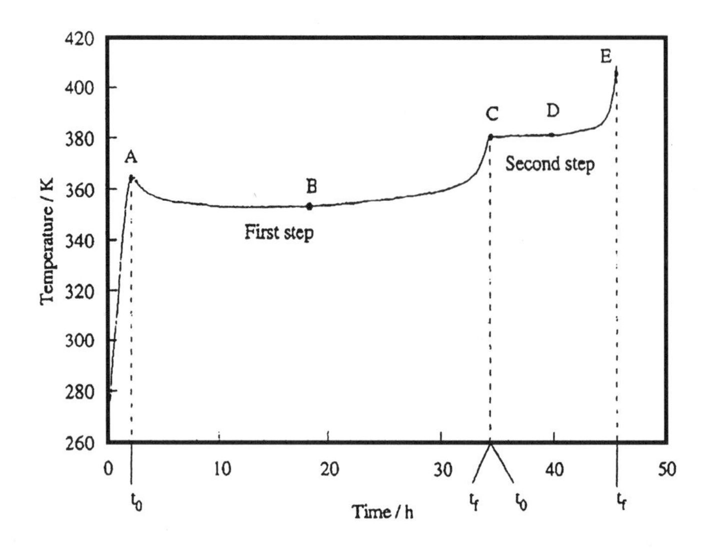

Figure 20. Dehydration of set plaster (100 mg; P = 900 Pa; reaction rate, 0.022 h^{-1}).[4]

The curve obtained at 900 Pa shows two steps between *(A)* and *(C)* at about 353°K (80°C) and between *(C)* and *(E)* at about 381°K (98°C) suggesting the existence of two dehydration steps. The length ratio of 3/1 between steps *(AC)* and steps *(CE)* suggests that the first step corresponds to the transformation of gypsum ($CaSO_4$•$2H_2O$) into calcium sulfate hemihydrate ($CaSO_4$•O•$5H_2O$) and that the second step corresponds to the transformation of calcium sulfate hemihydrate into anhydrite. The mass loss of 20.9% between point *(A)* and point *(E)* indicates a complete removal of the crystallization water.

The XRD analysis indicated the final product to be γ-$CaSO_4$ although diffraction peaks of calcium sulfate hemihydrate were observed, likely the result of spontaneous rehydration of γ-$CaSO_4$ into calcium sulfate hemihydrate.

An original result of this study is that depending of the microstructure of the initial sample of gypsum, the lattice of the intermediate calcium sulfate hemihydrate varies. There appears to be no transformation of the gypsum lattice in the case of micro-sized needles of gypsum. There is a partial transformation in the case of centimeter-sized single crystals of gypsum.

7.1 CRTA and Kinetic Modeling

Mass loss curves obtained using various CRTA modes (SIA, stepwise isothermal analysis; DRC, dynamic rate control; CDRC, constant decomposition rate control) are plotted in Fig. 21 for comparison with the conventional TG (designated TA) at 5°K (0.45°C)/minute. The sample gradually and continuously decomposes between 360° (87°C) and 410°K (137°C) using conventional TA. The curve reflects the removal of the two water molecules by a one-step reaction. The CRTA modes reflect the capability of separating this reaction into two steps. This is attributed to the ability of CRTA techniques to cause significant decomposition steps at low temperatures and over narrower temperature ranges than conventional TA. Reactions which occur at about the same temperatures can be conveniently studied by these methods. The CDRC method, with a constant rate of change of mass loss, indicates a linear dependence between the rate of reaction and time.

It can be shown that the following expression applies:

Eq. (6) $$\ln[f(\alpha)] = E/RT - \ln(A/C)$$

where $f(a)$ is a function depending on the kinetic model obeyed by the reaction and A is a pre-exponential factor in the Arrhenius expression for the rate constant, E is the apparent activation energy, R is the gas constant, and T is the absolute temperature.

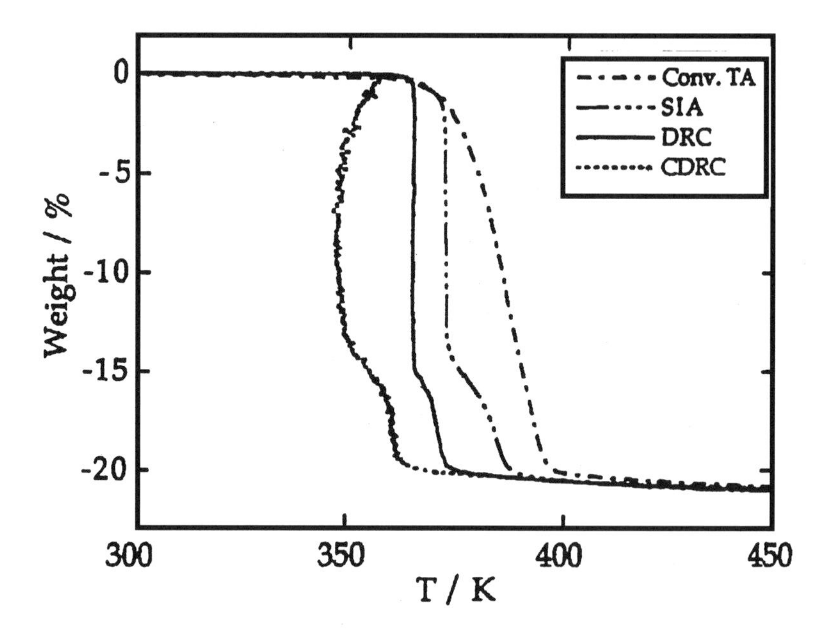

Figure 21. Comparison of mass loss curves for dehydration using conventional TG at 5°K min^{-1} with CRTA methods.[14]

Assuming the term $\ln(A/C)$ is a constant, plots of $\ln[f(\alpha)]$ versus I/T for various functions of a should be linear with the slope of E/R. The activation energy can then be obtained from a single CDRC experiment. It is apparent then that the general shape of the experimental curve, α versus T contains useful information concerning the actual dehydration mechanism. A number of model functions, for $f(\alpha)$ is given in Table 1.

Table 1. Kinetic Functions $f(\alpha)$ Describing Solid Rate Reactions

Model	Designation	$f(\alpha)$
One-dimensional diffusion	D1	$1/2\alpha$
Two-dimensional diffusion	D2	$1/[\ln(1-\alpha)]$
Three-dimensional diffusion (Jander)	D3	$3(1-\alpha)^{2/3}/2[1-(1-\alpha)^{1/3}]$
Three-dimensional diffusion (Ginstring-Broushtien)	D4	$3/2[(1-\alpha)^{-1/3}-1]$
Unimolecular decay	F1	$1-\alpha$
Phase boundary controlled	Rn	$n(1-\alpha)^{1-1/n}$; $1 \leq n \leq 3$
Nucleation and growth (Avrami/Enofeer)	Am	$m(1-\alpha)[-\ln(1-\alpha)]^{1-1/m}$; $½ \leq m \leq 4$

The theoretical curves can be divided into three categories as shown in Fig. 22. The curves were calculated by assuming the kinetic models cited in Table 1 and the following kinetic parameters: $E = 167$ kJ mol^{-1}, $A = 2 \times 10^9$ min^{-1}, and $C = 3 \times 10^{-2}$. The Avrami model for nucleation and growth leads to the curve with a temperature minimum and can be used for comparison with experimental data. The shape of the first stage profile in Fig. 22 can be attributed to this model. The letters in the figure legend are designated in Table 1. A similar kinetic analysis can be carried out for the second stage of dehydration, but is complicated by the possibility that the reaction is controlled by a diffusion or phase boundary mechanism. Analysis according to Eq. (6) is shown in Fig. 23. It is cautioned that, even though the correlation coefficients are close to unity, it is difficult to unambiguously select the correct mechanism. References 15 and 16 provide more details. It is clear, however, that controlled-rate thermal analysis has excellent potential for advancing understanding of the mechanism of gypsum dehydration.

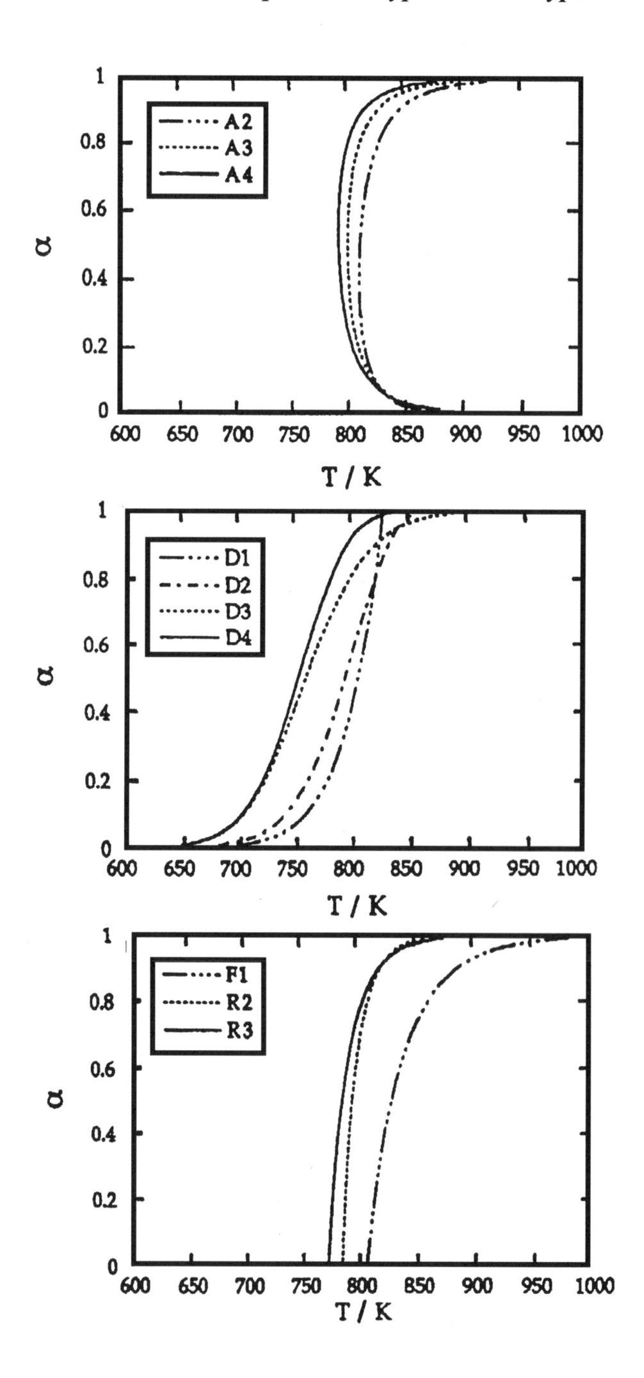

Figure 22. Shape of the theoretical CRTA curves corresponding to the kinetic models obtained assuming $E = 167$ kJ mol^{-1}, $A = 2.0 \times 10^9$ min^{-1}, and $C = 3 \times 10^{-2}$ min^{-1}.[14]

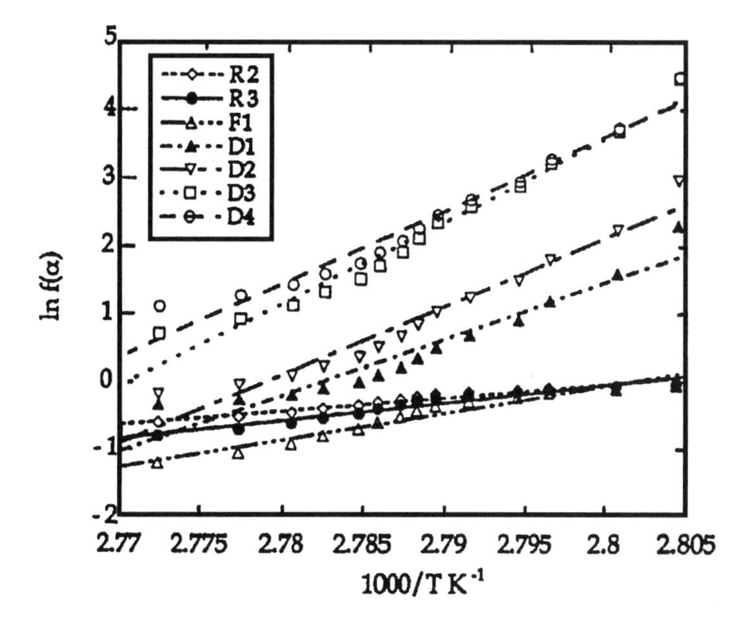

Figure 23. Results obtained from the representation of Eq. (6) for the second dehydration stage at a decomposition rate of *C*/2.7 in the various kinetic models.[14]

8.0 A THREE STEP GYPSUM DEHYDRATION PROCESS

Strydom and co-workers present evidence for a three-step gypsum dehydration process involving the formation of an intermediate hydrate, $CaSO_4{\bullet}0.15H_2O$.[17] The main dehydration step occurs between 95° and 175°C (heating rate 5°C/min). The extent of the dehydration ($\alpha < 0.1$) process proceeds with an activation energy of 392 ± 100 kJ mol^{-1}. For $0.1 < \alpha < 0.7$ (the second step), dehydration is an autocatalytic first order process with an activation energy value of 100.5± 12 kJ mol^{-1}. The third part of the reaction ($0.7 < \alpha < 1.0$) at temperatures up to 180°C has an activation energy value of 96 ± 15 kJ mol^{-1}. Dehydration appears to proceed through formation of the intermediate.

The thermal analysis evidence for the three-step process (referred to above) comprised the results of DSC and TG analysis. The largest mass loss (13.7 to 16.5%) occurred between 95° and 175°C (heating rate at 5°C/min). The enthalpy change values varied between 379 and 420 Jg^{-1}. The main product at 165°C was anhydrite (determined by XRD), but some hemihydrate was still present. No hemihydrate was observed at 172°C, but some $CaSO_4 \bullet 0.15H_2O$ and small amounts of $CaSO_3$ were observed. The reaction mixture (at 250°C) consisted of $CaSO_4 \bullet 0.15H_2O$, calcium sulfite, γ-$CaSO_4$, and probably some bassinite, $Ca_2(SO_4)_2 \bullet H_2O$. At temperatures up to 450°C, some $CaSO_4 \bullet 0.15H_2O$ and $Ca_2(SO_4)_2 \bullet H_2O$ are observed together with anhydrites $CaSO_3$, γ-$CaSO_4$, and possibly $Ca_3(SO_3)_2 \bullet SO_4$.

The activation energy was determined for heating rates varying from 0.5°C/min to 5°C /min. The following equation was used to estimate the activation energy:

Eq. (7) $$\ln \beta = \ln(AE/R) - \ln \gamma(\alpha) - 5.33 + 1.05\, E/RT$$

where β is the heating rate (°C min^{-1}), E is the activation energy, α is the degree of conversion, R is the gas constant, and T is the temperature.

The dehydration process occurs as follows. A plot of activation energy versus the degree of conversion is provided in Fig. 24 to facilitate discussion. The activation energy was determined for heating rates varying from 0.5°C/min to 5°C/min. It is apparent that the dehydration reactions do not occur in single steps. At least three stages are observed in Fig. 24. The activation energy increases in the state $0 < \alpha < 0.1$ and remains relatively constant (about 100 kJ/mol) until a sharp decrease occurs at $\alpha = 0.70$. A number of solid state decomposition kinetic models were fitted to the data. A diffusion model fits the data the best when $0 < \alpha < 0.1$. The equation is as follows: $[1 - (1 - \alpha)^{1/3}]^2 = kt$. It gives an activation energy value of 392 ± 100 kJ/mol. The correlation coefficient is marginally low with a value of 0.642. A much better fit for $0.1 < \alpha < 0.7$ is obtained using a first order reaction with an autocatalytic activation.[18] The activation energy value of 100.5 ± 1.2 kJ/mole correlates very well with the experimental values. A Sestak-Berggren equation[18] gives the best description of the dehydration during the last stage (correlation coefficient = 0.954) up to 180°C. An activation energy of 96± 15 kJ/mol is obtained using this model. The mass loss data are plotted in Fig. 25. It is apparent that the autocatalysis model provides a good correlation for $0.1 < \alpha < 0.7$. Other mechanisms appear to dominate at the beginning and end of the reactions. It is postulated that a second reaction could occur as follows:

Eq. (8) $CaSO_4 \cdot 0.5H_2O \rightarrow CaSO_4 \cdot 0.15H_2O + 0.35H_2O$

The calcium sulfate-water bonds appear to be relatively strong as they can be observed at temperatures up to 450°C. The possibility of parallel reactions after the first reaction is postulated. The reaction would be dehydration of hemihydrate directly to the anhydrite and dehydration of hemihydrate to an intermediate product containing 0.15 moles of water.

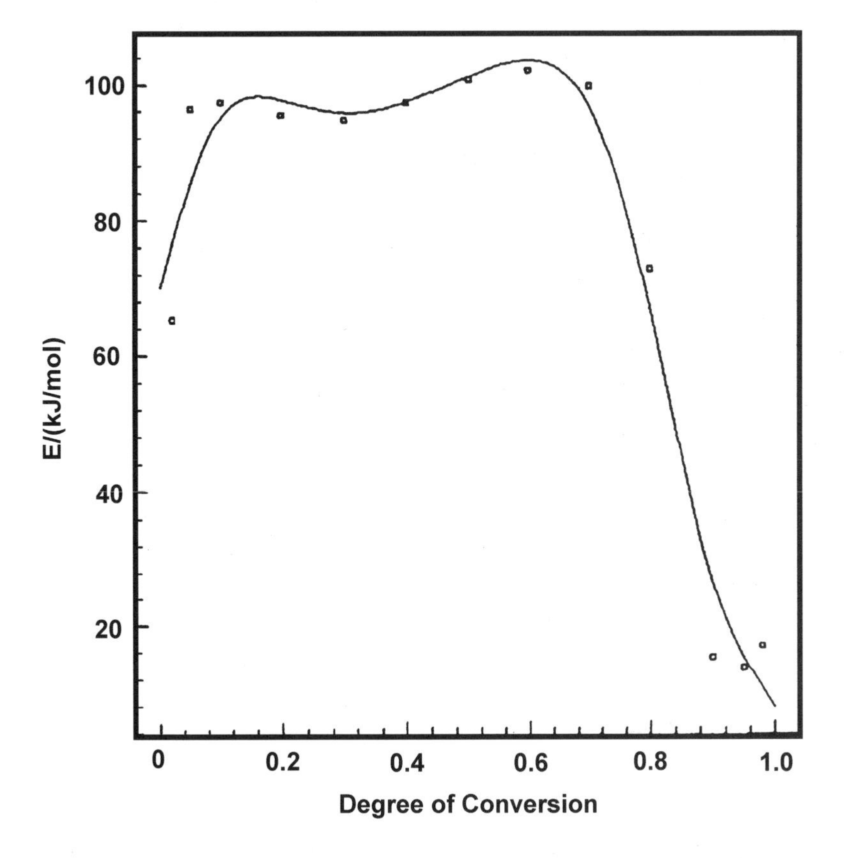

Figure 24. Activation energy values (E) at different degrees of conversion of gypsum.[17]

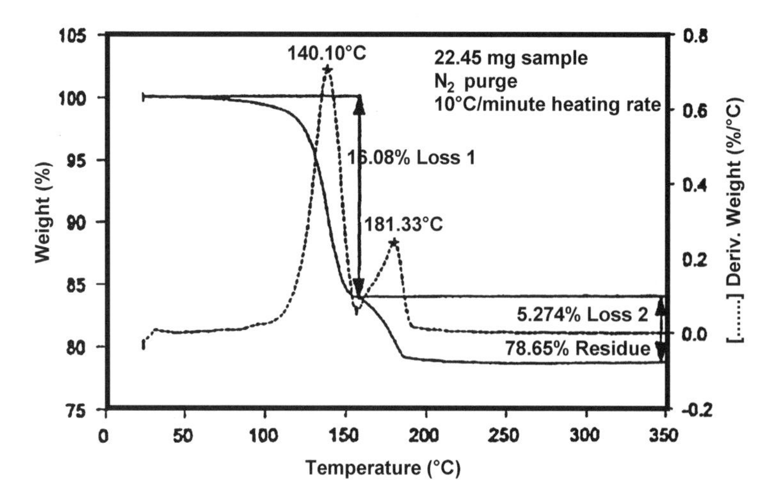

Figure 25. TGA of gypsum in a partially sealed container (pinhole in lid).[19]

9.0 INDUSTRIAL APPLICATIONS

9.1 Portland Cement and Stucco

The production of commercial building materials, e.g., portland cement and stucco, utilizes calcium sulfate in the form of the dihydrate or hemihydrate.[19] Gypsum is added in the production of portland cement to control set. Properties of cement can be adversely affected by the formation of hemihydrate during the grinding process. In the production of stucco, the hemihydrate is the preferred form. Consequently, investigators have adapted thermal analysis methods for estimating the quantities of each phase in these products.

The difficulty in determining the amounts of the product is associated with the separation of the two endothermic peaks which are not sufficiently resolved to allow quantification. This is generally the case when thermal analysis (e.g., DSC) is carried out in an open system (unsealed pan). The dehydration in a hermetically sealed pan improves the resolution of the DSC peaks dramatically (Fig. 26). This improved resolution results because the atmosphere of the water vapor generated by the first step retards the onset of the second dehydration and moves the peak to a higher temperature. Satisfactory resolution has been obtained using 5–8 mg samples in hermetically sealed aluminium pans using a 15°C/min heating rate. The resolution of the TGA mass loss profiles is similar to those for the DSC results. The successive mass losses are not adequately resolved (even using the derivative trace) using an open pan system to permit quantification. With the use of a sealed system (pan with a pinhole), the calcium sulfate hydrate peaks can be resolved (Fig. 26). A typical calculation involving the separated peaks in the DSC trace is as follows:

> The area associated with the low temperature DSC endotherm is measured. A calibration curve is used to determine the amount of gypsum in the cement. Using the peak area ratio for the dehydration of pure gypsum (found to be 3.3), the amount of heat in the second stage of the cement dehydration associated with gypsum is determined. The percentage of hemihydrate is then determined using a calibration curve for pure hemihydrate and the difference in peak area of the second peak, i.e., the area not associated with the original amount of gypsum in the material. A similar approach for the quantitative estimation of the calcium sulfate hydrates can be used with TGA analysis. The pinhole arrangement in the sample pan allows the initial water of dehydration to escape, but at a rate that retards the second hydration step. Using suitable reference standards, a calibration curve similar to that generated for DSC can be developed and subsequently used to determine the percent of gypsum at levels below 1% with an accuracy of 3 to 5%. Results of this quality requires a TG with a high sensitivity and low thermal drift because the presence of, for example, 0.5 % hemihydrate in a 2 mg sample, results in only a 160 μg mass loss.

Fujii and co-workers have reported the important factors governing the quantitative analysis of gypsum by DTA.[20] These included the following recommendations:

- Covering the sample with inert substances and bottom packing to improve the ability of reproduction of the DTA curves.

- Use of the fact that the peak area associated with the dihydrate is independent of the rate of temperature change.

- Estimation of hydrates based on the observation that the total peak area of the dihydrate is in proportion to the quantity of the sample dihydrate and that the area ratio of the first peak to the second peak is approximately 3.00.
- Establishment of a proportional relationship between the peak area of dehydration (second peak) and the quantity of the sample.

The following empirical relationship was developed to predict the relative proportion of calcium sulfate hydrates.

Eq. (9) $1/\gamma = 0.33 + 0.33R$

where γ is the area ratio of the first peak to the second peak and R is the mass ratio of hemihydrate to dihydrate.

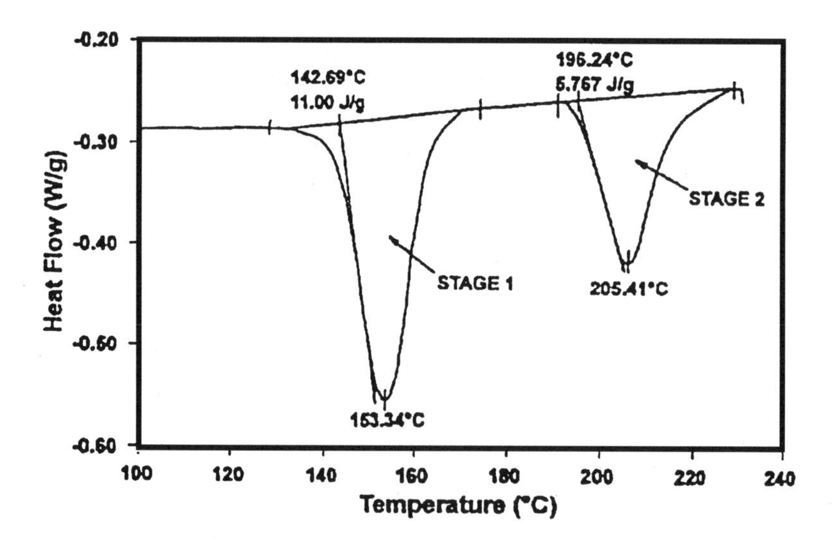

Figure 26. DSC curve for hydrated portland cement in a hermetically sealed pan.[19]

9.2 Gypsum–Based Cements

The utilization of gypsum-based binary and ternary blended cements has been shown to provide the desired characteristics of early hardening strength with improved durability.[21] This is illustrated in

Kovler's investigation of a blend of calcium sulfate hemihydrate (75% by mass), portland cement (20%), and silica fume (5%) containing 2% superplasticizer. The hydration of hemihydrate to form gypsum crystals in the binary cement-gypsum system can be conveniently followed by DTA methods (Fig. 27). Two characteristic peaks for gypsum and hemihydrate (150° and 200°C, respectively) clearly indicate the growth of the gypsum peak and the corresponding decrease in the hemihydrate peak. The analysis permits estimation of the amount of gypsum growth in the paste up to about 2 hours (not shown in the figure). It is thus possible to estimate the degree of reaction from DTA. Ten percent of the reaction occurs within 2 minutes, 50% within 5 minutes, and 90% within 10 minutes. Similar results are obtained for the ternary cement system containing silica fume. The DTA method is found to be a useful tool in providing the basis for the explanation of setting and hydration mechanisms influenced by the complex sequence of reactions in a ternary blend. It was apparent that strength development of the ternary blend was similar to that of pure gypsum. Long term improvement in the performance was the result of a different mechanism associated with the cement and silica fume hydration process.

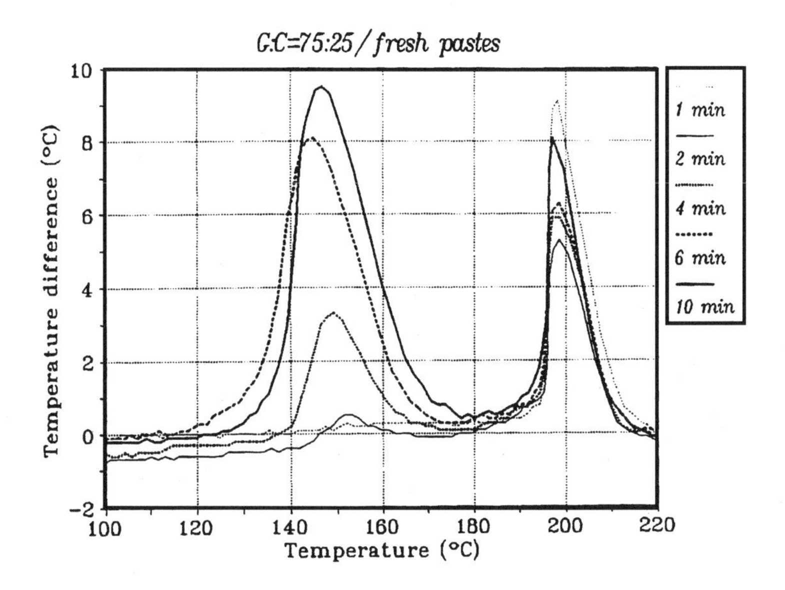

Figure 27. DTA thermograms for a gypsum-portland cement mix containing 2% superplasticizer (1 to 10 min after mixing).[21]

9.3 Sedimentary Rocks Containing Gypsum

A method for estimation of gypsum in sedimentary rocks by TG and DTG was proposed by Brigatti and Poppi.[22] The method involves equilibrating the material to a partial pressure of less than 0.01 in a flow anhydrous N_2. This generally requires about an hour. A second step involves isothermal heating to 80°C for an hour. TG is performed with a heating rate of 10°C per minute. The dehydration of gypsum associated with Ca-montmorillonite previously heated to 80°C for one hour is illustrated in Fig. 28. Curve (*a*) is the TG trace for Ca-montmorillonite and curve (*b*) is the trace of the gypsum-bearing montmorillonite. The DTG curves are designated (*a′*) and (*b′*). The estimation of gypsum with an error of about ±1% is possible.

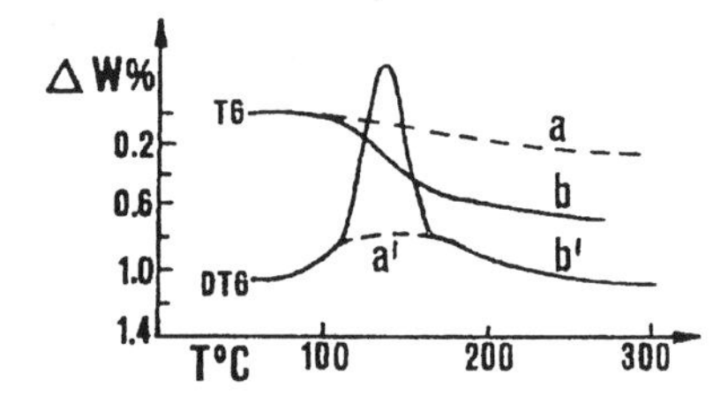

Figure 28. Dehydration of gypsum (solid line) associated with Ca-montmorillonite (dashed line) previously heated to 80°C for 60 min in a flow of N_2 gas.[22]

9.4 Quality Control of Commercial Plasters

Storage conditions of commercial plasters often result in partial hydration of the hemihydrate material. Dynamic or isothermal thermogravimetry can be used to estimate the primary phases and impurities present in commercial plaster.[23] In dynamic TG experiments at low heating rates, the hemihydrate and dihydrate contents can be estimated from separate mass change steps of the respective dehydration reactions shown in the resulting TG curves.

A typical TG curve of a commercial plaster (Fig. 29) shows two steps for mass loss. The first step is caused by the dehydration of calcium sulfate dihydrate. The second step is due to the dehydration of calcium sulfate hemihydrate. The second step is attributed to the hemihydrate present in the original sample and also to that formed during the first step of decomposition.

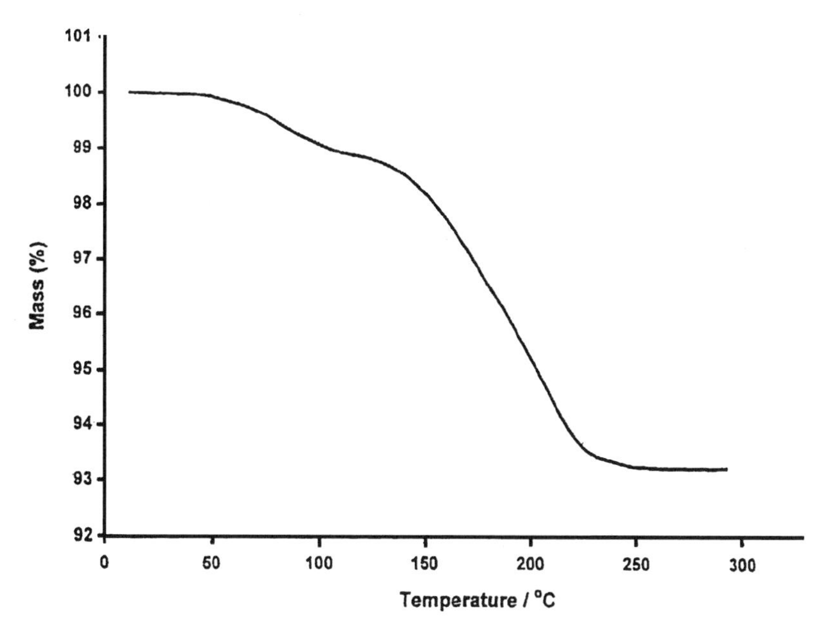

Figure 29. Typical dynamic TG curve of 2.5 g of plaster in air. Heating rate = 5°C min^{-1}.[23]

A typical calculation utilizing a TG trace (dynamic method) can be described as follows. The mass losses in steps 1 and 2 are designated ML1 and ML2, respectively. Further, H is the mass of the hemihydrate in the sample, D is the mass of dihydrate in the sample, and I is the mass of the impurities in the sample. The values of H, D, and I can be calculated by the following equations:

Eq. (10) $D = \mathrm{ML1}/0.1571\ I$

Eq. (11) $H = (\mathrm{ML2} - 0.062D)/0.062$

Eq. (12) $I = 100 - (D + H)$

The isothermal TG method is described as follows: The dihydrate and hemihydrate are decomposed simultaneously during isothermal tests when the temperature of the experiments is maintained at 250°C or higher. The isothermal TG curves show the cumulative mass loss of these reactions (Fig. 30). Two different analyses are required, one for the original sample

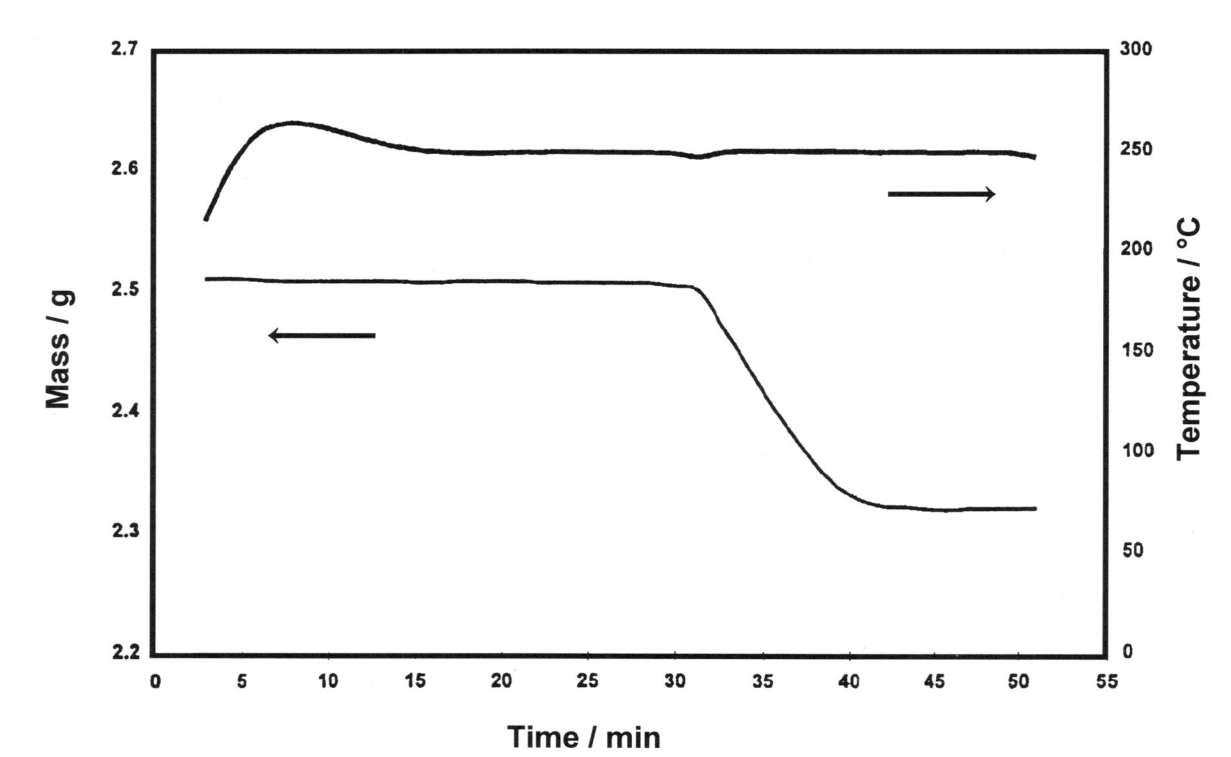

Figure 30. Typical isothermal TG curve of plaster at 250°C.[23]

and one for the modified sample. The following equations can be written for isothermal analysis of the original sample:

Eq. (13) $0.062H + 0.2091D + MT$

Eq. (14) $I = 100 - H - D$

where MT is the total mass loss in percent.

It is apparent that calculation of D, H, and I depends on the complementary data obtained from analysis of the modified sample. The modified sample is mixed with water (70% by mass). The fully hydrated sample is then dried at 45°C prior to being subjected to isothermal analysis. The apparent impurity content, I^*, can be calculated using the total mass loss, MT^*, from the following equation:

Eq. (15) $I^* = 100 - (100\, MT^*/20.91)$

The increase in the mass of the sample in this treatment is due to hydration of hemihydrate. The adjusted calculation of the impurities is given as follows:

Eq. (16) $I = I^* \bullet (1 + 0.001862H)$

The above system of equations can be solved for the hemihydrate content, H.

Eq. (17) $H = (MT^* - MT)/(0.18602 - 0.18602\, MT^*)$

D and I can then be calculated applying the Eqs. (13) and (14).

Thermogravimetry can be used for quality control of commercial plasters either by dynamic or isothermal methods. The partial hydration of the plasters, which have not been well stored, can be subjected to identification and estimation.

9.5 White Coat Plaster

Several investigators have reported that white coat plaster made from dolomite as the raw material and gypsum exhibits "popping" or "bulging" from 5–10 years after application. DTA and TG methods have been applied with success not only to determine the causes leading to the failure of plasters, but also to estimate the amounts of the ingredients in the finished plaster. The ingredients include gypsum, calcium hydroxide, magnesium hydroxide, magnesium oxide, and calcium carbonate.[24] Most failures could be explained by delayed hydration of MgO to $Mg(OH)_2$.

9.6 Expanding Cement

Normal concrete containing portland cement exhibits a linear shrinkage of about 0.05–0.1%. An expanding cement is produced by burning a mixture of bauxite, calcium carbonate, and gypsum and mixing it with portland cement. This cement causes expansion equal to or greater than the shrinkage that would occur otherwise. The expanding cement paste mainly comprises calcium sulfate, calcium aluminate, and calcium silicate. The cement may also be produced by burning a mixture of gypsum, alumina, and calcium hydroxide to a temperature of 1300°C. The reactions occurring during the synthesis of this cement can be followed by DTA.[3]

REFERENCES

1. Groves, A.W., *Gypsum and Anhydrite,* Overseas Geol. Surveys, Min. Res. Div., p. 108, Her Majesty's Stationery Office, London (1958)
2. Taylor, H. F. W., *Cement Chemistry,* p. 233, Academic Press, London (1990)
3. Ramachandran, V. S., *Applications of Differential Thermal Analysis in Cement Chemistry,* p. 308, Chem. Pub. Co., New York (1969)
4. Badens, E., Llewellyn, P., Fulconis, J. M., Jourdon, C., Veesler, S., Boistelle, R., and Rouquerol, F., Study of Gypsum Dehydration by Controlled Transformation Rate Thermal Analysis (CRTA), *J. Solid State Chem.,* 139:37–44 (1998)
5. Clifton, J. R., Thermal Analysis of Calcium Sulfate Dihydrate and Supposed α and β Forms of Calcium Sulfate from 25 to 500°C, *J. Res. Natl. Bur. Stands-A, Phys. and Chem.,* Vol. 78A (1972)
6. Khalil, A. A., Hussein, A. T., and Gad, G. M., On the Thermochemistry of Gypsum, *J. Appl. Chem. Biotech.,* 21:314–316 (1971)

7. Kelly, K. K., Southhard, J. C., and Anderson, C. F., Thermodynamic Properties of Gypsum and its Hydration Products, *U. S. Bur. Mines Tech.,* Paper 625, p. 71 (1941)

8. Khalil, A. A., and Gad, G. M., Thermochemical Behavior of Gypsum, *Build. Int.,* 5:145–147 (1972)

9. Elliott, C., Plaster of Paris Technology, *Chem. Trade J.,* 72:725–726 (1923)

10. Isa, K., and Oruno, H., Thermal Decomposition of Calcium Sulfate Dihydrate Under Self-generated Atmosphere, *Bull. Chem. Soc. Jpn.,* 55:3733–3737 (1982)

11. Lehman, H., and Rieke, K., Investigations of the System $CaSO_4$ – H_2O Under Special Considerations of Material and Experimental Parameters by Differential Thermal Analysis, *Proc. 4th Int. Conf. Thermal Analysis, Budapest,* 1:573–583 (1975)

12. Lehmann, H., and Rieke, K., Dr Einsatz der Differenzthermoanalyze zur Verfolgung der Ummwandlungsvorgange im System $CaSO_4$•H_2O, *Sprechsaal,* 106:(23):924–930 (1973)

13. Holdridge, D. A., and Walker, E. G., The Dehydration of Gypsum and the Rehydration of Plaster, *Trans. Brit. Ceram. Soc.,* 66:485–509 (1967)

14. Arii, T., and Fujii, N., Controlled Rate Thermal Analysis: Kinetic Study in Thermal Dehydration of Calcium Sulfate Dihydrate, *J. Anal. and Appl. Pyrolysis,* 39:129–143 (1997)

15. Criado, J. N., Ortega, A., and Gotor, F., Correlation Between the Shape of Controlled Rate Thermal Analysis Curves and the Kinetics of Solid State Reactions, *Thermochimica Acta,* 157:171 (1990)

16. Wiedenann, H. G., Sturzenegger, E., and Bayer, G., Nucleation Characteristics of Some Thermal Decomposition Processes, Thermal Analysis, in: *Proc., 4th ICTA, Budapest,* (I. Buzas, ed.), p 227 (1974)

17. Strydom, C. A., Hudson-Lamb, D. L., Potgieter, J. H., Dagg, E., The Thermal Dehydration of Synthetic Gypsum, *Thermochimica Acta,* 269/270:631–638 (1995)

18. Netzch Thermokinetic Analysis Multiple Scan, 2nd Ed., 5:2–11, Netzsch-Geratebau GmbH, Bayern, Germany (1993)

19. Blaine, R., Determination of Calcium Sulfate Hydrates in Building Materials Using Thermal Analysis, *American Laboratory,* 27:24–28 (1995)

20. Fujii, S., Hinota, M., and Seki, K., Quantitative Analysis of Gypsum by DTA, *Gypsum and Lime,* 86:23–28 (1967)

21. Kovler, K., Setting and Hardening of Gypsum-Portland Cement-Silica Fume Blends, Part 2: Early Strength, DTA, XRD and SEM Observations, *Cement Concr. Res.,* 28:523–531 (1998)

22. Brigatti, M. F., and Poppi, L., Quantitative Determination of Gypsum in Sedimentary Rocks by Thermal Analysis, *Miner. Petrogr. Acta,* 23:131–134 (1979)
23. Dweck, J., Lasota, E. I. P., Quality Control of Commercial Plasters by Thermogravimetry, *Thermochimica Acta,* 318:137–142 (1998)
24. Ramachandran, V. S., Sereda, P. J., and Feldman, R. F., Delayed Hydration in White Coat Plaster, *Mat. Res. Std.,* 4:663–666 (1964)

12

Clay-Based Construction Products

1.0 INTRODUCTION

The performance of clay-based construction products, e.g., bricks and roofing tiles, can be monitored using thermal methods. The types of raw materials, viz., clay and accessory minerals, and their reactions that occur during the firing process and the durability of clay products can be examined through the application of DTA, TG, TMA, and dilatometric methods. This is particularly important for quality control as physical and chemical behaviors are dependent on the raw material characteristics, e.g., composition, particle size, and morphology.

This chapter focuses on the application of thermal methods specific to raw materials suitable for industrial clay products that are suitable for brick-making. Details of thermal processes for the following materials are discussed: singular, binary, and ternary clay-based systems; structural ceramics; solid waste in clay bricks; significant archaeological clay products; and brick clay products from Central Europe.

The durability of clay bricks to freezing-thawing cycles is also presented. Emphasis is placed on the use of thermal methods, e.g., dilatometry to determine the factors that contribute to freezing-thawing resistance. A method for determining the firing temperature characteristics of clay brick is presented. The use of brick particulate additives to portland cement concrete for enhancement of freezing-thawing resistance and the applicability of thermal techniques to assess the bloatability of clays is also described.

2.0 THERMAL BEHAVIOR AND IDENTIFICATION OF CLAYS AND ACCESSORY MINERALS

2.1 DTA of Clay Minerals

Thermal methods (DTA, TG, TMA, and dilatometry) are well-established investigative tools in clay science and related industrial applications.[1][2] Clay brick manufacturers have employed these techniques to optimize their plant production procedures.

The successful firing of clays in industrial processes involves considerations related to volume change, phase changes, and crystallization phenomena. Drying shrinkage of clay minerals occurs due to pore water loss in the temperature range of 100–150°C. The oxidation of any organic material takes place in the range of 200–600°C. The oxidation of sulfides begins between 400 and 500°C. The hydroxyl water is removed from the clay minerals starting at temperatures somewhat below 500°C and continuing to temperatures approaching 900°C. The nature of the clay minerals in clays and the particle size influences the specific temperature and rate of hydroxyl loss.

Some of the general conclusions that can be drawn from the DTA studies of various clays may be summarized as follows.[3] An endothermic reaction below about 200°C usually indicates the presence of montmorillonite or illite. A clay containing these components will have high plasticity and shrinkage, will probably be non-refractory, and will burn red. In general, the larger this fraction the higher the plasticity and shrinkage. Endothermic reactions between 300 and 500°C usually indicate a hydroxide of alumina or iron. If the component is a hydroxide of alumina, the clay will be refractory and will have low shrinkage. A broad exothermic reaction between 200 and 600°C is due to the presence of an organic material. Clays yielding such thermal reactions will frequently be very plastic and will require careful burning to insure complete oxidation of the carbon. A sharp exothermic reaction between 400 and 500°C may indicate the presence of pyrite or marcasite. An intense endothermic reaction at about 600°C followed by a sharp exothermic reaction at about 975°C indicates the presence of kaolinite. A clay with a peak of low intensity at about 500 or 700°C followed by another endothermic reaction at about 900°C and then a final exothermic reaction may indicate the presence of illite or montmorillonite. A clay containing either of the two is not refractory or light firing and is apt to have a short vitrification range. If the component is mainly montmorillonite, it will also exhibit high plasticity and shrinkage. A small endothermic break at 575°C is typical of quartz which reduces plasticity and shrinkage of the clay. The presence of calcium carbonate is indicated by an intense endothermic peak at about 850°C. If it is present, the clay requires careful preparation and firing technique. It has to be emphasized that very careful study and experience are required to interpret the thermal curves. In some clays, thermal inflections cannot be easily interpreted unless additional analytical data are obtained with x-ray, chemical, and other methods of analysis.

Kaolinite is the most prevalent mineral in ceramic formulations.[4][5] It shows pronounced thermal effects on heating and generally has a more ordered structure than other clay minerals. Figures 1 and 2 illustrate typical DTA data for kaolinite, halloysite, and montmorillonite.[2] Kaolinite and halloysite lose their hydroxyls between 450 to 600°C. Variations within this range are attributed to differences in entrapped water vapor that is dependent on sample size and shape factors. The loss of hydroxyls from montmorillonites in the range of 450 to 650°C is typical for dioctahedral forms of these minerals. Dehydroxylation is more gradual for trioctahedral forms and can continue to temperatures up to 850°C.

The crystallization of new high temperature phases for fired kaolinite and holloysite occurs at about 950–1000°C. It has been shown that a significant variation in high temperature effects develop in montmorillonites with different exchangeable cations. The nucleation of the high temperature phase is often accompanied by a considerable release of energy and is shown on the thermogram by a sharp exothermic peak. Often the temperature must be raised above the nucleation temperature to enable the new structure to grow and develop. For example, mullite in a refractory brick produced from a kaolinite clay begins to form at about 1000°C, but only develops rapidly at about 1250°C. DTA methods are capable of indicating the formation of a new high temperature phase before it is detected by x-ray diffraction measurements. It shows pronounced thermal effects on heating and generally has a more ordered structure than other clay minerals. Kaolinite changes at 600°C with water evolution, becomes isotropic between 900 and 1000°C, takes on a granular appearance at 1250–1300°C and transforms into sillimanite above 1400°C.[6] Mellor and Scott[7] describe the thermal events as follows: completion of dehydration above 500°C; completion of alumina transformation accompanied by an exothermic reaction at 900°C; solid solutions development (sillimanite with $3Al_2O_3 \cdot 2SiO_2$) below 1200°C; formation of $3Al_2O_3 \cdot 2SiO_2$ above 1200°C in Georgia kaolin. γ-Al_2O_3 has also been identified after heating in the range of 960–1000°C.[8] Fluxing impurities present in brick clays tend to suppress the thermal reactions.[9] Some oxide additions accelerate the formation of mullite. The relevance of the exothermic peak at about 980°C has been the subject of considerable debate. Bradley and Grim[10] reported that the probable phases present include spinel-type phases as well as quartz, cristobolite, and mullite. The presence of the 980°C peak may reflect a series of reactions. The thermal effect was observed to be a maximum for coprecipitated gels of alumina at a SiO_2/Al_2O_3 ratio of between 0.67 and 1.00 (see Fig. 3).[11][12]

It is also instructive to examine factors that influence the high temperature behavior of montmorillonite. Two types of aluminous montmorillonite were identified by Grim and Kulbicki.[13] They noted that the high temperature phases formed depend on the type of this mineral.

Ceramists are particularly interested in reactions that occur in the firing range of ware, i.e., above 1200°C. It has been observed that in the range of 1200–1250°C, secondary mullite and amorphous silica are present.[14] The silica changes to cristobalite in the range of 1240–1350°C. Primary mullite is formed below 1200°C. The kaolinite-mullite transition involves metakaolin transformation to a spinel phase and then to mullite of possible composition $3Al_2O_3 \cdot 2SiO_2$.

The dehydration of allophanes can begin at about 100°C and continue up to about 900°C. More structural organization can also result in a rapid loss of hydroxyls in the range of 400–600°C.

Identification of Clay Minerals Through Clay Complexes. Basic dyestuffs are adsorbed by clay minerals by a cation exchange reaction. It has been found that clay mineral-dye complexes exhibit characteristic thermograms that can be used to identify clay minerals.[15] Certain clay minerals which show similar thermograms can be differentiated by the thermograms of their dye complexes.

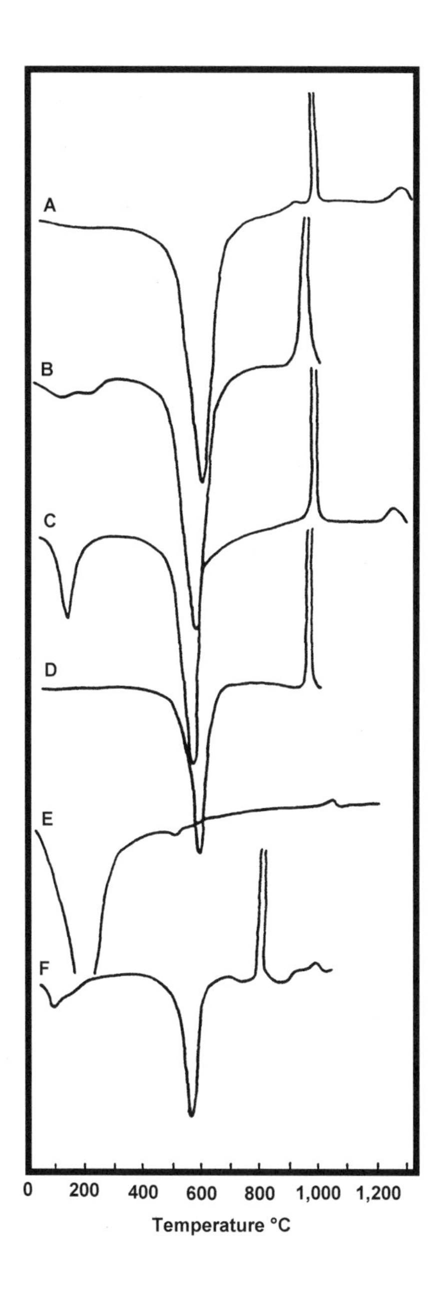

Figure 1. DTA curves: *(A)* kaolinite, Macon, Georgia, well crystalline; *(B)* kaolinite, Anna, Illinois, poorly crystalline; *(C)* hydrated halloysite, Bedford, Indiana; *(D)* anauxite, Ione, California; *(E)* allophane, Bedford, Indiana; *(F)* allophane, Iyo, Japan.[2]

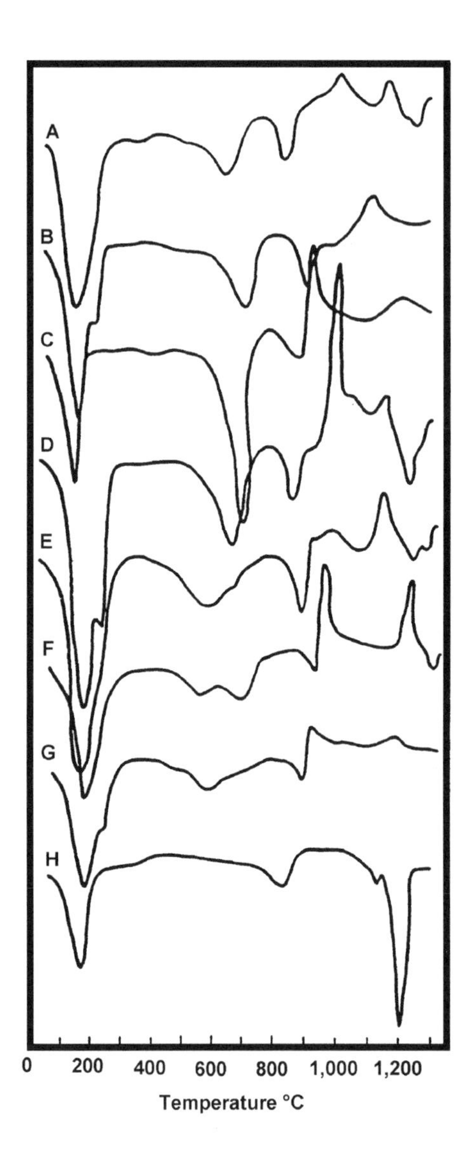

Figure 2. DTA curves: *(A)* montmorillonite, Otay, California; *(B)* montmorillonite, Tatatilla, Mexico; *(C)* montmorillonite, Upton, Wyoming; *(D)* montmorillonite, Cheto, Arizona; *(E)* montmorillonite, Pontotoc, Mississippi; *(F)* montmorillonite, Palmer, Arkansas; *(G)* nontronite, Howard County, Arkansas; *(H)* hectorite, Hector, California.[2]

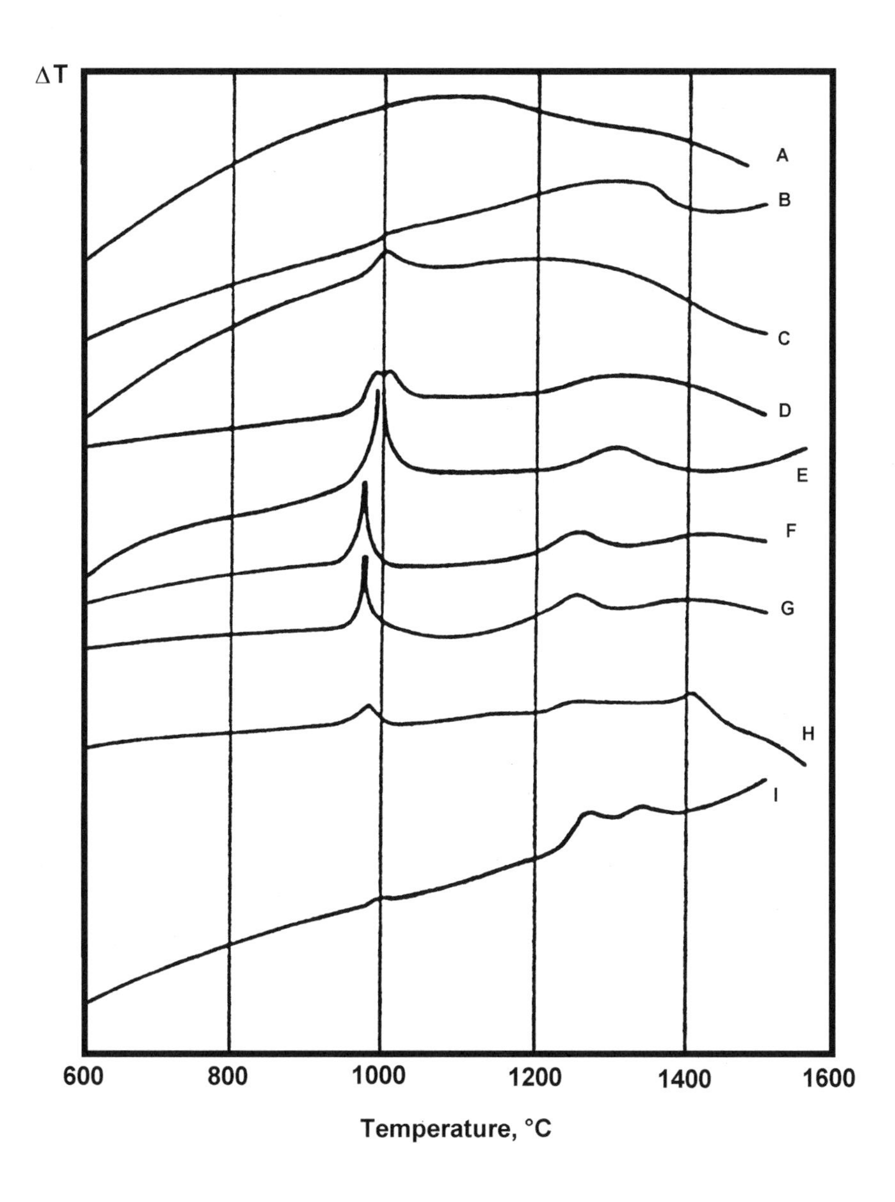

Figure 3. DTA curves for silica-alumina mixtures with Al_2O_3:SiO_2 ratios (by weight) of: *(A)* 0; *(B)* 0.0983; *(C)* 0.3452; *(D)* 0.6662; *(E)* 1.0152; *(F)* 2.002; *(G)* 2.927; *(H)* 10.041; *(I)* ∞.[12]

Thermograms of kaolinite, illite, nontronite, and montmorillonite with three basic dyestuffs, namely, malachite green, methylene blue, and methyl violet have been investigated. Thermograms of some of the clay minerals treated with dyes are shown in Fig. 4.[15] All the complexes exhibit low temperature endothermic peaks smaller than those of the untreated samples. The organic matter probably occupies the spaces normally held by the water molecules. Kaolinite-dye complexes show a small exothermal effect between 350 and 435°C due to the oxidation of the dye adsorbed by base exchange reaction. Malachite green-kaolinite complex has a much more intense exotherm than others. Illite-dye complexes (Fig. 4a) indicate two intense exothermic effects between 200 and 500°C. Nontronite and montmorillonite dye complexes (Fig. 4b) show three exothermic peaks, two having pronounced intensities. The second peak in illite complexes at 460–470°C is less intense than that of nontronite-complexes occurring at 600°C. Illites can be differentiated from nontronites by the second exothermic peak. In montmorillonite complexes, the higher temperature exothermic peak at 650–670°C serves to differentiate montmorillonites from nontronites and illites in which it occurs at 600 and 460–470°C, respectively.

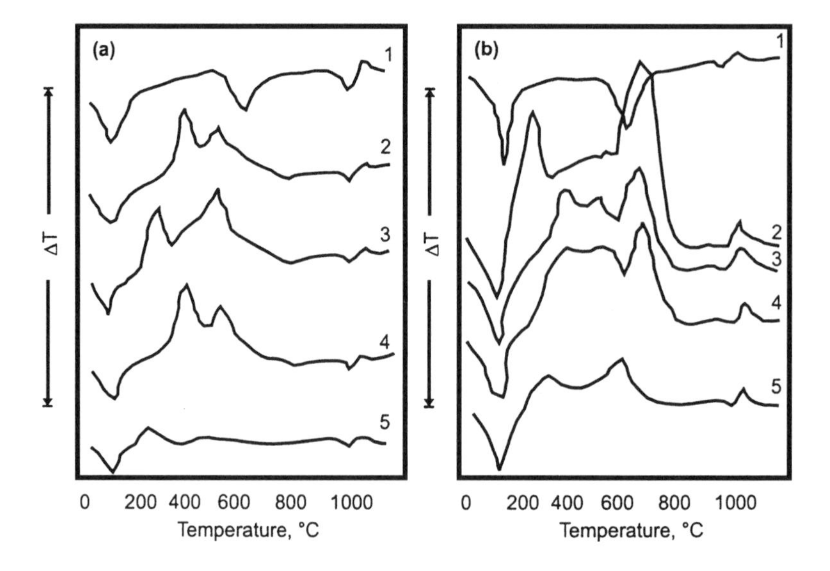

Figure 4. Thermograms of illite and nontronite-dye complexes. (*a*) Thermograms of illite and its complexes: (1) illite; (2) malachite green complex; (3) methylene blue complex; (4) methy violet complex; (5) piperidine complex. (*b*) Thermograns of nontronite and its complexes: (1) nontronite; (2) methylene blue complex; (3) malachite green complex; (4) methyl violet complex; (5) piperidine complex.[15]

Clay Mineral Mixtures. Natural clay is generally made up of a mixture of clay minerals. Some of the common clay minerals which occur in nature are kaolinite, nontronite, illite, and montmorillonite. The endothermal effects of kaolinite, illite, and nontronite occur in the range of 500–600°C, so it would be difficult to identify them on the basis of the endothermal effects. The exothermal crystallization at 900–1000°C shown by most clay minerals is not easy to use to differentiate one clay mineral from the other. It has been found that thermograms of kaolinite, illite, nontronite, and montmorillonite complexed with malachite green or methylene blue give characteristic exothermal peaks at 420, 460, 590, and 670°C respectively.[16] An example of the usefulness of the dye-clay complex study is illustrated with respect to illite-montmorillonite mixtures. In Fig. 5, thermograms of illite:montmorillonite mixtures formed in ratios of 1:4, 2:3, 3:2, and 4:1 are shown. Two endothermic peaks at 540–560°C and 670–700°C are evident in the mixtures. Although the second peak in all the mixtures is indicative of the presence of montmorillonite, the first may be due to kaolinite, illite, nontronite, or their mixtures. Figure 6 shows the thermograms of the methylene blue complexes of these mixtures. The first exotherm at 470–490°C is characteristic of illite and does not appear in kaolinite or nontronite. The second exotherm at 660–670°C is characteristic of montmorillonite. The exothermal peak intensity could be used to estimate the amount of the clay mineral.

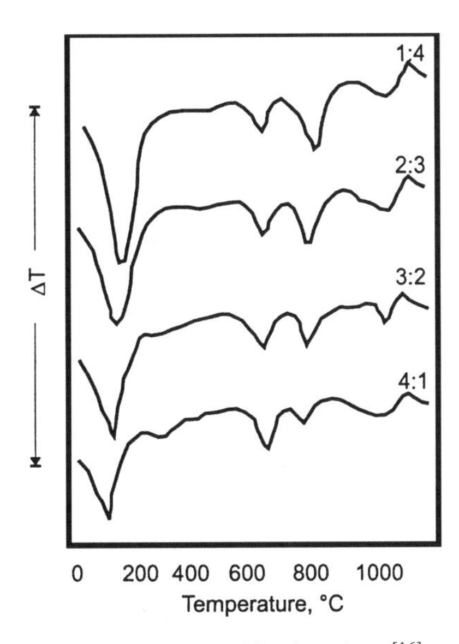

Figure 5. Thermograms of illite-montmorillonite mixes.[16]

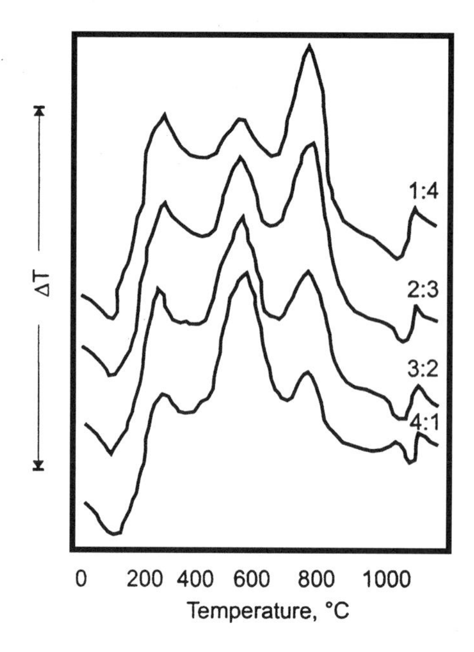

Figure 6. Thermograms of methylene blue complexes of illite-montmorillonite mixes.[16]

2.2 Other Thermal Methods

Application of multiple thermal techniques to clay mineral identification and behavior is more advantageous than applying any single technique.

Rai, et al.,[17] compared the relative merits in the use of DTA and DTG techniques for single, binary, and ternary systems containing clay minerals kaolinite, illite, nontronite, and monmorillonite. The techniques were found to be equally useful for single systems. DTA enabled semi-quantitative estimation of the heat values, DTG enabled a more accurate estimation of the weight losses due to desorption and dehydroxylation. For the binary systems, DTG was found to be more useful because it could resolve dehydration peaks more clearly than the DTA technique. In the DTG, the ternary systems showed three peaks for the dehydration, and, thus, the presence of all the clay minerals. It was also concluded that the dynamic weight loss values were slightly lower than those obtained by the static method. DTA was very advantageous in providing information on the exothermal effects.

Generally, the weight losses in TG showed correspondence with the endothermal effects in DTA. The weight losses were found to be very useful for quantitative measurements.[18] A comparison of the DTA and TG curves are illustrated in Fig. 7. The DTA curve shows an exothermic effect at 940°C for crystallization that is absent in TG.

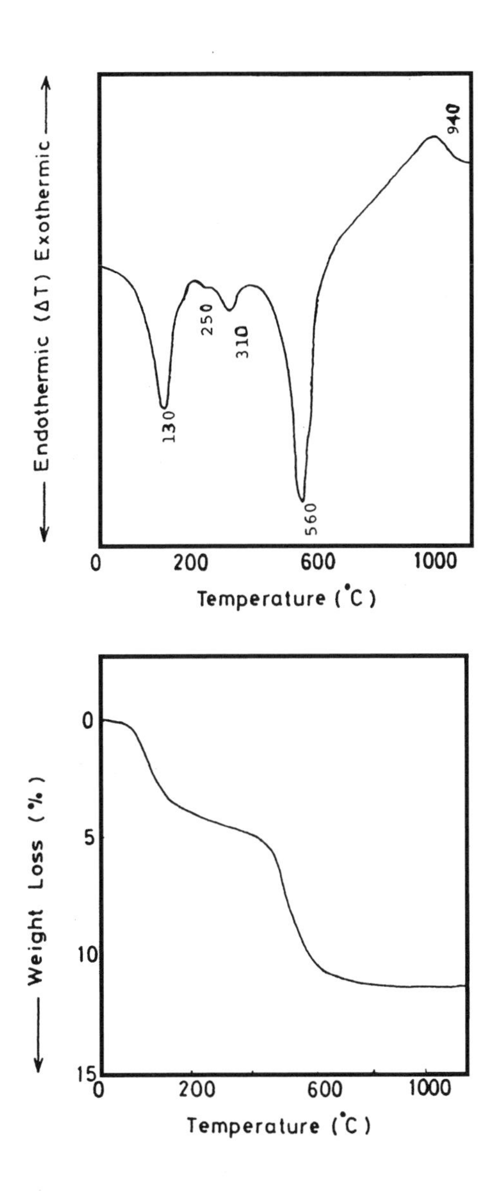

Figure 7. DTA and TG curves of a montmorillonite clay.[18]

Thermomechanical Analysis (TMA) permits determination of length changes occurring in a clay body that is heated from the ambient temperatures to the temperature that the body is subjected to in actual practice. The approximate shrinkage values for kaolinite, illite, and montmorillonite are 2–17%, 9–15%, and about 20% respectively. The total shrinkage in a clay body is determined by the shape, type, amount of clay minerals, and their constituents. For example, the presence of quartz will reduce shrinkage. Shrinkage may increase in the presence of alkali, iron, and alkali earths. The temperature-shrinkage curves not only are useful to identify the clay mineral, but they also give information on the desirable temperature to which clay should be fired. Ramachandran and Ramez applied various thermal techniques for investigating the plastic clays.[18] The TMA showed the following trends: Initial shrinkage occurred at 110–300°C due to the loss of adsorbed water from clay minerals. Maximum shrinkage occurred in samples containing montmorillonite. In the range 420–650°C, the shrinkage was slight although dehydroxylation took place. A large shrinkage, however, resulted in the range 680–800°C. This could be attributed to the vitrification and crystallization effects. Some samples exhibited expansion around 1000°C due to the bloating effect. Total shrinkage was higher in samples containing sodium chloride and montmorillonite. Illitic clays showed lower shrinkages. A TMA analysis of a clay to which 10% limestone is added is shown in Fig. 8.

Dilatometry (length change measurement at temperatures up to 1400°C) can provide useful information on the relevant phase changes that occur in fired clay products. This technique can be used to supplement information obtained using other thermal methods. Clay minerals generally shrink when fired. Shrinkage values (%) for kaolinite, illite, montmorillonite, and halloysite vary between 2 and 20%.[19] The magnitude of shrinkage is affected by the size and shape of the clay mineral particles as well as the compositional variations. The amount of shrinkage for kaolinites is dependent on particle morphology. Shrinkage is directly related to the percentage of very small plate-shaped particles. The presence of large masses of the mineral reduces shrinkage. A similar effect would be expected for illite. Also, the shrinkage generally increases as the amount of fluxing agent increases. There is, however, no apparent correlation between the content of hydroxyls in the clay mineral structure and the firing shrinkage.

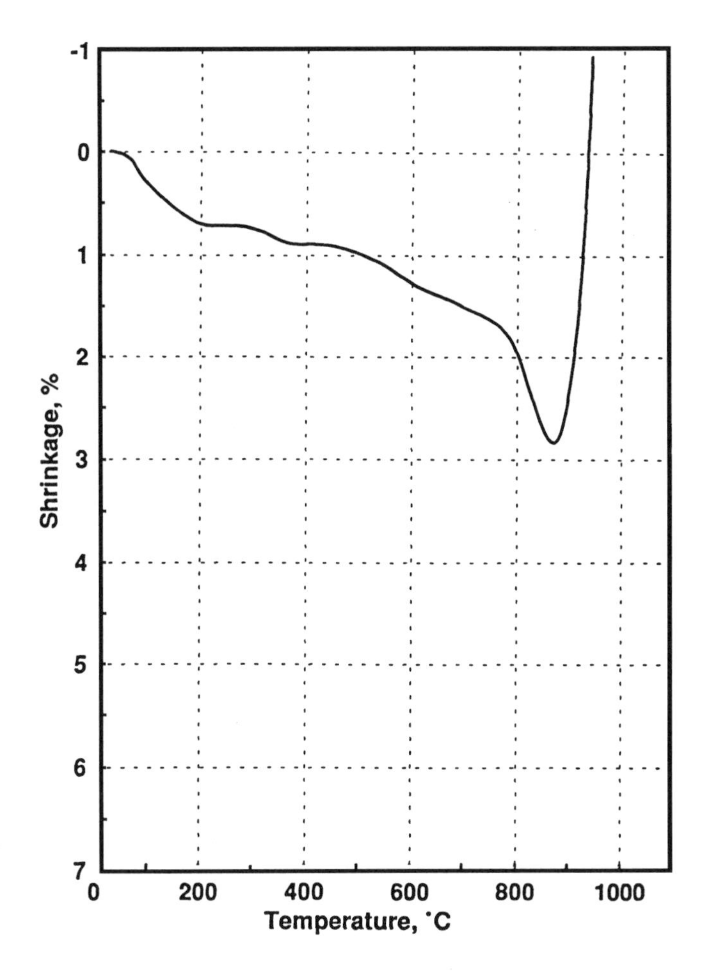

Figure 8. TMA of a clay containing 10% limestone.[18]

Dimensional changes occurring on firing kaolinite and illite are illustrated in Fig. 9. Kaolinite and holloysite exhibit a slight expansion up to about 500°C followed by a small amount of shrinkage accompanying the loss of hydroxyl water, up to about 900°C. At higher temperatures, larger shrinkage accompanies the formation of the high temperature crystalline phases. Shrinkage at about 1000°C has been attributed to the formation of mullite. Shrinkage at higher temperatures, to about 1200°C, is associated with the formation of cristobalite.

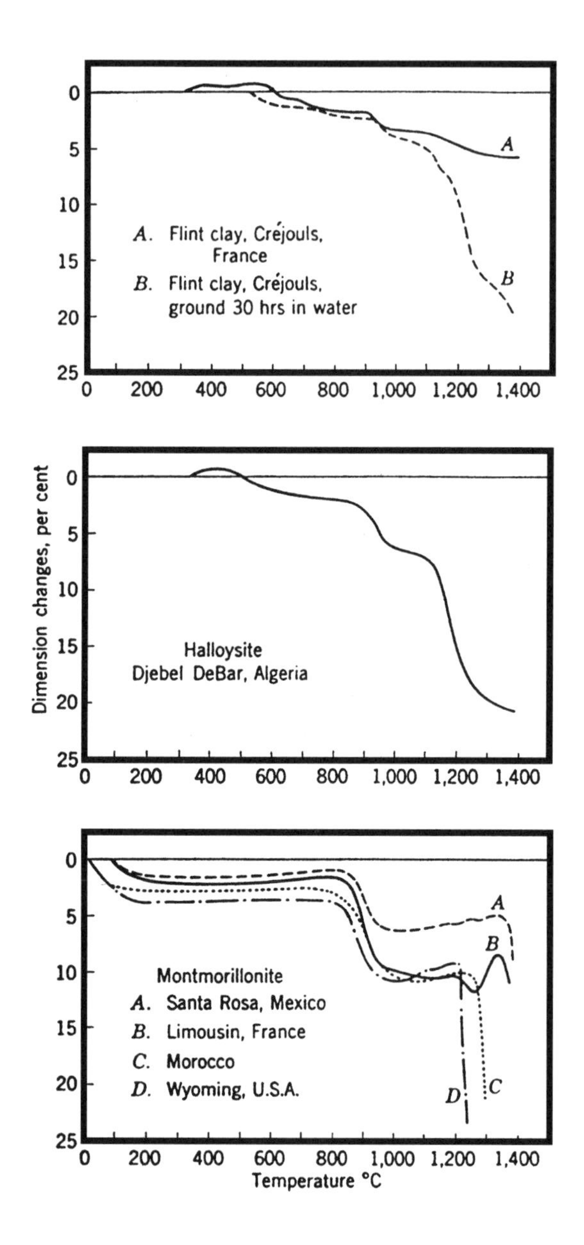

Figure 9. Dimensional changes on firing kaolinite (flint clay), halloysite, and montmorillonite.[21]

Data for illite (not shown in Fig. 9) indicate a slight expansion in the temperature range 450–800°C followed by a rapid continuous shrinkage.[20] Expansion is attributed to the development of the anhydrite structure, and the shrinkage to loss of structure of the phase. The high rate of shrinkage correlates with the relatively lower development of high temperature crystalline phases.

The aluminous montmorillonites (Fig. 9) are dimensionally stable between 200 and 800°C.[21] Above 800°C, there is a relatively rapid shrinkage to about 1000°C, a plateau to about 1200°C, and rapid shrinkage at higher temperatures. Dimensional change above 1000°C varies with composition and variations in the formation of the high temperature phases.

In general, the loss of hydroxyl water results in a slight shrinkage if there is some loss of crystalline structure. If the structure is not substantially altered, e.g., with montmorillonite, a slight expansion may be associated with the loss of hydroxyls. Pore space, textural variations, and formation of gases on firing can influence the magnitude of a shrinkage or expansion. Major shrinkage is related to formation of high temperature crystalline phases, and the rates are directly related to the development of glassy material. Formation of gases during vitrification can have a significant effect on the net dimensions of fired clay minerals.

2.3 Accessory Minerals

Many accessory minerals present in a clay influence its properties in the plastic and hardened states. Some of the organic and inorganic materials that may be present in clays include clarin, vitrain, marcasite, pyrite, quartz, gypsum, limestone, and dolomite. Their presence is indicated by endothermal peaks caused by decarbonation or exothermic peaks caused by oxidation. DTA curves of some of the accessory materials that may be present in a clay are shown in Fig.10.[3]

The DTA curve for a shale used in the manufacture of brick is shown in Fig. 11.[5] A single DTA curve is capable of identifying the following:

- Water (adsorbed as well as absorbed) below 150°C.

- An organic inclusion indicated by an oxidation effect extending over a temperature range from 200°C to nearly 600°C.

- A broad endothermal effect for the dehydration of constitutional water (partly being masked by the exothermal effect).
- An endothermal effect for the decomposition of the dolomite followed by an exothermal trend for the crystallization effect.

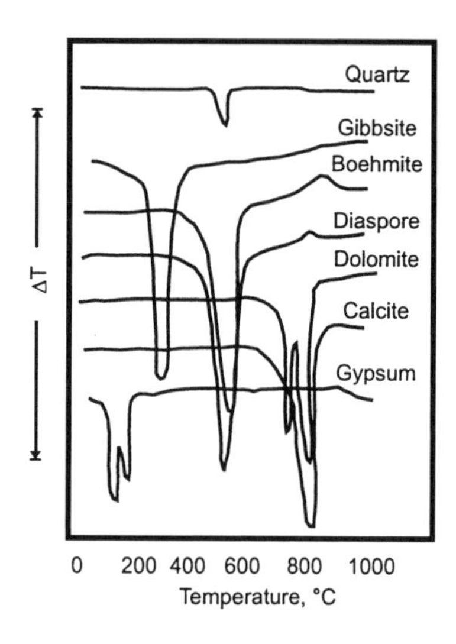

Figure 10. Some accessory minerals present in clays.[3]

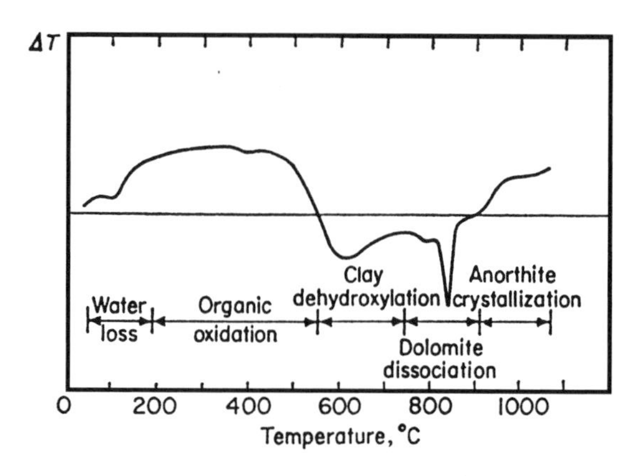

Figure 11. DTA curve for a shale used in brick manufacture.[5]

DTA and dilatometric curves for a clay containing about 20% dolomite and used in the manufacture of facing brick are shown in Fig. 12.[22] Above 700°C, the clay begins to shrink following the completion of dehydroxylation and commencement of dolomite dissociation. The residue from the dolomite reacts with the alumina and silica from the clay resulting in a large expansion at about 850–1000°C. It is a requirement that bricks made from this clay must be heated slowly to prevent cracking.

A vitrification range during which liquid phases are produced from the crystalline components exists in most clay materials. A long vitrification range is commercially desirable. Results from equilibrium systems of pure oxides may be significantly different from those produced from natural materials. It is, therefore, important to characterize raw materials at each plant location for most industrial applications.

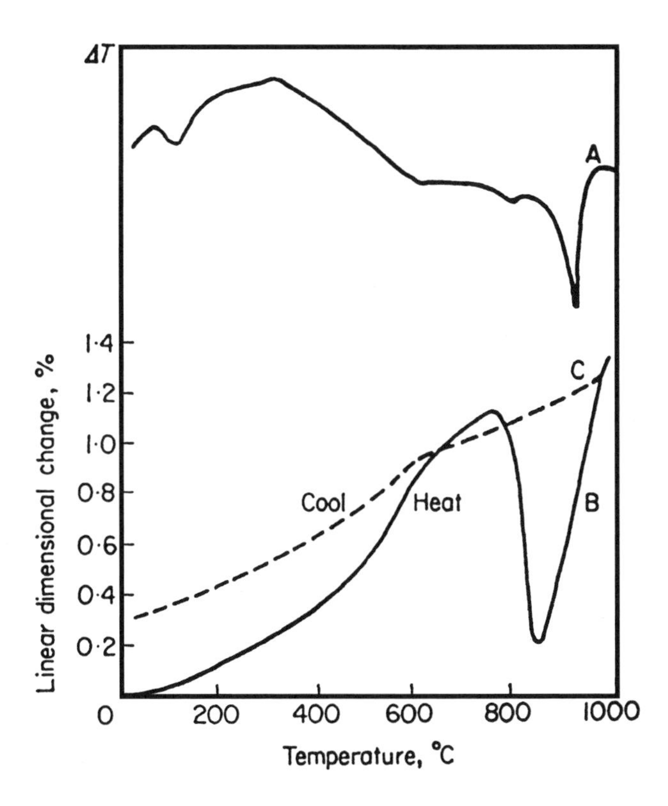

Figure 12. DTA (*A*) and dilatomeric, (*B*) on heating, (*C*) on cooling, curves for a clay containing about 20% dolomite.[5]

3.0 APPLICATIONS

3.1 Analysis of Brick Clays

The thermal techniques are routinely used for quality control purposes. A typical TG/DTA analysis is shown in Fig. 13 for a Thessalian brick clay.[23] None of these clays contains kaolin and the characteristic mullite peak is absent. The endothermic peak between 100 and 160°C is due to the removal of adsorbed water, and its size is dependent on the surface area and crystallinity of the clay. The peak (400–700°C) is attributed to the dehydration of the combined water (dehydroxylation of the silicate lattice) and decomposition of the clay. The third peak (800°C) indicates decomposition of the carbonates and other salts present in the clays.

3.2 Thermal Efficiency of Kilns

The heat of dehydration of clays caused by the water between 425 and 750°C accounts for nearly 10% of the total input in the thermal efficiency of kilns used for the production of structural clay products. The possibility of applying DTA to calculate directly the heat of dehydration of brick clays was examined by Ramachandran and Majumdar.[24] They tested six clays and a bentonite that showed endothermal effects for the loss of combined water (Fig. 14). Samples S-229, S-230, S-233, and S-234 exhibit double peaks, indicating that the clays contain a mixture of clay minerals. Heats of dehydration were determined by obtaining the areas of the endothermal peaks, using potassium sulfate as the calibrating agent. The values were in the range 8.9–69 calories/gram.

3.3 Dark Color of Soils

There are several types of soils that exhibit extensive cracks on drying. Such soils normally contain a montmorillonitic clay mineral. The black cotton soils of India contain about 0.5% organic matter and montmorillonite as the predominant clay mineral. The causes leading to the black color have implications for brick making and in agriculture. The dark color has been attributed variously to titaniferrous magnetite, magnetite, a high C:N ratio, type of cations, clay character, organic matter, etc. Ramachandran, et al.,[25] found that a sensible reduction in organic matter by any method invariably was followed by a consequent reduction of the dark color. The DTA technique was applied to investigate the effect of various treatments on the exothermal effect occurring due to the organic matter. Treatment with 10% sodium hexametaphosphate at the boiling point, or with H_2O_2 at 80°C, or the dichromate was effective in removing the color and so was heating the clay at 400°C. It was concluded that the oxidation of the organic matter was responsible for the dark color, and it occurred in stages at 325° and 410°C.

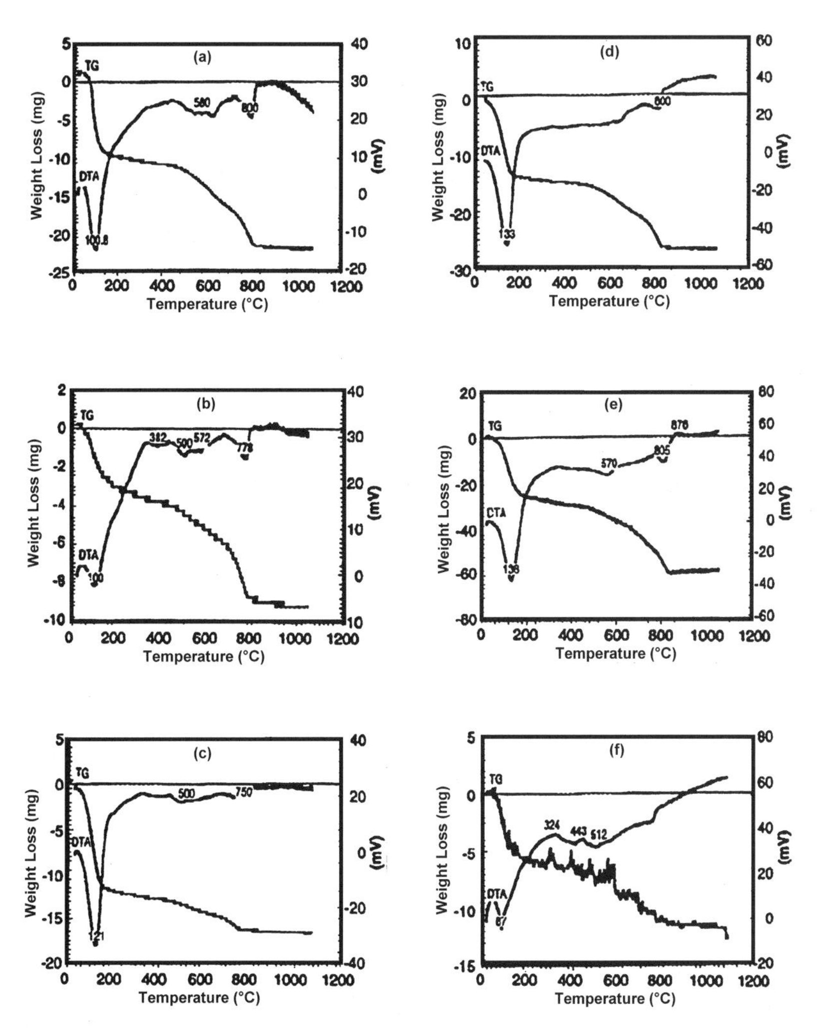

Figure 13. TG-DTA curves for brick-clay powders: *(a)* 1; *(b)* 2; *(c)* 3; *(d)* 4; *(e)* 5; and *(f)* 6.[23]

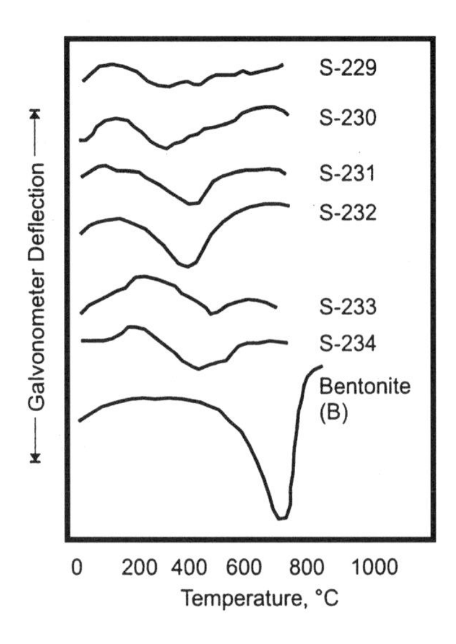

Figure 14. DTA of some clays.[24]

3.4 Bloatability of Clays

Clays form an important group of raw materials for the production of lightweight concrete. In practice, the clay pellets are fired to a temperature 1000–1300°C for a short period when the pellets suddenly increase in volume, called bloating. Various materials present in the clay are reported to be responsible for the bloating phenomenon. The gas-producing agents are variously given as hematite, pyrite, carbonates, sulfates, alkali earths, graphite, lignite, and micaceous minerals. The gases responsible for bloating are oxygen, sulfur, sulfur dioxide and trioxides, carbon dioxide, water, hydrocarbons, and entrapped air either alone or in combination with others. Ramachandran, et al.,[26] mixed several potential bloating agents with montmorillonite and a synthetic aluminate silicate corresponding to illite. The agents, such as hematite, calcite, molasses, pyrites, gypsum, siderite, magnesite, dolomite, saw dust, fly ash, talc, and muscovite, were mixed with the clay fired at 50°C intervals from 1100 to 1300°C. Figures 15 and 16 show the DTA curves of some of the bloating agents. The temperature, intensity, and the endothermal/exothermal nature of the peaks were used to determine the mechanism and the efficiency of bloating of various agents.

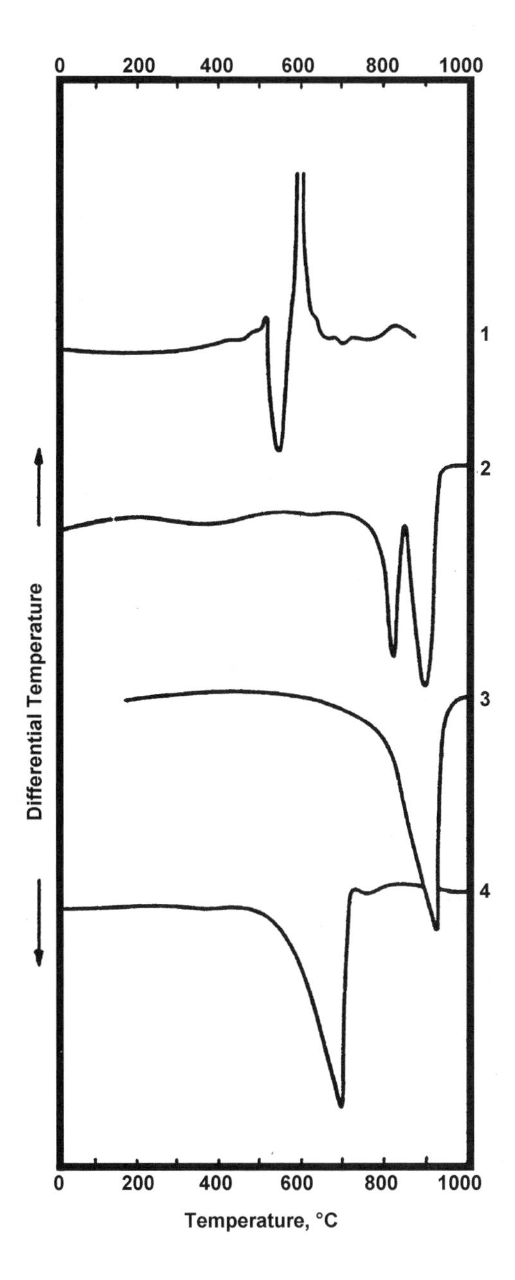

Figure 15. Thermograms of some gas-producing agents: *(1)* siderite; *(2)* dolomite; *(3)* calcite; *(4)* magnesite.[26]

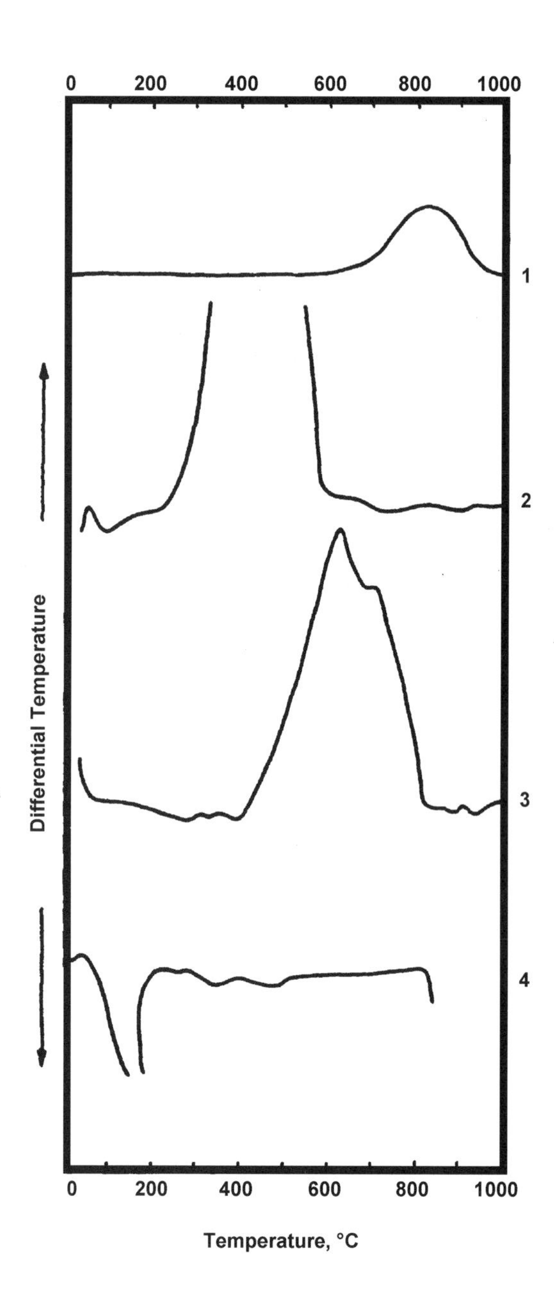

Figure 16. Thermograms of some gas producing agents: *(1)* graphite; *(2)* sawdust; *(3)* fly ash; *(4)* sodium carbonate.[26]

3.5 Weathering of Roofing Slates

Slates are related to clays. The DTA technique has been applied to study the weathering quality as well as an aid to the identification of slate of unknown origin. Figure 17 shows the differential thermal analysis of samples of roofing slate of three different qualities.[3] The first curve represents a slate of excellent durability. The peaks at 610° and 850°C may be due to some type of chlorite. The inflection at 575°C is caused by the presence of quartz. In the calcined material (2nd curve), the presence of quartz is more clear. The third curve is obtained with a slate that was found in practice to delaminate slowly on roofs under conditions of low atmospheric pollution. The poor durability may be caused by the slow oxidation of pyrite in the slate and subsequent reaction between the oxidation product and calcite to form calcium sulfate. The presence of pyrite and calcite is indicated by the peaks at 450° and 770°C, respectively. The fourth curve of slate of pure quality is dominated by the large calcite peak. The exothermic peaks at 930° and 420°C suggest the presence of chlorite and a small amount of pyrite, respectively.

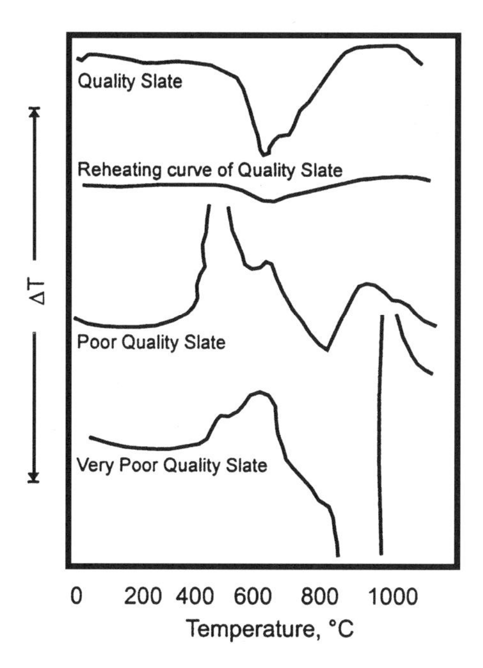

Figure 17. DTA of roofing slates.[3]

3.6 Soil Stabilization

Multi-component clay-based chemistries involving reactions between clays and lime and pozzolans are of interest in the area of soil stabilization. Thermograms[27] for kaolinite and montmorillonite treated with lime are presented in Fig. 18. Addition of lime results in the gradual diminution of the primary kaolinite dehydroxylation peak (500–600°C) to a greater extent than can be accounted for by dilution alone. All samples have a small peak at about 130°C and a broad endothermic peak at about 210°C. The decomposition of carbonated lime is associated with endothermic reactions at 700–800°C.

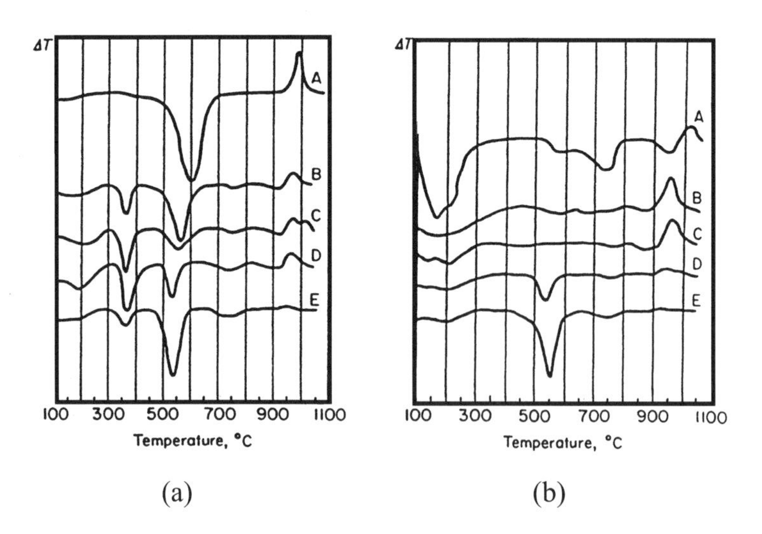

Figure 18. (*a*) DTA curves for kaolinite reacted with various proportions of lime at 35–40°C for 8 weeks: *(A)* 0% lime; *(B)* 25% lime; *(C)* 40% lime; *(D)* 60% lime; *(E)* 75% lime. (*b*) Montmorillonite reacted with various proportions of lime at 35–40°C for 8 weeks: *(A)* 0% lime; *(B)* 25% lime; *(C)* 40% lime; *(D)* 60% lime; *(E)* 75% lime.[27]

3.7 Structural Ceramics

Structural ceramics are generally brittle composite systems with a coarse microstructure consisting of two or more phases. Kaolinitic clays are used in the industrial production of structural ceramics. Thermograms from TG and DTA tests are shown in Fig. 19 for kaolin, a pottery mixture, brick clay, and fly ash.[28] The possible transformation reactions are shown on the graphs. The large endothermic peak (575°C) is associated with the dehydroxylation of the silicate lattice and decomposition of the clay. This is accompanied by a significant weight loss from combined water (about 12%) calculated from the TG curve. The exothermic peak at 996°C is attributed to the crystallization of mullite and/or γ-alumina and spinels. There remains considerable disagreement regarding the cause of the exothermic peak and the composition of the spinel phase. The pottery mixture exhibited a small endothermal peak due to the removal of adsorbed water at 96°C. The peak at 553°C is also associated with dehydroxylation. A very small third peak at 985°C may be the result of mullite formation. Impurities in the kaolin contribute to the lack of definition for this peak. The clay brick mixture displayed an endotherm at 100°C (adsorbed water) and two peaks (600–800°C) indicating decomposition of illite. No new crystal phases were formed up to 1100°C. The TG and DTA curves for fly ash showed mainly water loss and gas evolution.

Determining the relationship of ceramic properties to structure and sintering temperature is facilitated by thermal analysis. The contribution of mullite relative to glass may be related to the strong interface between mullite and glass when mullite needles are formed and grown in a glass matrix. The improvement of mechanical properties of brick clays with sintering temperature suggests that glass formation is beneficial to the microstructure. Increase of sintering temperature for kaolin specimens from 1000 to 1100°C increased the flexural strength substantially. A second steep increase, almost doubling the strength, occurred when the sintering temperature increased to 1400°C. This was followed by a decrease at higher temperatures suggesting that 1400°C was an optimum temperature. The pottery mixture sintered from 900 to 1300°C showed similar mechanical behavior to that of kaolin. An increase in temperature from 900 to 1150°C increased strength by a factor of about 6. At temperatures giving complete vitrification, kaolin and the pottery mixture had similar strength, toughness, and other mechanical characteristics.

The main characteristic differentiating brick clay from the other two materials was the presence of sand particles and the absence of mullite even after sintering at high temperatures. The sand particles were large enough to act as initiating sites of catastrophic fracture without any substantial preceding crack growth. The addition of fly ash to kaolin ceramics resulted in formation of cracks and deterioration of properties.

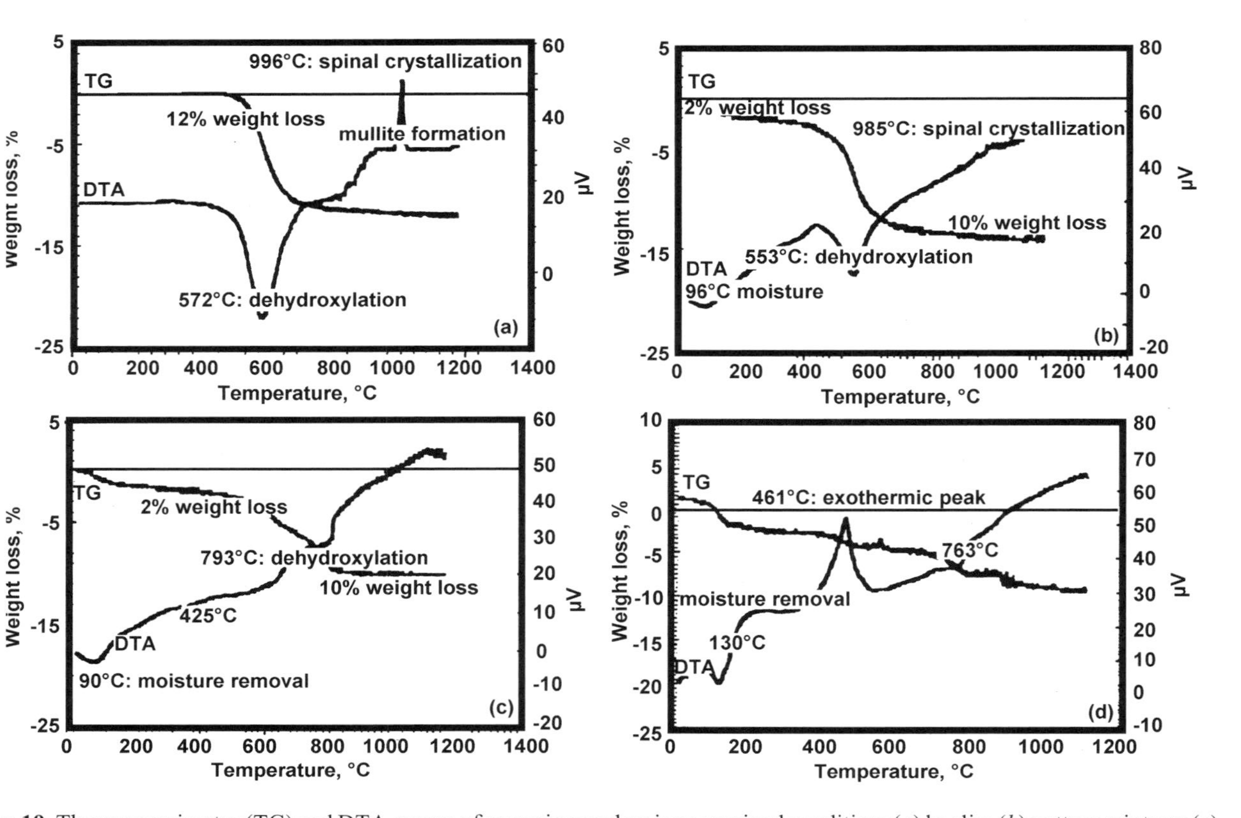

Figure 19. Thermogravimetry (TG) and DTA curves of ceramic powders in as received condition: (*a*) kaolin; (*b*) pottery mixture; (*c*) brick clay; (*d*) fly ash.[28]

3.8 Solid Waste in Clay Bricks

The possibility of using Aswan clay combined with industrial waste products, e.g., blast-furnace slag and air-water cooled converter slag, has been investigated with the assistance of thermal analysis.[29] The results show that the substitution of 10% clay by blast-furnace slag fired at 900 and 1000°C improves the compressive strength and the bulk density, but the substitution by converter slag decreases these properties. The results of DTA provided confirmation of the mineralogical composition of the starting materials, Fig. 20. The thermograms show two endothermic reactions with peaks at about 120 and 560°C, respectively. The first broad peak is due to the removal of the moisture and the interlayer water loss from kaolinite and illite. The second sharp peak is related to the dehydroxylation of lattice water of clay minerals. The exothermic peaks at 930°C occur due to the recrystallization of the dehydrated constituents of metakaolinite into spinel.

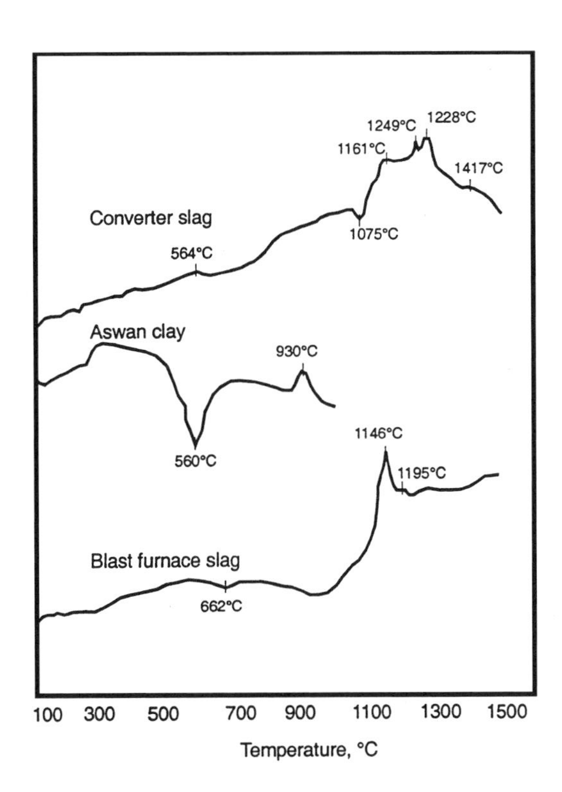

Figure 20. DTA thermograms of blast-furnace slag; Aswan clay and converter slag.[29]

3.9 Archaeological Investigations

The thermal analysis of different archaeological samples of bricks, terra-cotta, and local clays provides useful information on the chemical composition of historical samples and can provide confirmation of specific sites of origin.[30] The thermal characterization of local clays is valuable to assist the restoration of historical pottery materials.

The temperature at which ancient ceramics, terra-cotta, and pottery were fired varies from 500–1300°C and depends on the type of clay. The clay minerals are the main materials for the production of bricks and terra-cotta figures. The clays contain kaolinite, feldspar, quartz, aluminite, gehlenite, hydromagnesite, calcium, magnesium and iron carbonates and hydroxides, hygroscopic and bound water, and organic substances. TG and DTG thermograms for Leharu clay are presented in Fig. 21. The weight-loss peak around 100°C is due to the removal of adsorbed water. Peaks at 200 to 350°C are attributed to bound water. The dehydration of hydromagnesite (250–280°C) and brucite (350–420°C) are followed by the decomposition of calcium hydroxide (400–520°C). Magnesium carbonate decomposes in the 450–520°C range, and calcium carbonate in the 700–900°C range. The dehydroxylation of clay minerals is, thus, considered to occur over the wide temperature range 125–555°C.

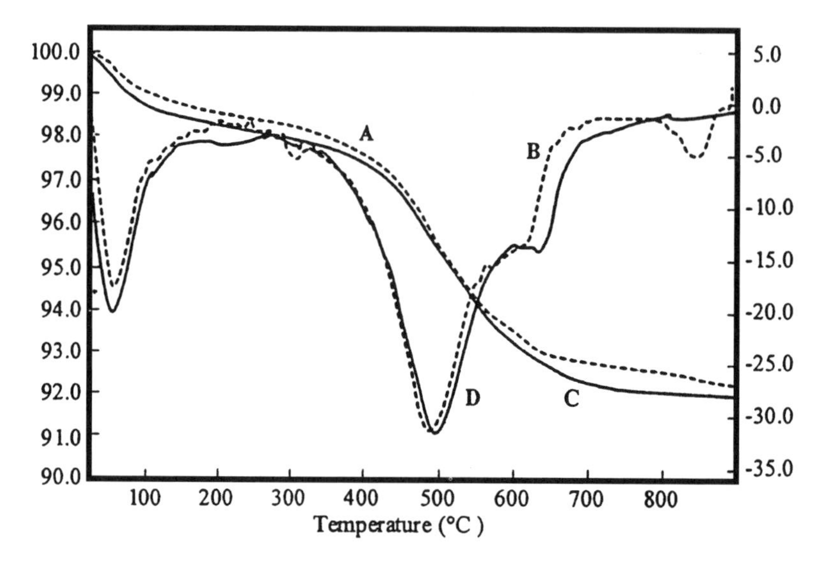

Figure 21. TG-DTG analysis of Leharu clay in air and nitrogen atmosphere: *(A)* TG curve in air; *(B)* DTG curve in air; *(C)* TG curve in nitrogen; *(D)* DTG curve in nitrogen.[30]

4.0 DURABILITY OF CLAY BRICKS

4.1 Dimensional Changes

Chemical, mineralogical, and physical characteristics of raw materials used for brick manufacture are important indicators of potential frost durability. The following discussion pertains to raw materials containing illites mixed with chlorites.

Thermal dilatometric curves obtained for six bricks extruded from different raw materials are shown in Fig. 22.[31] Curves for those manufactured from similar raw materials, but made by different forming methods, i.e., soft mud, dry press, and stiff extrusion, are shown in Fig. 23. From room temperature up to 500°C, the dimensional change is approximately linear and is due to thermal expansion. The abrupt increase in expansion (500–800°C) is due to phenomena such as quartz inversion, exfoliation of the illitic, and chloritic minerals due to dehydroxylation, and, possibly, the escape of CO_2 under pressure.

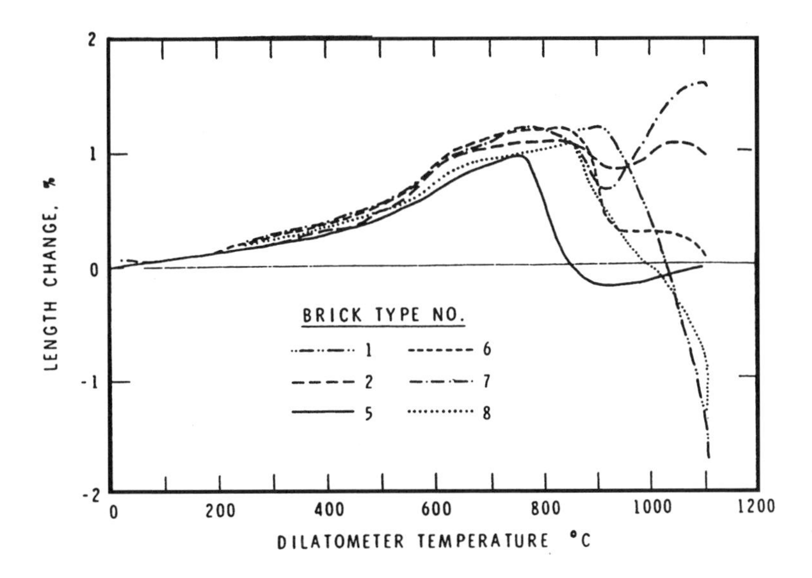

Figure 22. Thermal dilatometric curves of bricks extruded from different raw materials.[31]

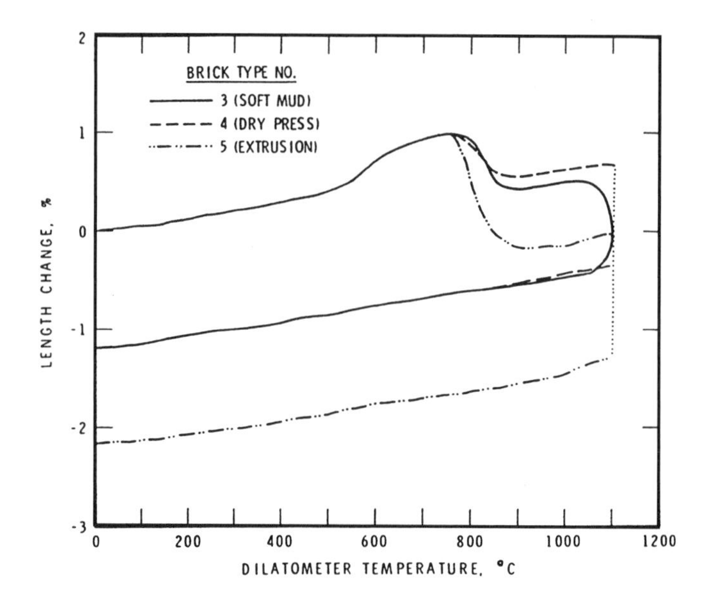

Figure 23. Thermal dilatometric curves of bricks made by different forming methods using similar raw materials.[31]

At higher temperatures, the exfoliated layers of the clay minerals collapse and carbonates decompose resulting in shrinkage and vitrification in the brick matrix. The temperature at which the collapse begins (800–900°C) is typical of illites and seems to be a function of the geology of the raw materials. Three distinct behaviors above 900°C (Fig. 22) were observed. Bricks labeled number (1) and (8) display abrupt shrinkage caused by continued vitrification until melting and a substantial glassy phase may be developed. This type of behavior is common among clays and shales low in minerals containing calcium. Bricks labeled (5) and (6) illustrate the second kind where further collapse is prevented by the formation of a crystalline matrix, consisting mainly of calcium silicates, such as gehlenite, diopside, and anorthite. The third type is characterized by a secondary expansion after the first shrinkage (bricks 2 and 7). This secondary expansion is related to the reaction between fine clay particles and calcite and is chemically linked to the crystallization and growth of the calcium silicates characteristic of some shales.

The general dilatometric behavior of bricks made by different forming methods using common raw materials (Fig. 23) is similar. The difference in the amount of shrinkage at temperatures higher than 800°C may be attributed to the differences in particle size distribution and packing density of the green bricks. The brick-making materials were primarily mixtures of illites and chlorites. The properties of the laboratory-fired samples were similar and representative of the plant-burnt bricks for the types studied with the exception of brick (8) which contained the highest surface area raw material.

4.2 Saturation Coefficient

The absorption value of the burnt brick is a function of the calcium carbonate content in the raw material (within the normal brick-firing temperature range). This can be determined by thermal analysis. The different forming techniques alter this relationship. The saturation coefficient (standard CAN3-A82.2-78) seems to be proportional to the amount of fine particles. It is emphasized that both absorption and saturation coefficients should be used in evaluating brick durability.

The relationship between saturation coefficient and absorption reflects both the quality of the brick and the history of the brick, e.g., the starting raw material composition (clay minerals, quartz, calcite) and the firing condition (Fig. 24). These factors are conveniently identified and estimated by thermal techniques as already discussed. For each raw material, a rational durability index should be defined in terms of absorption and saturation coefficient. The appropriate index can be determined for quality control as the raw material changes.

4.3 Firing Temperature of Clay Brick

The freezing-thawing durability of brick made of the same clay (undergoing the same process) depends essentially on the degree of firing or vitrification in their manufacture. Litvan argued that a test that could predict the firing temperature of bricks would also enable an assessment of the freezing-thawing durability of the product in addition to the performance of the kiln.[32] Dilatometry was shown to be a potentially useful technique for this application.

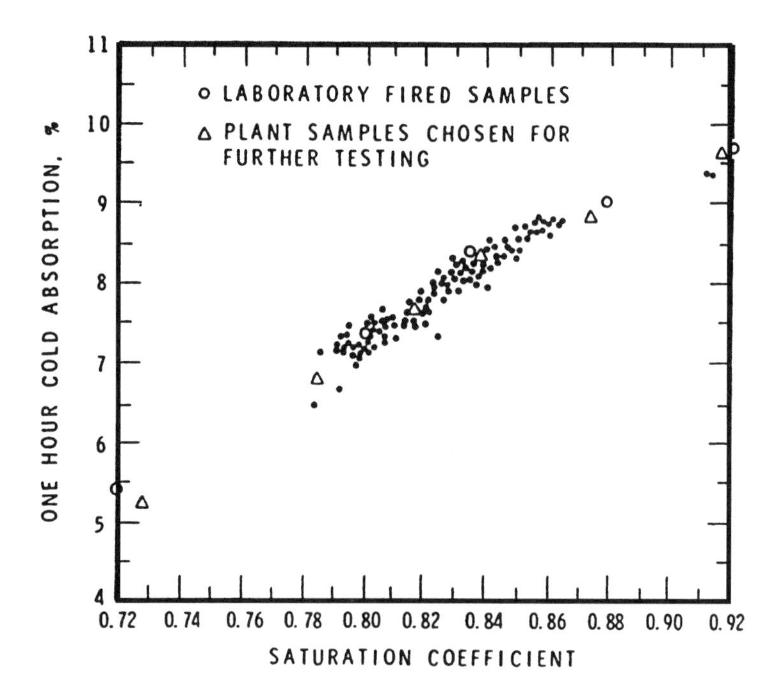

Figure 24. Relation between one hour cold absorption and the saturation coefficient of laboratory and plant-fired clay brick specimens.[31]

Length changes for a Canadian brick fired at various temperatures between 900 and 1100°C are shown in Fig. 25. All curves show a marked hysteresis and a length anomaly around 575°C (due to α-quartz - β-quartz inversion). These parameters have a magnitude that is inversely proportional to the temperatures at which the specimens had originally been fired.

Residual length changes produced by the temperature cycle are plotted in Fig. 26 as a function of firing temperature. Values range from 0.4% shrinkage (900°C firing temperature) to 0.06% expansion (1100°C firing temperature). Clay brick shrinks during firing and undergo additional shrinkage during the refiring test. The previously well-fired brick (1100°C) expanded slightly instead of shrinking as a result of re-heating.

The integral of the peak in the length change curve (at approximately 575°C) was shown to decrease systematically with firing temperature. This peak suggests the α-quartz to β-quartz inversion. This inversion, unlike the quartz-tridymite conversion, is rapid and instantaneously reversible. A decrease in the anomaly with increased firing temperature is likely a result of a more complete conversion of quartz to tridymite above 870°C.

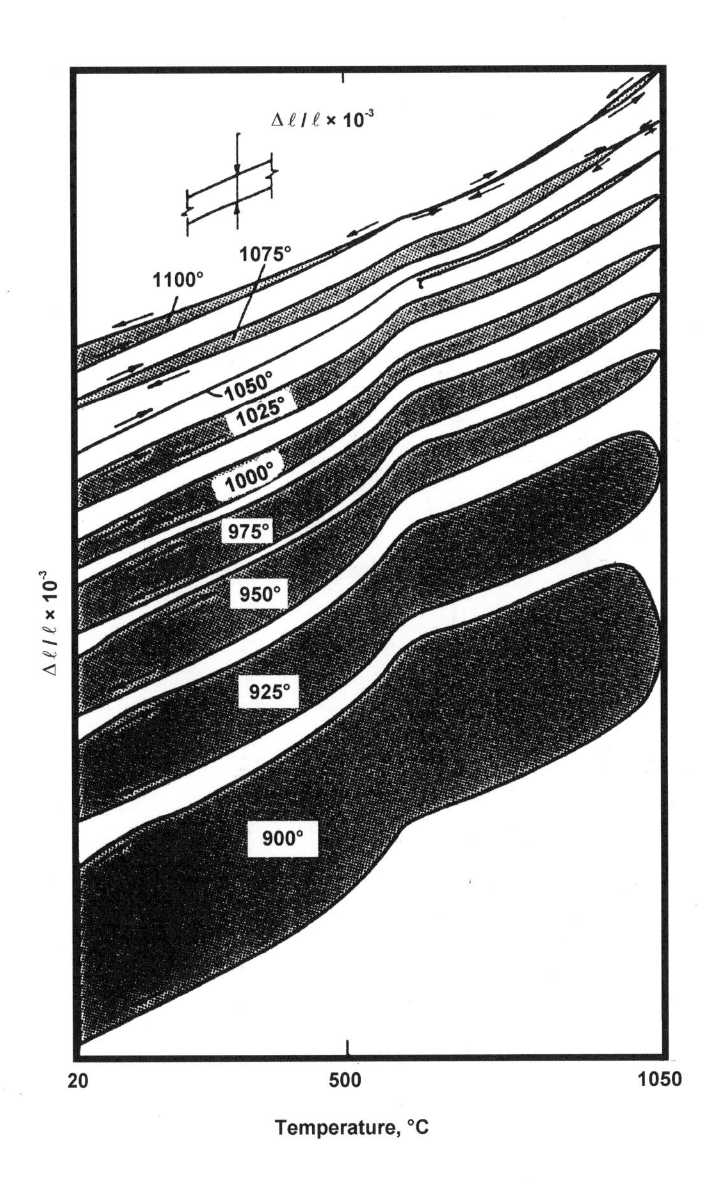

Figure 25. Length changes of brick specimens fired at temperatures indicated during a temperature cycle; heating and cooling rate, 3°C/min.[32]

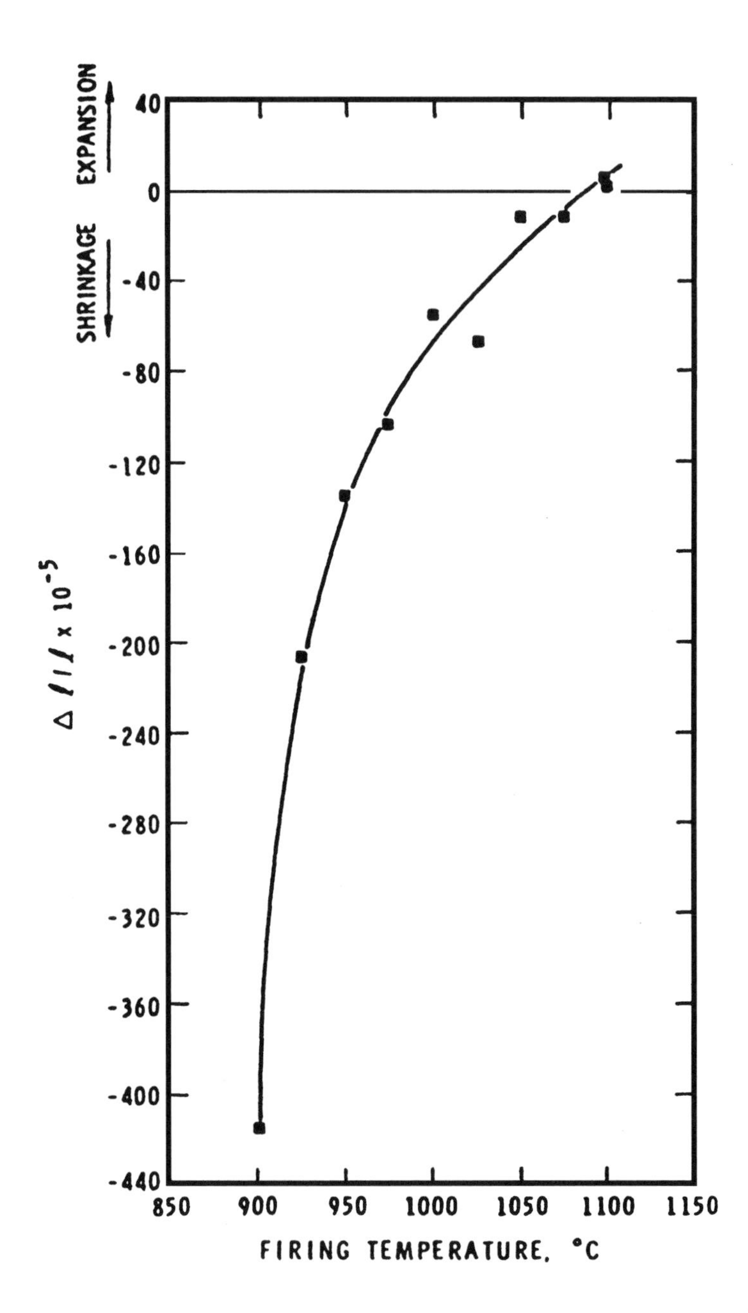

Figure 26. Permanent length changes of bricks fired at temperatures produced by the temperature cycle of a dilatometric experiment.[31]

The DSC thermograms for brick specimens fired at various temperatures are given in Fig. 27. The heat liberated per unit mass of the specimen (calculated by arbitrarily defining the limits of the peak area) as a function of the firing temperature is indicated in the figure. An inverse proportionality is indicated. The results are dependent on clay composition and other characteristics and will be different for clays of different origin. However, the information can be of value in brick manufacture at an individual plant for quality control of the finished product and for the firing process.

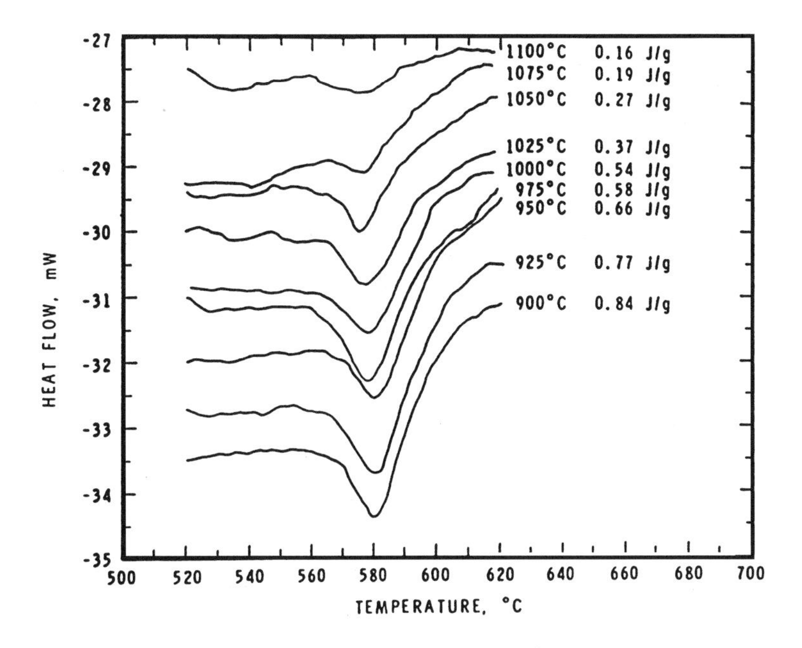

Figure 27. Thermograms of brick specimens fired at temperatures indicated.[32]

The DTA technique has been applied to examine the fire clay products for underfiring. An example has been provided by Honeyborne.[3] In Fig. 28, thermograms of a brick from a certain locality are given. *(A)* refers to the locality, *(B)*, *(C)*, and *(D)* show the thermograms of the same clay fired to 600, 800, and 1000°C. The increase in the firing temperature modifies the thermograms, and at 1000°C, only the peak due to the inversion of quartz (570°C) remains. Curve *(E)* refers to a sample of facing brick with extensive decay. Comparison of the thermograms suggests that this brick was made from the locality from which *(A)* was taken, and, further, it was fired to a temperature between 600 and 800°C. The large endothermal peaks in the vicinity of 150–160°C denote gypsum which was present in the original clay, or was formed by the reaction between the calcium carbonate in the brick and sulfur gases from the atmosphere. Its crystallization in the pores may have contributed to the disintegration of the poorly bonded brick particles.

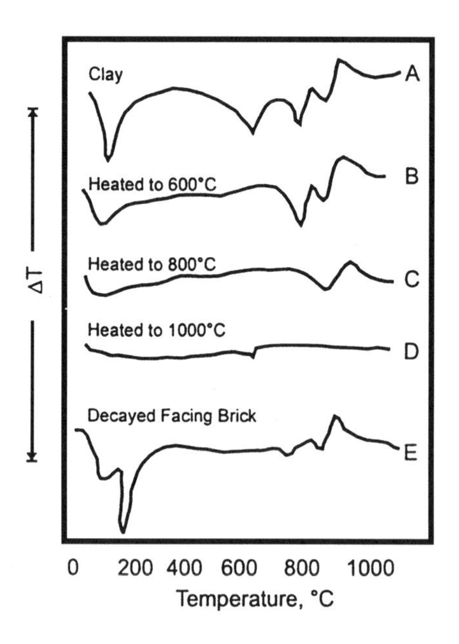

Figure 28. Brick clays heated to different temperatures.[3]

4.4 Brick Particulate Additives for Concrete

Thermomechanical Analysis (TMA) enables the measurement of length change that occurs on cooling and warming saturated porous materials. The magnitude of these length changes can be used to assess the durability of building materials to freezing and thawing. Incorporation of porous brick particles, with at least 30% total porosity, (pore diameters primarily between 0.3 and 2 micrometers) into cement paste and concrete mixes can significantly improve their freezing-thawing resistance.[33] The concentration necessary to achieve a given level of frost resistance depends on the physical characteristics of the material (mineralogy and clay mineral content) and also the temperature to which the brick is fired. Thermal techniques may be applied for the optimization of firing temperature and the retention period. A cement paste (w/c = 0.50) can withstand in excess of 1000 freezing-thawing cycles when it contains about 16 mass percent of brick particulate (0.5 ± 0.08 mm in size, 36% total porosity).

The spacing of brick particulates (distance between particle surfaces) is linearly related to the particle diameter for a given volume concentration. Freezing-thawing cycles expressed as the number of cycles required to produce 0.2% residual expansion is plotted versus spacing in Fig. 29 for cement paste, w/c = 0.50 and wet cured for 28 days. There is a dramatic increase in the number of cycles when the spacing is reduced below 0.8 mm.

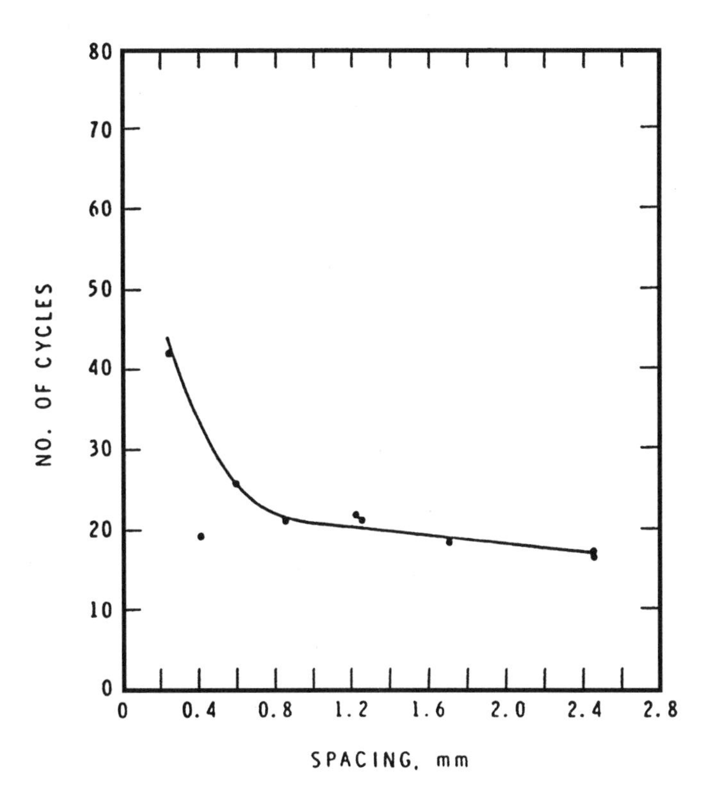

Figure 29. Freezing-thawing resistance, expressed as the number of cycles required to produce 0.2% residual expansion, as a function of spacing of neat cement paste samples.[33]

Residual length changes of concrete specimens subjected to freezing-thawing cycles and containing brick particles (total porosity, 16%) are recorded in Table 1. The basic concrete mix had a cement:aggregate:sand ratio 1:2.75:2.25 and a water:cement ratio of 0.58. The spacing factors for the concretes listed in Table 1 were 0.046, 0.013, 0.033, 0.029, and 0.033 respectively.

Table 1. Residual Length Changes of Concrete Containing Brick Particulates[26]

	Length Change, %		
Sample	**50 Cycles***	**160 Cycles**	**360 Cycles**
Plain	0.117, 0.103	0.702, 0.478	1.908, 1.202
5% brick	0.009, 0.006	0.015, 0.018	0.026, 0.025
10% brick	0.011, 0.010	0.108, 0.034	0.218, 0.158
20% brick	0.003, 0.005	0.012, 0.015	0.022, 0.022
Air entrained	0.007, 0.005	0.010, 0.009	0.025, 0.019

*ASTM C-666 Procedure "B"

The two types of brick described above (36% and 16% porosity) demonstrate that the larger porosity sample is more effective in protecting the sample against the effect of freezing and thawing. Apparently the larger porosity and its particular pore size distribution are more effective in relieving internal stress. The mean pore diameter is about 1 μm. The lower porosity brick has a greater number of pores in excess of 2 μm and is more restrictive regarding moisture movement across particle boundaries.

It is, therefore, apparent that incorporation of suitable porous brick particles in the plastic mix increases the freezing-thawing resistance of hardened concrete without the serious shortcomings of the conventional air-entrainment method, i.e., instability of bubbles and strength reduction.

REFERENCES

1. Grim, R. E., *Applied Clay Mineralogy,* p. 422, McGraw-Hill, New York (1962)
2. Grim, R. E., *Clay Mineralogy,* p. 384, McGraw-Hill, New York (1953)
3. Ramachandran, V. S., and Patwardan, N. K., Differential Thermal Analysis of Building Materials, *J. Natl. Bldg. Org.,* 2:1–9 (1957)
4. Brindley, G. W., and Nakahita, M., Kinetics of Dehydroxylation of Kaolinite and Halloysite, *J. Am. Ceram. Soc.,* 40:346–350 (1957)
5. Mackenzie, R. C., (ed.), *Differential Thermal Analysis: Applications,* 2:607 Academic Press, New York (1972)
6. Satoh, S., A Study of the Heating and Cooling Curves of Japanese Kaolinite, *J. Am. Ceram. Soc.,* 4:182–194 (1921)
7. Mellor, J. W., and Scott, A., The Action of Heat on Kaolinite and Other Clays I, *Trans. Br. Ceram. Soc.,* 23:322–329 (1923–24)
8. McVay, T. N., and Thompson, C. L., X-ray Investigation of the Effect of Heat on China Clays, *J. Am. Ceram. Soc.,* 11:829–841 (1928)
9. Gruver, R. M., Henry, E. C., and Heystek, H., Suppression of Thermal Reactions in Kaolinite, *Am. Miner.,* 34:869–873 (1949)
10. Bradley, W. F., and Grim, R. E., High Temperature Thermal Effects of Clay and Related Materials, *Am. Miner.,* 36:182–201 (1951)
11. Demediuk, T., and Cole, W. F., A New Approach to the Investigation of the Cause of Moisture Expansion of Ceramic Bodies, *Nature,* 182:1223–1224 (1958)
12. West, R. R., and Gray, T. J., Reactions in Silica-Alumina Mixtures, *J. Am. Ceram. Soc.,* 41:132–136 (1958)
13. Grim, R. E., and Kulbicki, G., Montmorillonite: High Temperature Reactions and Classification, *Am. Miner.,* 46:1329–1369 (1961)
14. Gerard-Hinne, J., and Meneret, J., Les Réactions Thermique à Haute Temperature des Kaolins et Argiles, *Bull. Soc. Franç. Céram.,* 30:25–33 (1956)
15. Ramachandran, V. S., Kacker, K. P., and Patwardan, N. K., Basic Dyestuffs in Clay Mineralogy, *Nature,* 191:696–698 (1961)
16. Kacker, K. P., and Ramachandran, V. S., Identification of Clay Minerals in Binary and Ternary Mixtures by Differential Thermal Analysis of Dye-Clay Complexes, *9th Int. Ceram. Congr., Brussels,* p. 483–500 (1964)
17. Rai, M., Kacker, K. P., and Ramachandran, V. S., Identification of Clay Minerals in their Mixtures by Differential Thermal Analysis and Differential Thermogravimetry, *Indian J. Technol.,* 4:93–97 (1966)

18. Ramachandran, V. S., and Ramez, M., Shale Brick Production, Prepared for IDRC of Canada, p. 114 (1993)
19. White, W. A., The Properties of Clays, MS Thesis, U. Illinois, p. 121 (1947)
20. Kiefer, C., Importance et Influence de la Nature Minéralogique des Constituants sur la Comportement des Masses Céramique, *Bull. Soc. Franç. Céram.,* 17:14–33 (1952)
21. Kiefer, C., Propriétés Dilatométriques des Minéraux Phylliteux entre Oct 1400°, *Bull. Soc. Franç. Céram.,* 35:95–114 (1957)
22. Brady, J. B., Bell, K. E., and Zengals, L. K., Some Methods of Investigating Firing Problems, *J. Canadian Ceram. Soc.,* 34:9–20 (1965)
23. Papargyris, A. D., Botis, A. I., Papapolymerou, G., Spiliotis, X., and Kasidakis, D., Thermal Analysis of Brick Clays, *Bull. Am. Ceram. Soc.,* 73:53–55 (1994)
24. Ramachandran, V. S., and Majumdar, N. C., Differential Thermal Analysis as Applied to Study of Thermal Efficiency of Kilns, *J. Am. Ceram. Soc.,* 42:96 (1961)
25. Ramachandran, V. S., Ahmad, F. U., and Jain, L. C., The Dark Colour of Black Cotton Soils, *Proc. Indian Acad. Sci.,* 50:314–322 (1959)
26. Ramachandran, V. S., Kacker, K. P., and Garg, S. P., Effect of Various Gas Producing Agents on the Bloatability of Illite and Montmorillonite, *7th Conf. Silicate Ind., Budapest,* pp. 183–197 (1963)
27. Croft, J. B., The Pozzolanic Reactivities of Some New South Wales Fly Ashes and Their Application to Soil Stabilization, *Proc. Austr. Road Res. Bd.*, 2(2):1144–1168 (1964)
28. Papargyris, A. D., and Cooke, R. D., Structure and Mechanical Properties of Kaolin Based Ceramics, *Br. Ceram. Trans.,* 95:107–120 (1996)
29. Elwan, M. M., and Hassan, M. S., Recycling of Some Egyptian Industrial Solid Wastes in Clay Bricks, *Ind. Ceram.,* 18:1–6 (1998)
30. Paama, L., Pitkanen, I., and Peramaki, P., Analysis of Archeological Samples and Local Clay Using ICP-AES, TG-DTG and FTIR Techniques, *Talanta,* 51:349–357 (2000)
31. Kung, J. H., Frost Durability Study on Canadian Clay Bricks III. Characterization of Raw Materials and Burnt Bricks, *Durability of Bldg. Mater.,* 5:125–143 (1987)
32. Litvan, G. G., Determination of the Firing Temperature of Clay Brick, *Bull. Am. Ceram. Soc.,* 63:617–618 (1984)
33. Litvan, G. G., and Sereda, P. J., Particulate Admixture for Enhanced Freeze-Thaw Resistance of Concrete, *Cement Concr. Res.,* 8:53–60 (1978)

13

Introduction to Organic Construction Materials

1.0 INTRODUCTION

There are many types of organic-based materials used in construction, such as adhesives, sealants, paints and coatings, asphalts, and roof covering materials. Different types of polymers constitute such materials. The behavior of these materials during fabrication and exposure to natural elements may be investigated by various physico-chemical techniques. In recent years, thermoanalytical techniques have been used with some success for such studies. In this chapter, various types of organic-based materials and polymers that are used in the construction industry are discussed. Subsequent chapters present information on the application of thermal techniques for the investigation of these materials.

Macromolecules are very large molecules with atoms linked together by covalent bonds. They may be linear chains, branched, or three-dimensional networks.[1] Some natural macromolecules, such as those present in biological construction materials, have a very complex structure.

However, many synthetic macromolecules have a relatively simple structure being formed by a chemical process called polymerization, which consists in joining together repeating (*poly*) small units (*mers*) to form a macromolecule called a *polymer*.

Polymers with long linear chains and a high degree of symmetry are called *linear* (e.g., thermoplastics), and those with side branches are called *branched* polymers. If the macromolecules are highly interconnected through chemical cross-linking of linear polymers by means of chemical reactions, they become a three-dimensional network (e.g., thermosets). They are, therefore, called three-dimensional polymers. The mechanical behavior of cross-linked molecules is different from those without cross-linking. Examples of polymer types are as follows:

- Linear polymer, e.g., poly(vinyl chloride)

$$—(CH_2)_n—CHCl$$

- Branched polymer, e.g., natural rubber (polyisoprene)

$$\begin{array}{l} —CH_2 = CH—C = CH_2 \\ \quad\; | \\ \quad CH_3 \end{array}$$

- Network polymers, e.g., silicones, epoxies, and urethane foams

$$\begin{array}{ccccccc} & R & & O^- & & & \\ & | & & | & & & \\ — & Si & — & O & — & Si & —O— \\ & & & | & & & \\ & & & O & & & \text{silicone} \\ & & & | & & & \\ & & — & Si & — & O— & \\ & & & | & & & \\ & & & R & & & \end{array}$$

According to Kravelen,[1] the fundamental characteristics of a polymer are the chemical structure and the molecular mass distribution pattern. The former includes the nature of the repeating units, end groups, composition of possible branches and cross-links, and defects in the structural sequence. The molecular mass distribution, which depends upon the synthesis method, provides information about the average molecular size and its irregularities. These characteristics are responsible, directly or indirectly, for the polymer properties. They are directly responsible for the cohesive force, packing density and potential crystallinity, and molecular mobility (with phase transitions). Indirectly, these properties control the morphology and relaxation phenomena (behavior of the polymer).

There are three factors that affect the nature of a polymeric material and determine whether it is glassy, rubbery, or fiber-forming under a given set of conditions.[2] These factors are: polymer flexibility, the magnitude of intermolecular forces, and the stereoregularity of the polymer. They influence: the ability of the polymer to crystallize, the melting point of the crystalline regions, and the glass transition temperature. Discussion of each of these factors is beyond the scope of this chapter.

There are organic polymers, semi-organic polymers, and inorganic polymers.[3] Organic polymers contain carbon, hydrogen, oxygen, nitrogen, sulfur, and halogen atoms. Examples of these are poly(vinyl chloride) (PVC) and polyurethane:

$$–CH_2–CHCl–CH_2–CHCl–CH_2 \qquad \text{(PVC)}$$

$$–OC–NH–(CH_2)_n–NH–CO–O–(CH_2)_n–O– \qquad \text{(polyurethane)}$$

Semi-organic polymers include materials with heteroatoms directly bonded to the carbon backbone of the polymer, but do not contain nitrogen, sulfur, oxygen, or a halogen. One example is poly(dimethyl siloxane)

$$–(CH_3)Si–O–Si(CH_3)–O–$$

Inorganic polymers do not contain any carbon in their chains, and they can be formed by the elements of the group IV of the periodic table. Cements, polysilanes, and polysilicates below are typical examples of these types of polymers:

H_2–Si–Si–H_2– (polysilane)

–Si–O–Si–O– (polysilicate)

Polymeric silicates may be lamellar or have three-dimensional crystalline structure. Talc and crysotile asbestos are lamellar silicates. Asbestos fibers are based on a chain of Si, O, and lateral groups with MgO.[3] Other important inorganic polymeric compounds are inorganic cement pastes, which consist mainly of various silicates.[4] Organic and inorganic polymers are considered the essential engineering materials in the building/ construction industry.

Polymeric materials are easy to synthesize. Therefore, the final products are substantially more economical than other types of building materials. Synthetic polymers are widely used in the construction industry. Some of the organic polymer materials used in the building industry are fibers, rubber, plastics, adhesives, sealants, coatings, and caulking compounds. Polymers are also used as modifiers to improve the properties and performance of some construction materials, such as roofing and road asphalt.

According to Feldman,[3] more than 100×10^6 tons per year of synthetic polymers are produced in the world in a variety of products for different applications. Building materials consume close to 3.6×10^6 tons. The use of polymeric building materials is greater in Europe than in North America. This is because raw materials, such as wood, are scarce and the cost of steel, aluminum, and other metals is usually high. Moreover, construction tends to be more innovative in Europe.

There are different methods of classifying polymers. Engineers and materials scientists classify polymers based on the physical properties as:

(a) Thermoplastics

(b) Thermosets

(c) Elastomers or rubbers

(d) Elastoplastics and

(e) Thermostable polymers[5]

Another method is to use their response to thermal treatment and classify them as thermoplastics and thermosets.[2] According to Young,[6] a useful way is to categorize the polymers into three groups displaying similar properties, which has the advantage of their underlying molecular structure. These groups are thermoplastics, rubbers, and thermosets. The above classifies polymeric materials based on their properties rather than the chemistry used in their manufacturing process.

Thermoplastics, referred to as plastics, are obtained from the polymerization of long chain molecules. In such materials, the molecules are not interconnected. Hence, they can be melted on heating, molded, and remolded by conventional methods. The thermoplastics can be subdivided in two groups: crystalline and non-crystalline polymers.[6] Thermoplastic building materials include poly(vinyl chloride) (PVC), thermoplastic polyolefins (TPO), and polyethylene (PE).

Thermoset polymers are solid products, which cannot be softened, will not melt on heating, and will decompose at a high temperature. Their molecules are substantially cross-linked in an extensive three-dimensional network of covalent chemical bonding.[2] Examples of thermosets are EPDM polyesters and epoxies.

Rubbers are polymeric materials consisting of long-chain molecules coiled and twisted in a random manner, which display elastomeric properties. They are able to completely recover from very large deformation once the applied stress is released. Uncured elastomers are unable to completely recover from large deformation due to the sliding of the molecules over each other during stress. After curing, the molecules are cross-linked and held together preventing sliding of the molecules without changing the coiled form.[1] Rubber properties include resilience, low temperature flexibility, resistance to oils, greases, ozone, and many acids and bases.

Thermoplastic rubbers (TPR) are a type of rubber produced by polymerization of a mixture of substances, such as butadiene and styrene. They have some advantages over conventional rubbers. The advantages are: (a) the material can be melted and easily molded; and (b) the chemical composition can be varied to improve or to produce a material with new properties with a wider range of use. Styrene-butadiene-styrene (SBS) polymer is commonly used as a modifier in the manufacturing of asphalt-based roofing materials.

Thermoplastic, thermoset, rubber, and rigid foam polymeric materials are of great commercial importance in the building industry. However, they cannot always be used as engineering materials without improving properties such as strength. Polymer properties can be enhanced by adding reinforcing fillers. They are solid, chemically inert substances added to a resin to modify their properties and, most of all, reduce cost. Most fillers are minerals; they reduce cost, provide speed of cure; add strength; and provide special electrical, mechanical, and chemical properties.[3] The combination of the polymer (thermoplastic, thermoset, elastomer) and reinforcing-fiber materials such as glass or carbon results in a material known as a *composite.* A wide range of crystalline and amorphous materials can be used as the fiber. Carbon black and glass fiber are commonly used in the fabrication of construction materials, e.g., roofing.

There is a strong interaction between the three major construction sectors: the building industry, the building materials manufacturers, and the industrial mineral producers who supply many of the raw materials to make the building products. Therefore, any changes in the construction industry and technological advances in building products and construction methods could have a potential impact in any of these sectors.[7]

Most of the construction materials are derived from four major sources: forest products, metal fabrication, chemical products, and industrial minerals. Wood, as a forest product, has many applications, e.g., as a fuel, construction materials, in weapons, and transportation. As the science of engineering developed, wood was replaced by other materials. However, there has been a tremendous increase in the number and variety of products made either directly or indirectly from wood.[8] There is a diversity of products used in construction, which are derived from industrial minerals such as portland cement, calcium carbonate, and fillers. Similarly, stone in one form or another has been an important construction material from early time. For example, internationally, the great ages of stone construction have reflected the wealth of societies. Modern construction uses cut natural stone for the structural fabric and decoration of many buildings.

Natural chemical products and polymer-based (e.g., asphalt) materials have been part of building construction. Their applications range from the reinforcement given by natural fibers to the use of resins as adhesives. As a result of new trends in the construction industry, there has been a major increase in the use of synthetic polymers such as adhesives, sealants, paints, coatings, insulation materials, and plastics. Since these polymers contain carbon, they are also called *organics*. Examples of their applications are given Table 1.

The large increase in the use of polymer-based materials in the construction industry is due to the fact that they have certain advantages over mineral materials. For example, they are lightweight; can be welded; cut and glued in rapid, automated processes; and are unlike cement-based materials which need a long-curing time. However, as with other materials, polymer-based and plastic building materials have some disadvantages. Many of them have low thermal stability as well as a low melting point. Hence, emission of volatiles is of great concern due to environmental regulations. Furthermore, there is a potential for the lethal release of hazardous fumes by some plastics during a fire.

Table 1. Polymer-Based Building Products[7]

Product Type	Application
Glues and resins	Adhesives
Foams, beads, sheeting	Thermal insulation
Putty, caulking, weather stripping	Sealants
Coatings	Paint and varnishes
Sewer and wastewater pipes, rainwater control	Drainage
Electrical conduits, drinking water pipes	Conduit
Trims and fixtures	Door and window frames, baseboards, lighting fixtures
Shingles, roofing membranes, siding, vapor barriers, ceiling tiles	Cladding
Tiles, linoleum, asphaltic and polymer concrete, carpeting	Flooring and paving
Composite materials used as grouts, glazing, manufactured stones, sanitary ware, etc.	Miscellaneous
Tapes	Electrical insulation

2.0 ADHESIVES AND SEALANTS

The adhesives and sealants represent a $22 billion a year global industry. The market for adhesive products spans a diverse range of industries including: electronics, packaging, appliances, automotive, medical and dental products, construction and building materials, industrial assemblies, furniture, paper, and aerospace.[9] For example, nearly 60% of all gun-grade joint sealants produced globally are used in the construction industry. In 1996, this amount was of the order of 420,000 tons at an installed cost of about $30 billion. A survey, carried out in the UK in 1990, showed that only 5% of building joint sealants have lasted more than 10 years and 55% failed, due to different causes, within ten years of service.[10]

2.1 Adhesives

Adhesive is defined by the American Society for Testing and Materials (ASTM)[11] as a substance capable of holding materials together by surface attachment. Kinloch[12] defines an adhesive as a material which, when applied to a substrate surface, can join them together and resists separation. Adhesion is the holding together of two surfaces by interfacial forces, which may consist of molecular attraction (chemical bonds) or mechanical adhesion (interlocking action). The chemical bond is the active force that holds the surface together whereas the mechanical adhesion is passive and not very effective until an outside force is applied.

Adhesives have been used since ancient times. For example, Egyptians utilized gum Arabic from the acacia tree, egg, glue, semi-liquid balsams, and resins from trees as binders. Wooden coffins were decorated with pigments bonded with a mixture of chalk and glue (gesso).[13]

Vegetable and animal glues have been used as adhesives since the beginning of history. They supply a good bond under dry conditions, but fail if the joints become damp. Moreover, they can only be used on certain materials such as wood and paper. The advent of casein and blood glues was an improvement, but these protein materials are not resistant to either water or fungal attack.[3]

Except for the introduction of rubber and pyroxylin cements over a century ago, there was little advance in the technology of adhesives until well into the twentieth century. In the last few decades, the quality of natural adhesives has been improved, and many synthetics have emerged from laboratories into the market.

Adhesives have a wide range of applications in packaging, sports, shoes, furniture, appliance and houseware assembly, rug backing, fabric combining, surgical, millwork manufacture, electronics, automotive, aircraft, space, transportation, and the building construction industries. Discussion of the application of adhesives in this chapter will be limited to the construction industry.

Most adhesives are organic. The traditional glues and gums from natural sources are either proteins or polysaccharides.[3] There are different types of adhesives. Some of the classifications are based on the end-use, such as metal-to-metal adhesives, paper and packaging adhesives, and general purpose adhesives. They can also be classified according to the chemical composition, methods of application, physical form, and suitability. Landrock[14] classifies adhesives depending on their function, chemical composition, physical form, and application.

Depending on the physical form, the adhesives can be classified as liquid, paste, tape and film, and powder or granule adhesives. According to the mode of application (viscosity), there are sprayable, brushable, or trowelable adhesives.[14][15] Based on the origin and chemical composition of the main component, adhesives can be classified as natural and synthetic.[3] Natural adhesives include three groups: animal (e.g., shellac,animal glue, casein); vegetable (e.g., natural rubber, linseed oil), and mineral (e.g., waxes). Synthetic adhesives can be classified as thermoplastic, thermosetting, elastomeric, or a combination of these types (alloys).[14][15][16]

According to the adhesive function, there are two principal types of adhesive—structural and non-structural. *Structural adhesives* are bonding agents used for transferring required loads between adherends exposed to a service environment, as is typical of a structure.[15] They are materials based on phenolic, epoxy, or acrylic pre-polymers that on curing provide a highly cross-linked adhesive[3] of high performance and strength.[14] Structural adhesive bonds must be capable of transmitting structural stress, without loss of structural integrity, within limits.[17] They are stronger over a wider temperature range than other adhesives. Their primary use is for high strength and high temperature where the highest and strictest end-use conditions must be met regardless of cost (e.g., military applications).[14][15] *Non-structural adhesives* are materials which do not support substantial loads, but hold lightweight materials in place. They are sometimes called "holding adhesives," e.g., adhesive/sealants (sealing adhesives) which are primarily intended to fill gaps.[15]

Thermoplastic adhesives soften and melt when heated (provided that they do not decompose). In other words, they do not cross-link during cure. These adhesives are single-component and harden upon cooling by evaporation of a solvent or water vehicle. They have poor creep resistance, fair peel strength, and are used mostly in unstressed joints and designs with caps, overlaps, and stiffeners. The most common application is to bond nonmetallic materials such as wood, leather, cork, and paper.[15][16] In general, thermoplastic adhesives are not used for structural applications or at temperatures above 66°C (150°F). Thermoplastic adhesives include:[13][14]

- Cellulose acetate
- Cellulose acetate butyrate
- Cellulose nitrate
- Polyvinyl acetate
- Vinyl vinylidene
- Polyvinyl acetals
- Polyvinyl alcohol
- Polyamide
- Acrylic
- Phenoxy

Thermosetting adhesives set or harden when heated due to curing, which takes place by chemical reaction (cross-linking) at room or elevated temperature. They are supplied as one- or two-component systems. The one-part (or component) system usually requires an elevated-temperature cure and has a limited shelf-life. The two-part systems have longer shelf lives and can be cured slowly at room temperature or somewhat faster at moderately higher temperature. However, they need careful metering and mixing before application. Examples of these adhesives are cyanoacrylates, polyesters, urea-formaldehyde, epoxies, polyimides, and acrylic acid-diesters.[15] Thermosetting adhesives are resistant to heat and solvents, have good creep and fair peel strength. Their main application is for stressed joints at somewhat elevated temperature (93–260°C)[14] because they show little elastic deformation under load.

Elastomeric adhesives are natural or synthetic polymers with superior toughness and elongation. Examples of elastomeric adhesives include natural rubber, reclaimed rubber, butyl rubber, polyisobutylene, nitrile rubber, styrene-butadiene-rubber, etc.[15] Elastomeric adhesives are supplied as solvent solutions, latex cements, pressure sensitive tapes, and single- or multi-component nonvolatile liquid or pastes.[14] However, they are usually supplied in liquid form. Most are solvent dispersions or water emulsions. The service temperature is up to 204°C (400°F). They never melt, have excellent flexibility, but low bond strength. The main application of elastomeric adhesives is on unstressed joints on lightweight materials (e.g., joints in flexure). Hence, they are not considered structural adhesives.[14]

Alloy adhesives are obtained from the combination of resins of two different chemical groups selected from the thermoplastic, thermosetting, and elastomeric groups. This type of adhesive includes:[15]

- Epoxy (phenolic, polysulfide, nylon)
- Nitrile-phenolic
- Neoprene-phenolic
- Vinyl-phenolic
- Polyvinyl acetal-phenolic

The construction industry is now applying structural bonding, a method of structural fabrication used in the aircraft industry. It has been used for manufacturing prefabricated panels where brick slips, or other decorative facings, are bonded to epoxide resin "concrete" backings. In the construction industry, wood, concrete, and other building materials are also bonded with adhesives. Table 2 shows examples of adhesives used in the construction applications.

The adhesives are used in the construction industry that includes residential and commercial buildings, roads, bridges, and other elements of the infrastructure.[18] They are utilized for manufacturing a number of structural products, such as softwood plywood, laminated timbers, laminated paper-based panels, etc. More recently, adhesives have been used to replace mechanical fastenings in the assembly operation of building construction. These are called *construction adhesives*. The compounds must be capable of producing adequate bonds in poorly fitting joints with thick and variable thickness bondlines.[14]

Table 2. Some Adhesive Types Used in Construction Applications[13]

ADHESIVE	APPLICATION
Polyvinyl acetate (PVAC), starch	Gypsum board
Epoxy, PVAC, acrylic, styrene-butadiene-styrene (SBS)	Concrete
Polyurethane rubber (PUR)	Recreational surfaces
Phenolic adhesives, asphalt, neoprene, polyethylene (PE), polypropylene (PP), sodium silicate	Glass fiber and rock wool insulation
Asphalt, SBR/asphalt, neoprene, butyl	Roofing
SBR, acrylic, epoxy, furan, silicone	Grouts and mortar for ceramic tiles
Resin, acrylic	Installation of ceiling tiles
SBR, acrylic, polyvinyl alcohol (PVA), PVAC	Ceramic tiles-additives to cement
SBR, acrylic, neoprene	Stud and framing
Dextrin, starch, acrylic, carboxy-methyl cellulose	Wall covering
Asphalt, PVAC, epoxy, SBR	Vinyl flooring tile
SBR, PVAC, linoleum paste, epoxy	Vinyl flooring sheet
Polyvinyl chloride (PVC), ABS, chlorinated PVC	Pipe-joint cements

Table 2. *(Cont'd.)*

ADHESIVE	APPLICATION
Casein, PVAC, urea-formaldehyde (UF)	Wood bonding
Neoprene, epoxy	Metal bonding
SBR, PUR	Plywood floors
UF, SB, acrylic	Asphalt roofing
Epoxy, neoprene, PUR phenolic, PVAC	Thermal/sandwich panels: e.g., interior/exterior partitions, cold storage, and thermal-insulating panels
SBR	Parquet floor installation
Acrylic, SBR	Decorative brick installation
Asphalt, SBR	Asphalt tiles installation
PVAC, PUR, neoprene	Mobile homes
PVAC	Binding of fiber in cellulose insulation
PF(phenol-formaldehydes), UR, PUR	Manufacturing of plywood
Epoxy-polysulfide	Wood, ceramic, bonding of concrete floors, airport runways, bridges, and other concrete structures especially for outdoor extreme temperatures (freeze-thaw cycles)

In general, adhesives are utilized in construction for many structural and decorative applications, e.g., for installing vinyl flooring, carpeting, ceramic tile, wall coverings, making doors, gluing floors, etc. The construction industry is one of the largest outlets for adhesives, utilizing some 40 types of adhesive in about 30 different applications. Adhesive bonding has many advantages to offer to the building industry. No other method of attachment is as versatile. For example, one cannot consider nailing a ceramic wall tile into position. Their use began with the finishing trades. Flooring materials, wallpaper and roofing cements were the first volume applications. The bulk of the volume consists of thermoplastic and water-based adhesives.[3][13]

The use of adhesives offers many advantages over more conventional methods of joining two surfaces such as brazing, welding, riveting, bolting, screwing, etc. Among these advantages are:[3][19]

- The ability to join dissimilar materials (e.g., plastics and rubbers)
- The ability to join thin sheet materials efficiently
- An increase in design flexibility
- An improved stress distribution in the joint, which leads to an increase in fatigue resistance of the bonded component
- A convenient and cost effective technique

Floor covering such as wood, resilient and rigid surfaces (ceramic tile, slate, and brick), and soft fabric materials, such as carpeting can be installed with the proper adhesive.[13] Until a few years ago, asphalt was used as the adhesive to install wood floors. However, it posed a hazard to the workers due to the material flammability. Although the cost was low, the hazard prompted users to switch to nonflammable adhesives for this application. Most adhesive-installed wood floorings today utilize chlorinated solvent systems, polyvinyl acetate, and epoxy resins.[13]

Solvent-based styrene-butadiene-styrene (SBS) and nitrile formulations as well SBS latex emulsions are commonly used as adhesives for ceramic or ceramic mosaic tiles. It is important to choose a fast drying formulation that allows grouting the tile within a 24–48 hour period. Hence, the solvent systems are favored over the slower drying latex systems.[13]

There are a number of adhesives, glues, and cements with unique properties which are used in the construction industry for special applications. For example, cyanoacrylates, referred to as "super glues" are adhesives of great strength and quick-setting characteristics. Moreover, nitrile rubber (acrylonitrile-butadiene rubber) has some use in curtain wall construction.[13]

Adhesives are also used in the manufacturing of prefabricated sections or assemblies. Examples include: complete wall sections with the sheathing bonded to wood frames; roof sections with sheathing bonded to joists and rafters to provide T-beam action; and box beams and other types of composite beams for joists or rafters.[14] Phenolic adhesives are used in the fabrication of laminated beams and trusses. Epoxies are also widely used for bonding and the repair of concrete.[18]

Adhesive performance is determined not only by the degree to which the adhesive forms intimate contacts with both surfaces, but also by its ability to resist separation from these surfaces. Therefore, the adhesive must possess the rheological properties that promote bonding, but resist debonding. The dual set of properties can be achieved by changing the state of adhesives from wetting to setting or by manifesting its viscoelastic properties.[20]

In the selection of adhesives, applications and performance characteristics are most often considered simultaneously. The most commonly used performance criteria are:[3]

- Minimum tensile, shear, or peel strength at specific temperatures
- Minimum and maximum service temperatures
- Glue line resilience
- Resistance to water, moisture, humidity, and light
- Resistance to oils, greases, and solvents
- Resistance to acid, alkaline, and chemical attack
- Aging and weathering
- Color suitability

Adhesive joints should be designed to take advantage of the desirable properties of adhesives and to minimize their shortcomings. Since the purpose of the joint design is to obtain maximum strength for a given area, the joint's adhesive bonding must be designed for the use of adhesives and not for another method of fastening. The strength of an adhesive joint is determined primarily by:[14]

- The mechanical properties of the adherent and adhesive
- Residual internal stresses
- The degree of true interfacial contact
- The joint geometry

Each of the above four factors has a strong influence on the joint performance. Therefore, the joint must be designed to eliminate stress concentrations, which are not always apparent and may occur as a result of differential thermal expansion of adhesive-adherend, and shrinkage of adhesive on curing. Air can also become entrapped at the interface if the adhesive does not flow easily during curing or does not wet the substrate. The types of stress found in adhesive joints are: compression, shear, tension, peel, and cleavage (see Fig. 1).[14]

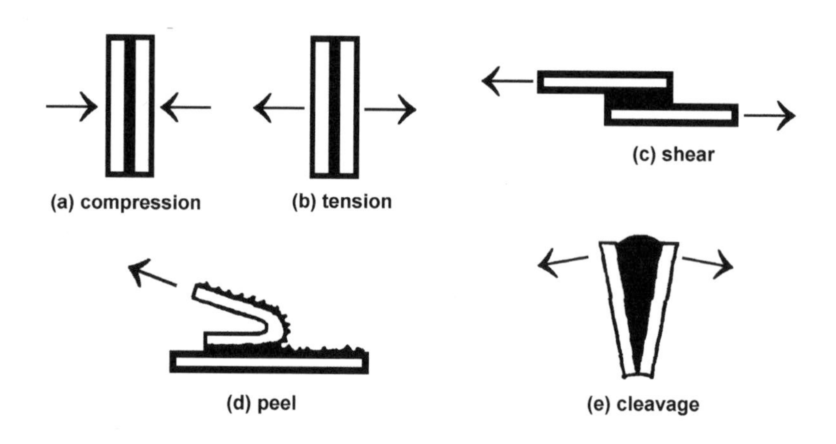

Figure 1. Types of stresses in adhesive joints. *(Reprinted with permission.)*[14]

Compression results from loads in pure compression, but compression joints are limited in application. *Shear loading* imposes an even stress across the bonded area, using the joint area to the best advantage, and providing an economical joint that is most resistant to joint failure. In *tension*, the load is distributed over the joint area as in shear. However, other stresses may be present. With any slight offset of the load, the joint is more likely to fail. In *peel,* a very high stress is applied to the boundary line of the joint and unless the joint is wide or the load small, bond failure will occur.

Cleavage is similar to peel and occurs when forces at one end of a rigid bonded assembly act to split the adherend. The stress in this case is not evenly distributed, but concentrated on one side of the joint.[14]

The ability of the adhesive to support the design loads under service conditions for the expected life of the structure is the first thing to be considered in adhesive selection. Thus, the mechanical properties, durability, and environmental resistance of the bonded structure are also important.[18]

2.2 Sealants

The American Society for Testing and Materials, ASTM C 717-00a, defines a *sealant* as a material that has the adhesive and cohesive properties to form a seal, which in turn is defined as a barrier against the passage of liquid, solids, or gases, and is the name given to the material that has, in general, replaced putties and caulk.[3][21] They are elastomeric materials, which are used to seal cracks and joints in window frames or between panels; to prevent rain, air, and dust from passing through the joint; and even to improve the thermal performance of a wall. The largest variety of sealants fall into the classification of solvent release and are composed of three parts:[22] the liquid portion of the component or non-volatile vehicle, the pigment, and the solvent or thinner to provide ease of application. The non-volatile component varies from a vegetable oil (e.g., linseed) to synthetic elastomers. The pigment introduces color to the material and assists the flow control. The solvent provides a reduced viscosity to the material and enables easy application of the sealant.

Sealants are available in a non-cured, pourable, or extrudable state for easy application. Upon curing, they are transformed into solid elastomeric material. According to Feldman,[3] the characteristics that define a sealant are that they:

- be deformable during application and in service
- absorb cyclic movement without permanent distortion
- have recovery properties
- do not flow from the joint
- adhere to a wide range of surfaces
- have properties that do not vary greatly across the service temperature range
- be durable

Sealants can be marketed in three main forms:

- putty-like mastics
- non-cured tapes, ribbons, and beads
- cured gaskets

The general composition of a sealant consists of a base polymer and additives. However, there is an inclination to classify sealants in terms of their chemical composition or physical properties. Based on this, the main sealant groups are:[3]

- silicones
- polyurethanes
- polysulfides
- polymercaptans
- chlorosulfonated polyethylenes
- polyacrylics
- polychloroprenes
- butyl rubbers
- halogenated butyl rubbers
- polyisobutylenes
- polybutylenes
- drying and non-drying oil-based caulks

A brief description of the different types of sealant is given below. More information is contained in Refs. 3 and 23.

Silicones. They are produced from linear polymers based on polydimethylsiloxane (see reaction) by partial cross-linking. Silicone elastomers consist of a polyorganosiloxane (usually polydimethylsiloxane due to its low cost and wide range of attainable properties), additives (which provide strength and other desired properties) and a cross-linking system.[3]

$$\underset{\text{dimethyldichlorosilane}}{Cl{-}Si(CH_3)_2{-}Cl} \rightarrow \underset{\text{polysiloxane}}{-[-Si(CH_3)_2{-}O{-}Si(CH_3)_2{-}O{-}Si(CH_3)_2{-}O{-}]_n-}$$

The high flexibility of the polymer backbone contributes to the ability of silicones to remain elastic down to low temperatures. They have high recovery and are highly stable in both cured and uncured states.

Polyurethanes. The main raw materials for urethanes are organic diisocyanates and various active hydrogen components, such as polyether polyols, hydroxyl-terminated polyesters, hydroxyl-terminated urethane polyethers, vegetable oils modified by alcohol or some diamines.[23]

One-component sealants are based on isocyanate-terminated prepolymers with an isocyanate equivalent of 1000–2000, which are made from isocyanate and active hydrogen materials with an excess of isocyanates. The curing relies on the atmospheric moisture and proceeds with the formation of urea linkages:[23]

$$\text{NCO–R–NCO} + H_2O \rightarrow \underset{\text{urea linkage}}{\text{NCO–R–NH–CO–NH–R–NCO}} + CO_2\uparrow$$

Curing of the one-component system can be enhanced by introducing a tertiary amine or a metallic catalyst.

The two-component sealants are cured by a chemical reaction between a free isocyanate group of part (A) and the active hydrogen group of part (B). To accelerate the curing, metallic or tertiary amine catalysts can be incorporated in part (B). The curing of the two-component systems is via the formation of a urethane linkage:[23]

$$\text{R–NCO} + \text{–R}'\text{–OH (A)} \quad \text{and} \quad \underset{\text{urethane}}{\text{–R–NH–CO–OR (B)}}$$

Polyurethanes have many valuable sealant properties. They are high recovery sealants with only CO_2 as the by-product of their production. Their low modulus of elasticity is important especially for their use under severe conditions such as continuous cycling between expansion and contraction. However, they suffer from adhesion problems upon immersion in water (hydrolysis) and tend to yellow upon exposure to UV radiation.

Polysulfides. Polysulfides are based on liquid polysulfide polymers containing a small amount of cross-linking in their structure. They are prepared from the reaction between an inorganic polysulfide and di- or polyfunctional halogenated hydrocarbons.[23] The degree of cross-linking is determined by the amount of the polyfunctional that reacts. Their basic structure is $(–R–S_x–)_n$ with n ranging from 2 to 4.[3]

Polysulfide sealants are two-component systems based on liquid polymers compounded with additives as part (A) and an oxidizing curing agent as part (B). They have elastomeric properties and are distinguished by outstanding resistance to oil and solvents. The resistance increases with the increasing content of sulfur. They are also resistant to oxygen and ozone attack, but are susceptible to strong oxidizing acids and strong alkali.

Polymercaptans. Polymercaptans have the same functional thio group as the polysulfides. Liquid linear polymercaptans have the same polyoxyalkylene backbone as the polyether polyols, which constitute an important group of urethane polymers.

Polymercaptans occupy an intermediate position between urethanes and polysulfides. They exhibit the good elastic properties of urethanes and some rheological characteristics of the polysulfides. They are also available as one- and two-component systems. The one-component consists of basic liquid polymer with an incorporated additive, which can be activated by water or oxygen. Part (A) of the two component system consists of a liquid polymercaptan with additives, and part (B) is the oxidizing curing agent. Curing is initiated by metal or organic peroxides, anionic oxidants, or cationic oxidizing agents:[23]

$$-O(CH_2)_3-SH + O_2 \rightarrow -O(CH_2)_3-S-S-(CH_2)_3O- + H_2O$$

Chlorosulfonated Polyethylenes. Chlorosulfonated polyethylenes are made from the reaction of polyethylene chains with various chlorine and sulfonyl chloride groups. The degree of substitution and the ratio of the types of groups depend on the reaction conditions. They can be represented as follows:[3]

$$[-(CH_2)_6-CHCl)_{12}-CH-SO_2Cl-]_n$$

Chlorosulfonated PE is used as a rubber, sealant, and in areas such as flooring, lining for chemical plants, roofing, rollers, etc., where resistance to weathering, corrosion, or abrasion is required.

Polyacrylics. Polyacrylic adhesives and sealants are formulated from functional acrylic monomers, which achieve excellent bonding upon polymerization.[3] Alkyl esters of acrylic or methacrylic acids up to 80,000 molecular weight constitute the main bonds of acrylic sealants.[23]

$CH_2=CCH_3-CO-OR$
alkyl ester of methacrylic acid

$CH_2=CH-CO-OR$
alkyl ester of acrylic acid

$$-\left[\begin{array}{cc} -CH_2-CCH_3 & -CH_2-CH- \\ | & | \\ RO-C=O & RO-C=O \end{array}\right]_n-$$

acrylic copolymer

where R = $-CH_3$, $-CH_2-CH_3$, $-CH_2-(CH_3)_2$.

The adhesive properties of polymers produced by solvent polymerization are better than those produced from emulsion polymerization. Acrylic sealants set by solvent evaporation, have good adhesion, excellent flexibility, good heat and chemical and ozone resistance.[23] However, they shrink upon evaporation of the solvent and upon aging. Acrylic sealants are not recommended for joints exposed to contraction and expansion.

Polychloroprenes. The structure of the polychloroprene chain can be represented as shown below:[23]

$$(-CH_2-CCl{=}CH-CH_2-)_n-$$

Polychloroprene is commercially produced by emulsion polymerization and exists as a one- and two-component system. The one-component polychloroprene sets to touch in 4–6 hours and cures slowly (3–7 days).[23] The two-component system is a solvent-free, non-shrinking, slump resistant sealant which sets quickly, but cures slowly.[3] Polychloroprene sealants have poor color retention (usually are pigmented with black pigments), and high shrinkage.

Polyisobutene and Butyl Rubber. Polyisobutene, referred to sometimes as polyisobutylene, is considered the precursor of butyl rubber and is processed by low temperature cationic polymerization of isobutylene. The chemical structure can be represented as:[3]

$$\cdots\cdots CH_2-C(CH_3)_2-[CH_2-C(CH_3-)_2]_n-CH_2-C(CH_3)_2\cdots\cdots$$

Since this polymer is 100% saturated, it cannot be converted to a practical elastomer nor vulcanized by normal means. To overcome the lack of unsaturation, isobutene is copolymerized with a small amount of a diene monomer. The resulting polymer is vulcanized by sulfur-based systems.

$$\cdots\cdots-[-CH_2-C(CH_3)_2-]_n-CH_2-CCH_3{=}CH-CH_2-\cdots\cdots$$

Butyl rubber sealants can be obtained using butyl crude rubber, which can be dissolved in an appropriate organic solvent to give a solvent-based sealant. Cured butyl rubbers have exceptional resistance to aging by oxidation, ozone, heat and corrosive chemicals. Hence, butyl rubber is used for weather seals and gaskets in buildings. However, organic liquids have a similar swelling action as that of other rubbers and sealants.

Halogenated Butyl. Butyl elastomers have a low density of unsaturation, which results in a low cure rate when conventional vulcanization systems are used. Butyl copolymers containing a small amount of combined chlorine or bromine are vulcanized more quickly than normal butyl rubber because the halogen atoms provide additional sites for the cross-linking process. Butyl rubber has poor adhesive properties to metals and other rubbers because of the lack of polar groups. This is the reason for using halogenation.

Drying Oil-based Caulks. The oils used in sealant formulations are triglyceride esters of long chain unsaturated fatty acids. Oil-based caulking compounds were employed almost exclusively for sealing joints in buildings as lining against the infiltration of air, dust, and water during the first half of the twentieth century.[3]

Presently, they are supplied in gun grade or knife grade formats. The gun grade is a viscous semi-liquid, which is suitable for applications by air or hand-operated guns. The knife grade caulking compounds are more viscous with a consistency similar to mortar, but used to a lesser extent than the gun grade.

$$
\begin{array}{l}
CH_2\text{–}OOR_1 \\
| \\
CH\text{–}OOR_2 \\
| \\
CH_2\text{–}OOR_3
\end{array}
$$

triglyceride

where R_1, R_2, and R_3 are straight chain hydrocarbons.

There are three main factors to be considered in selecting a sealant: joint movement, durability, and cost.[3]

Since most sealants are used in exterior applications, they must be deformable, have good recovery properties, and should have good overall elastic properties. Many sealants used in the construction industry are required to maintain functional performance characteristics over many years. While in service, sealants are exposed to many environmental factors such as UV radiation, water, oxygen, and thermal cycling. Depending on which face of the building the sealants are placed on, the type of substrate, and geographic region, they may be exposed to extreme environmental conditions, stress, and strain gradients. Hence, sealants are susceptible to weather-induced degradation. In general, degradation involves both chemical and physical processes with the chemical reactions usually preceding the physical process.[24]

3.0 PAINTS AND COATINGS

According to ASTM D 16-00, *paint* is a pigmented coating and a *coating* is a liquid, liquefiable, or mastic composition that is converted to a solid, protective, decorative, or functional film after application as a thin layer.[25] A paint can also be defined as a fluid material that when spread over a surface in the form of a thin layer, will form a solid, adherent, and cohesive film.[26]

Paints have been used for decorative purposes for many centuries. They were used about 25,000 years ago by primitive men who used them in caves to outline the shape of animals they hunted.

The paintings produced in ancient Egypt have been shown to incorporate gum arabic, gelatin, egg white, and beeswax. In classical Greece, paints were used extensively in sculpture (for hair, lips, and eyes of statues), architecture, and in painting ships as well as in interior decoration.[27]

By the late eighteenth century, demands for paints of all types had increased such that it became worthwhile for businesses to make paints and varnishes for widespread use.[28] The industrial revolution had a major effect on the development of the paint industry due to the demand for paint to protect machinery, and this marked the start of modern paints. An acceleration of the rate of scientific discovery had a growing impact on the development of paints from the eighteenth century to the present day. Prussian blue, the first artificial pigment with a known chemistry, was discovered in 1704.[28]

Paints were based mainly on drying oils, and this type remained in common use until about the end of the first quarter of the twentieth century when oleoresinous varnishes and, later, alkyd resins gradually replaced the oils. British Standard specifications for oil-type paints were still current in 1982.[27] The purpose of paints and surface coatings is two-fold. They may be required to provide the solution to aesthetic or protective problems, or both.

Since World War II, rapid developments have taken place in the polymer field leading to new types of resin suitable for use in paints. These have enabled the paint technologists to satisfy the demand for high-performance coatings with rapid drying or curing times to meet modern productions.

Fluid paints contain three major ingredients, viz., pigment, binder or film former, and solvent (thinner). The relative proportions of the ingredients can be varied to produce films with the desired physical properties and applications characteristics.

Paints have a discontinuous phase (pigment) and a continuous phase (vehicle). The vehicle is formed by the polymer resin or binder and the solvent or diluent. The pigment is formed by the primary pigment and extender.[28]

Pigments are finely divided solids used to provide color and enhance film durability and hardness. They can be classified as inorganic or organic.

Inorganic pigments can be sub-classified as natural or manufactured. Both can be either true pigments or extenders. In finishes, they contribute to durability. A pigmented film is more weather resistant than an unpigmented film of the same binder. *Extenders* are inorganic solids that differ from "true" pigments in their behavior when dissolved in organic media and are practically transparent. Unless very impure, they do not make any contribution to colors or opacity. Extenders are used in certain types of paint such as undercoats, primers, and some low-gloss finishes to modify or control physical properties. The majority of natural pigments are iron oxides or hydroxides, but may contain appreciable quantities of clay or siliceous matter.[27]

The organic or manufactured pigments cover the entire spectrum range, but the brilliance and opacity vary considerably. In general, they are brighter than the inorganics, but show a much greater variation in opacity and in light fastness. Organic pigments derived from plant and animal sources are no longer used by the paint manufacturers because they lack permanence.

Binders or resins (film former) are continuous phases in the paint that bind the pigment particles into a coherent film that adheres to the substrate. In modern paints, the mechanical and resistance properties of the film are contributed largely by the binder.

The majority of binders are organic materials such as oleoresinous varnishes, resins containing fatty acids from natural oils (alkyd, epoxy esters, urethane oils), treated natural products (cellulose nitrate, chlorinated rubber), and synthetic polymers. There are two general types of organic binders, convertible and non-convertible.[27]

The *convertible binders* undergo chemical reaction in the film. The drying and hardening of the film is catalyzed by "driers" with oxygen absorption. Hence, there is a limit to the thickness of the film, which will dry. The two-pack materials, notably epoxies and polyurethanes, cure by chemical reaction between two components in the film with no oxygen absorption.

Another type of organic binder is the thermosetting known as "thermohardening" or "stoving" type.[27] In these coatings, film hardening is the result of three-dimensional linking of the polymer molecules due to heat.

The *non-convertible binders* do not depend on chemical reaction for film formation. The process is mainly solvent evaporation. These materials are often called "lacquers."[27] Some of the inorganic binders are pre-hydrolyzed ethyl silicate, quaternary ammonium silicate, and alkali silicates (sodium and lithium) which are pigmented with zinc dust to give primers required for steel work.

Solvents (thinners) are volatile organic liquids which dissolve the binder or film former. They are used to liquefy the pigment/binder mixture sufficiently to allow the formation of a uniform film. The most commonly used thinners are organic solvents with the exception of benzene (e.g., hydrocarbons, alcohols, esters, and ketones) and water (in latex emulsions and water-soluble paints). The choice of solvent is not only based on solvency, but also on other important factors such as toxicity, odor, evaporation rate, flammability, and cost.[26][27]

Additives are added to paints as dryers, anti-skinning and anti-settling agents, fungicides, or bactericides to assist in the dispersion of the pigment.[26] In practice, a pigment, dispersed in a binder, carried in a solvent or non-solvent liquid phase is rarely satisfactory. Usually some defects arise from limitations in both chemical and physical terms that require elimination or mitigation before the paint can be considered satisfactory for the market. Some of these problems or defects are depressions in the film, surface shriveling (wrinkled film surface), sagging (uneven coating from excessive flow), floating (color difference after application due to separation of components), and flooding or brush disturbance (permanent color change after application due to shear). Therefore, to overcome these defects, additives are available for most paints from specialist manufacturers.[27]

Thousands of coatings and paint products are sold for industrial applications, making it impossible to memorize their names, attributes, properties, and limitations. Therefore, a classification method is needed based on some similarities. The most common method of classification is to refer to the chemical attribute—most often the resin type—that is unique to a group of coatings. This is the most useful classification system because coatings of the same generic type have similar handling and performance properties. This is based on the binder (resin) used in the formulation.[29]

The secondary generic classification is by curing mechanism or some other compositional element. For example, vinyl and epoxy coating types are names based on the resin. Another generic method is to use systems designed to indicate even broader classifications. For example, inorganic zinc-rich coating indicates that high loadings of zinc dust are part of the formulation, while the resin is only broadly classified by its general chemistry as inorganic or organic. Table 3 shows the most commonly used generic types of coatings.

Paint and coating manufacturers use over 40 different types of extenders, fillers, and pigments, in construction, perhaps more than in any other industry.[7] These materials are added to lower the cost, except for pigments and minerals added for gloss or flatness of finishes. Titanium dioxide pigment is the most important mineral in paints. Other types of fillers include alumina hydrate, barite, calcium carbonate, kaolin, mica, silica, and talc.[7]

Due to environmental regulations in the 1980s and 1990s, the industry of solvent-based paints has slowly been driven toward water-based systems.[7] Furthermore, new technologies have been developed including powder coatings, water reducible systems, and high solid and UV cured systems.

Table 3. Generic Coating Types Classified by Resin[29]

Name	Characteristics
Oil-based and alkyd	Based either on natural fish, plant oils, or synthetic resins (alkyd); all are organic; cure by air oxidation; moderate moisture vapor transmission rates; good exterior durability in non-aggressive atmospheres; topcoats should have good sunlight resistance; relatively expensive.
Chlorinated rubber and vinyl	Organic resins; cure by solved evaporation; must be applied by spray because of the curing mechanism; dry quickly; easy to topcoat or repair; very low moisture vapor transmission rates; form good chemically resistant barrier coats; poor solvent and heat resistance; high content of volatile organic compounds (VOC); difficult to formulate to meet regulations; good gloss retention; chlorinated rubbers are good as swimming pool coatings; unmodified chlorinated rubber vinyl resin has poor adhesion to steel.
Bituminous	Derived (totally or in part) from distillation of crude oil or coal; coal tar and asphaltic materials are relatively inexpensive; used for waterproofing, protecting buried structures, and lining tanks; poor resistance to sunlight.
Epoxy	Contain epoxide resin; cures by chemical reaction; two-part system (hardener and resin); excellent adhesion to most substrates, tough and durable films; good resistance to solvents, water, chemicals, and abrasion; poor resistance to sunlight (chalking breakdown process) which limits outdoor use to primers and intermediate coats; most commonly used are polyamide- and amine-cured; modified epoxies can be used in steel; concrete, flooring and lining applications.
Silicones	Contain silicone rather than carbon; good gloss retention and temperature resistance; one of the few used for protection of high temperature surfaces.
Phenolic	Based on inorganic resin phenol-formaldehyde; pure resins are cured by heat; good resistance to chemicals, solvents, water, boiling water, and steam; can be reacted with drying oils to give air-dried, oil-based coatings (considered as oil-based coatings).

Table 3. *(Cont'd.)*

Name	Characteristics
Polyesters and vinyl esters	Based on organic unsaturated polyester dissolved in unsaturated monomer; cured by free radical reaction catalyzed by peroxide; used mainly for lining due to their good acid and chemical resistance.
Water-borne acrylic (or acrylic latex)	Resin is dispersed in water to form water emulsion; acrylic is major resin type; one-component materials; cure by solvent evaporation followed by coalescence of the resin particle; used as protective and architectural coatings in the form of primers, intermediate coats, and topcoats; high moisture vapor transmission rates; good to protect wood or concrete where moisture must be allowed to pass through the coating; cannot be used on steel without anticorrosive pigments, but are used in mild weather to protect steel; used as overcoats for oil-based or alkyd coatings because of low shrinkage.
Urethanes	*(Chemically cured):* The isocyanate (–N=C=O) group is used to cross-link the resin (curing type); cured PU available in two separate containers; chemically cured PU has good water and chemical resistance; aliphatic urethanes have good gloss and color retention.
	(Moisture cured): Isocyanate group reacts with water to form an amine, which reacts with isocyanate polymers to form the film; available in one-part container; limited pot life—moisture reacts with isocyanate in can; used for cold weather applications, poor color and gloss retention; strong performance properties as steel coatings.

Table 3. *(Cont'd.)*

Name	Characteristics
Cementitious	Based on inorganic materials, e.g., portland cement; they are a mixture of cement, aggregate, and water; high compressive strength and good chemical resistance (depending on cement and aggregate type); good for fireproofing and high temperature service.
Inorganic zinc-rich	Contain high loading of metallic zinc dust; use silicate resin; cured by heating, application of a curing solution, evaporation, or reaction with moisture depending on the type of coating; they are used as primers or one-coat system on steels for corrosion protection; they are unaffected by most organic solvents; poor resistance to acid or alkalis without topcoat; resistance to high temperature up to ~538°C (1000°F) without topcoat; abrasion resistance; long service life, especially in marine environments.
Organic zinc-rich	Contain high loading of metallic zinc dust in organic binders; tend to have better compatibility with topcoat than inorganic zinc-rich coatings; more tolerant to variations in surface preparation; long service life, especially in marine and other environments.
Hybrid	Results from combination of other categories; intermediate characteristics between technologies used.

4.0 ASPHALT - BITUMINOUS MATERIALS

Asphalts are dark brown or black solids or semi-solids found in the natural state and are also produced by the refining of petroleum. Raw asphalt is a complex mixture of bitumen and low-volatile organic chemicals that are recovered from petroleum refinery residues or natural deposits. Bitumen is a generic name applied to various mixtures of hydrocarbon. The American Society for Testing and Materials has the following standard definitions for asphalt.[30]

1) Relating in general to bituminous materials: *Bitumen*—a class of black or dark-colored (solid, semi-solid, or viscous) cementitious substances, natural or manufactured, composed principally of high molecular weight hydrocarbons, of which asphalts, tars pitches, and asphaltites are typical.

2) Relating specifically to petroleum or asphalts: *Asphalt*—a dark brown to black cementitious material in which the predominating constituents are bitumens which occur in nature or are contained in petroleum.

Asphalt can be generally classified as natural or artificial. *Natural asphalts* include bituminous materials laid down in natural deposits, such as those in Trinidad, and as gilsonites and grahamite bitumens, which are completely soluble in carbon disulfide. *Artificial asphalt* includes mainly petroleum-derived asphalts and, to a lesser extent, coal tar, water-gas tars, and their pitches.[31][32] There are types of asphalt products obtained from *straight-run asphalt* (refined naphtha-based crude oils): hot, cutback, and emulsion asphalt.[8]

Hot asphalts can be softened with heat and have good resistance to water and water vapor transmission when applied to dry surfaces under controlled heating, but have poor bond adhesion to wet surfaces. In addition, they have relatively poor flexibility, are brittle at low temperatures, and oxidize upon exposure to solar radiation.

Cutback asphalts comprise four types in the US (I–IV) and three types (I–III) in Canada. The No. I type consists of straight-run asphalt, and solvent and either a small amount of fiber or none; the No. II type is obtained by adding a large amount of fibers and filler to the asphalt cut with solvent; and No. III is primer-type in solution without any fillers or fibers.

These types of asphalt have poor adhesion to wet surfaces, but some have a damp-bonding ability. Type II asphalt has excellent vapor-barrier properties and better weather resistance than types III or IV. Contrary to type

I and II, type III asphalt is thin enough to penetrate masonry, wood, and paper to provide a bond for other bitumen applications. It is also used to wet surfaces (e.g., metal). Depending on requirements, type III can be made from soft, ductile asphalt as well as harder-base asphalts for application to dense, metal, or porous surfaces.[8]

Emulsion asphalts are easier to handle than other bituminous products. Water is added in just the necessary amount to decrease viscosity. Drying involves, primarily, loss of water due to evaporation. Their application does not require heat. Furthermore, they have good bonding properties to either damp or wet surfaces, and are more weather resistant than the other bitumen types.

The properties and behavior of asphalts are critically dependent on the nature of the constituents, which consist of hydrocarbon and heterocyclic or nitrogen-, sulfur-, and oxygen-containing compounds. Separation of the various asphalt fractions (Table 4) is usually based on their different boiling points, molecular weights, and solubilities in solvents of different polarities.[31]

Table 4. Solvent Fractions of Asphalt and Related Carbonaceous Materials[31]

Fraction			
No.	**Designation**	**Solubility**	**Remarks**
1	gas oil	propane soluble	saturated and aromatic hydrocarbon
2	resin	propane insoluble	combined 1 and 2 are called maltene or petrolene
3	asphaltene	pentane insoluble benzene soluble	ASTM uses CCl_4 instead of benzene
4	carbene	CS_2 soluble benzene insoluble	ASTM uses CCl_4 instead of benzene
5	carboid	CS_2 insoluble pyridine soluble	combined 4 and 5 are referred to as *asphatol*
6	mesophase	pyridine insoluble	seldom found in ordinary asphalt, can be generated by heat treatment

Asphalt can be considered as a colloidal system, similar to petroleum, with the lighter molecules removed. The carbon types in petroleum can be classified as paraffinic (C_P), naphthenic (C_N), and aromatic (C_A). As fractions become heavier, the proportion of C_P, C_N, and C_A becomes larger.[31]

The physical properties of asphalt are directly related to the quantity of the dispersed phase (asphaltenes); the size of the micellar structures, which depends upon the degree of adsorption of resins; the nature of the dispersion medium; and maltenes (oils and resins). In heavier fractions of petroleum, single molecules contain C_P, C_N, and C_A portions. The aromatic protons near the aromatic ring system can participate in electrophilic reactions, such as nitration, sulfonation, halogenation, and others. The interaction between the molecules will determine the chemical properties of asphalt.

The rheological properties of asphalt are concerned with the study of physical or mechanical stability of the colloidal system under varying conditions of time, temperature, pressure, load or stress. Hence, a study of its rheological properties is essential for practical applications such as pavement, roofing, coating, and other types of applications.[31]

Asphalt products have a variety of applications in the construction industry:

- Buildings (conduit, insulation, paint composition, papers, roofing membranes, shingles)
- Roofing (building papers, BUR adhesive, felts, primers, caulking compounds, cement waterproofing compounds, laminated roofing, shingles, liquid roof coatings, plastic cements)
- Floors (damp-proofing and waterproofing, tiles, coverings, insulating fabrics, step treads)
- Walls, siding, ceilings (acoustical blocks, felts, bricks, brick siding, building blocks, masonry coatings, plaster boards, putty, asphalt, soundproofing, stucco base)
- Paints, varnishes (acid-proof enamel, acid-resistant coatings, anti-oxidants and solvents, boat deck sealing compound, lacquers)
- Electrical products (tapes, wire coatings, junction box compounds, molded conduits)

The applications of asphalt products in the industrial world are immense. Considerable amount of work has been published on the chemistry and application of asphalt.[31]–[36][38]–[55]

5.0 ROOF COVERING MATERIALS

Polymeric materials and asphalt-based products have a variety of applications in the construction industry. Some of these applications are as roof coverings. Table 5 shows the various types of related material used as roof coverings.

Table 5. List of Various Roof Covering Materials

Material Type	Acronym
Built-up Roof	BUR
Modified Bitumen	MB
Styrene-Butadiene-Styrene Modified Bitumen	SBS
Atactic Polypropylene Modified Bitumen	APP
Ethylene-Propylene-Diene Monomer	EPDM
Poly(Vinyl Chloride)	PVC
Thermoplastic Oligomers	TPO
Polyurethane Foam	PUF
Metal	

Today, millions of square meters of residential and non-residential roofs are covered with water shedding or weatherproofing assemblies. The 2000 US roofing market was approximately $34 billion US dollars.[37] A roof assembly contains a roof system over a roof deck. The *roof system* is defined as having the elements which cover, protect, and insulate the roof surface of a structure against the external environment. Roof systems vary from the traditional types [e.g., shingles for sloping roofs and built-up roofing (BUR) for flat roofs] to the non-traditional roofing materials (e.g., polymer-based single-ply or modified bitumen). They also vary in the method by which the covering has been put down. The conventional method has the covering (e.g., membrane) above the insulation and exposed to the environment. Alternatively, in a protected system, the covering is directly above the deck and is covered with the insulation.

Irrespective of the system, new materials have been introduced as alternatives to the older ones (see Table 5). For example, some shingles are made of modified bitumen instead of blown bitumen. Asphalt-saturated organic felt is increasingly being replaced by asphalt-impregnated glass fiber mat in the manufacture of shingles and as felt plies in BUR. In the new systems, more factory prefabrication and less work atop the roof is involved than in the traditional roofs.

According to the 1994 Canadian Roofing Contractors' Association (CRCA) *Project Pinpoint* survey, the three main roof membrane systems (for flat roofs) in Canada (and their market shares) were as follows: BUR (54.7%), polymer-modified bitumen (33.4%), and polymer-based single-ply (10%). In the USA, the 2000 National Roofing Contractors Association (NRCA) survey[37] for new construction showed that the asphalt-based systems (BUR and polymer-modified bitumen) occupied ~41.5% of the market and the polymer-based systems were almost as popular. The market share was as follows: BUR (22.4%), modified bitumen (19.1%), single-ply (38.1%), metal (4%), and 16% others (tiles, PUF, liquid-applied, asphalt singles). The survey showed that in Canada the market was SBS-modified bitumen (50.4%), BUR (25.2%) and single-ply (14.9%).

Some find this choice overwhelming and confusing. Many years of experience have accrued in the application of BUR, but not so with the newer materials. As a result, the knowledge acquired in the past was of little use for explaining the problems experienced with the new generation of roof systems.

Asphalt has been used for thousands of years as a waterproofing material. In North America, asphalt has been used for approximately 150 years as a roofing material. More specifically, BUR has been used for over 100 years. It is still the single biggest type of roofing system installed. The new materials introduced as alternatives to BUR are products of different chemical formulations. Although no panacea, they do provide a wide range of options that meet required performance characteristics. The first generation of these materials suffered some set-backs due to lack of design and performance criteria and lack of experience. However, improvements in their compositions, reinforcing, and lap joint techniques have resulted in a second generation of products with better and progressively improved performance characteristics. The development and promotion of new materials were prompted by the following factors:

- The energy crisis of the early 1970s resulted in an increase in the cost of petroleum-based products. The unpredictability of the sources of oil supplies meant that the quality of asphalt was not consistent. This, in turn, affected the quality of roofing materials.
- Energy-induced inflation raised the cost of labor-intensive BUR, thus making the alternatives economically more viable.
- Advances in polymer chemistry and technology resulted in the development of many polymer-based synthetic materials that could be used for roof coverings.
- During the 1960s, new structural design principles gave rise to lightweight structures that caused problems for conventional roofing assemblies owing to their increased structural movement.
- Highly insulated roofs and decks with unusual architectural configurations allowed innovative designs of roof systems that only the new materials could meet.
- The aesthetics of roofs in terms of color and pattern presented the architect with attractive alternatives that would complement the architecture of other elements, like flooring and carpeting, in a building as well as other exterior elements of the building.
- Better corrosion-resistant metals.

As a consequence, literally hundreds of new roofing materials have appeared on the market. Most of them are polymeric in nature. They are reinforced with a variety of woven and non-woven fabrics of synthetic and glass fibers.

5.1 Polymers

In the field of polymer chemistry, there are many terms related to internal structure such as monomers (as in EPDM), co-polymers as in vinylidene chloride (Saran wrap), block co-polymers (as in SBS), and interpolymers (as in EIP), etc. Polymers are giant molecules of different chemicals. A polymer or a macromolecule is made up of many (*poly*) molecules (*mers*) or monomers linked together like wagons in a train, for example, poly(vinyl chloride), polyethylene, etc. Monomers may have the same or different chemical compositions.

Elastomers are a group of polymers that stretch under low stress to at least twice their original length and recover after the removal of the stress. The formation of elastomers depends on the system. For example, EPDM is formed by adding sulfur to the mixture of ethylene, propylene, and diene monomer. The sulfur forms linkages between the polymer molecules. This process is known as *vulcanization*. Similarly, esters constitute a family of chemicals whose macromolecules are known as polyesters. They are used in synthetic fibers, filaments, threads, fabrics, etc. They also are used in reinforcing roofing sheets and membranes. Polyurethanes include rigid, semi-rigid, flexible, and integral skin foams used in interiors of automobiles and many everyday products.

As can be seen in Fig. 2, there are essentially two families of polymer-based roofing membranes: thermosets (TS) and thermoplastics (TP).

Types of Covers. In addition to BUR, there are many different types of new membranes that are mostly prefabricated sheets or liquid-applied materials that cure to form waterproof sheets or closed cell foams. They are made from a wide variety of synthetic organic materials (polymers) with various chemical compositions and additives. In some cases, natural materials such as bitumen, organic fibers, etc. are compounded with them.

A brief generic list of roofing membranes is given in Table 5. In each type, there is a long and growing list of products. No two products are identical even if they consist of the same predominant polymer. A manufacturer may have a number of different products: unreinforced, or reinforced with different fabrics, for protected or exposed application, with different seaming techniques, different attachment methods, availability in different colors, etc.

The term sheet and membrane tend to be interchangeable. By definition, a *membrane* refers to the finished built-up waterproofing layer comprising one or more prefabricated sheets. As such, a sheet becomes a membrane in a single-ply application and a modified bituminous membrane may have two sheets, base and cap.

Most of these membranes are composed of mastic, which is the waterproofing component, and reinforcing fabrics which give the membranes the desired physical properties.

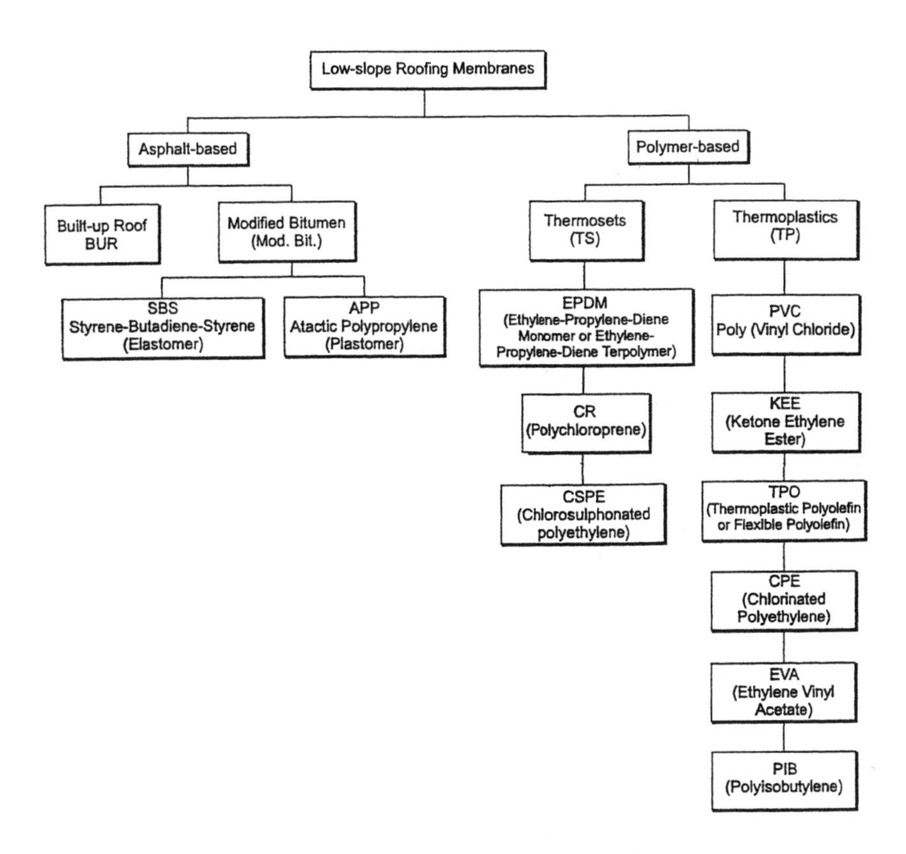

Figure 2. Two families of polymer-based roofing membranes: thermosets (TS) and thermoplastics (TP).

Type of Reinforcements. Conventional roofing felts are made of organic fibers and are impregnated with asphalt. They act as binders and as reinforcement in the waterproofing component in the BUR. In recent years, non-woven fabrics have made significant inroads in roofing in addition to many other fields, such as the geotextile and medical fields.

Most of the reinforcements used in roofing are woven fabrics or mats and scrims of non-woven glass fibers and synthetic fibers. They are placed within the body of the membrane. In some cases, a lightweight reinforcing mesh is incorporated to act as a carrier during manufacture. Some exposed modified bituminous membranes without granule surfacing have a light glass mat embedded in the top surface to make it crack resistant.

5.2 Membrane Characteristics

Bituminous Built-up Roofing (BUR). Bituminous materials used in BUR include *(a)* asphalt obtained in petroleum processing, and *(b)* a product extracted from coal known as coal-tar pitch (CTP) (also referred to as tar or pitch). In Canada, CTP is generally not available. Asphalt used for saturating organic felts is commonly called No. 15 because the earlier types weighed 15 lb/100 ft.[2] This and asphalt-impregnated glass fiber felts are used as plies in the construction of BUR. Heated asphalt is mopped on each felt layer to bind them together into the finished multi-ply membrane.

Many problems in BUR, like blistering, are related to moisture and air voids in the membrane. The organic felts absorb moisture from improper storage, and the lack of protection from rain during installation can result in loss of strength due to wetness. Sometimes moisture is trapped where there are skippings in the mopping of asphalt under the felt. Also, synthetic fiber from rags sometimes gets included in the felts during manufacture, resulting in poor asphalt saturation. Glass fiber felts are quite porous and provide good adhesion. Their use in flashing is not recommended. During installation, walking over felt on hot asphalt causes the asphalt to squeeze out, which could result in a void and lack of adhesion in the membrane, a potential source of moisture problems.

A very common source of problems is overheating of asphalt. This apparently makes the mopping easy, but it hardens the asphalt and reduces both its softening point and coefficient of linear thermal expansion. The surface causes shrinkage cracks and alligatoring. Some other problems in BUR include shrinkage of the membrane that pulls flashings away, caused by a lack of adhesion of the roofing system to the deck. Membrane slippage occurs if the softening point of the asphalt is too low with respect to the roof slope, or the amount of interply asphalt is excessive.

Modified Bituminous (MB) Sheets. This class of sheets is made from bitumens and modifying polymers (synthetic rubbers or plastic materials) together with fillers and special-property additives. Since the process is basically mixing components, the amount of modifier can be varied according to the required characteristics. The two most widely used bitumen modifiers are SBS (styrene-butadiene-styrene) and APP (atactic polypropylene). The average SBS content in the formulation is 12–15%. Generally, more SBS means greater low temperature flexibility and fatigue resistance as well as a higher softening point and wider temperature use. There are about a dozen different SBS grades that accentuate one or the other property required for processing and performance of the membranes. APP is a by-product of the manufacture of IPP (isotactic-polypropylene). It comprises 25–35% of the modified compound, its primary function is to improve the mechanical properties of the finished membrane. The APP-modified product has higher strength and lower elongation compared to the SBS-modified type. A small quantity of filler provides rigidity to the compound, but large quantities reduce flexibility and adhesion. Consequently, the best products have the least filler.

Proper modification of bitumen results in a product whose performance characteristics are far superior to those of normal bitumen. Various types of reinforcements, particularly glass and polyester composites incorporated in the membrane, provide improved properties. Granules protect the surface from the degrading effect of UV. In some membranes, a light glass mat laminated to the surface protects the surface from cracking and acts as a replacement for granules. The number of reinforcing fabrics and their positioning depends on the design of the product. The sheets are up to about 5 mm thick.

These membranes are frequently applied by torching (open-flame melting) the underside as the sheet is being unrolled. Others have self-adhesive backing, or they may be adhered with a mopped-on adhesive. Since open-flame torching is considered a fire hazard, some manufacturers have introduced an electric heat-welding process. Overheating modified bitumen degrades the mastic and leads to poor adhesion and weak lap joints.

Hot-Applied Rubberized Asphalt. Hot-applied rubberized asphalts consist of proprietary blends of asphalt, mineral fillers, elastomers (natural, synthetic, or a blend of both), virgin or reclaimed oil, and a thermoplastic resin. It is applied hot in such a manner as to form an impermeable monolithic membrane over the surface to be waterproofed, which may be concrete, gypsum board or wood. Improved versions of this type of system consist of two coats of rubberized asphalt with a polyester mat in between, called the fully reinforced or two-ply system.

Both use bitumen (asphalt or tar) for waterproofing the systems and organic fibers in rags, cloth or cellulosic felt saturated with bitumen for reinforcing. Mineral materials such as granules and gravel are applied on surfaces to protect bitumens from ultraviolet (UV) radiation.

Polymeric Sheets. Single-ply polymer roofing membranes have been available in North America since the 1950s. Many new products were introduced to the market such as polychloroprene (CR), chlorosulfonated polyethylene (CSPE), EPDM, and PVC. Unfortunately, since these products were different from the traditional asphalt, new installation techniques had to be developed. Many companies were unaware of the critical role that installation techniques would play and, as a result, many failures have occurred.

Today, there are many fewer companies involved in the manufacturing of single-ply membranes. The advantages of such systems include speed of installation and no requirement for open flames or heated asphalt. However, they must be installed by properly-trained and manufacturer-approved installers. Most single-ply manufacturers claim that their products have a service life of at least fifteen years. The membranes are formulated to resist UV, heat, bacterial attack, etc.

The nomenclature used in the industry for these single-ply systems is based on the main chemical ingredient (e.g., PVC, EPDM, etc.). This is convenient for discussion purposes, but it must be remembered that all of these membranes contain additives which are required to impart the desired properties such as flexibility and weatherability. In general, there are two categories of polymeric sheets: elastomeric and thermoplastic.

Elastomeric Sheets. There are many types of elastomers or synthetic rubbers used in roofing, including EPDM (which is naturally flexible), neoprene, CSPE (also known as Hypalon™), butyl, nitrile, etc. They are compounded with polymers and ingredients such as fillers, anti-degradants, processing oils, and processing aids to impart the required properties. Polymers provide the muscle, and fillers provide bones to the materials. Stabilizers (e.g., anti-degradants) improve weathering properties of the membranes. The most commonly used elastomer in roofing is EPDM. The compound contains 30–50% polymer (ethylene-propylene-diene monomer), 20–30% carbon black, and 30–50% extender oil, sulfur, accelerator, and antioxidant. Sheets are produced by laminating two plies with or without reinforcing.

Vulcanization is the process of converting a raw rubber to a cross-linked network. This is generally an irreversible process. The most common method of vulcanization is with the addition of sulfur (and metal oxides) although other cross-linking agents can be used. Chemically, the sulfur forms bridges (or bonds) between the rubber molecules. The end result is increased stiffness and reduced sensitivity to solvent swell as well as other enhanced physical properties.

The non-vulcanized or uncured rubber sheets that are self-curing are gradually cured on the roof by heat from the sun. Once they are cured, their behavior is similar to that of cured elastomers. If they are not self-curing, they remain uncured and exhibit properties similar to thermoplastics during their service life. In general, elastomeric sheets have good tensile and other mechanical properties and excellent resistance to UV, ozone, many oils, and solvents.

Field seaming of some vulcanized sheets, known to cause some problems, is progressively being improved. Doing the maximum possible seaming in the factory reduces the amount of field seaming and the probability of problems. Proper choice of adhesives, care in the preparation of the seam area, skillful application, and adequate curing time could result in a durable joint. Work is currently underway at the National Institute of Standards and Technology(NIST) to study the use of tapes instead of liquid adhesives for seaming.

Thermoplastic Sheets. As the name implies, *thermoplastic polymers* soften when heated, and, thus, can be easily extruded or molded. They are distinguished from thermosets by the fact that there is no cross-linkage or vulcanization of the molecules. Welding together using heat or a solvent is easy and creates new molecular linkages during service life.

PVC. Poly(vinyl chloride) is one of the most versatile thermoplastics in use today. Its use in roof covering started in the sixties. In a PVC sheet, the compounded plastic is the key element that determines the final characteristic of the product and acts as a binder for the system. The plasticizers impart flexibility to the sheet and improve processing. Fillers and extenders (such as calcium carbonate) are used primarily to lower the raw material cost of the compound. They also improve processing and affect other mechanical properties, such as the hardness and dimensional stability of the finished product. Stabilizers protect PVC against heat during processing and against ultraviolet radiation during service. Pigments are added to color the plastic material.

Loss of plasticizers was once a major concern, as it caused embrittlement in the PVC sheets. This is now considerably improved by using high molecular weight plasticizers that have less of a tendency to volatilize or migrate out of PVC resin.

PVC sheets are produced by three basic methods, calendering, extruding, and coating. There are three types of sheets: unreinforced, lightly reinforced with fibers or fabrics that act as carriers, and reinforced sheets that contain glass and/or polyester fibers or fabrics. The carrier facilitates manufacturing and adds to the dimensional stability of the sheet. Reinforcement provides enhancement of tensile strength and other properties.

PVC sheets have good resistance to industrial pollutants, bacterial growth, and extreme weather conditions. Minor damage to the sheet during installation or in service can be easily repaired by patching the hole using heat or solvent.

TPO. Defining "thermoplastic polyolefin (*TPO*)" is difficult. *Thermoplastic* is a generic term in polymer science; it encompasses a class of polymers that soften when heated in a reversible process. The term *olefin* is generic, being an old chemical name for any molecule containing carbon-carbon double bonds (also known as alkenes). Any polymer chain formed by chemically linking up many olefin molecules is termed a *polyolefin*. These polyolefins only contain carbon and hydrogen atoms. While TPO roofing membranes are relatively "new" (appearing in the last 10 years or so), the polymer technology has been around for some time. Today, polyolefins are used extensively in the automotive industry, plastic bags, bottles, etc. The polymer for TPO roof membranes is reported, from one source, to be an alloy (i.e., a blend) of ethylene and propylene (two monomers that are joined to create a uniform polymer during synthesis).

Cold-Applied Liquid Compounds. This category of material comprises a number of different products in the market. They consist of emulsions and solutions of *(a)* various resins or elastomers such as polyurethanes, silicones, acrylics, etc., and *(b)* bitumens and modified bitumens. The material left after the exploration of volatiles (water or organic solvents) forms the waterproofing layer. Their surface coatings may contain white pigment, aluminum flakes, or they may be vinyl films for protection from solar radiation. These liquids are generally applied by spraying or with rollers. The emulsions cure slowly at low temperatures and they cannot be applied below water's freezing temperature. The solution forms a film much faster under these conditions.

Also, the cutbacks (solutions) and emulsions (water dispersions) of asphalt and coal tar pitch are used in various types of cold applications of BUR. Polyester mats are used as alternatives to conventional felts for plies. The market share of cold-applied BURs is small. From the point of view of economy and availability, asphalt is more commonly used in this type of roofing product.

Polyurethane Foam. Polyurethane rigid foams (PUF) were first developed in the late thirties and used during the war to strengthen aircraft wings. Commercial use in different industries started only in the late fifties. The sprayed-in-place PUF roofing system was introduced in the early 1960s.

This system is made up of three components: PUF, a protective cover, and a vapor barrier. PUF forms a closed cell waterproofing barrier and provides insulation. The foam is made from the combination of two materials, a resin (containing polyol, catalyst, blowing agent, surfactant, etc.) and a polyisocyanate component. Their combination during application from a two-head spray gun produces a polymeric structure and a vapor that forms bubbles before the foam becomes rigid. During the chemical reaction, it expands 20 to 30 times from its original volume within seconds. Minimum thickness of the foam layer is 25 mm.

The use of PUF in roofing faced many problems in early years that were related to the ambient temperature variations. On hot days, the foam reacted too rapidly leaving a rough texture, while on very cold days it did not react leaving the material in a liquid form. Accordingly, attention to environmental conditions (temperature, wind, moisture on deck) is necessary. PUF, once considered only as a re-roofing alternative to BUR, is now being used in a wide range of new construction projects.

Since urethane foam is very sensitive to UV radiation, it must be protected in some manner. Various elastomeric coatings and latex paints have been used for this purpose. In some cases, mineral roofing granules are sprinkled into the coating when wet. They improve abrasion resistance, weathering characteristics, and fire resistance. Coatings must have high tensile strength, elongation, and water transmission resistance, since water is the foam's prime enemy.

REFERENCES

1. Van Krevelen, D. W., *Properties of Polymers, Their Correlation with Chemical Structure; Their Numerical Estimation and Prediction from Additive Group Contributions,* 3rd Ed., Elsevier Science Publ. Co. Inc., The Netherlands (1990)
2. Nicholson, J. W., *The Chemistry of Polymers,* Royal Society of Chemistry Paperbacks (1991)
3. Feldman, D., *Polymeric Building Materials*, Elsevier Science Publ. Co., Inc., New York (1989)
4. Meyer, K. H., and Mark, H., *Makromolekulare Chemie,* 3rd Ed., Akademishe Verlagesellschaft Geest and Patin K. G., Leipzig (1953)
5. Turi, E. A., *Thermal Characterization of Polymeric Materials,* Ch. 3, Academic Press, Inc., New York (1981)
6. Young, R. J., *Introduction to Polymers,* Chapman and Hall Ltd., New York (1981)

7. Industrial Mineral Background Paper 13, *Developments in Building Products: Opportunities for Industrial Minerals,* Ontario Ministry of Northern Development and Mines, by Institute for Research in Construction (National Research Council Canada), Ottawa, ON and MATEX Consultant Inc., Mississsauga, ON (1990)
8. Smith, R. C., Materials of Construction, 3rd Ed., Ch. 11, Gregg Division, McGraw-Hill Book Co. (1979)
9. Commercial Adhesives: Environmental Evaluation and Proposed Environmental Standard, Draft for Public Comment Prepared by Green Seal, 1001 Connecticut Ave., NW, Suite 827, Washington, DC 20036–5525, (Feb. 11, 2000) (http://www.greenseal.org/adhesdraft.htr)
10. Wolf, A. T., Review of Modern Analytical Techniques Used in the Study of Aging-induced Changes in Sealants, *Durability of Building Sealants,* (A. T. Wolf, ed.), Ch. XVII, pp. 323–342, RILEM Publ. S.A.R.L., The Publ. Co. of RILEM (1999)
11. Standard Terminology of Adhesives, *Am. Soc. for Testing and Mater.,* ASTM D 907–00 (2000)
12. Kinloch, A. R., *Mater. Sci.*, 15:2141–2166 (1980)
13. Skeist, I., ed., *Handbook of Adhesives,* 3rd Ed., Van Nostrand Reinhood, New York, USA (1990)
14. Landrock, A. H., *Adhesives Technology Handbook,* Noyes Publ., Park Ridge, NJ (1985)
15. Petrie, E. M., *Handbook of Plastics and Adhesive,* (C. H. Harper, ed.), Ch. 10, McGraw-Hill, NY (1977)
16. Merriam, J. C., Adhesive Bonding, *Materials in Design Engineering,* 50(3):113–128 (1959)
17. Shields, J., *Adhesives Handbook,* 2nd Ed., Newnes-Butterworths, London (1976)
18. Hartshorn, S. R., ed., Structural Adhesives, Chemistry and Technology, *Topics in Applied Chemistry,* p. 18, Plenum Press, New York (1986)
19. Laaly, H. O., *The Science and Technology of Traditional and Modern Roofing Systems*, Vol. 1, Ch. 33, Laaly Scientific Publ., Los Angeles, CA (1992)
20. Cheremisinoff, N. P., ed., *Elastomer Technology Handbook,* p. 836, CRC Press, Inc., USA (1993)
21. Standard Terminology of Building Seals and Sealants, ASTM C717-00a.
22. Hollaway, L. C., ed., *Polymer and Polymer Composites in Construction,* Ch. 11, p. 250, Thomas Telford Ltd., Thomas Telford House, London (1990)
23. Damusis, A., ed., *Sealants,* Reinhold Publ. Corp., NY (1967)

24. Blaga, A., Durability of Building Materials and Components, ATM STP691, (P. J. Sereda and G. G. Litvan, eds.), *Am. Soc. for Testing and Mater.,* p. 827, Washington, DC (1980)

25. Standard Terminology for Paint, Related Coatings, Materials, and Applications, *Am. Soc. for Testing and Mater.,* ASTM D 16-00

26. Morgans, W. M., ed., *Outlines of Paint Technology, Vol. 2:Materials,* Charles Griffing and Co., Ltd., London (1982)

27. Morgans, W. M., (ed.), *Outlines of Paint Technology, Vol. 1: Materials,* Charles Griffing and Co., Ltd., London (1982)

28. Lambourne, R., (ed.), *Paint and Surface Coatings: Theory and Practice,* Ellis Horwood Ltd. Publ., Chichester (1987)

29. Smith, L. M., *J. Protective Coatings and Linings,* 12(7):73–82 (1995)

30. Standard Terminology Relating to Materials for Roads and Pavements, ASTM D8-97 (1997)

31. Mark, B. F., Bikales, N. M., Overberger, C. G., Menges, G., (Editorial Board), *Encyclopedia of Polymer Science and Engineering,* 2nd Ed., Index Volume Asphaltic Materials to Processing Aids, John Wiley & Sons, Inc., (1990)

32. Barth, E. J., *Asphalt, Science and Technology*, Gordon and Breach Sci. Publ., New York (1962)

33. Traxler, R. N., *Asphalt, Its Composition, Properties and Uses,* Reinhold Publ. Corp., New York (1961)

34. Pfeiffer, J. P., *The Properties of Asphaltic Bitumen*, Elsevier Science Publ. Co., New York (1950)

35. Bestougeff, M. A., and Byranhee, R. J., (T. F. Yen and G. V. Chilingarian, eds.), *Asphaltenes and Asphalt,* Vol. 1., Elsevier Scientific Publ. Co., Amsterdam (1989)

36. Yen, T. F., *Chemistry of Asphaltenes*, (J. W. Burger and C. N. Li, eds.), 195:37–51, Am. Chem. Soc., Washington, DC (1981)

37. Good, C., Eyeing the Industry, *Professional Roofing,* pp. 28–30 (April 2001)

38. Laaly, H. O., *The Science and Technology of Traditional and Modern Roofing Systems,* Vol. 1, Chap. 14, Laaly Scientific Publishing, Los Angeles, (1992)

39. Abraham, H., *Asphalt and Allied Substances,* 5th Ed., Vol. 1, *Raw Materials and Manufactured Products,* D. Van Nostran Co., Inc., New York (1945)

40. Hoiber, A. J., *Asphalt - Encyclopedia of Chemical Technology*, Kirk-Othimer, 2nd Ed., 2:762–806, HOLB Author Index, HOL:17-03.

41. Hoiber, A. J., *Bituminous Materials - Encyclopedia of Polymer Science and Technology*, pp. 402–437, HOL:17-03 PUBL HOLB Author Index
42. Kraus, G., Modification of Bitumen by Butadiene/Styrene Block Copolymers, *ISRR - Second International Symposium on Roofs and Roofing, Brigthon, England*, HOL:17-03, HOLB#03
43. Rodriguez de Sancho, I., Modification de Betunes Asphálticos por Adición de Estereoisómeros de Polipropileno (Spanish) (Modification of Bitumen by Addition of Polypropylene Stereoisomers), Central Laboratory of Materials and Structures CEDEX, Madrix, Spain (July 1990)
44. Performance Testing of Roofing Membrane Materials, Recommendations of the Conseil International du Batiment pour la Recherche l'Etude et la Documentation (CIB) W.83 and Réunion Internationale des Laboratoires d'Essai et de Recherche sur les Matériaux et les Constructions (RILEM) 75-SLR Joint Committee on Elastomeric, Thermoplastic and Modified Bitumen Roofing, RILEM, Paris, France (Nov. 1988)
45. Gorman, W. B., and Usmani, A. M., Application of Polymer and Asphalt Chemistries in Roofing, presented at a meeting of the Rubber Division, Am. Chem. Soc., Philadelphia, PA (May 2–5, 1995)
46. Meynard, J. Y., Modified Bitumens: What to Look For, The Roofing Industry Educational Institute, 6851 S. Holly Circle, Suite 250, Englewood, CO 80112, Reprinted from *The Roofing Spec* (June 1992)
47. Lu, X., and Isacsson, U., Characterization of Styrene-Butadiene-Styrene Polymer Modified Bitumens-Comparison of Conventional Methods and Dynamic Mechanical Analysis, *J. Testing and Evaluation,* JTEVA, 25(4):383–390 (1997)
48. Penchev, V., and Stojanova, M. A., Study of the Heavy Fractions and Residues from West Siberian Crude Oil by Thermogravimetry and Differential Thermal Analysis, *J. Thermal Analysis,* 35:35–45 (1989)
49. Kaki, L., Masson, J. F., and Collins, P., Rapid Bulk Fractionation of Maltenes into Saturates, Aromatics, and Resins by Flask Chromatography, *Energy & Fuels,* 14:160–163 (2000)
50. Michon, L. C., Netzel, D. A., and Turner, T. F., A ^{13}C NMR and DSC Study of the Amorphous and Crystalline Phases in Asphalt, *Energy & Fuels,* 13:602–610 (1999)
51. Fernández, J. J., Figueiras, A., Granda, M., Bermejo, J., and Menéndez, R., Modification of Coal-Tar Pitch by Air-Blowing, I: Variation of Pitch Composition and Properties, *Carbon,* 33(3):295–307 (1995)
52. Gahvarl, F., Effects of Thermoplastic Block Copolymers on Rheology of Asphalt, *J. Mater. Civil Eng.,* 9(3):111–115 (1997)

53. Masson, J. F., and Polomark, G. M., Bitumen Microstructure by Modulated Differential Scanning Calomerity, *Thermochimica Acta*, 374:105–114 (2001)

54. Memon, G. M., and Chollar, B. H., Glass Transition Temperature of Asphalts by DSC, *J. Thermal Analysis,* 49:601–607 (1997)

55. Lu, X., and Isacsson, U., Rheological Characterization of Styrene-Butadiene-Styrene Copolymer Modified Bitumens, *Const. and Bldg Mater.,* 11(1):23–32 (1997)

14

Sealants and Adhesives

1.0 INTRODUCTION

Most sealants are used in exterior applications, hence, they must be deformable, have good recovery properties, and should have good overall elastic properties. Many sealants used in the construction industry are required to maintain functional performance characteristics over many years. While in service, sealants are exposed to environmental factors such as UV radiation, water, oxygen, and thermal cycling. Depending on which face of the building sealants are placed, type of substrate, and geographic region, they may be exposed to extreme environmental conditions, stress, and strain gradients. Hence, sealants are susceptible to weather-induced degradation. In general, degradation involves both chemical and physical processes with the chemical reactions usually preceding the physical process.[1]

Weathering of construction sealants is an issue of importance not only to sealant formulators, manufacturers, and specifiers, but also to contractors and users. Sealants can fail due to weather exposure and aging. This type of failure is often characterized by discoloration and crazing and/or stiffness of the sealant surface and is a result of the individual or combined effects of solvent evaporation, ozone attack, migration of plasticizers, UV radiation, water immersion, etc.

To prevent failure and promote certain performance characteristics, additives, such as adhesion promoters, fillers, pigments, plasticizers, and thixotropic agents, can be introduced into the polymer. Typical fillers are carbon black, calcium carbonate, talc, clays, and ground silica. Primers, release agents, and backup materials are used to promote better performance. Primers provide better adhesion of the sealant to the substrate. The more elastic the sealant, the greater the need for a primer.[2]

Adhesive performance is determined not only by the degree to which the adhesive forms intimate contact with both surfaces, but also by its ability to resist separation from these surfaces. Therefore, the adhesive must possess the rheological properties that promote bonding, but resists debonding. The dual set of properties can be achieved by changing the states of adhesives from wetting to setting or by manifesting its viscoelastic properties.[3]

Since an adhesion joint is expected to perform satisfactorily under the expected service for the lifetime of the bonded structure, it is important to know or predict the changes in the properties of the materials due to environment, fatigue, temperature, loading rate, and age. Therefore, knowledge of the chemical and physical properties of both adhesive and adherends is required.[4]

Testing of adhesives is necessary in order to determine the level of performance and/or predicted durability. Some of the tests provide information on the working properties of the adhesive, such as viscosity, which affects mixing, application, and spreadability as well as wetting and penetration of the substrate. Other test methods measure the amount of resin present (solid content). This not only influences the viscosity of the adhesive, but also the performance of the bonded assembly.[5]

2.0 TEST METHODS

Adhesives and sealants may undergo progressive deterioration with time as a result of environmental factors. This may be the result either of chemical changes, for example, thermal degradation, polymer chain hydrolysis, or physical changes such as crazing and cracking caused by diffusion of environmental species into the polymeric materials.[6] All these factors lead to deterioration and, hence, failure. Because of this failure, the adhesives and sealants industry urgently needs test methods for generating long-term performance data rapidly and with assured reliability. Accelerated laboratory aging experiments are the most promising methods for acquiring durability information in a short time. However, reliable test methods for conducting and interpreting the results of the experiments are needed.[7] The following section will address the use of thermoanalytical techniques as a promising characterizing tool for adhesives and sealants.

The mechanical properties of polymeric materials, such as adhesives and sealants, have to be considered in all applications. The mechanical behavior involves the deformation of a material under the influence of applied forces. The force can be applied either in compression (pressure), tension, or shear modes. The latter is applied tangentially to the material, and is very important in structural adhesives.

For example, the physical properties of cured adhesives can be evaluated by measuring "stress-strain" properties. This can be done with a testing machine as shown in Fig. 1. The machine records force applied to a specimen of defined shaped as a function of either separation of the jaws of the specimen holder or the elongation of the sample.

A curve, such as the one shown in Fig. 1, can be obtained. The tensile strength is defined as the maximum tensile force divided by the cross-sectional area (i.e., width × thickness) of the specimen while the ultimate elongation is the ratio between the final gauge separation and the initial gauge length in percentage as measured by a device (extensiometer).

This test gives the "tensile stress" property of the material, which can be represented by the following equation:[6]

Eq. (1) $$\sigma = F/A$$

where σ is the stress, F is the force applied to the sample, and A is the cross-sectional area of the sample at its narrowest point.The tensile strength (ε) of the specimen is given by:

Eq. (2) $$\varepsilon = l/l_o$$

where l is the length of the specimen as a function of the tensile stress (σ) and l_0 is the original length of the specimen. As shown in Fig. 2, the breaking point of the materials *(a,b)* occurs at the point at the maximum tensile force where the material can no longer resist the applied force (stress).

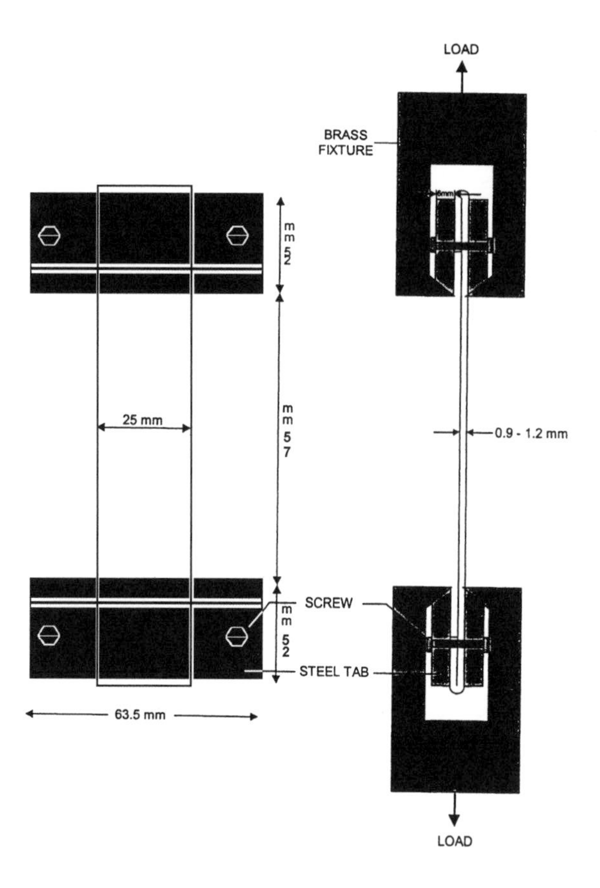

Figure 1. Loading assembly for tensile tests conducted at low temperatures.

The dashed line *(c)* in Fig. 2 corresponds to a typical stress (force) versus strain (displacement) curve for a material which obeys Hooke's Law (elastic) and curve *(b)* represents a viscoelastic behavior. It can be observed from the figure that the curve from the thermoplastic material is almost linear for a force (stress) below 300 N. In the linear region of the curve, the material behaves like an elastic material and obeys Hooke's Law. The curve in Fig. 2 can be represented by the following equation:[6]

Eq. (3) $$\sigma = E\varepsilon$$

where σ is the stress, ε is strain, and E is the slope of the curve (resistance to tensile forces). E is also called Young's modulus of the material. It can be seen from Eq. 3 that the higher the E value, the higher resistance of the material to tensile forces.

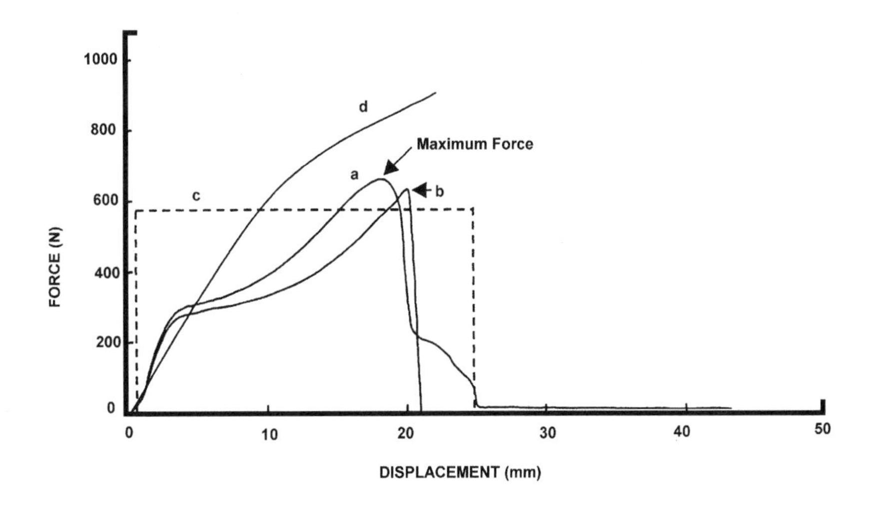

Figure 2. Typical force-displacement curve for a thermoplastic roof membrane.

Scientists have long sought a practical and accurate method for predicting the useful lifetime of polymeric materials such as adhesives and sealants. The proliferation of materials has given scientists and engineers the freedom to select the best material for the desired application. Before selecting the material, the required performance properties such as rigidity, strength, impact resistance, creep, and the environment in which the product will operate must be defined. Then the life expectancy of the adhesive or sealant can be determined. Only then does the material selection begin.[8]

Traditional evaluation procedures generally are laborious, time-consuming, expensive, and, in general, more empirical than analytical. They require fabrication of prototype parts and testing under actual end-use or simulated-service conditions. Therefore, the results may sometimes be questionable.[8]

Thermoanalytical techniques can be used to monitor a wide array of material properties including enthalpy, heat capacity, weight loss, thermal stability, thermal expansion coefficient, and glass transition temperature. They have been widely used to study the aging behavior of roofing construction materials such as roofing membranes and sealants.[9]–[15]

3.0 APPLICATIONS

3.1 Sealants

The wall cladding of a building constitutes a major function in the building's life cycle. An effective system is essential in contributing to a successful weathertight building envelope and preventing the infiltration of air and water. The high cost of energy for heating and air conditioning of inefficient building systems has a long-term financial impact on building owners. Many sealants in high-rise buildings, for example, are reaching the limit of their life span, hence, owners of many buildings in cities are anticipating repairing or replacing sealants in the near future.[16]

The most common types of sealants used for joints in wall cladding include silicones, polyurethanes, polysulfides, and acrylics. Because of this variety, the suitability of a sealant for a particular situation has to be carefully assessed.[16] Therefore, it is necessary to have an in-depth understanding of the causes of failure of the sealants.

The causes of sealant failure depend on the types of sealant used, installation, and service conditions. A sealant material can fail due to age and weather exposure. In the presence of environmental factors, such as ultraviolet light (UV), water, heat, oxygen, and micro-organisms, a polymeric sealant can degrade due to oxidation, photochemical and chemical reaction, hydrolysis, microbial attack, or to chain scission due to induced stress. Regardless of the type of failure or degradation mechanism, it is important to have the proper test to evaluate the new and in-service sealants as well as the failure or degradation mechanism.

Usually, sealants and adhesive materials for construction applications are evaluated by looking at the engineering side, but not the chemistry of the material. As a result, only tests that measure the mechanical properties are used. Most of the studies on the viscoelastic properties use traditional tests such as tensile testing to obtain data, which can be used in complicated mathematical equations to obtain information on the viscoelastic properties of a material. For example, Tock and co-workers[17] studied the viscoelastic properties of structural silicone rubber sealants. According to the author, "the behavior of silicone rubber materials subjected to uniaxial stress fields cannot be predicted by classical mechanical theory which is based on linear stress-strain relationship. Nor do theories based on "ideal elastomers" concepts work well when extensions exceed

30%. For these reasons, there are few design specifications related to the use of silicone rubber sealants for structural applications."

They presented a theoretical approach to predict the behavior of silicone rubber under uniaxial stress. The model is based on the concept of the classical Maxwell treatment of viscoelasticity and stress relaxation behavior, and the Hookean spring component was replaced by an ideal elastomer component. From the test data, the substitution permits the new model estimation of the cross-link density of the silicone elastomer and allows a stress level to be predicted as a complex function of extension, cross-link density, absolute temperature, and relaxation time. Tock and co-workers[17] found quite good agreement between the experimental behavior based on the new viscoelastic model. By using dynamic mechanical analysis (DMA), the authors would have been able to obtain similar information on the silicone elastomer.

It is of paramount importance to have knowledge of both the physical and chemical properties of polymeric adhesives and sealants for a proper understanding of their behavior as construction materials. Thermoanalytical techniques can provide that information but, although thermoanalytical techniques have been widely used for characterizing polymeric adhesives and sealants formulations, applications to polymeric construction material did not really start until the early 1990s. The advantage of thermoanalytical techniques is that they require little sample preparation and a small sample size. Therefore, they are very suitable for the characterization of in-service sealants or adhesive materials. As stated by Wolf:[7] "it is surprising that these techniques have not been used more widely in field studies, either in an effort to better characterize the sealant material or to study the effects of aging." In the last decade, however, these techniques have gained popularity as an analytical tool.

For example, Leonard, et al.,[18] studied the relation between rheology, artificial weathering, and urethane sealant performance after 3500 hours of total QUV exposure (QUV uses fluorescent UV lamps to simulate the effects of sunlight). The authors used a mechanical spectrometer to measure the viscoelastic properties of elastomeric joint sealants at cold temperatures and a constant stress rheometer to generate all other rheological data. The results of the study showed that at temperatures below -40°C the silicone sealants were stiffer than urethanes. This was attributed to the crystallinity melting of the polydimethylsiloxane polymer. The authors concluded that, in general, all silicone polymers would show this fundamental shift.

The rheological data showed that, after 600 hours of exposure, the urethane sealants (A) and (B) did not meet the requirements of an ideal sealant which would have an E' independent of temperature and thermal history. On the contrary, in both cases, the E' values decreased dramatically at high temperature. Furthermore, at sub-ambient temperature, the sealants became unacceptably stiff. However, two other sealant samples (C and D) studied were stable across the entire range. The author did not interpret the reason for changes in modulus. It would be reasonable to expect that the E' will not be independent of temperature and thermal history. Sealants are neither ideal solid nor ideal viscous materials, they are polymeric materials, and, as such, they will show viscoelastic behavior. The drastic drop in the E' values at high temperatures may be explained by the fact that the material has reached the rubbery plateau region. For temperatures above this region, melting or decomposition may occur. In fact, decomposition of polyurethane (PU) sealants may start near 200°C.[11][13][15] Derivatives of TG (DTG) curves (Fig. 3) of polyurethane sealants heated at 20°C/min, indicate that decomposition of PU sealants starts near 200°C whereas decomposition of silicone sealant starts above 300°C. For low heating rates, decomposition may even begin at a lower temperature because the sample is allowed to reach equilibrium. This may be the case since the heating rate used was 1°C/min with 1 minute soak time. For high heating rates, as in the work done by Paroli and co-workers[11][14] at 20°C/min, the sample temperature will lag behind the furnace temperature.

After 1000 and 3500 hours exposure, sealants (A) and (B) continue to deteriorate, losing their ability to withstand movement at high temperature. Again, the authors reported that results showed that the sealants (C) and (D) did not undergo rheological degradation. This indicated that the thermal and UV stresses from the QUV procedures have minimal impact on the storage modulus E'.

It was concluded that the technique used was a powerful rheological technique because it provided quantitative evidence that, as indicated by the xenon and QUV tests, there are significant differences between urethanes (A), (B), and (C) sealants. Also, at least one silicone sealant was much stiffer below -35°C (-31°F) than the polyurethanes studied. One of the four multicomponent urethane sealants showed similar performance to silicone sealants under severe test condition. Finally, the authors concluded that the current test incorporated in ASTM C 920 Specification for Elastomeric Joint Sealants does not identify the critical performance differences between sealants and must be updated to accomplish this critical task.

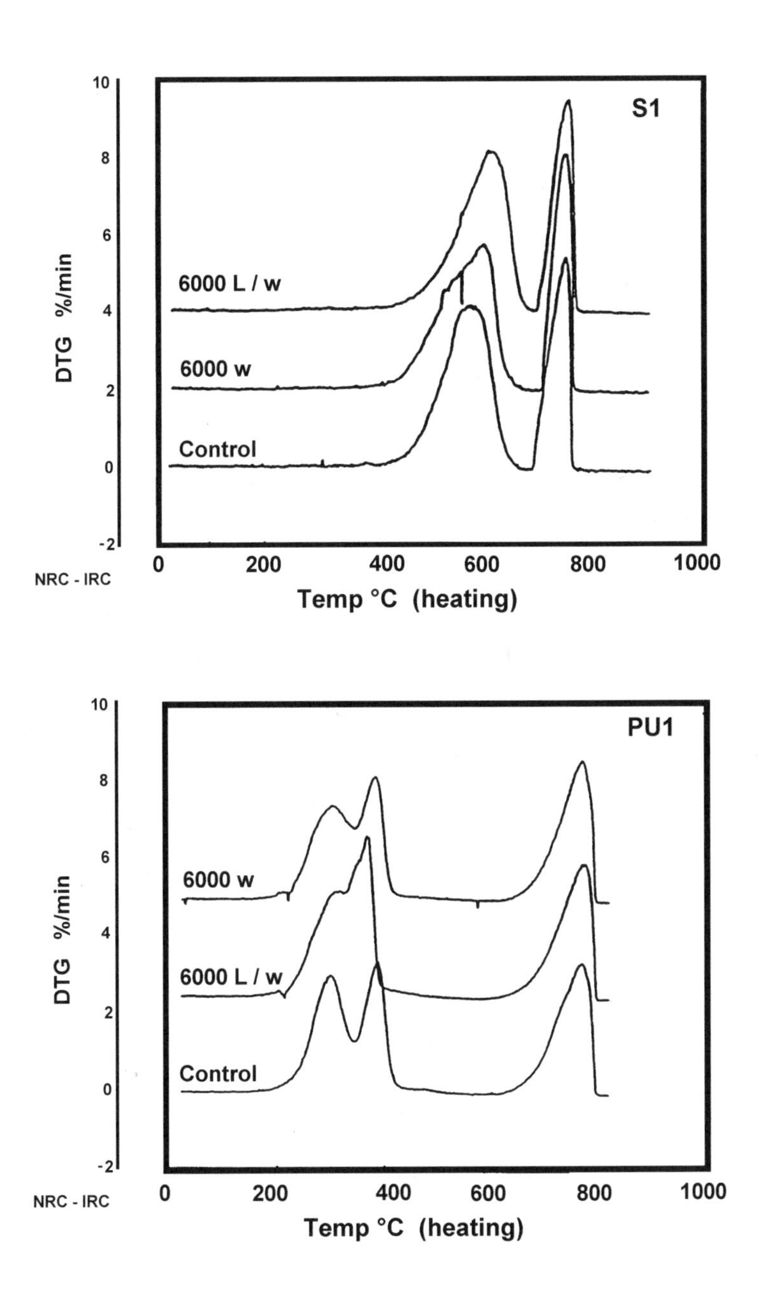

Figure 3. DTG of silicone and polyurethane sealants.

Using thermogravimetry (TG), Hugener and Hean[19] studied the field aging and laboratory aging-induced changes in four polymer-modified bitumen-based bridge deck sealants. The reference (non-exposed), field, and laboratory-exposed samples were analyzed by TG and other characterizing techniques (e.g., ring and ball softening point, gel permeation chromatography). The TG showed only small differences between the reference samples, the field exposed, and laboratory-aged samples. The authors did not provide any TG results because the differences observed were in the same range as the uncertainty of the technique, hence, they considered evaluation of the results unreasonable. The authors did not show any DTG curves, which sometimes better show the changes taking place. Bituminous materials are difficult to analyze by TG due to the different volatility of bitumen components. Paroli, et al.,[11][13][15] studied the weatherability of polyurethane and silicone sealant. By using STA (simultaneous thermal analysis, TG/DTA), they were able to identify the filler level and type in laboratory-cured silicone and polyurethane sealants. They also followed the changes that the sealants underwent after 8,000 hours of accelerated exposure in a xenon-arc Weather-O-Meter.

The DTG curves (Fig. 4) showed that the silicone sealants only underwent minor changes after exposure when compared to the control (unexposed) samples whereas the changes for the polyurethane sealants were more pronounced especially for two of the series, PU1 and PU2. The weight loss in the 200–400°C regions, shown in Fig. 4, was attributed to the polymer decomposition. As can be observed from the DTG curves, the control samples in both PU1 and PU2 display two well-resolved peaks in this region. However, the DTG curves for the exposed samples showed unresolved peaks, regardless of the type of exposure (water or both UV light and water). No attempts were made to interpret the degradation mechanism of the changes observed by STA. Paroli and co-workers[13] also used TG-FTIR to study the induced changes in the chemical composition of the same series of sealants. The IR results supported the changes observed by STA alone. Therefore, it was concluded that this hyphenated technique could be useful in identifying the decomposition products of construction sealants.

In an attempt to correlate aging-induced changes on the mechanical properties (modulus, tensile strength, and elongation at break) of silicone sealants, Lacasse and Paroli[20] used the same series of sealants. The results from both the mechanical properties and the STA (TG/DTA) showed changes in the silicone sealants. However, the authors were not able to correlate the results obtained from the mechanical test with those from the STA.

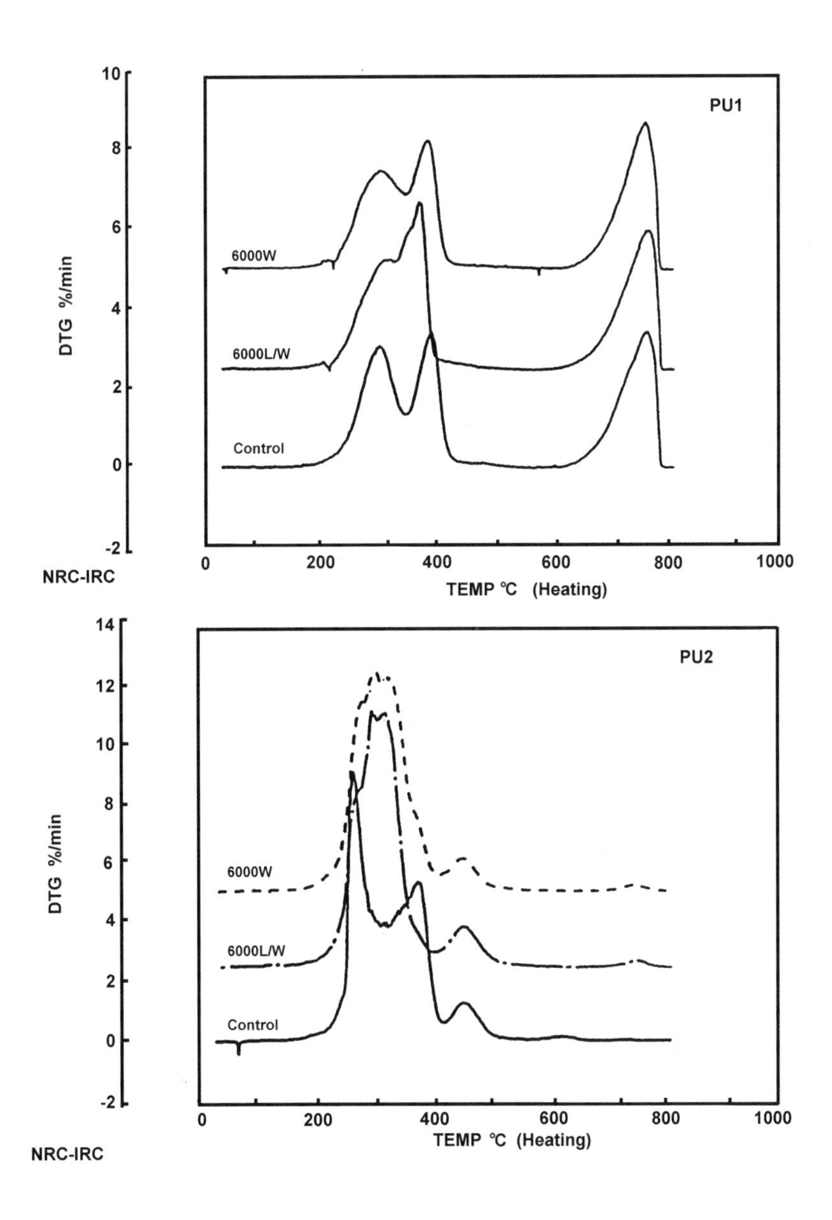

Figure 4. DTG curves for control and exposed polyurethane sealants PU1 and PU2.[13]

Dynamic mechanical analysis (DMA) can be considered one of the most suitable analytical techniques to study the viscoelastic properties of polymeric materials such as adhesives and sealants. According to Rogers, et al.,[21] it is more meaningful to correlate the performance test with the sealant viscoelastic properties to explain their behavior. In a proposed test protocol for selecting concrete pavement joint sealants, the authors recommended DMA as a technique to initially evaluate the curing of sealants and to study the effect of mechanical fatigue by measuring, in both cases, the glass transition. In another study[22] on polyurethane sealants for the same application, the authors concluded that the viscoelastic properties of the sealants, as determined by DMA, were able to explain sealant behavior observed in adhesion-in-peel and shear fatigue tests and that the correlation was conclusive for a variety of test conditions (e.g., high and low temperature, thermal cycling, and water and chemical exposure).

Malik[23] used dynamic mechanical analysis to study the effect of accelerated weathering on the storage and loss moduli of commercially available polyurethane (PU) and silicone construction sealants. Three multicomponent PU and one-component silicone sealants were used in the study. The sealants were exposed to 8 hours of UV exposure at 65°C alternating with 4 hours of condensation at 50°C for 600 and 1000 hours. The moduli G', G'', tan δ, and the dynamic viscosity (η) were measured in the temperature range of 25° to 125°C. The temperature was increased to 125°C, held for 2 hours and cooled to 25°C.

The study showed that silicone sealants exhibit very high dynamic moduli (10^9 dyn/cm^2) compared to PU (10^7 dyn/cm^2). Moreover, silicone is stiffer than PU sealants below -30°C (Fig. 5a–b). Malik reported that silicone and PU sealants behave similarly only above this temperature and that the modulus of the silicone sealant falls in the acceptable window. According to the author, the storage modulus for a good construction sealant should not exceed 10^8 dyn/cm^2 at -40°C. Therefore, he indicated that silicone sealants may not be suitable for temperatures below -30°C because they are too rigid.

It was reported that a comparison of the DMA data for two of the polyurethane samples (PU-A and PU-C) showed remarkable differences in the viscoelastic parameters of the two samples. The curves for G', G'', tan δ and η for PU-C (Fig. 5c) show almost a plateau whereas the same curves for PU-A display a transition between 50° and 70°C.

From his work, Malik concluded that contrary to the conventional view by sealant manufacturers and end use, silicone sealants are a lot stiffer than most of the PU sealants at temperatures below -35°C. Also, some PU sealants show decomposition, softness, and reversion at high temperatures and stiffness at lower temperatures after 600 and 1000 hours of exposure to UVB. Furthermore, PU sealants can show performance similar to commercial silicone sealants under severe test conditions. Finally, he concluded that rheological additives (used for sag resistance) may play an important role in the overall performance. Although, the author hints at the usefulness of DMA as a tool to characterize sealants and to predict their performance, interpretation of the results is difficult because no data was reported for unexposed sealant samples.

In another study, Jones, et al.,[24] utilized dynamic mechanical thermal analysis (DMTA) to examine the effects of movement during cure development on bulk joints sealants. Movement parameters were imposed upon the joints during cure. These parameters included a combination of temperature cycling over the relevant amplitudes of ±7.5% and ±12.5% at temperature ranges of 30° and 60°C, respectively. The joints were subjected to an elevated temperature cycle during compression (35° at -7.5% and 50°C at -12.5%) and to a low temperature during tension (5°C at + 7.5% and 50°C at +12.5%). Both one and ten cycles a day were imposed on the joints. Specimens were also cycled mechanically, but without temperature cycling. Mechanical testing was performed on these joints. The results indicated that cyclic movement during cure reduced significantly the performance of tensile adhesion joints prepared from a one-part system. However, the effect for two-part systems was minimal. Similar joint materials and configurations were analyzed using DMTA.

The analysis was carried out in the temperature range of -80° to 40°C at a frequency of 1 Hz, using a shear sandwich arrangement and a nominal peak-to-peak displacement of 23 μm. Samples from inner and outer sections of the bead after 1, 3, and 7 days of cure were tested together with uncured sealant. The curing of the sealants was monitored by observing the shifting of the primary peak to a higher temperature on the tan δ versus temperature curves, accompanied by a decrease in the height of the very broad secondary transition (Fig. 6).

Jones, et al.,[24] reported that DMTA results showed that movement during cure increases the rate of cure for a one-part sealant material, but had little effect on the actual mechanism of cure as shown in Fig. 7. For the two-part sealants, it was observed that movements during curing did not affect the shape and maximum of the tan δ. Therefore, they identified that the actual mechanism of curing for a two-part sealant is unaffected by movement, and this has no effect on the rate of cure.

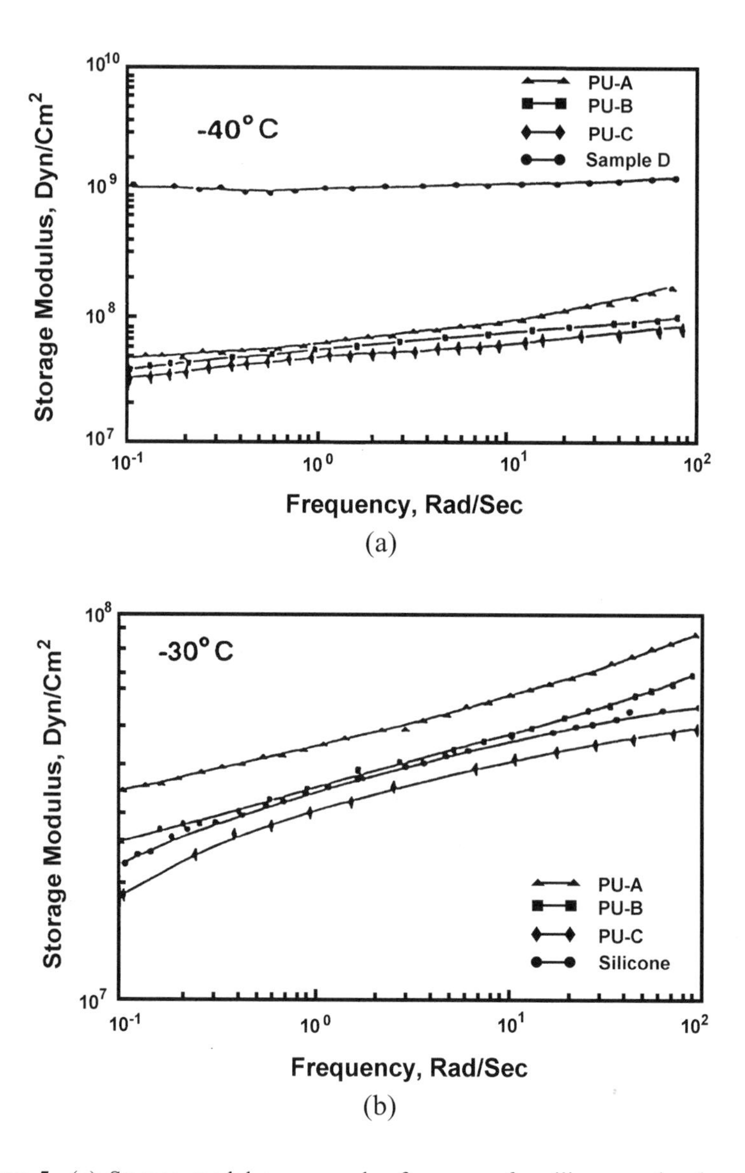

Figure 5. (*a*) Storage modulus vs angular frequency for silicone and polyurethane sealants at -40°C. (*b*) Storage modulus vs angular frequency for silicone and polyurethane sealants at -30°C. (*c*) Dynamic mechanical data for silicone sealant at 1 rad/sec. *(Reproduced with permission.)*[31] Note: 1 Dyn/cm^2 = 0.1 Pascal (Pa).

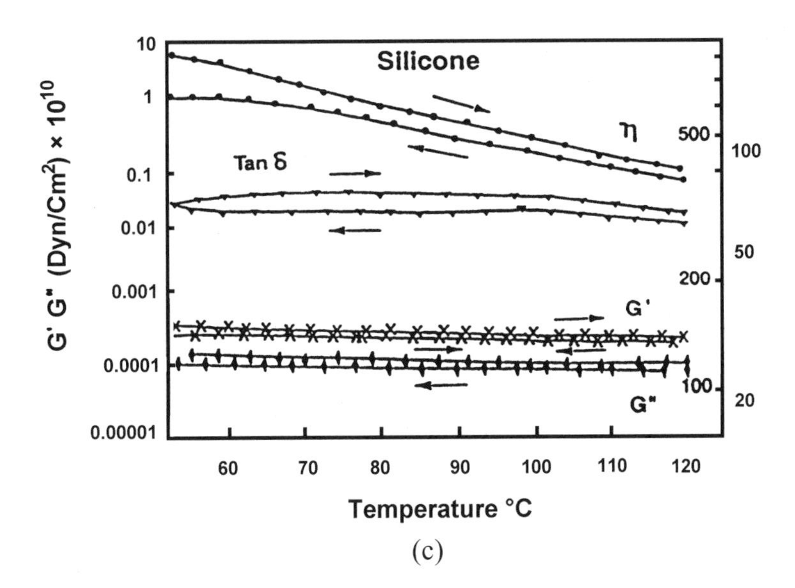

(c)

Figure 5. *(Cont'd.)*

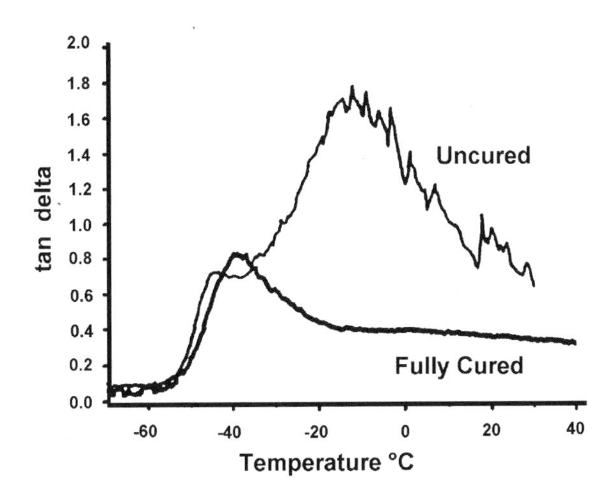

Figure 6. Curing of a typical one-part polyurethane sealant. *(Reprinted with permission.)*[7]

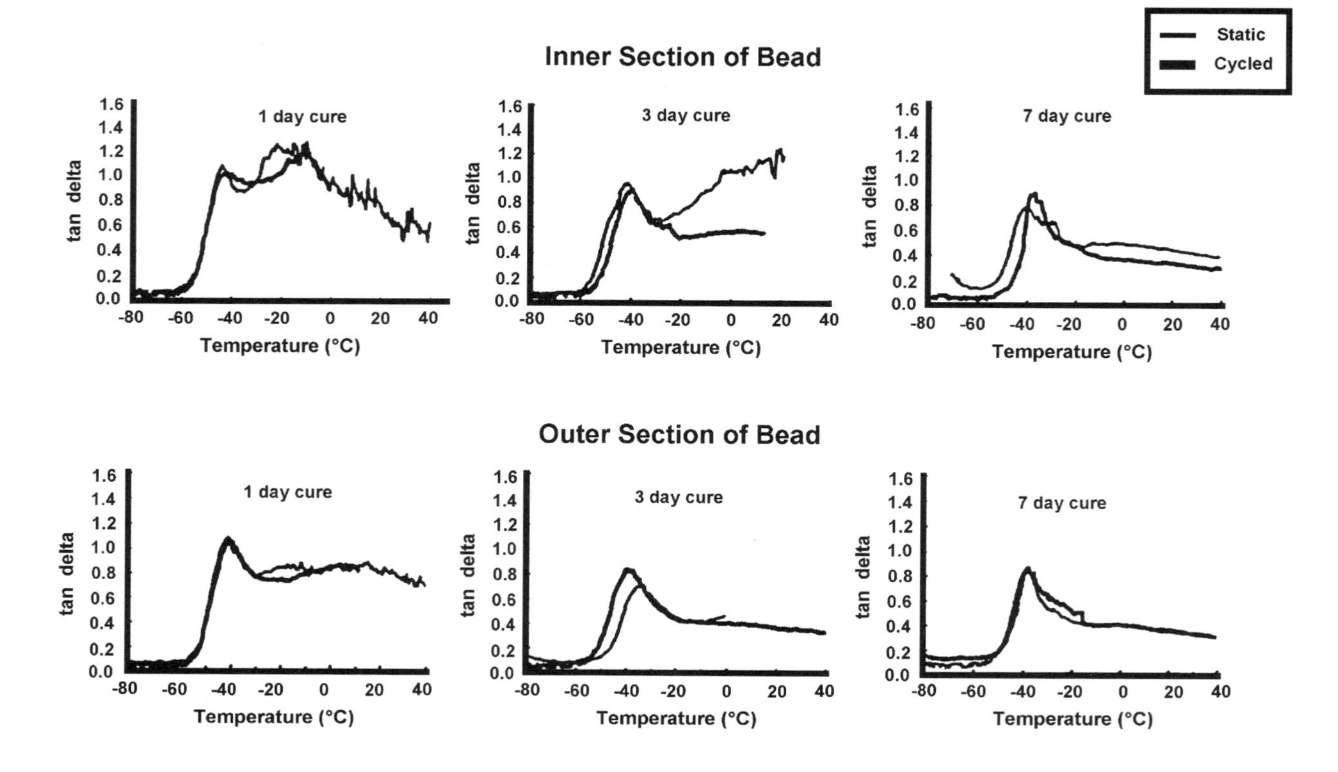

Figure 7. Effect of movement on bead sections of a one-part sealant at various stages throughout the cure. *(Reprinted with permission.)*[7]

As early as 1982, Riesen and Bartels[25] demonstrated the usefulness of dynamic load thermomechanical analysis (DLTMA) for the characterization of commercial joint sealants for building construction. They used DLTMA to study the viscoelastic behavior of four different sealant types: polysulfide, polyurethane, silicone, and polyacrylate. The difference in viscoelastic behavior of the sealants can be seen from Fig. 8.

From the DLTMA curves, the authors reported that besides the likely relaxation effect observed at -50°C (see Fig. 8), the silicone rubber behaves as an ideal rubber-elastic solid over the whole temperature range. The polysulfide and polyurethane sealants, on the other hand, exhibit viscous flow at temperatures higher than 150°C. The polyacrylate sealant will soften above 50°C indicating a pseudoplastic behavior.

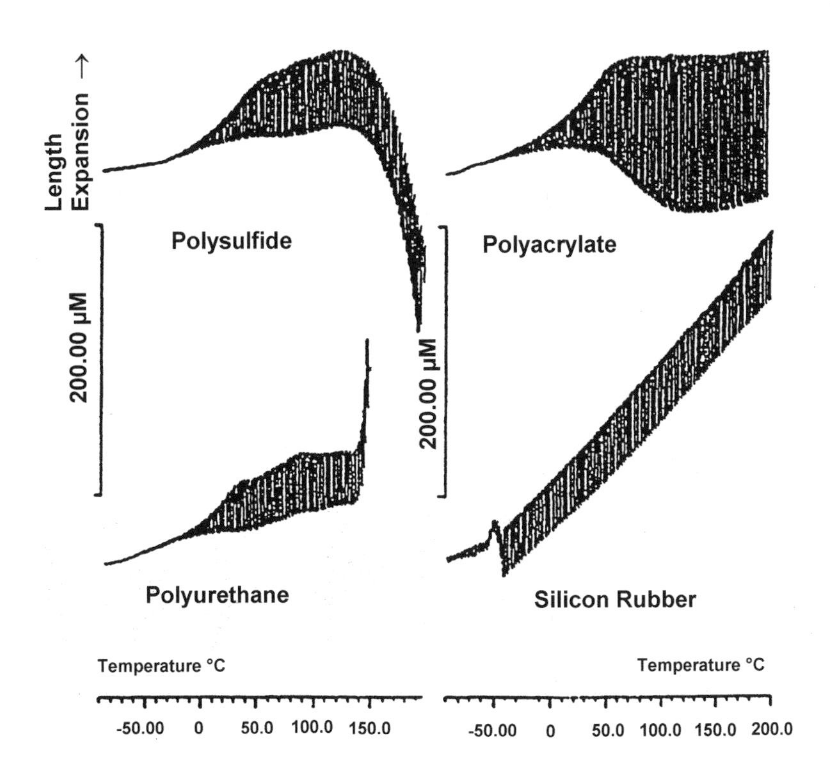

Figure 8. Typical DLTMA curves of four different elastomeric sealants (sealing compounds) in the temperature range of -90° and 200°C. *(Reprinted with permission.)*[7]

The length change due to the change in DLTMA load allows determination of Young's modulus. As shown by Riesen and Bartels' work, the use of thermoanalytical techniques for the characterization of sealants began almost 20 years ago. Wolf[7] stated that "studies in this direction had begun by Bartels but the results were never published due to the death of one of the authors."

In 1995, Squire[26] reported results of TG and DSC analysis on aging-induced changes in silicone, polysulfide (PS), and polyurethane (PU) sealants. The three types of sealants were exposed to both artificial and natural weathering regimes. The artificial weathering included exposure to UVB radiation, UVB/condensation cycling, heat aging, and water immersion. The natural weathering consisted of exposure to hot, dry, hot humid, and temperate climates. The TG analysis of the weathered sealants showed changes in the amount of unbound polymer present in the artificially weathered PU and PS sealants. The variable heating rate method, which provides information on kinetic parameters, estimated lifetimes, etc., was used to examine changes in the samples aged for twenty-eight and forty weeks in the laboratory. Changes observed in the thermal decomposition profiles were attributed to a combination of factors, including plasticizer loss, chain scission, and cross-linking. The author also used high temperature differential scanning calorimetry (DSC) to analyze the PU sealants. He reported that distinct fusion endotherms were obtained and that their position and intensities were altered by the accelerated aging regimes as well as by natural weathering. However, no attempts were made to assess aging-induced changes in the glass transition temperature of the sealants.

Rama Rao and Radhakrishnan[27][28] studied the degradation kinetics and mechanism of liquid polysulfide polymers and of ammonium dichromate cured polysulfide using thermogravimetry (TG) and pyrolysis-gas chromatography-mass spectrometry. Although the authors did not investigate the aging-induced changes in the sealants, this study showed that thermoanalytical techniques combined with other analytical techniques can provide significant information on degradation mechanisms of polymers.

Boettger and Bolte[29] conducted a multi-year durability study on a series of high-performance artificially- and naturally-aged sealant products. The physical and mechanical properties of seven series of sealants (polyurethane, silicone, polysulfide rubber, polychloroprene, and EDPM rubber) were evaluated after 6000 hours exposure to natural and artificial aging. The artificial aging consisted in exposure to UV radiation at four different temperatures and condensation at 50°C. Samples were also exposed to xenon-arc radiation, heat and water spray for three weeks, heat in a ventilated oven, and to acid dew and fog. Furthermore, artificial weathering was combined with extension and compression. The properties of the exposed and reference sealants were evaluated every 1000 hours by using DSC.

They found that DSC was the most suitable technique to investigate changes in the glass transition temperature (T_g) for elastomeric sealants. Figure 9 displays the DSC thermal curves for a polysulfide rubber sealant (B9) before and after 6000 hours of artificial exposure in an UVCON (80/50). As can be observed, the T_g of the exposed shows a 4°C increase. From the DSC results, the authors were able to establish a loose correlation between mechanical properties and T_g. They also reported that the data agree with the changes observed in the mechanical properties of the B9 sealant.

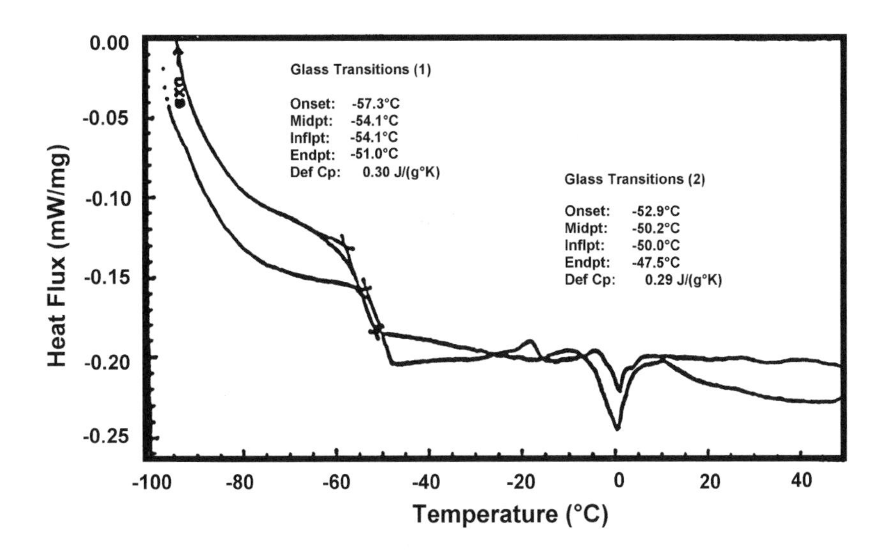

Figure 9. DSC thermograph for Material B9 subjected to 6000 hours of exposure to artificial aging in a UVCON (80/50) apparatus. Determination of the glass transition temperature is shown in the figure. *(Reprinted with permission.)*[7]

Feldman and Lacasse[30] studied the morphology of blended lignin-polyurethane (L-PU) sealants. The objective of the work was to establish the viability of blending lignin (L), a naturally occurring, readily available polymer resource, with polymeric based sealants. The L-PU blends were exposed to accelerated and natural weathering. The control, pure L, and exposed blends were analyzed by DSC. The control specimens were subjected to 23°C and 50% RH. The accelerated weathering regimen performed in a weathering chamber, consisted of 170 cycles, each between temperatures of -30° and 40°C, four times every twenty-four hours with twelve hours daily UV-radiation exposure. Samples were continuously artificially weathered for 180 days. The L-PU specimens were exposed to exterior weather conditions in central Montreal for thirty-seven weeks between July 15, 1985 and March 31, 1986. During this period, temperature varied between +32.6° and -26.7°C.

DSC curves for L and L-PU blends (unexposed and exposed) are displayed in Fig. 10. The authors attributed the transition near -52°C to the T_g of the soft PU segments and that at about +9°C was attributed to the microcrystalline melting of the soft segments. Similarly, the transition observed at ~ 90°C was probably due to the breakup of short-range ordered hard segments. The transitions for blended specimens occur in the same temperature interval as for the neat PU. The glass transition temperature, barely detectable for L at 125°C, indicates that the L-PU blend exists in two distinct phases which, in turn, confirms their observation by microscopy. From the DSC results, the authors concluded that DSC confirms the immiscibility of L with PU at blend ratios used in the study.

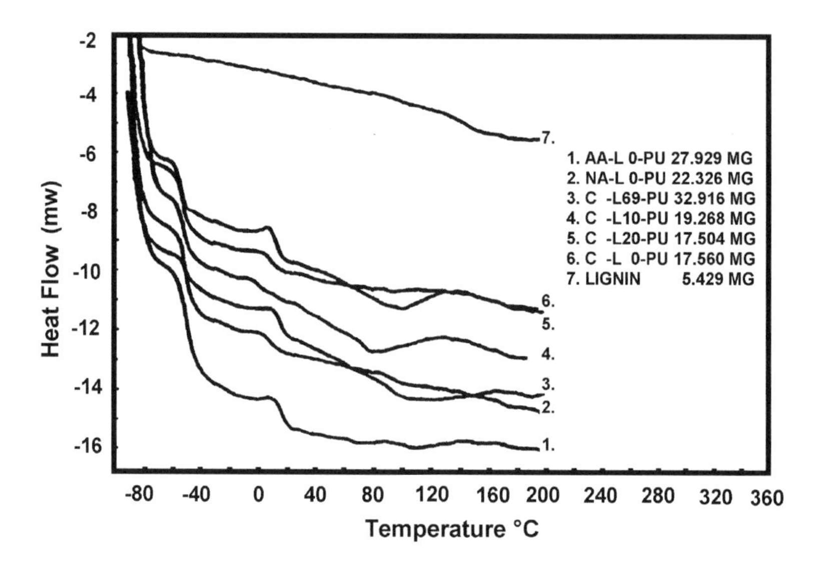

Figure 10. DSC curves of lignin and L-PU polyblends. *(Reprinted with permission.)*[30]

3.2 Adhesives

Dynamic mechanical analysis (DMA) has been used to study the flow behavior of hot-melt adhesives.[31] Brummer and co-workers[32] used DMA to study the viscoelastic behavior of adhesives. They found that dynamic mechanical measurements in adhesives provided insight in the macromolecular mobility of the polymer or rubber system studied. The viscoelastic behavior at various temperatures can be correlated with standard measurements such as adhesive force, shear strength, and tack. The authors concluded that three-dimensional DMA plots from frequency-temperature sweeps provide a complete overview of the frequency and temperature dependence of the adhesive. Foster, et al.,[33] characterized the hot-tack differences in hot-melt adhesives using DMTA.

Although thermoanalytical techniques have been widely used in the characterization of different types of adhesives, little work has been published on the use of these to characterize construction adhesives.

Onic and co-workers[34] reported the results of their work on the use of DMTA to investigate the viscoelastic behavior of wood joints bonded with cross-linking thermoset adhesives. The resins studied were phenol formaldehyde (PF), melamine-urea-formaldehyde (MUF), resorcinol-formaldehyde (RF), tannin-formadehyde (TF), and tannin-hexamethylenetetramine (TH). The curing of the adhesives was followed by observing the changes in the storage modulus E' and tan δ. The authors reported that the value of the joint E' increases as the adhesives pass from the liquid to the rubbery state and then finally to the glassy state. From this, three distinct zones were observed.

The first zone was observed at very low values of E' where the adhesive behaves as a liquid (T_{gel}). The second zone starts at E'_{min} characterized by an increase in the E' value up to E'_{max}•E'_{min} (the elastic modulus of the joint occurring at the adhesive gel temperature (T_{gel}). E'_{max} occurs at T_f where the hardened adhesive network is likely to become tighter (vitrification). The third zone was identified as that where the value of the E' decreases because of physical and chemical changes in the properties (e.g., degradation/softening) of the materials in the joint.

Onic, et al.,[34] considered that the decrease in modulus of the bonded joint could not just be attributed to degradation of the adhesive because all of the adhesives used, with exception of MUF resin, were thermally stable in the temperature range used. It was rather due to the wood substrate and delamination of the joint. Therefore, they concluded that the

DMTA results can only be interpreted as the behavior of the total joint rather than the adhesive glueline alone.

Dynamic mechanical thermal analysis (DMTA) has also been used by Brinson, et al.,[35] to determine the suitability of the technique for evaluating damage in the adhesive bond from the viscoelastic properties of bonded beams and for evaluating the effects of various environmental conditions and various surface treatments. The authors considered that if the bond becomes damaged (either adhesive and/or interphase) due to excessive load, fatigue, moisture, or corrosion, it would seem likely that dissipation mechanisms or loss modulus and tan δ would change. Therefore, they used DMTA to measure the viscoelastic properties of beams with simulated flaws and beams taken from lap specimens, which had been exposed to humidity and/or corrosion for extended periods.

The DMTA data for the simulated flaw obtained at the two frequencies used (1.0 and 10 Hz) showed a substantial change in storage and loss moduli (E' and E'') with the inclusion of simulated flaws from no flaw, 0%, to a flow the length of the beam, 100%. Also, the glass transition as given by E'' appears to decrease with increasing simulated flaw length. The authors also reported that the significant changes were observed with increasing exposure to humidity. Specimens exposed for over a month had higher storage moduli, E', than some of those exposed for shorter periods of time. Furthermore, exposure to corrosive conditions showed to be more complicated than for humidity. However, initial changes in the dynamic mechanical properties are essentially the same as in humidity exposure.

From the results of the work, it was concluded that it does appear that DMTA studies on bonded beams may allow the determination of progressive damage due to fatigue, moisture, corrosion, or, perhaps, to other environmental parameters. But it may be necessary to build special DMTA equipment more sensitive to small changes in damping behavior.

Differential scanning calorimetry (DSC), DMA and TG were used by Tabaddor and co-workers[36] to investigate the cure kinetics and the development of mechanical properties of a commercial thermoplastic/ thermoset adhesive, which is part of a reinforced tape system for industrial applications. From the results, the authors concluded that thermal studies indicate that the adhesive was composed of a thermoplastic elastomeric copolymer of acrylonitrile and butadiene phase and a phenolic thermosetting resin phase. From the DSC phase transition studies, they were able to determine the composition of the blend. The kinetics of conversion of the thermosetting can be monitored by TG. Dynamic mechanical analysis measurements and time-temperature superposition can be utilized to describe the development of the mechanical properties of the material over several decades of time. The results of the study also demonstrated the power of the various thermoanalytical tools for increasing understanding of industrial polymers and controlling compositional variability in applications.

Another example of the applications of thermoanalytical techniques for adhesives characterization, in general, is the work done by Ludbrook and Whitwood.[37] They have used DSC and DMA in conjunction with lap shear strength testing to relate the degree of cure of several toughened epoxy adhesives to the build-up of mechanical strength. DMA was also used to measure the T_g of adhesives and its changes with curing.

It was concluded that DSC and DMA permit understanding of and defining the properties of curable adhesive formulations by relating mechanical performance to the extent of cure. This allows the isolation of the changes of mechanical properties that result from incomplete cure from those due to other factors such as aging. Since it may be necessary to screen many potential formulations to achieve the desired mechanical properties and cure characteristics during adhesive development, DMA is a very useful tool for broad characterization of the mechanical properties as a function of temperature. Finally, they concluded that alternative mechanical test methods, such as lap shear strength, only gives a pass/fail criterion whereas DMA enables the influence of compositional changes to be correlated with the T_g and allows the influence of such changes to be studied as part of an overall experiment design.

Allsop, et al.,[38] found that DMTA and modulated DSC (MDSC) could be used to identify useful secondary components for adhesive formulations. Thermal analysis techniques have been successfully applied by the Naval Aviation Depot, North Island, in the evaluation and repair of advance composite components.[39] Differential scanning calorimetry (DSC) has been used to detect a residual exotherm in undercured FM300 film adhesive and the amount of residual exotherm has been correlated to a reduction in the lap shear strength of the adhesive. A similar correlation has been obtained between lap shear strength and the reduction in the glass transition temperature (T_g) as measured by DMA. A reduction in the T_g of thermally exposed AS4/3501-6 laminates has been detected using TMA. The authors correlated the T_g to the reduction in the mechanical properties of the material.

Eastman[40] published a paper on "The Anatomy of Hot Melt Adhesives by Thermal Analyses." By using DSC, TMA, and TG, the author attempted to show relationships between measurable thermal properties and physical characteristics of different hot melt systems. From the results, it was concluded that modern thermoanalytical techniques, individually or combined, can be invaluable tools in quantifying the total energy aspects of hot melt adhesives and sealant compositions. The author also reported that the effect of crystalline content of the different components on both properties and performance characteristics can be better understood by use of DSC. Furthermore, thermomechanical analysis (TMA) offers a reproducible, quantitative means to study the deformation characteristics, either expansive or compressive, of components of a fully formulated system during controlled thermal conditions. Also, it is ideally suited to characterize the heat resistance properties of hot melt adhesives. On the other hand, TG offers many ways to better understand the effects of components on thermal stability, particularly when combined with infrared spectroscopy.

In another study, Lu, et al.,[41] used DSC and TG to investigate the thermal characteristics of the Larch (*Larix gmelini*) tannin-phenol-formaldehyde adhesive, and its application in plywood and hardboard manufacturing. From their investigation, it was concluded that Larch tannin extracted from *Larix gmelini* bark is suitable to partially replace the phenolic component in adhesives for plywood and hardboard applications. Moreover, they concluded that both the TG and DSC methods can be utilized to directly study the thermal curing reaction of the Larch. Thermogravimetric analysis has also been used to investigate the degradation of holt melt adhesive materials used in disposable diapers.[42]

Thermomechanical analysis (TMA) was used to investigate the curing behavior of resorcinol-formaldehyde (RF) and phenol-resorcinol-formaldheyde (PRF) adhesive resins used for structural glued laminated timber and veneer lumber.[43] Samples were heated from room temperature to 200°C and the storage modulus of the adhesive was measured. The TMA results showed a rapid increase in the storage modulus of the commercial RF between 60° and 80°C and remain stable above 85°C, the same as the synthetic PRF adhesives. Furthermore, RF and PRF curing periods at room temperature conditions required more than thirty days.

The thermal curing behavior of phenol-formaldehyde (PF) adhesives was also studied by DSC[44] using three different scanning methods (single-heating, multi-heating rates, and isothermal methods). The results showed that the single-heating rate method is fairly rapid and produces abundant reaction kinetic parameters, which are useful for comparison of different PF resins. However, it produced greater activation energy compared to the multi-heating rate method. This method was applied successfully to characterize thermal behavior of power resins displaying two exothermic peaks. Finally, the isothermal method can provide a means of determining a reaction type.

Yin, et al.,[45] studied the use of TMA to determine the variation of physical properties of UF, MUF, and PMUF thermosetting resins in-situ in wood joints. The TMA analysis was carried out on a three-point flexion from 25° to 220°C at 10°C/min heating rate and a constant applied force of 0.3 N or at a dynamic force of 0.2 N. Some of the tests were performed at 60, 80, 100, and 120°C. Depending on the rate of the adhesive hardening, the measured times varied between 5 and 80 minutes.

They reported that the three adhesives/wood joints behave in a similar manner. The relative elastic modulus versus temperature of the joints are shown in Fig. 11. According to the authors, three distinct zones are observed on the TMA curves. The low temperature zone (1st zone) is characterized by a low relative elastic modulus, the middle zone shows a sudden and marked increase of the modulus value, and in the higher temperature zone (3rd), a slow decrease is observed. It was concluded that before gelling the adhesive behaves as a liquid, hence, a low elastic modulus as it cannot transfer the stress between the two wood layers. On gelling, a tri-dimensional structure started to form, transforming the adhesive from a liquid to a rubbery state increasing the relative elastic modulus of the adhesive and, thus, the wood joint. The beginning of the second zone can be considered as the gel temperature.

The results indicated that the final hardening appears like a second order transition and that the definition of hardening temperature is complicated because hardening occurs within either a wider or a narrower temperature range, which varies with the technique. Therefore, Yin, et al.,[45] defined the hardening temperature (T_V) as the one at which the rate of increase of the elastic modulus attains its maximum.

From the TMA results, Yin, et al.,[45] concluded that non-isothermal and isothermal TMA analysis is a useful tool for the determination of the variation of different chemical properties of thermosetting wood adhesives directly in the wood joint during the cross-linking and hardening process of the adhesives. Furthermore, an increase in the mechanical resistance of a bonded joint during adhesive hardening can be clearly correlated with the degree of resin conversion. In this respect, gelling, hardening, vitrification temperatures, and the temperature at which the modulus attains its maximum can be determined by TMA.

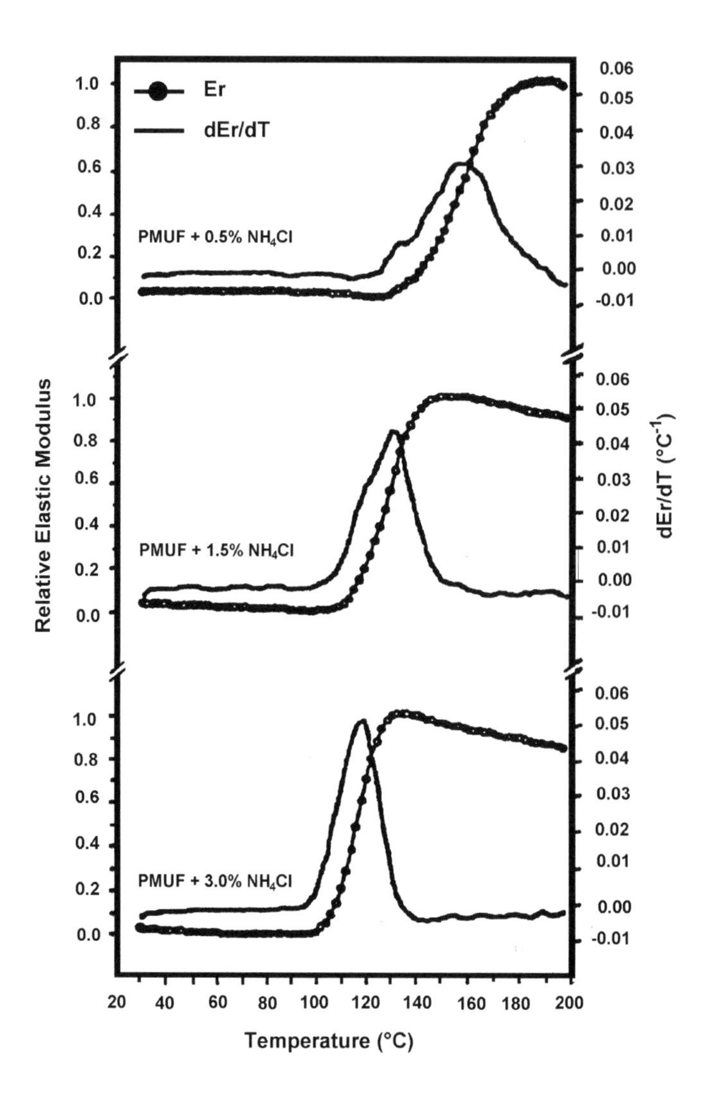

Figure 11. (*a*) Evolution of relative modulus and of its first derivative of PMUF/wood joints as a function of temperature and hardener; (*b*) Evolution of relative modulus and of its first derivative of MUF/wood joints as a function of temperature and hardener. *(Reprinted with permission.)*[45]

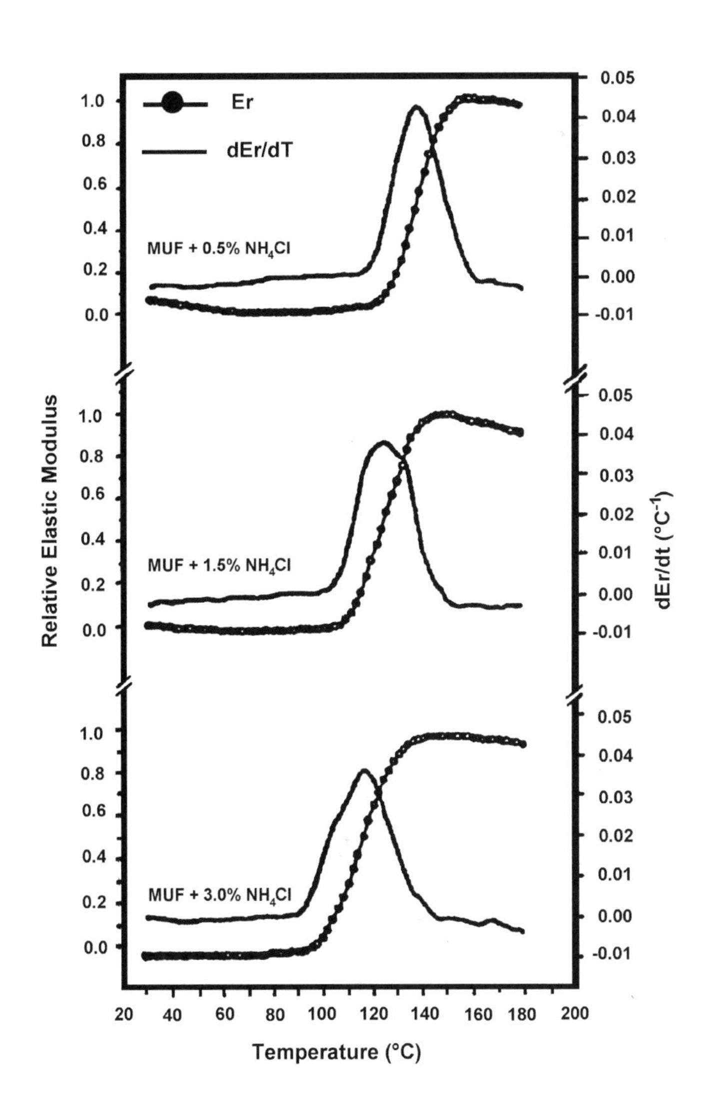

Figure 11. *(Cont'd.)*

Development of thermal analytical techniques for materials characterization is always in progress. Thermal wave microanalysis for testing of adhesive bonding has been reported in the literature.[46] This technique is a reliable tool for non-contact, non-destructive testing of adhesive bonding of thin-walled materials. The spatial distribution of the adhesive under a metal foil can be visualized during all stages of the curing process.

REFERENCES

1. Blaga, A., Durability of Building Materials and Components, (P. J. Sereda and G. G. Litvan, eds.), p. 827, ATM STP691, American Society for Testing Materials (ASTM), Washington, DC (1980)
2. Feldman, D., *Polymeric Building Materials*, Elsevier Sci. Publ. Ltd., UK (1989)
3. Cheremisinoff, N. P., *Elastomer Technology Handbook*, p. 836, CRC Press, Inc., USA (1993)
4. Landrock, A. H., *Adhesives Technology Handbook,* Noyes Publ., Park Ridge, NJ (1985)
5. Laaly, H. O., *The Science and Technology of Traditional and Modern Roofing Systems*, Vol. 1, Ch. 33, Laaly Scientific Publ., Los Angeles, CA (1992)
6. Hartshorn, S. R., Structural Adhesives, *Chemical and Technology Topics in Applied Chemistry,* p. 18, Plenum Press, New York (1986)
7. Wolf, A. T., Review of Modern Analytical Techniques Used in the Study of Aging-induced Changes in Sealants, *Durability of Building Sealants*, pp. 323–342, RILEM Publ. S.A.R.L., Ch. XVII, The Publ. Co. of RILEM (1999)
8. Sichina, W. J., A Practical Technique for Predicting Mechanical Performance and Useful Lifetime of Polymeric Materials, Reprints from *Am. Lab.* (Jan. 1998)
9. Paroli, R. M., Delgado, A. H., Rossiter, W. J., and Flueler, P., Using Thermoanalytical Techniques to Characterize Roof Membrane Materials, in: *Proc. 4th Int. Symp. on Roofing Technol.,* pp. 58–65, Gaithersburg, MD (Sept. 17–19, 1997)
10. Oba, K., and Roller, A., Characterization of Polymer-Modified Bituminous Roofing Membranes Using Thermal Analysis, *Mater. Struct.,* 28:596–603 (1995)

11. Paroli, R. M., Delgado, A. H., and Cole, K. C., Applications of Thermogravimetry and PAS in the Characterization of Silicone Sealants, *Canadian J. Appl. Spectroscopy,* 39:7–14 (1994)

12. Paroli, R. M., Cole, K. C., and Delgado, A. H., Weatherability of Polyurethane Sealants: An Spectroscopic Approach, *Polymeric Mater. Sci. Eng.,* 71:435–436 (1994)

13. Paroli, R. M., and Delgado, A. H., Applications of Thermogravimetry-Fourier Transform IR Spectroscopy in the Characterization of Weathered Sealants, pp. 129–148, ACS Symp. Series, No. 581 (1994)

14. Paroli, R. M., Cole, K. C., and Delgado, A. H., Evaluating the Weatherability of Polyurethane Sealants, pp. 117–136, ACS Symp. Series, No. 598 (1995)

15. Paroli, R. M., and Delgado, A. H., Applications of Thermogravimetry in the Characterization of Silicone Sealants, *Polymeric Mater. Sci. Eng.,* 69:139–140 (1993)

16. Chew, M. Y. L., Joint Sealant for Wall Cladding, *Polymer Testing,* 19:643–651 (2000)

17. Tock, W., Dinivahi, M. V. R. N., and Chew, C. H., Viscoelastic Properties of Structural Silicone Rubber Sealants, *Advances Polymer Technol.,* 8(3):317–324 (1988)

18. Leonard, N. E., and Malik, T. M., The Relation Between Rheology, Artificial Weathering, and Urethane Sealant Performance, Science and Technology of Building Seals, Sealants, Glazing, and Waterproofing, (J. C. Myers, ed.), 3:21–32, ASTM STP 1254, ASTM, Philadelphia (1994)

19. Hugener, M., and Hean, S., Comparison of Short-term Aging Methods for Joint Sealants, in Durability of Building Sealants, *Proc. Int. RILEM, Symposium,* (J. C. Beech and A. T. Wolf, eds.), pp. 37–47, E & FN Spon, London (1996)

20. Lacasse, M. A., and Paroli, R. M., Evaluating the Durability of Sealants, in: *Sciences and Technology of Building Seals, Sealants, Glazing and Waterproofing,* (D. H. Nicastro, ed.), 4:29–48, ASTM STP 1243, ASTM, Philadelphia (1995)

21. Rogers, A. D., Lee-Sullivan, P., and Bremner, T. W., *J. Mater. Civil Eng.,* 11(4):302–308 (1999)

22. Rogers, A. D., Lee-Sullivan, P., and Bremner, T. W., *J. Mater. Civil Eng.,* 11(4):309–316 (1999)

23. Malik, T., Sealant Rheology and Its Practical Measurements, *Polymeric Mater. Eng. Sci.,* 75:291–292 (1996)

24. Jones, T. G. B., Hutchinson, A. R., and Atkinson, K. E., Effects of Early Movement on the Performance of Sealed Joints, in: *Proc. Int. RILEM Symp. on Durability of Building Sealants,* (A. T. Wolf, ed.), pp. 117–132, E. & F. N. Spon, London (1999)

25. Riesen, R., and Bartels, W., Rapid Quantitative Characterization of Elastomers by Dynamic Load Thermomechanical Analysis, in: *Proc. 7th Int. Conf. on Thermal Analysis,* 2:1050–1056, John Wiley & Sons, NY (1982)

26. Squire, R., Chemical Changes in Roof Membranes and Building Sealants Building Research Establishment Occasional Paper, OP 60, pp. 29–48, Bldg. Res. Establishment, Garston, UK (1995)

27. Rama Rao, M., and Radhakrishnan, T. S., Thermal Degradation of Liquid Polysulfide Polymers: Pyrolisis-GC-MS and Thermogravimetry Studies, *J. Appl. Polymer Sci.,* 30:855–873 (1985)

28. Radhakrishnan, T. S., and Rama Rao, M., Kinetics and Mechanism of the Thermal Degradation of Polysulfide Polymer Cured with Ammonium Dichromate/Pyrolysis-Gas Chromatography and Thermogravimetric Studies, *J. Anal. Appl. Pyrolysis,* 9:309–322 (1986)

29. Boettger, T., and Bolte, H., Results from the University of Leipzig Project Concerning the Long-Term Stability of Elastomeric Building Sealants, in *Science and Technology of Building Seals, Sealants, Glazing, and Waterproofing,* (J. M. Klosowski, ed.), 7:66–80, ASTM STP 1334, ASTM, West Conshohocken, PA (1998)

30. Feldman, D., and Lacasse, M. A., Morphology of Lignin-polyurethane Blends, *Mater. Res. Soc. Symp. Proc.,* 153:193–198 (1989)

31. Rohn, C. L., Flow Behavior of Holt Melt Adhesives, *Thermal Trends,* 3(1):18 (Winter 1996)

32. Brummer, R., Hetzel, F., Harder, C., and Germany, H., Correlation of Polymer Properties with Dynamic Mechanical Measurements, *Appl. Rheology,* 7(4):173–178 (1997)

33. Foster, B. W., and Barrret, A. E., Characterizing Hot-tack Difference in Hot-melt Adhesives Using DTMA, *Tappi Journal,* 74(9):135–138 (1991)

34. Onic, L., Bucur, V., Ansell, M. P., Pizzi, A., Deglise, X., and Merlin, A., Dynamic Thermomechanical Analysis as a Control Technique for Thermoset Bonding of Wood Joints, *J. Int. Adhesion & Adhesives,* 18:89–94 (1998)

35. Brinson, H. F., Dickie, R. A., and Debolt, M. A., Measurement of Adhesive Bond Properties Including Damage by Dynamic Thermal Analysis of a Beam Specimen, *J. Adhesion,* 55:17–30 (1995)

36. Tabaddor, P. L., Aloisio, C. J., Bair, H. E., Plagianis, C. H., and Taylor, C. R., Thermal Analysis Characterization of a Commercial Thermoplastic/ Thermoset Adhesive, *J. Thermal Anal. Calorimetry,* 59:559–570 (2000)

37. Ludbrook, B. D., and Whitwood, R. J., The Use of Thermal Analysis to Investigate the Cure of Reactive Adhesives, *Int. J. Adhesion Adhesives,* 12(3):138–144 (1992)

38. Allsop, N. A., Bowditch, M. R., Glass, N. F. C., Harris, A. E., and O'Gara, P. M., Thermal Analysis in the Development of Self Validating Adhesives, *Thermochimica Acta,* 315:67–75 (1998)

39. Marien, M. E., Perl, D. R., and Gaenzle, K., Applications of Thermal Analysis in the Repair of Advance Composite Structures, *39th Int. SAMPE Symp. and Exhibition,* 39:639–650, Anaheim, CA (April 11–14, 1994)

40. Eastman, E. F., The Anatomy of Hot Melt Adhesives by Thermal Analyses, *Polymer, Laminations, and Coatings Conf.,* pp. 391–425, Nashville, TN (1996)

41. Lu, Y., Shi, Q., and Gao, Z., Thermal Analysis and Application of Larch Tannin-based Adhesive, *Holz als Roh- und Werkstoff,* 53(3):205–208 (1995)

42. Ryan, C. M., Investigating the Degradability of Holt Melt Adhesive Formulae, *Adhesive Age,* pp. 30–34 (July 1990)

43. Taki, K., Yamada, M., and Hoshida, H., Thermal Analysis and Bonded Strength of Thermosetting Resin, *Euradh'98: WCARP1, 4th Adhesion-Europe Conf.,* 2:167–168, Garmisch Partenkirchen, Germany (1998)

44. Park, B. D., Riedl, B., Bae, H. J., and Kim, Y. S., Differential Scanning Calorimetry of Phenol-formaldehyde (PF) Adhesives, *J. Wood Chem. Technol.,* 19(3):265–286 (1999)

45. Yin, B. S., Deglise, X., and Masson, D., Thermomechanical Analysis of Wood/Aminoplastic Adhesives Joints Cross-linking - UF, MUF, PMUF, *Holzforschung,* 49(6):575–580 (1995)

46. Netzelmann, U., Thermal Wave Micro Analysis for Testing of Adhesion Bonding, *Materialprufung, Materials Testing, Materiaux Essais et Recherches,* 38(6):246–249 (1996)

15

Roofing Materials

1.0 INTRODUCTION

The use of polymers as a roofing material is continuously increasing. The function of a roof is to protect the building from environmental factors such as light, wind, rain, snow loads, temperature changes, hail, and storms. Therefore, the material used on a roof must withstand those factors for many years.

The performance of a material or system depends on the environment and the degrading effects to which it is exposed. Until the early 1990s, methods of measuring the properties and predicting the durability of products were not well known, hence, much of the knowledge about building materials was based on experience from long-time use.

An estimate of 90% of all flat roofs in Canada used in industrial, commercial, and public buildings are protected by bituminous roofing, and essentially all sloping residential roofs are covered with shingles having a bituminous-felt base. In recent years, changes in building practices have produced roofs of unusual design for which these materials were not suitable. This led to the development of new rubber and plastic roofing material, which, in turn, led to the development of rubber- and plastic-modified bitumen and bitumen-modified rubbers and plastics.[1]

In 1993 and 2000, the commercial low-slope roofing market in the USA consisted of:

Type	1993[2]		2000[3]	
	New construction	Re-roofing	New construction	Re-roofing
Built-up roofing	29.0%	31.0%	22.4%	27.8%
Single ply roofing	38.8%	33.3%	38.1%	33.6%
Modified bitumen	17.0%	21.7%	19.1%	23.8%
Metal	3.5%	1.6%	4.3%	2.5%
Other types*	11.7%	12.4%	16.1%	12.3%

* Other types = tiles, PUF, liquid-applied, asphalt shingles, metal

The NRCA's 2000–2001 Annual Market Survey[3] reported that in 2000, the low-slope roofing market accounted for 64.1% of the total roofing market, a slight decrease from 68.7% in 1999. The roofing contractors predicted that the low-slope roofing market would be 63.3% in the year 2001.

Most types of roofing materials are bituminous and synthetic (polymeric) roofing membranes. The most commonly used roofing and waterproofing membrane is made by combining asphalt or coal tar pitch (bitumen) with felts or mats, or fabrics of organic or inorganic fibers.

2.0 BITUMINOUS ROOFING MATERIAL

The built-up roof membrane consists of bitumen, reinforced with roofing felts, and aggregates which protect the bitumen from the UV radiation and oxidation. Bituminous materials have been used since 3500 BC. Because of their waterproofing, preservative, and binder characteristics, they were utilized by the ancients for the construction of houses and roads.[4] Bituminous materials were also used by ancient civilizations such as Egyptians for construction, mummification, waterproofing, preservatives, and binders.

The asphalt is a complex mixture of organic and inorganic compounds and their complexes. Some of the organic compounds are aliphatics, aromatics, polar aromatics, and asphaltenes. Polar aromatics are responsible for the viscoelastic properties of asphalts.[5]

Over the years, multilayers of tar-based waterproofers replaced the hot asphalt used in roofing. In early 1900, asphalt became available from petroleum refining, and it was followed by oxidized bitumen interlaid with roofing felt and then alternated with a mineral base sheet.[4] Asphalt-based materials are used extensively as binders, sealants, and waterproof coatings in diverse applications because of their low cost, inherent cohesive nature, weather-resistant properties, and ease of processing in the molten state.[6]

Despite its natural viscoelastic properties, asphalt cannot be used as such in roofing applications because of its inherent limitations, such as brittleness at low temperature and flow properties at high temperature. Therefore, studies have been done to improve the properties of bitumen. Combining bitumen with natural or synthetic rubbers or lattices, new materials with higher elasticity, low temperature flexibility, higher strength, and better fatigue resistance can be obtained.

Polymer-modified bituminous membranes were developed in Europe in the mid-1960s and have been in use in North America since 1975. The polymeric systems have varied from natural rubber to more complex synthetic systems such as block copolymers and polymer blends. Most common polymers used as modifiers are polyisobutylenes, polybutadienes, polyisoprenes, styrene-butiene-monomer, styrene-butadiene-rubber, butyl rubber, ethylene-vinylacetate (EVA), atactic polypropylene (APP) as well as natural rubber. Polymers, such as atactic polypropylene or styrene-butadiene-styrene (SBS), impart flexibility and elasticity, improve cohesive strength, resist flow at high temperatures, and toughness.[7] They are the most widely used modifiers of bitumen-based roofing materials.

3.0 SYNTHETIC ROOFING MEMBRANES

Polymers such as poly(vinyl chloride) (PVC), ethylene-propylene-diene monomer (EPDM), chlorosulfonated polyethylene, ketone ethylene ester (KEE), reinforced polyurethane, butyl rubbers, and polychloroprene (neoprene) have proven to be suitable for roofing membranes.[4] In the last ten years, a new synthetic roofing material (thermoplastic polyolefins)

(TPO) has entered the market. Polymeric membranes have advantages and disadvantages.[8]

Advantages:

- Ultra-lightweight (for unballasted, adhered membranes)
- Adaptable to irregular roof surfaces
- Isotropic physical behavior
- Superior architectural quality (color)
- White membranes have general superior heat reflectivity (for cooling-energy conservation)
- Better elongation (up to 800% at 21°C, 70°F)
- Superior performance at sub-zero temperatures [flexibility down to at least -40°C, (-40°F) for some materials]
- Easier application
- Fewer weather restrictions on application
- Less hazard from moisture entrapment during installation
- Easier repair of punctures, splits, tears
- Easier flashing applications at corners and irregular surfaces where stiffer built-up roof materials are difficult to form
- Possible greater reliance on factory-manufactured material quality

Disadvantages:

- Greater requirement for good workmanship
- More limited range of suitable substrates
- Less puncture resistance. Hence, greater vulnerability to traffic damage
- Lack of performance and design criteria comparable to those available for built-up roofs

4.0 APPLICATIONS

Regardless of the chemical composition of the materials, in a number of applications, such as in roofing, the materials are exposed to a wide range of temperature, wind, and load conditions. For example, in northern climates, the temperature of an asphalt or black roof membrane could be as high as 100°C on a hot summer day and as cold as -40°C in winter. Moreover, the material has to sustain the stresses generated by thermal expansion and contraction. Therefore, a combination of high-temperature and low temperature performance is required from end-use consideration. Critical properties such as high softening points, good tracking and flow resistance at elevated temperatures (60–70°C for modified bitumens), good impact and crack resistance, good cohesive strength and high elastic modulus are required.[6]

Exposure of the material to temperature, solar radiation, water, wind, environmental pollution, and stabilizer failure, in addition to roof traffic, poor workmanship, and lack of maintenance will induce either physical or chemical changes or both that affect the properties of the membrane. Therefore, proper evaluation of the membrane is required to determine the type of change.

Polymeric roofing membranes are evaluated using various test methods developed for assessment of durability. Mechanical properties of polymeric materials have two facets: one is related to the macroscopic behavior and the other to the molecular behavior.[9] Engineers are concerned only with the description of the mechanical behavior (physical properties) under the design conditions. Evaluation of the mechanical properties of roofing membranes (tensile properties at different temperatures, load at break, elongation at break, and energy to break) provides information about the material structural failures and how it can be improved, but does not offer an explanation. If the failure is related to molecular activity, additional information is necessary to comprehend the problem fully.

Current laboratory procedures used to evaluate the durability of roofing membranes utilize artificial weathering devices, which attempt to simulate the primary weathering agents, namely: solar radiation, temperature, ozone, and moisture. After the use of the artificial weathering device, the physical properties are usually compared with an unaged or "original" sample with the results stating "Retains x% of the original physical properties after aging."[10]

Differential scanning calorimetry (DSC), TG, DTA, TMA, and DMA have proven to be useful in the characterization of materials. DSC provides information on the glass transition temperature, vulcanization reaction, and oxidative stability. Thermogravimetry (TG) is applied for the quantitative analysis of the material components. The changes in sample dimensions as a function of time or temperature under a nonoscillatory load are measured by TMA, whereas DMA or DLMTA measures the rheological properties. These methods can also provide information about the thermal stability of polymers, their lifetime or shelf life under particular conditions, phase changes in the polymer, glass transition, the influence of additives, kinetics, oxidation stability, and many others.[11]–[20] Thermoanalytical techniques bridge the gap between the traditional engineering evaluation and chemistry. Thermoanalytical methods have long been used to characterize construction materials,[21]–[28] but they are not widely used to characterize roof membrane materials.

In 1988, an international roofing committee, working under the auspices of CIB/RILEM (Conseil International du Bâtiment pour la Recherche, l'Étude et la Documentation/Réunion Internationale des Laboratoires d'Essais et de Recherches sur les Matériaux et les Constructions),[29][30] recommended that thermoanalytical methods be added to the inventory of test methods currently used to characterize roof membrane materials.

Since little research had been reported on the application of thermal analysis (TA) methods to roofing, the committee recommended that more research be carried out to provide the technical basis for this application. The recommendation was based on research by Farlling[31] and Backenstow and Flueler.[32] These authors used TG, DSC, and DMA to characterize EPDM, PVC, and polymer-modified materials. Backenstow and Flueler reported the application of torsion pendulum analysis to characterize the above membrane materials. They concluded that TA techniques were useful for membrane characterization and should be investigated as methods for incorporation into standards. Previous work published by Cash[33] on the use of DSC to characterize neoprene, chlorinated polyethylene (CPE) and PVC had shown that DSC could be used to identify not only the components in a single-ply sheet and the manufacturer, but also to differentiate between new and exposed materials.

In 1990, Gaddy, et al.,[34] conducted a study to provide data on the feasibility of using thermoanalytical methods to characterize roofing membrane materials. The authors used TG, DSC, and DMA to analyze white and black EPDM before and after laboratory exposure to heat, ozone, UV, and outdoor exposure. The results were compared to changes in load-elongation measured as per ASTM D 412 for rubber sheet materials. The TG, DSC, and DMA results showed only slight changes in the white and black membrane materials after exposure. Such changes (e.g., T_g of ± 10°C) were close to the limits suggested by the CIB/RILEM[35] Committee. On the other hand, the percent elongation values displayed relatively large changes compared to changes obtained by TA techniques. In addition, the study showed that the DSC method was less practicable than other methods such as TMA because of analysis time. Based on the results of the study, Gaddy, et al., concluded that the TA methods were more appropriate for determining changes in bulk properties induced during exposure than the elongation measurement because the latter is, to a great extent, influenced by surface characteristics. They also concluded that further research was necessary to integrate TA methods into ASTM standards for EPDM roofing membrane materials.

Since previous research[31]–[34] had shown that TA methods can detect changes in the properties of roofing membrane materials, Gaddy, et al.,[36] investigated the applicability of thermomechanical analysis (TMA) for the characterization of white and black EPDM roofing membrane materials before and after laboratory (heat, ozone, and UV) and outdoor exposure and compared the results with those from load-elongation. The TMA results were similar for the black and white EPDM. The maximum change in glass transition (T_g), as measured by TMA was 12°C. The authors reported that the UV and ozone exposures generally produced greater changes than heat. In some cases, the black material showed slightly greater T_g changes than the white when exposed to UV conditions which was unexpected since it had been previously reported that the titanium dioxide and colored pigments used for white EPDM gave much less UV protection than carbon black.

The results of the EPDM study showed that, in general, the changes observed were relatively small and representative of the commercially available white EPDM roofing sheets. Also, it was found that the TMA procedure was readily applicable to the analysis of EPDM roofing materials, more practicable, used small specimens, and was reproducible. According to the authors, the use of small specimens makes the TMA technique very attractive for the analysis of in-service membrane since small specimens may be taken from roofs with minimal disruption to the waterproofing integrity of the membrane. From the results of the study, it was concluded that TMA can be applied to the characterization of EPDM membrane materials and yield results using small specimens:

a) The changes in glass transition (T_g) measured by TMA for the exposed specimens from different manufacturers allowed identification of the various products

b) Changes in T_g after exposure of 14°C or less, were considered to be minor (compared to ±8°C recommended by CIB/RILEM)

c) No differences between the black and white EPDM were observed for the same exposure conditions

d) The changes in the T_g of the EPDM obtained from TMA were not as large as those in the percent elongation found by load-elongation measurement

As in their previous paper, Gaddy, et al.,[34] attributed this to the influence ofthe surface characteristics and that TMA may be appropriate for determining bulk properties and possible changes which result from exposure.

In March 1990, a workshop on the "Applicability of Thermal Analytical Techniques to the Characterization of Roof Membrane Materials"[37] was convened to address the issue of incorporating thermoanalytical methods for characterizing roof membrane materials. The consensus of the workshop was that TA methods are valuable tools in the laboratory for research studies and troubleshooting as well as for tracking manufacturing processes to ensure that they are in control. However, participants concluded that thermoanalytical techniques or methods did not have immediate use in roofing standards because they could not alone provide the in-service performance of a product. In addition, the high cost of thermoanalytical equipment may prohibit general use of the techniques or make their incorporation in standards unattractive.

The fact that previous research[31]–[34][36] on the use of thermoanalytical techniques to characterize roof membrane materials showed little change in the materials analyzed does not indicate that the thermoanalytical techniques were not a useful tool. If the materials such as EPDM selected for the study were good, no changes would be observed. This was probably the case for the research carried out by Gaddy, et al.[34][36] Research carried out by Paroli, et al.,[38] on the effect of heat-aging on three EPDM roofing membranes showed that thermoanalytical techniques can provide some insight as to why some roofing materials fail more prematurely than others. The objective of Paroli's, et al., study was to demonstrate the utility of TG and DMA in establishing the durability of roofing membranes and correlating the thermoanalytical results with the mechanical properties. Three commercially available non-reinforced EPDM roofing membranes were selected for the study. They were heated in air-circulating ovens to 100°C and 130°C. Specimens were removed from the oven after 1, 7, and 28 days.

Thermogravimetry (TG) was used to measure the weight loss of the unheated and heated specimens. The glass transition temperature was measured by DMA and tensile testing was used to measure the mechanical properties of the specimens before and after heating. The TG results showed that degradation of the unheated (control) and heated EPDM specimens from two of the manufacturers occurred in one step (52–54%) in the 300–500°C range, which is well above the aging temperatures used in the study. Hence, no significant changes were observed between the unheated and heated specimens. This supported the findings reported by other researchers and discussed above.

The EPDM specimens from the other manufacturer showed, however, two-step degradation with 20% of the unheated (control) material being volatized between 150° and 430°C while the polymer decomposition (22.5%) occurred in the 440–550°C range. The lower temperature weight loss was attributed to the loss of oils and/or plasticizers of low volatility, which are typical ingredients in EPDM roof membranes. It was observed by TG that the specimens heated at 100°C and 130°C displayed a decrease in the oil content with exposure time. For example, after 28 days at 100°C, only 4.9% was detected (~75% of oil was lost during the heat treatment) compared to 13.2% (~ 34% oil lost during heat-aging) after 1 day of the same heat-treatment. Similarly, after 28 days of heating at 130°C, only 1.6% (~92% oil lost) was detected versus 5 % of oil (75% oil was lost during heat aging) after 1 day of heating. The TG results also showed (Fig. 1) that there was an increase in the polymer loss with decreasing oil loss for specimens from this manufacture. An increase in hardness was also observed.

The dynamic mechanical analysis results for the EPDM roof membrane samples from the three different manufacturers (S1, S2, S3) showed a similar trend to that observed by TG. The glass transition temperature (T_g) of the unheated specimens from S1 and S2 manufacturers was -49° and -46°C, respectively. The T_g of heated specimens did not change significantly (-48° to -42°C range) even after 28 days of exposure at 100° and 130°C. As predicted by TG, the EPDM specimens from the S3 manufacturer showed an increase in T_g with the number of days of heating. For example, the T_g changed from -61°C for the unheated to -46°C after 28 days at 100°C and to -43°C after 28 days at 130°C.

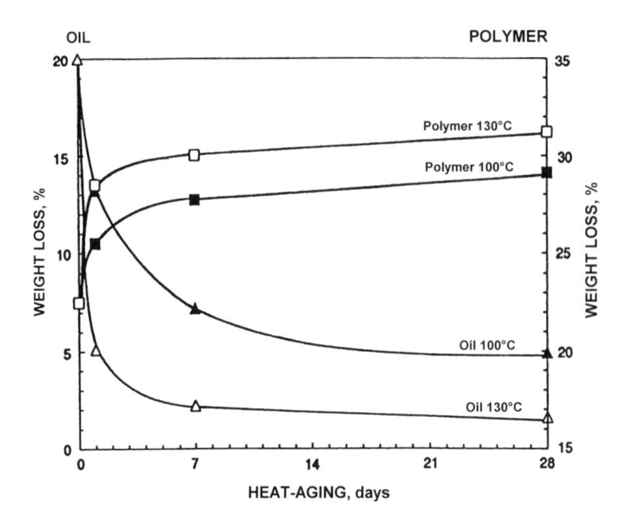

Figure 1. Relationship between the weight loss for the oil and polymer components for specimen S2.

A comparison between the TG and DMA results with those from the tensile and elongation tests indicated that as shown in Table 1, the strength of the membranes, which was not significantly affected by the heat exposure (S1, S2), remained fairly constant or decreased slightly after exposure. However, for the other sample (S3), it was observed that the strength at break increased with heat-aging, indicating that the sample became more brittle. Similarly, the elongation-at-break data for this sample (unheated) was lower than for the other two non-heated membrane materials.

From the results of the study, Paroli, et al.,[38] concluded that both TG and DMA can be used to evaluate the properties of various roofing membranes. The use of dynamic mechanical analyzers facilitates the determination of the glass transition temperature, which can be used to explain the change in mechanical properties of roofing membranes. Thermogravimetry provides information on the volatiles and polymer components of the membrane. Moreover, these thermoanalytical techniques can be used to assess the validity of the accelerated aging process that exposes membranes to elevated temperatures and thus reduces the time required to evaluate the effect of heat on the mechanical properties of polymer-based roofing materials. Finally, the equivalency of material response at different temperatures could help in establishing standard requirements.

Table 1. TG, DMA, and tensile strength and elongation values of (*a*) S1 EPDM Membrane, (*b*) S2 EPDM Membrane, and (*c*) S3 EPDM Membrane

Heat-aging Schedule		S1				
Temp. (°C)	Days	T_g (°C)	Oils loss** (%)	Polymer loss*** (%)	Strength (kNm^{-1})	Elongation (%)
22*	0	-49		52.5	10.6	522
100	1	-48		54.0	10.9	433
100	7	-49		53.4	11.6	396
100	28	-50		54.5	10.7	324
130	1	-48		54.0	9.9	348
130	7	-49		53.6	9.7	307
130	28	-48		51.6	8.0	217

(a)

Heat-aging Schedule		S2				
Temp. (°C)	Days	T_g (°C)	Oils loss** (%)	Polymer loss*** (%)	Strength (kNm^{-1})	Elongation (%)
22*	0	-61	20.0	22.5	8.7	285
100	1	-58	13.2	25.5	8.8	266
100	7	-50	7.2	27.7	9.5	211
100	28	-46	4.9	29.1	10.5	132
130	1	-50	5.1	28.5	9.2	180
130	7	-46	2.2	30.0	10.9	88
130	28	-43	1.6	31.2	11.6	38

(b)

Heat-aging Schedule		S3				
Temp. (°C)	Days	T_g (°C)	Oils loss** (%)	Polymer loss***(%)	Strength (kNm^{-1})	Elongation (%)
22*	0	-46		52.5	11.4	307
100	1	-45		51.8	10.8	314
100	7	-42		51.9	11.3	297
100	28	-44		51.6	10.6	270
130	1	-46		52.8	9.3	273
130	7	-45		52.4	9.8	227
130	28	-44		51.8	9.7	159

(c)

* Unheated control sample.

** Percent based on the integrated area of peak between 200 and 400°C (not available for S1 or S3).

*** Percent based on the integrated area of peak between 400 and 550°C. In the case of S1 and S3, the area most probably includes both oil and polymer loss.

The fact that no significant changes were observed on the materials characterized by thermoanalytical techniques in earlier sudies,[31]–[34][36] did not mean that the techniques were not suitable. Changes were not observed because the materials studied were not affected by the exposure conditions. Therefore, no changes would be observed regardless of the technique used for characterizing changes. The study carried out by Paroli, et al., is a clear example. They used three different techniques, TG, DMA,[38] and tensile testing. All show that out of three materials, two performed well after exposure whereas the other did not stand the exposure.

The applicability of thermoanalytical techniques for characterizing roof membrane materials was supported by studies carried out in the early 90s.[39]–[46] In 1993, Paroli, et al.,[47] also published the results of a study on the effects of accelerated heat aging on three poly(vinyl chloride) (PVC) roofing membranes using TG, DMA, and tensile testing. The results of the study showed that the three techniques provide similar information about the behavior of the PVC samples. The authors concluded that TG and DMA were able to not only record the changes incurred by the PVC samples due to heat-treatment, but also to provide a ranking of the heat stability of the various materials. The order of the stability determined by TG and DMA corresponded to that observed from actual records obtained from in-service roofs. Thermoanalytical techniques are important prognostic tools for predicting the behavior of single-ply roofing membranes and may also be used for developing new formulations of roofing materials.

By 1993, the shattering of unreinforced PVC roofing membranes (7.5–16 years old) had been reported in the USA. As a result, the National Roofing Contractor Association in collaboration with Paroli, et al.,[48] at the Institute for Research in Construction, National Research Council, Canada carried out a study to investigate the causes of failure. The glass transition temperature of the samples was measured by DMA. The results of the study showed that in nearly all cases, the glass transition temperature shifted to much higher temperatures. For example, specimens taken from different areas of the same sample showed changes in the glass transition (ΔT_g) of -3°C, + 15° to + 23°C as compared to -34°C. The specimens taken near the shattered line always showed the highest ΔT_g.

The authors also reported that for shattered samples not exposed to the environment (area under the seam or bottom) the change in T_g was negligible when compared to the control. It would appear that only factors related to exposure to the outside environment affected the samples. From the results of the study, Paroli, et al., concluded that glass transition temperature may be useful in explaining the shattering of unreinforced roofing membranes. The T_g can also be used to predict the behavior of unreinforced roofing membranes. Dynamic mechanical analysis is a valuable tool in characterizing reinforced and unreinforced PVC roofing membrane. Based on the result of the study, some recommendations to reduce the possibility of shattering of existing unreinforced PVC membranes were given. Some of them were to avoid rooftop traffic when the ambient temperature was below approximately +8°C (+50°F), or consult the membrane manufacturer for recommendations regarding minimum temperature. They also recommended semi-annual roof inspections.

In response to the CIB/RILEM recommendations,[29] RILEM 120-MRS/CIB W.83 Committee on Membrane Roofing Systems, representing 22 countries worldwide was formed in 1989. Following the recommendations of its predecessor, the new committee organized two task groups. One, on methods and the other on codes of practice. The purpose of the thermoanalytical group was to conduct interlaboratory testing to examine the reproducibility of thermoanalytical techniques when applied to typical roofing membrane materials, and to investigate further the feasibility of using these methods for detecting changes, which occur in the aging process.

In examining the application of thermoanalytical methods to the characterization of roofing membrane materials, the task group defined three criteria for selecting test procedures:

- Testing be done under laboratory conditions
- Results be available within a reasonable period of time
- Results be applicable to demonstrating the stability of materials and to comparing performance of materials under laboratory and outdoor exposure conditions

The members of the task group chose samples of EPDM, PVC, and polymer-modified (SBS and APP) bituminous roofing membrane materials to conduct the study. Thermogravimetry (TG), dynamic mechanical analysis (DMA), torsion pendulum analysis (TPA), and differential scanning calorimetry (DSC) were the techniques selected. The roofing materials were tested "as received" (unaged) and after exposure in the laboratory to heat and water submersion, and to natural weathering in climates of four countries. The variability within and between laboratories was investigated. Five, two, one, and four laboratories participated in the TG, DMA, TPA, and DSC analysis respectively.

In September 1993, the task group issued a preliminary report[49] describing its activities and reporting aging conditions and experimental parameters for each method and the interlaboratory results of unaged roofing membrane materials. Based on their study and results from previous research,[31]–[48][50][51] the Committee presented its final report in December 1995[52] reporting the results of studies and included recommendations for conducting thermoanalytical methods under controlled test conditions.

The results of the inter-laboratory study showed that the exposures produced only minor changes in the measured thermoanalytical properties. The materials tested were generally stable under the outdoor and laboratory exposure conditions. In the case of TG analysis of the pyrolyzable organic constituents, it was found that the method showed rather good within- and between-laboratory variability. The number of participants in the dynamic mechanical/TPA analysis was limited; hence, no conclusion was made regarding the variability within and between laboratories. In the case of

DSC analysis of the glass transition, it was found that the method is not universally applicable to the analysis of all roofing membrane materials. Difficulties were encountered in determining T_g of the polymer-modified bitumens. Consequently, it was recommended that although DSC can have utility for certain roofing membrane materials and specific investigation, it is not a universal method for the type of material under investigation.

During the study, it was also found that some of the recommendations of the 1988 report on TG analysis needed to be revised. For example, it was recommended in 1988 that "thermoanalytical methods should be applied to characterizing roofing membrane materials. Thermogravimetry (TG) analysis of new and aged materials should be reported. The change in organic constituents after aging should not exceed±2% (if a material is to be considered stable)." Although the findings of the study supported the previous recommendation on TG, the results did not support the recommendation that "change in the organic constituents after aging should not exceed ±2%." The data showed that variations of less than 2% (absolute) were achievable for replicates within a laboratory. However, between laboratories, average determinations within 2% were not always possible to achieve, particularly for bituminous materials. Therefore, the following recommendation was made:

> *Thermogravimetry (TG) is an acceptable method for characterizing new and aged roofing membrane materials. It is complementary to other mechanical, chemical, and physical methods used for membrane material characterization. The change in the mass of the organic constituents after exposure (i.e., aging) should not exceed ±3 percent (absolute) if a material is to be considered stable to the exposure. Changes greater than ±3 percent may indicate changes in the material resulting from the exposure.*

The Committee concluded that DMA/TPA were readily applicable to roofing membrane materials and the test parameters selected were suitable for roofing membrane materials except that the temperature range of -100° to +100°C was not necessary. Therefore, it was recommended that DMA and TPA are acceptable methods for evaluating new and aged roofing materials and that they are complementary to other mechanical, chemical, and physical methods used for membrane characterization. The Committee also recommended that if the material is to be considered stable to exposure conditions, the change in T_g should not exceed 8°C and change in the storage modulus (E'/G') at T_g ± 10°C should not be more than a factor of ten.

Changes greater than these values indicate changes in the material resulting from the exposure. A standard test procedure for DMA/TPA was also recommended.

The conclusions from the DSC analysis were that the technique is not universally applicable to the analysis of roofing materials. Since the variability of the method was not investigated, recommendations were not made. The use of newer techniques such as modulated DSC (MDSC) and oscillating DSC (ODSC) to characterize the T_g range of roof membrane materials was beyond the scope of the study. Therefore, the Committee did not recommend the use of DSC as a universal method. However, it was recommended that the applicability of MDSC and ODSC for determining glass transition of roofing membrane materials be investigated.

The applicability of thermoanalytical techniques for the characterization of roofing membrane materials is gaining popularity in the roofing industry. Several papers have been published in the last five years.[53]–[59] For example, more and more manufacturers, contractors, consultants, and building owners are turning to these techniques to solve in-service roofing membrane problems as well as to evaluate new roofing membrane materials. A typical example is the paper on the problem of EPDM membrane shrinkage.[55] The study was conducted in 1995 to investigate the causes of EPDM shrinkage. The performance of EPDM roofs, typically, had been quite good, however, instances of flashing problems attributed to membrane shrinkage had occurred. The laboratory study included TG, DMA, tensile strength, elongation, and thermally induced load. The TG results showed that the oil loss of the EPDM membrane overlapped with the region where the polymer degraded. Therefore, it was difficult to measure the total oil content for each sheet.

Since the EPDM rubber polymer is relatively stable and not expected to decompose when exposed to natural weathering, Oba and Paroli, et al.,[54][55] compared the difference in oil content between the top and bottom layers by subtracting the total weight loss below the polymer decomposition region (500°C) for the top sheet from that of the bottom sheet. It was found that for some membranes, the oil loss during exposure to natural weather was as high as 8% when compared to the bottom sheet. In addition, the TG results showed that for one sample taken from near a flashing, the weight loss was not typical of an EPDM membrane.

The glass transition measured by DMA also showed similar results. An infrared spectrum taken from the sample showed that the sample taken from the flashing was neoprene-based. These thermoanalytical techniques not only provided insight into the shrinkage/contraction problem of some of the membranes, but also showed that they can easily be used to differentiate between the various membranes on this roof. Based on the thermoanalytical and mechanical testing results, it was concluded that it was difficult to attribute the cause of shrinkage to changes in material property. Although some of the roof membrane had shown a difference in weight loss of 8% (between top and bottom layer) and changes in glass transition >3°C, they did not exhibit problems. Nearly all samples passed the ASTM criterion for elongation. However, all of the techniques indicated property changes for three of the samples. It was also concluded that membrane shrinkage did not appear to be the cause of the flashing problem on one of the other roofs. The damage was probably related to temperature-induced loads since it was installed over an unreinforced PVC membrane, which shattered. The work resulted in some recommendations about existing and new EPDM roofs.

In the last ten years, thermoplastic polyolefins (TPO) have entered the roofing market. Since this is a relatively new product only a few studies using thermoanalytical techniques have been published; one on the investigation of the long-term performance of in-service thermoplastic olefin (TPO) roofing membranes[60] and the other on the effects of welding parameters on seam strengths.[61]

As a result of the previous studies, both the American Society for Testing and Materials (ASTM) and the Canadian General Standards Board (CGSB) have initiated the development of standards that will include thermoanalytical techniques to evaluate roofing membranes. The work and effort of the thermoanalytical task group,[49]–[52] supported by recent publications,[53]–[61] on the use of thermoanalytical techniques to characterize roofing membrane materials, made possible the incorporation of TG and DMA methods into standards to characterize roof membrane materials. An ASTM standard (ASTM 6382-99) was published in 1999. As a result, roofing manufacturers, building owners, and contractors have started to recognize the usefulness of thermoanalytical techniques to characterize roofing membranes.

REFERENCES

1. Baker, M. C., Roofs, Design, Application and Maintenance, sponsored by the Natl. Res. Council of Canada, Polyscience Publ. Ltd., Quebec, Canada (1980)
2. LaValley, A. L., NRCA Market Survey Shows Golden Days of Roofing are Fading, *Professional Roofing,* pp. 16–19 (April 1993)
3. Good, C., Eyeing the Industry, *Professional Roofing,* pp. 28–32 (April 2001)
4. Feldman D., *Polymeric Building Materials*, Elsevier Science, New York (1989)
5. Gorman, W. B., and Usmani, A. M., Application of Polymer and Asphalt Chemistries in Roofing, Paper No. 34 presented at the Rubber Division, American Chemical Society (ACS), Philadelphia, PA (May 2–5, 1995)
6. Nadkarnl, V. M., Shenoy, A. V., and Mathew, J., Thermomechanical Behavior of Modified Asphalts, *Ind. Eng. Prod. Res. Dev.* 24:478-484 (1985)
7. Rodriguez, I., Dutt, O., Paroli, R. M., and Mailvaganam, N. P., Effect of Heat-Aging on the Thermal and Mechanical Properties of APP- and SBS-Modified Bituminous Membranes, *Mater. Struct.* 26:355–361 (1993)
8. Griffin, C. W., *Manual of Built-up Roof Systems,* pp. 247–275, McGraw-Hill, New York (1982)
9. Dutt, O., Paroli, R. M., Mailvaganam, N. P., and Turenne R. G., Glass Transitions in Polymeric Roofing Membranes: Determination by Dynamic Mechanical Analysis, Building a Worldwide Roofing Community: *Proc. 1991 Int. Symp. on Roofing Technology,* National Roofing Contractors Association, pp. 495–501 (1991)
10. Farlling, M. S., Using Thermal Techniques to Evaluate Durability of Roofing Membranes, *Rubber World* (Jan. 1988)
11. Pielichowski, J., and Pielichowski, K., Applications of Thermal Analysis for the Investigation of Polymer Degradation Processes, *J. Thermal Analysis,* 43:505–508 (1995)
12. Judd, M. D., The Use of Thermo-Analytical Techniques in Materials Evaluation, *Proc. Int. Symp. on Spacecraft Material in Space Environment, Toulouse, France,* jointly organized by CNES, ESA & CERT (June 8–12, 1982)
13. González, V., Thermo-Oxidation of Elastomers by Differential Scanning Calorimetry, *Rubber Chem. and Technol.*, 54:134–145 (1980)
14. Sircar, A. K., Galaska, M. L., Rodrigues, S., and Chartoff, R. P., Glass Transition of Elastomers Using Thermal Analysis Techniques, Rubber Division Meeting, ACS Louisville, KY (Oct. 1996)

15. Maurer, J. J., Advances in Thermogravimetric Analysis of Elastomers Systems, *J. Macromol, SCI.-Chem.*, A8:73–82 (1974)

16. Gibbons, J. J., Application of Thermal Analysis Methods to Polymer and Rubber Additives, *Am. Lab.,* p. 33 (Jan. 1987)

17. Laird, J. L., and Liolios, G., Thermal Analysis Techniques for the Rubber Laboratory, *Am. Lab.,* p. 46 (Jan. 1990)

18. Sircar, A. K., Analysis of Elastomer Vulcanizate Composition by TG-DTG Techniques, Paper presented at a Rubber Division Meeting, ACS, Toronto, Ontario, Canada (May 21–24, 1991)

19. Staub, F., Applications of TA to Elastomers, *Am. Lab.,* 18:56–63 (Jan. 1986)

20. Knappe, S., and Urso, C., *Thermochimica Acta,* 227:35–42 (1993)

21. Ramachandran, V. S., and Patwardhan, N. K., Differential Thermal Analysis of Building Materials, *J. Natl. Bldg. Org.,* II(I-II) (1957)

22. Abdelrazig, B. E. I., Bonner, D. G., Nowell, D. V., and Dransfield, J. M., Estimation of the Degree of Hydration in Modified Ordinary Portland Cement Paste by Differential Scanning Calorimetry, Part II, *Thermochimica Acta*, 168:291–295 (1990)

23. Bushnell-Watson, S. M., Winbow, H. D., and Sharp, J. H., Some Applications of Thermal Analysis of Cement Hydrates, *Anal. Proc.* 25:8–10, London (1988)

24. Valenti, G. L., and Cioffi, R., Quantitative Determination of Calcium Hydroxide in the Presence of Calcium Silicate Hydrates, Comparison Between Chemical Extraction and Thermal Analysis, *J. Mat. Sci. Letters*, 4:475–478 (1985)

25. Whitehead, M. B., and Russel, G. A., Determination of Hydrated Cement Content of Portland Cement Concrete by DTA, *Am. Lab.*, 11:37–38, 40, 42–43 (1979)

26. Bhatty, J. I., A Review of the Application of Thermal Analysis to Cement-Admixture Systems, *Thermochimica Acta*, 189:313–350 (1991)

27. Ramachandran, V. S., Lowery, M. S., Wise, T., and Polomark, G. M., The Role of Phosphates on the Hydration of Portland Cement, *Mat. and Struct.,* 26:425–432 (1993)

28. Ramachandran, V. S., Sarkar, S. L., Xu, A., and Beaudoin, J. J., Physico-Chemical and Microstructural Investigation of the Effect of NaO on the Hydration of $3CaO{\cdot}SiO_2$, *Il Cemento*, 89:1–17 (1992)

29. Performance Testing of Roofing Membrane Materials, Recommendations of the CIB W.83 and RILEM 75-SLR Joint Committee on Elastomeric, Thermoplastic, and Modified Bitumen Roofing, RILEM, Paris, France (Nov. 1988)

30. Elastomeric, Thermoplastic, and Modified Bitumen Roofing: A Summary Technical Report for CIB W.83 & RILEM 75-SLR Joint Committee, *Materiaux et Constructions*, 19(112):323–329 (July–Aug. 1986); Warshaw, R. I., Summary of the RILEM/CIB Report on Performance Testing of Roof Membrane Materials, *Int. J. Roofing Technol.,* 1(1):10–11 (Spring 1989)
31. Farlling, M. S., New Laboratory Procedures to Evaluate the Durability of Roofing Membranes, Appendix D in Performance Testing of Roofing Membrane Materials, Recommendations of the CIB W.83 and RILEM 75-SLR Joint Committee on Elastomeric, Thermoplastic, and Modified Bitumen Roofing, RILEM, Paris, France (Nov. 1988)
32. Backenstow, D., and Flueler, P., Thermal Analysis for Roofing Characterization, *Proc. 9th Conf. on Roofing Technol., Rosemento, IL,* pp. 85–90, Natl. Roofing Contractors Assoc. (May 1989)
33. Cash, C. G., Thermal Evaluation of One-Ply Sheet Roofing, Single-Ply Roofing Technology, (W. H. Gumpertz, ed.), pp. 55–56, ASTM STP 790, American Society for Testing Materials (ASTM), Philadelphia, PA (1982)
34. Gaddy, G. D., Rossiter, W. J., Jr., and Eby, R., Application of Thermal Analysis Techniques to the Characterization of EPDM Roofing Membranes, *Roofing Research and Standards Development*, (T. J. Wallace and W. J. Rossiter, eds.), pp. 37–52, 2nd Vol., ASTM STP 1088, ASTM, Philadelphia (1990)
35. Minutes of the April 28–29, 1988 Meeting, CIB/RILEM Committee on Elastomeric, Thermoplastic and Polymer Modified Bituminous Roofing, RILEM, Paris, France (April 1988)
36. Gaddy, G. D., Rossiter, W. J., Jr., and Eby, R. K., The Use of Thermal Mechanical Analysis to Characterize Ethylene-Propylene-Diene Terpolymer (EPDM) Roofing Membrane Material, Materials Characterization by Thermomechanical Analysis, (A. T. Riga and C. M. Neag, eds.), pp. 168–175, ASTM STP 1136, ASTM, Philadelphia (1991)
37. Rossiter, W. J., Jr., and Gaddy, G., Applicability of Thermal Analysis to Characterization of Roof Membranes Materials: A Summary of the March 28–29, 1990 Workshop, Army Corps of Engineers, Const. Eng. Res. Lab., USACERL Special Report M-91/92 (Sept. 1991)
38. Paroli, R. M., Dutt, O., Delgado, A. H., and Mech, M. N., The Characterization of EPDM Roofing Membranes by Thermogravimetry and Dynamic Mechanical Analysis, *Thermochimica Acta*, 182:303–317 (1991)
39. Paroli, R. M., Rodriguez, I., and Fernandez, M., Characterization of Bitumens Using High Performance Gel Permeation Chromatography and Thermal Analysis for Improving Performance of Bitumen/Polymer Blends Used in Waterproofing, *VIII Intl. Roofing Waterproofing Congr.,* Madrid, Spain, pp. 418–430 (1992)

40. Paroli, R. M., and Dutt, O., Dynamic Mechanical Analysis Studies of Reinforced Polyvinyl Chloride (PVC) Roofing Membranes, *Polymeric Mater. Sci. Eng.*, 65:362–363 (1991)

41. Paroli, R. M., Dutt, O., and Delgado, A. H., The Characterization of Roofing Membranes Using Thermal Analysis and Mechanical Properties, *Poster Book CIB World Bldg. Congr., Montreal, Quebec, Canada,* pp. 122–123 (May 18–22, 1992)

42. Delgado, A. H., Paroli, R. M., and Dutt, O., Thermal Analysis in the Roofing Area: Applications and Correlation with Mechanical Properties, Presented at CIB World Bldg. Congr., Montreal, Quebec, Canada (May 18–22, 1992)

43. Penn, J. J., and Paroli, R. M., Evaluating the Effects of Aging on the Thermal Properties of EPDM Roofing Materials, *Proc. 21st N. Am. Thermal Analysis Soc. (NATAS) Conf.,* pp. 612–617 (1992)

44. Penn, J. J., and Paroli, R. M., Evaluating the Effects of Aging on the Thermal Properties of EPDM Roofing Materials, *Thermochimica Acta,* 226:77–84 (1993)

45. Paroli, R. M., and Dutt, O., A New Approach to Assessing Roofing Problems, Construction Technology Update, Supplement: *Building Management and Design, The Canadian Architect, Canadian Consulting Engineer* (Feb. 1993)

46. Oba, K., and Bjöerk, F., Dynamic Mechanical Analysis of Properties of Single-Ply Roof Coverings for Low-Slope Roofs and the Influence of Water, *Polymer Testing,* 12:35–56 (1993)

47. Paroli. R. M., Dutt, O., Delgado, A. H., and Stenman, H. K., Ranking Polyvinyl Chloride (PVC) Roofing Membranes Using Thermal Analysis, *J. Mater. Civil Eng.*, 5:83–95 (1993)

48. Paroli, R. M., Smith, T. L., and Whelan, B. J., Shattering of Unreinforced PVC Roof Membranes: Problem, Phenomenon, Causes and Prevention, *NRCA/NIST 10th Conf. on Roofing Technol.,* pp. 93–107 (April 22–23, 1993)

49. Thermal Analysis Testing of Roofing Membrane Materials: Interim Report of the Thermal Analysis Task Group, RILEM 120-MRSCIB W.83 Joint Committee on Membrane Roofing Systems (Sept. 1993)

50. Paroli, R. M., and Penn, J. J., Measuring the Glass-Transition Temperature of EPDM Roofing Materials: Comparison of DMA, TMA, and DSC Techniques, *Assignment of the Glass Transition,* pp. 269–276, ASTM STP 1249 (1994)

51. Paroli, R. M., Dutt, O., Smith, T. L., and Whelan, B. J., Enhancement of the Standard for PVC Roof Membrane Material with the Use of Mechanical and Chemical Analysis, *Roofing Research and Standards Development,* (T. J. Wallace and W. J. Rossiter, Jr., eds.), 3:116–122, ASTM STP 1224, ASTM, Philadelphia (1994)

52. Rossiter, W., Paroli, R. M., Flueler, P., Beech, J., Cullen, W., Lobo, O., Oba, K., Puterman, M., Saunders, G., Tanaka, K., Vandewynckel, J., and Burn, S., Thermal Analysis Testing of Roofing Membrane Materials, Final Report of the Thermal Analysis Task Group RILEM 120-MRS/CIB W.83 Joint Committee on Membrane Roofing Systems, p. 68 (Dec. 1, 1995)

53. Paroli, R. M., Dutt, O., Delgado, A. H., and Lei, W., A Novel Approach to Investigating the Durability of Roofing Membranes, *Proc. ACS Div. of Polymeric Mater. Sci. Eng.,* 72:378–379 (1995)

54. Oba, K., and Roller, A., Characterization of Polymer Modified Bituminous Roofing Membranes Using Thermal Analysis, *Mater. Struct.* 28:596–603 (1995)

55. Paroli, R. M., Delgado, A. H., and Dutt, O., Shrinkage of EPDM Roof Membranes: Phenomenon, Causes, Prevention and Remediation, *Proc. 11th Conf. on Roofing Technol., Gaithersburg, MD,* pp. 90–110 (Sept. 21–22, 1995)

56. Delgado, A. H., and Paroli, R. M., Influence of Gases, Glow Rates and Heating Rates on TMA Measurements of Roofing Membranes, *Proc. 24th N. Am. Thermal Analysis Soc., (NATAS),* pp. 554–560 (Sept. 1–13, 1995)

57. Paroli, R. M., and Delgado, A. H., Evaluating the Performance of Polymeric Roofing Materials with Thermal Analysis, *Rubber World,* 214(4):27–32 (July 1996)

58. Paroli, R. M., and Delgado, A. H., Analysis of Asphalt-based Roof Systems Using Thermal Analysis, Preprints of Papers Presented at the 211th Am. Chem. Soc. Natl. Meeting, New Orleans, LA, pp. 38–42 (March 24, 1996)

59. Sichina, W. J., DMA as Problem-Solving Tool: Characterization of Asphalt Roofing Shingles, Application Note AN-43, Haake/Seiko, Thermal Analysis

60. Simmons, T. R., Paroli, R. M., Liu, K. K. Y., Delgado, A. H., and Irwin, J. D., Evaluation of In-Service Thermoplastic Olefin (TPO) Roofing Membranes, 4:19–42, ASTM Special Techn. Publ. 1349, 12/6/98, Roofing Research and Standards Development, Nashville, TN (Sept. 24, 1999)

61. Simmons, T. R., Liu, K. K. Y., Paroli, R. M., Delgado, A. H., Irwin, J. D., and Runyan, D., Effects of Welding Parameters on Seam Strength of Thermoplastic Polyolefin (TPO) Roofing Membranes, *N. Am. Conf. on Roofing Techn., Toronto, Ontario,* pp. 56–65 (1999)

16

Paints and Coatings

1.0 INTRODUCTION

Regardless of the type of paints or coatings and the field applications, a study of their mechanical and chemical properties is necessary to understand their behavior and predict the performance. For example, *paints* (as discussed in Ch. 13) are dispersions of one or more liquid or solid phases in a liquid or solid matrix. Such materials exhibit very complex responses to the application of quite small forces. Some paints (thixotropic) look like solids or very viscous liquids when at rest in the container and become thin when stirred. They will recover their original appearance when allowed to rest.[1] These complex responses are related to the rheological properties of the material. Resistance to weather is an important performance aspect, which affects durability. This is critical for paints applied to exteriors of houses and buildings, ships, chemical plants, agricultural implements, and cars.

The mechanical properties of paints and coatings can be studied by application of various standard mechanical test methods, whereas chemical characterization can be achieved using a variety of analytical techniques such as gas chromatography (GC), gel permeation chromatography, high performance liquid chromatography (HPLC), infrared spectroscopy (IR), nuclear magnetic resonance spectroscopy (NMR), surface analysis, UV-visible spectroscopy, microscopy, x-ray diffraction, and thermoanalytical techniques. This chapter will deal with the applicability of thermoanalytical techniques for characterizing paints and coatings.

Generally, the basic chemical components of paint and coating materials can be considered the same regardless of the field of application. Considerable work has been published on the use of the thermoanalytical techniques for characterizing coatings and paints for many applications.[2]–[38] However, these have not been extensively applied to paints and coatings for construction applications.

2.0 PAINTS

Thermoanalytical techniques are potentially useful in studying paints for any application. One of the applications of thermoanalytical techniques, for example, is the use of DSC and DTA to study the influence of impurities or differences in crystallinity during pigment formation. The use of these techniques for purity evaluations will depend on the material and the impurities.[2]

Another important aspect of coatings technology is the heat of reaction or mixing. This can provide essential information in calculating heat balance for processing coating materials since the evolution or absorption of a substantial amount of heat could change the temperature of the material enough to interfere with its flow. Odlyha, et al.,[3] used DSC to evaluate the oxidation stability of paint media, particularly as drying oils represent the main material other than egg yolk which has been used as a binding medium in paintings from the fourteenth century in Northern European countries and Italy. The authors studied the effects of age, composition (e.g., oil, oil-protein), and oil and pigment type on the oxidative degradation of paint media. They also calculated kinetic parameters to determine whether the observed phenomena can be explained by a classical Arrhenius activation energy approach.

Odlyha[4] also used DSC to study samples from three paints during treatment in the Conservation Department of the Tate Gallery. Knowledge of the type of paint media used is of interest both to the conservator who needs to select the appropriate material for cleaning and to the art historian for documentation of artists' techniques. DSC provided the means to characterize the paint media at the microscopic level. Furthermore, DSC, TG, and TMA were used by Wingard[5] to study hypalon paint coatings for the solid rocket booster of the space shuttle.

DTMA has even been used in the evaluation of deacidification treatment, as well as the effects of environmental conditions, and the preventive conservation treatment of painting canvases.[6][7] From the results of both studies, it was concluded that DTMA is a suitable technique for the evaluation of the effect of treatment.

Dielectric thermal analysis (DEA) is another thermoanalytical technique that has been used to characterize paints. The technique measured the electrical characteristics (capacitance and conductance) of a material as a function of time, temperature, and frequency. Odlyha, et al.,[8] used DEA to study variations in the relative humidity in canvas paintings. The study showed that DEA allowed an assessment of the effects of water vapor on samples of primed canvas. It was concluded that the ability to quantify the degree of water binding within works of art in a non-invasive manner assisted optimization of conservation treatment such as rehumidification of surface paint layers to remove surface deformations and, where necessary, to allow adhesion of flaked paint fragments. Thermal stress analysis is another promising technique used by Perera and coworkers[9] to study the cure of stoving paints. The technique allows, amongst other things, measurement of the glass transition temperature of a coating.

The final properties of a paint, leveling characteristics and durability, depend on the cure conditions. Uncured coatings present unsatisfactory resistance to water and chemical reagents and insufficient hardness. Overcured coatings can be hard yet fragile and present signs of thermal degradation such as a decrease of adhesion, gloss and change in color. To solve this problem, it is necessary to know the mechanism of film formation or factors affecting it. Hence, techniques capable of evaluating cure characteristics are essential. This can be achieved by using a combination of thermoanalytical techniques. For example, DSC, DMA, TG, and TMA can be used to study the reaction kinetics and determine a number of basic coating characteristics sensitive to the degree of curing.

The use of combined thermoanalytical techniques for materials characterization has increased in the last few years. For example, differential scanning calorimetry and thermally stimulated current (TSC) have also been used to study polyurethane and mono- and bilayer paints.[10] TSC (considered a relaxation technique) is based on the degree of chain and branch mobility present in a non-conductive system. In TSC, a DC voltage is applied to the sample heated above the transition temperature of interest. This will cause orientation of the dipole(s) with the field. The sample is then cooled to freeze the dipole into its reoriented configuration. The field is removed and a heating ramp is applied. This causes depolarization of the dipoles, creating a small current, which is a function of the molecular mobility. Figure 1 illustrates a typical thermally stimulated current measurement.

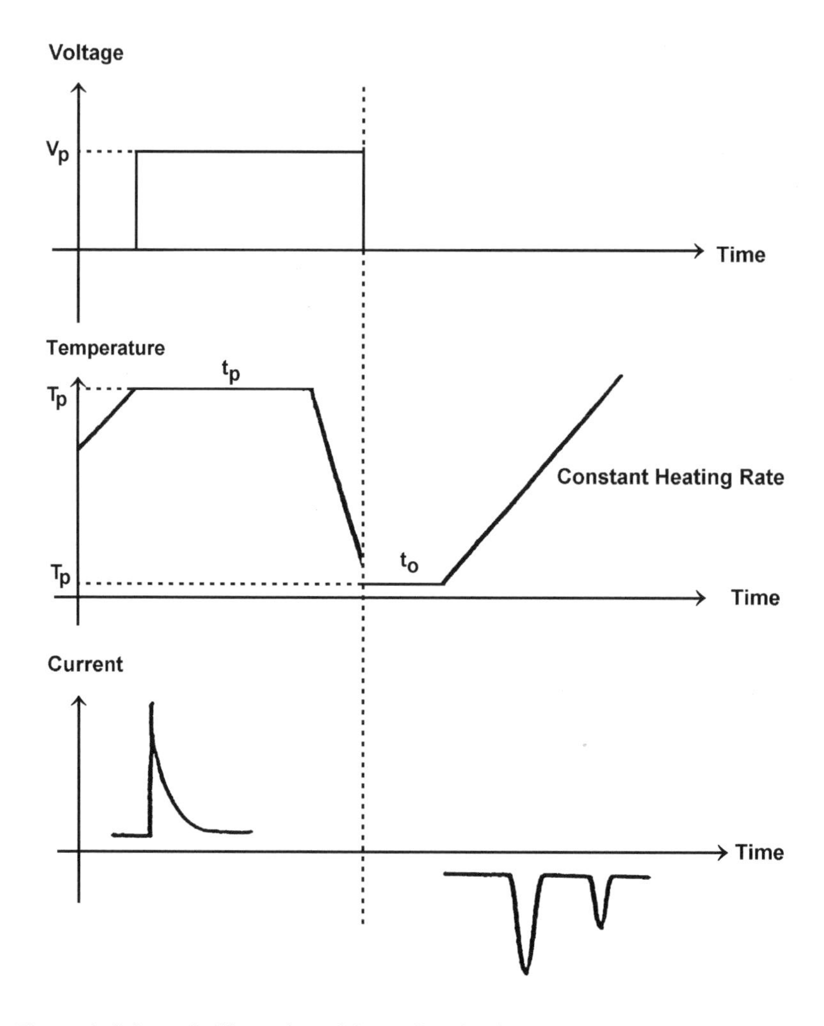

Figure 1. Schematic illustration of thermally stimulated current measurements.

The results from DSC and TSC have shown that both techniques can be used to investigate the molecular mobility in polyurethane coatings as well as to study the aging procedures in polymeric coatings.

Clausse, et al.,[11] used calorimetry and DSC to characterize the thermal transitions of thermosetting paints containing organic solvents. They concluded that the study provided a correlation between the T_g, specific heat (C_p), and the degree of cross-linking of the mixture. The degree of cross-linking increased as the C_p decreased.

Sebastian[12] studied paint-shop waste disposal from metallurgical factory painting houses by DTA and TG. Figure 2 shows the DTA and TG curves of the dry and wet wastes. It was reported that characteristics of the curves were in good agreement with the physico-chemical compositions of the wastes. The enthalpy of reactions was also obtained by integrating DTA curves.

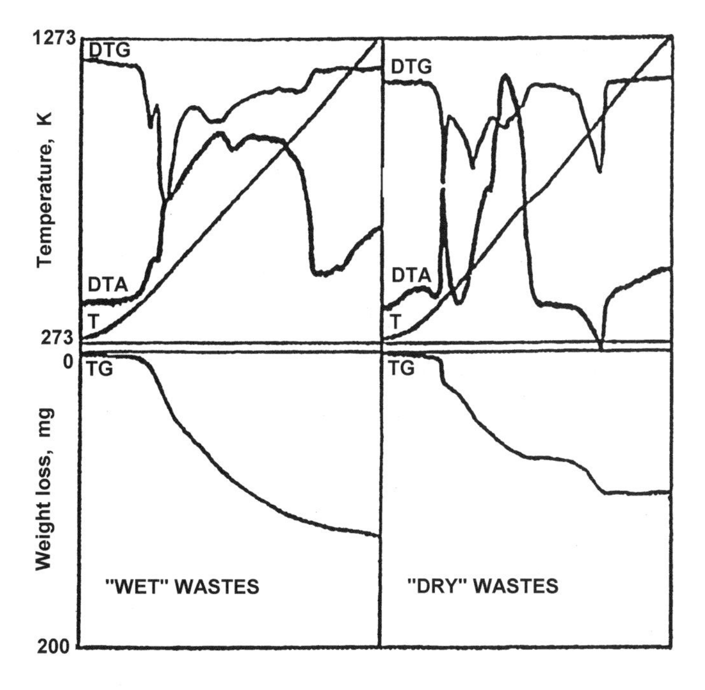

Figure 2. Thermal decomposition of paint-shop wastes. *(Reprinted with permission.)*[12]

The DTA curve for the wet wastes shows an exotherm effect in the range of 160–790°C, which corresponds to a major weight loss in the TG curve between 170° and 350°C. A sharp exotherm is observed in the DTA curve of the dry wastes in the initial stage for the first weight loss (below 200°C), which may be caused by oxidation of the low-boiling components of the painting material.[12] A second exotherm is also observed between 240° and 550°C due to another weight loss, followed by an endotherm which was attributed to the decarbonation of calcium carbonate (44.6% weight loss). From the results, Sebastian reported that the DTA curves showed that the thermal behavior of the mixture depends on the decomposition of the raw-paint waste, the main component. It was concluded that the decomposition process, the final temperature, the corresponding weight loss, and the calorific value curves provide valuable data on the suitable mode of preparation of wastes prior to their combustion.

Thin film analysis (TFA) is another technique that has been applied to the study of the drying of curable and latex-based coatings.[13] This novel approach, based on a simple force-sensing device capable of carrying various probes, measures quantitative changes in mechanical and rheological properties of drying films in-situ on a test panel. By imposing a thermal gradient along a heating block under the sample, measurements on a series of temperatures can be obtained in a single experiment. The technique has been used by Bahra, et al.,[13] to monitor changes in the viscosity during film drying of household paints and automotive basecoat films. Figure 3 displays the viscosity measurements obtained with the TFA for two types of household paints, a water-based emulsion and a white spirit-based gloss paint. The viscosity values for both paints remains low for an initial period less than four minutes, after which a sharp increase is observed. The initial period is longer for the gloss paint, which was attributed to the greater resistance to brushmarking experienced with this type of paint.

Bahra, et al.,[13] also used TFA to monitor the buildup of the scratch resistance of coatings. The reported results can be related to the extent-of-cure in cross-linking resins and to phenomena such as vitrification. Figure 4 shows an example of scratch resistance of a drying film for an alkyd-resin gloss paint at two different measuring speeds (made lengthwise at a constant temperature) as a function of time. The scratch resistance builds up sooner when the needle is moved with a fast stroke rather than with a slow stroke. Measurements from scratches along the x-direction were also performed at temperatures between 20° and 60°C after 0, 15, and 45 minutes. The results showed a general increase in the scratch resistance with time (Fig. 5), which indicated an increase in the degree of curing with temperature and time. The small decrease observed at higher temperatures between 30 and 45 minutes was attributed to a film softening (polymer approaching T_g). They concluded that the TFA technique can be used to provide a rapid empirical assessment of cure rates at different temperatures and to estimate the T_g of the coating at different times.

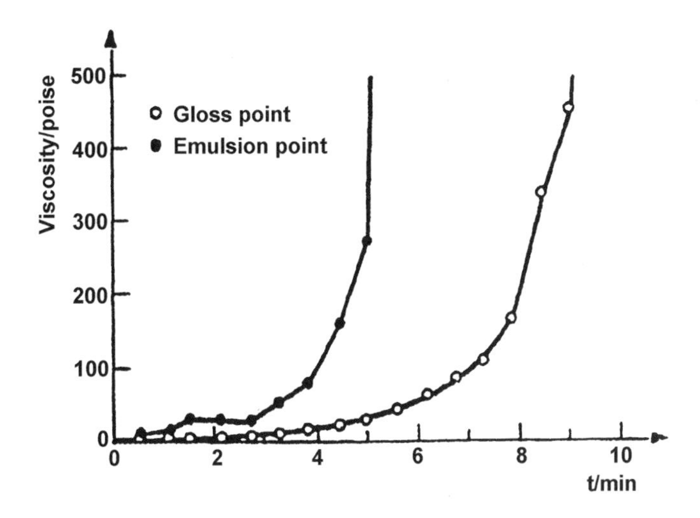

Figure 3. Comparison of the viscosity-time profiles of drying films of an emulsion paint and a gloss paint. (*Reprinted with permission.)*[13]

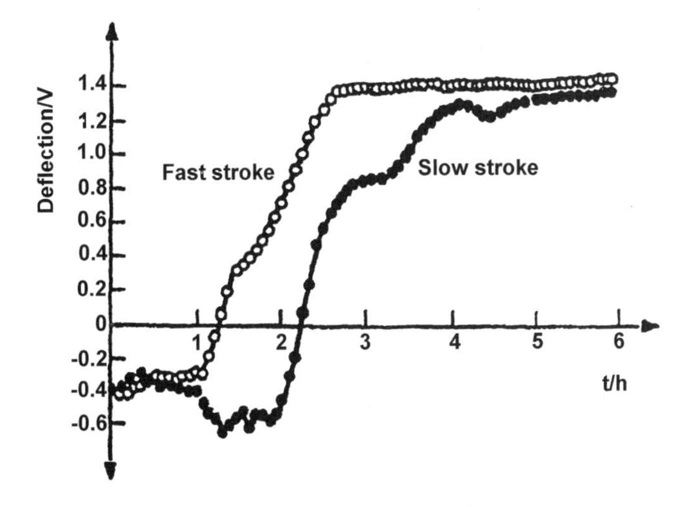

Figure 4. The build up of scratch resistance for fast and slow strokes, seen with a drying film of a gloss paint. (*Reprinted with permission.)*[13]

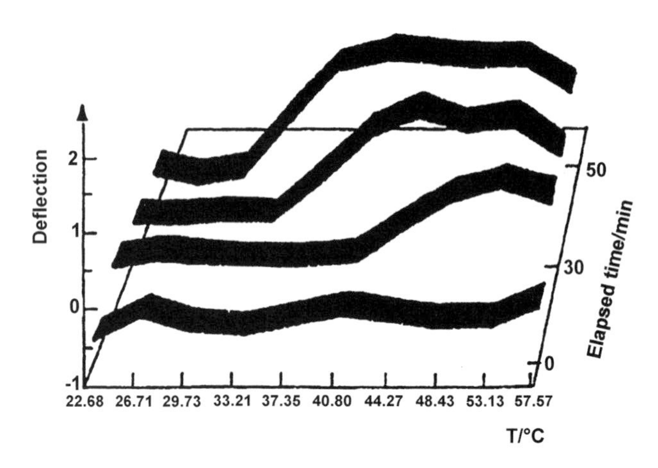

Figure 5. 3-D plot showing the results of a temperature scanning scratch test with a gloss paint. (*Reprinted with permission.*)[13]

3.0 COATINGS

3.1 Intumescent Coatings

Pagella, et al.,[14] used differential scanning calorimetry to study intumescent coating systems. *Intumescence* is the property of a coating enabling it to 'swell up' and form a carbonaceous char useful for the protection of the underlaying substrate from fire. They are formulated with several active compounds that react as temperature increases to form a char and to evolve gases. These gases cause the coating to bubble, foam, and ultimately expand to more than 100 times its original volume as a solid charred material. An insulating barrier that protects the substrate from a rapid increase in temperature is formed. Such a property is useful for increasing the time to collapse under fire for steel structures or to upgrade the fire resistance of walls or ceilings of various materials, avoiding diffusion of fire, smoke, and temperature on the opposite side of the wall.

Single reactive fillers of a typical intumescent system comprising pentaerythritol, melamine, and ammonium polyphosphate, as well as mixtures of these active components, were studied by DSC. The tests on binary and ternary mixtures were performed in weight ratios of 1:1 and 1:1:2, respectively, and on a dry paint sample with fillers ratio of 1:1:2. A sample mass of 50 mg pure substance and <20 mg for mixtures or dry coating material were heated from 20° to 600°C at 5°C/min under a flow rate of 200 mL/min of air or nitrogen depending on the experiment.

The phenomena observed represented decompostions of the pure substance or self-cross-linking reactions. Both binary and ternary mixtures were considered to detect interaction between the components and identify thermal aspects of the chemical reaction involved in the complex phenomenon of intumescence. The test performed on both mixtures shows a combination of these effects and interactions between components in every binary mixture, including the effect of the "blowing agent" (melamine). The overall intumescence reaction can be seen from the thermal point of view in the ternary mixture. DSC has an interesting potential as a tool for studying *thermally active* coating systems, such as the intumescent systems studied, which perform by means of chemical reactions occurring inside the coating film when a change in temperature occurs.

Fire retardant intumescent coatings are being frequently used to protect buildings containing structural steel from the effects of exposure to high temperatures caused by fire. The strength of structures decreases, especially those made of concrete and steel, with increasing temperatures of the fire, reaching a critical point at approximately 550°C. Below this temperature and without the benefit of an intumescent coating, a steel beam reaches its critical point in 17 minutes.[15] However, with the insulating effect of an intumescent coating, such time can be extended to approximately 70 minutes and up to 2 hours by using the appropriate coating thickness.[16] Therefore, it is important to study the thermal stability of coating materials for this type of application.

Simultaneous thermal analysis (STA) (TG/DTA) is a useful tool to study intumescent coatings because it provides information on both the thermal stability and the reaction type occurring with temperature in a single experiment. For example, Trehan and Kad[15] studied cellulose and ammonium polyphosphate intumescent coatings. They used the technique to elucidate the mechanism of the development of intumescent chars during the heating of an intumescent coating system based on ammonium polyphosphate (APP) as the acid source and cellulose as the carbonic source with a blowing agent. The completed coating was obtained by mixing APP, cellulose, chlorinated paraffin, and TiO_2 at different ratios by weight. The thermal behavior of pure APP and cellulose was studied first.

Figure 6 shows the TG/DTA curves for pure APP and pure cellulose. Trehan and Kad[15] reported that the TG/DTA curves for APP indicated that the mechanism of decomposition of APP occurs by two possible cross-linking reactions with the formation of P-O-P and P-N-P links after loss of ammonia and water as was also reported by Camino, et al.[17] The first loss in the TG curves was attributed to the competing cross-linking reactions responsible for the jagged appearance of the DTA signal in the 200°–400°C region. The second weight loss observed in the TG curve above 500°C was attributed to the decomposition of the complex phosphate polymer into P_2O_5.

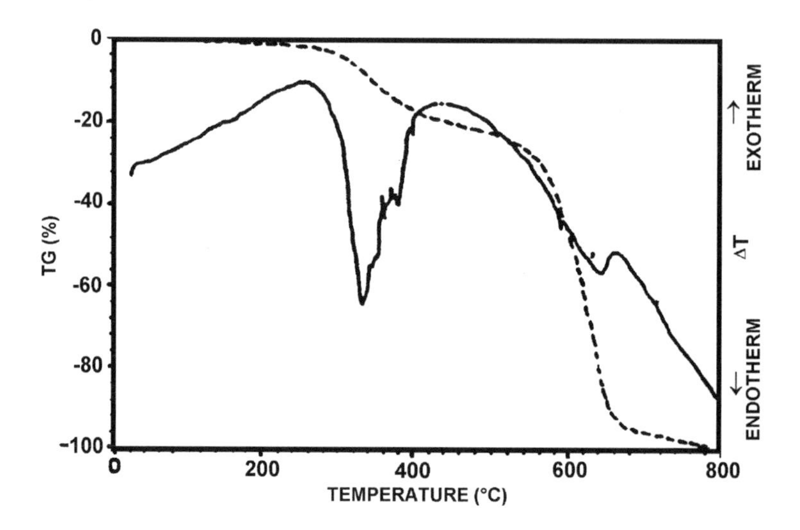

Figure 6. TG/DTA of APP. *(Reprinted with permission.)*[15]

The TG/DTA curves for cellulose are displayed in Fig. 7. An exotherm occurs at 322°C due to auto-oxidation of the carbonyl groups and C-H bonds. The endotherm observed at 336°C is associated with the pyrolysis of cellulose into levoglucosan, water, carbon monoxide, and carbon dioxide as reported in the literature.[18] The DTA curve, however, does not show clearly the two-stage reaction.

The TG/DTA curves for the APP/cellulose phosphate mixture (Fig. 8) show that the reaction between components results in an effective increase in the thermal stability of the mixtures. The enhanced thermal stability can be attributed to reactions between cellulose and APP.

From the results of their study, Trehan and Kad[15] concluded that the STA data showed clearly that the series of reactions between cellulose and APP led to an increase in thermal stability of a mixture. This increase indicates expansion of carbonaceous char formed prior to a final ceramic insulating layer. Cellulose not only acts as a blowing agent, but also helps to provide valuable thermal stability. Moreover, the addition of titanium dioxide and chlorinated paraffin further increase the quality of the coating.

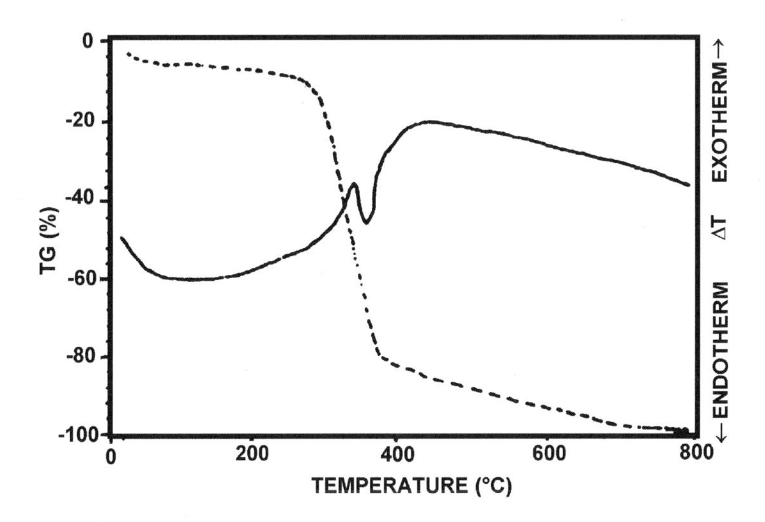

Figure 7. TG/DTA of cellulose. *(Reprinted with permission.)*[15]

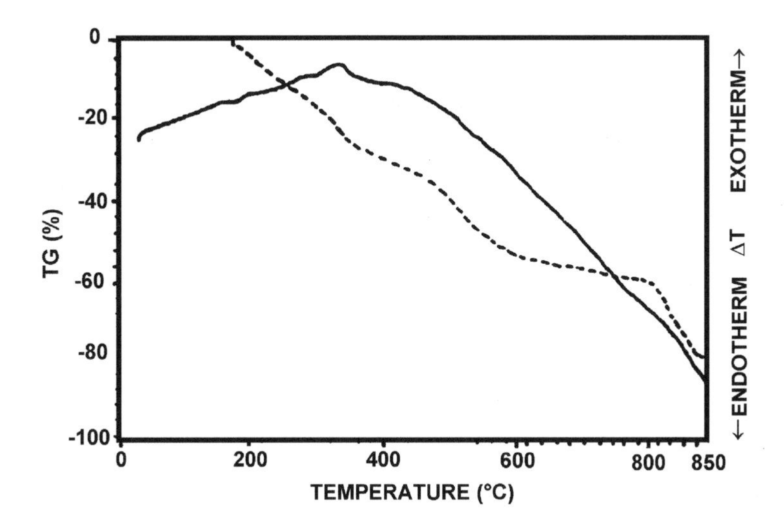

Figure 8. TG/DTA of APP/cellulose-phosphate. *(Reprinted with permission.)*[15]

Intumescent coatings have also been studied by Taylor[19] using STA. The mechanism of thermal decomposition of three single components [ammonium polyphosphate (APP), pentaerythritol (PER), and melamine] then binary, and ultimately ternary mixtures were investigated by STA and evolved gas analysis. The DTA traces for ammonium polyphosphate showed that decomposition steps involved different processes as previously reported.[15][17]

The STA traces for the pentaerythritol also showed a two-stage reaction as shown in Fig. 9.[19] The first one corresponds to the endotherm at 185°C due to the melting of PER. The second one was characterized by an exotherm under air and an endotherm under nitrogen. Therefore, it was suggested that the burning or decomposition process is dependant upon the oxygen present in the gas atmosphere. The melamine traces were also obtained by STA. They indicated a sublimation process at 291°C. The STA curves for the binary systems (APP/PER and APP/melamine) showed that the two components undergo several reactions following more than one mechanism (e.g., cross-linking), giving rise to a complex random cross-linked polymeric structure.

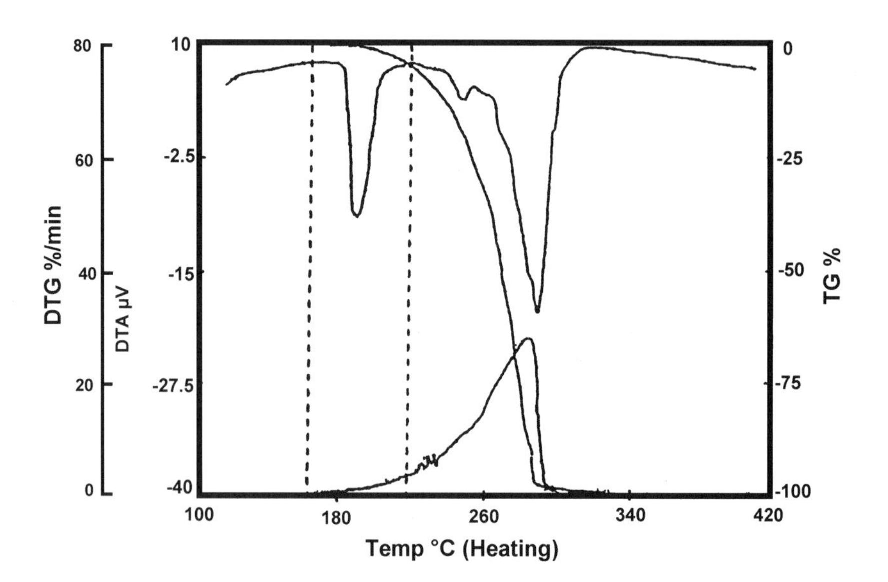

Figure 9. TG/DTA traces for pentaerythritol. *(Reprinted with permission.)*[19]

Taylor[19] concluded from the thermal analysis study, that the understanding of the mechanism of intumescent coatings increased, and that it was possible to optimize the mixing ratios in terms of minimizing weight loss. Furthermore, the results proved that the melamine was present not only as a blowing agent (spumific), but also as a contributor to the chemical composition of the char.

3.2 Silicone Coatings

Modulated temperature differential scanning calorimetry (MDSC), a modification to the linear temperature DSC, was used by Meyer and co-workers[20] to demonstrate the advantage of MDSC over DSC when analyzing coatings. Two batches of product (one manufactured in the pilot plant and one in the lab) were heated at 2°C/min (linear temperature ramp) and a modulation of 1°C amplitude with a period of one minute. The DSC traces showed that the change in C_p at T_g is obscured by kinetic events. The traces from the amplitude of the temperature difference between the sample and the reference showed clearly the difference between the two batches and T_g can easily be assigned. However, it was concluded that the MDSC offers a tremendous advantage over the DSC mode for determining the T_g of two batches of coatings. In addition to the increased sensitivity, MDSC allows separation of the kinetic and non-kinetic (C_p) components.

Cross-link density in silicone release coatings has important consequences on both the release as well as adhesion performance of pressure-sensitive adhesive constructions. The degree of recrystallization of silicone release coatings with different cross-link densities has been determined by Chang[21] using DSC and DMA. The recrystallization peak was obtained from DSC and DMA, and it was found that as the cross-link density of the release coating increased, the recrystallization peak of the silicone diminished sharply, indicative of the increasing constraints imposed by the high cross-link density which reduces the recrystallizability of the silicone. Also, an increase in the T_g of silicone together with lower recrystallization and melting temperatures was observed.

The temperature dependence of the tan δ curve for the silicone materials obtained by DMA from -150° to 0°C displayed in Fig. 10 for samples A–C shows that Sample A (under-cured) has a major transition at about -123°C with a shoulder at -116°C. Chang[21] assigned them to the T_g and recrystallization respectively. Sample B (medium-cured) shows peaks at -123°C, -116°C and -96°C. He reported that the three peaks suggested a heterogeneous chemical environment. Sample B had a low amount of cross-linker and the lowest catalyst content. It was possible that it contained heterogeneous regions of low and high cross-link densities responsible for the two lowest peaks. The peak at -96°C was due to recrystallization. The well-cured sample (C) showed only one major transition at about -116°C, attributed to high amounts of amorphous and high cross-link density material.

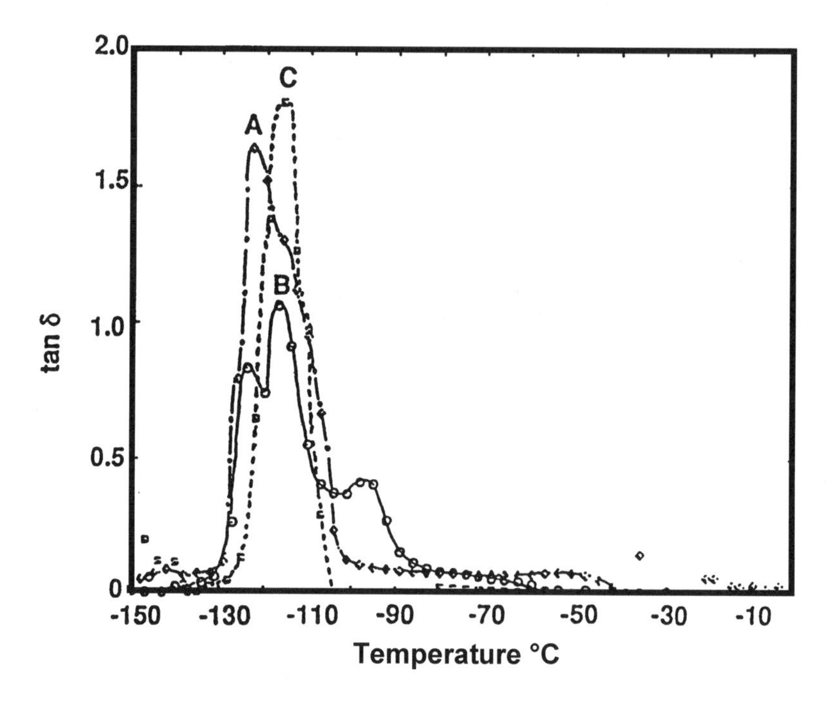

Figure 10. Temperature dependence of tan δ for release coating samples with different levels of cure. *(Reprinted with permission.)*[21]

The recrystallization behavior of the samples was confirmed by DSC. Figure 11 shows the DSC curves for samples A–C. Sample C shows the highest T_g and no detectable recrystallyzation. Samples A and B display not only well-defined recrystallization peaks but also melting endotherms not observed in Sample C. From the results of the study, it was concluded that the DSC and DMA methods can be used for monitoring the cure levels in the silicone release coatings.[8]

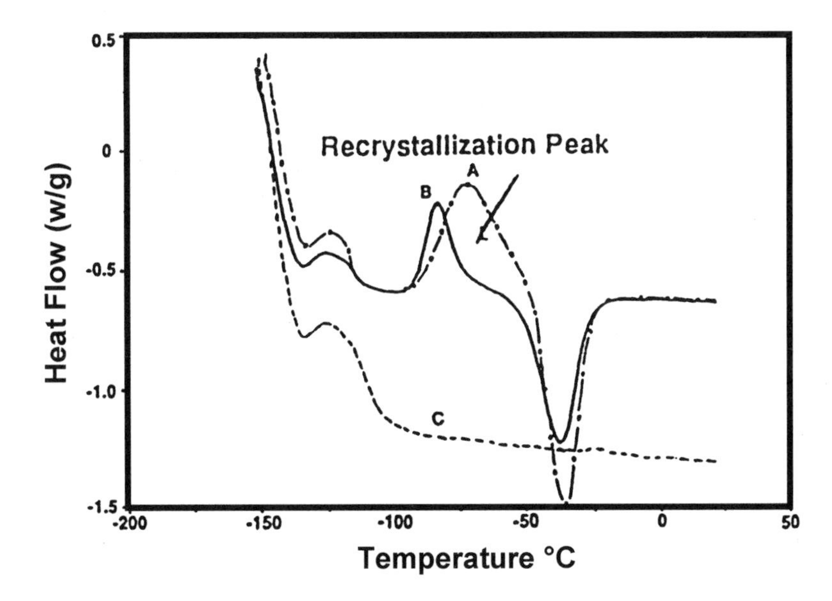

Figure 11. DSC curves showing the different glass transition, recrystallization melting transition, temperatures, and heats of recrystallization and melting for silicone samples with different levels of cure. *(a)* under-cured sample; *(b)* medium-cured sample; and *(c)* well-cured sample. *(Reprinted with permission.)*[21]

3.3 Organic Coatings Degradation (Service-Life)

The thermal stability of formulations may be important due to possible exposure conditions of the coating. The nature of failure depends on the material, but a measurable thermal effect is inevitable. This would yield the stability limit for short-term exposure and provide a limit for long-term exposure studies. Both thermal stability limits may be quite different because the possible reactions involved are slow. The degradation process responsible for the failure of the coatings may involve entirely different reactions. Hence, both kinds of data may be needed.[2]

The weight loss of a material as a function of temperature is used to obtain information on thermal decomposition characteristics. The use of TG is naturally limited to those decompositions or other reactions in which there is a gain or loss in weight. The technique is useful in providing information on the thermal stability of coatings and paints.

Differential thermal analysis can provide insight into the nature of some of the reactions. For example, cure temperatures for thermosetting resins can be determined by DTA. As the formulation is heated, it will reach a temperature at which the reaction is spontaneous and self-sustaining. Similarly, the degree of cure can be estimated from the heat effect observed on heating a partially-cured resin.[2] The evaluation of adhesion energies would require interpretation of differences in curves arising from the degradation of the bond between the coating and the substrate surface. It is well known that additional energy due to strain, to mechanical history, or to thermal history can affect thermal decomposition.[2]

Failure of coatings due to thermal degradation is a concern because their applications may involve service at elevated temperature. It is necessary to have analytical methods that establish a product's useful life expectancy at temperatures encountered in the field. Thermogravimetric analysis is an analytical tool that provides information on degradation processes and the kinetics involved at a given temperature. TG data obtained at different temperatures can be used to construct master curves by relating them to a single reference temperature. This can be done through a single Arrhenius factor that accounts for changes in the time scale. The method of building the master curve is called *Time-Temperature Superposition* (TTS). The information obtained helps in predicting service-life at elevated temperatures. Neag and co-workers[22] reported the application of TG to the service-life prediction of coatings. They studied coating formulations with different levels of additive. The coating preparations were aged at 140°, 160°, 180°, and 200°C in a forced-air oven. Formulations were checked daily for the onset of change in color, and film blackening was reported as the end of the degradation. Formulations were studied by TG under isothermal and non-isothermal conditions. The isothermal conditions consisted of scans at 100°, 110°, 120°, and 130°C whereas the non-isothermal ones consisted of scans performed at 1°, 2°, 5°, and 10°C/min from room temperature to 600°C under a gas purge of 50 mL/min. The TG/DTG curves (Fig. 12) obtained at 1°C/min show the failure and several other features of the degradation of the formulation.

The Time-Temperature Superposition (TTS) method[23] for monitoring cure behavior of coatings was adapted to predict the rate of degradation between 60° and 93°C and to obtain the degree of degradation curves. From the isothermal data, the endpoints between 100° and 130°C of the DTG curve in Fig. 12 can identify the onset and endpoint of the "catastrophic" failure process. By using Flynn's, et al.,[24] and Toop's[25] methods, the activation energy and estimated life-time of the material from the non-isothermal data can be calculated. From the results of the study, Neag, et al.,[22] concluded that the results from both the TG and oven studies showed that the time-to-failure of the formulation increased with increasing additive level. Film blackening obtained from the oven studies corresponds to a rapid change in weight after reaching about 2.5% weight loss. It was also found that the onset temperature and the weight loss derivative peak correlate well with the time to degradation in forced air ovens. Furthermore, the predicted time-to-failure values obtained using the TTS method were in excellent agreement with failure times obtained from oven-aging studies.

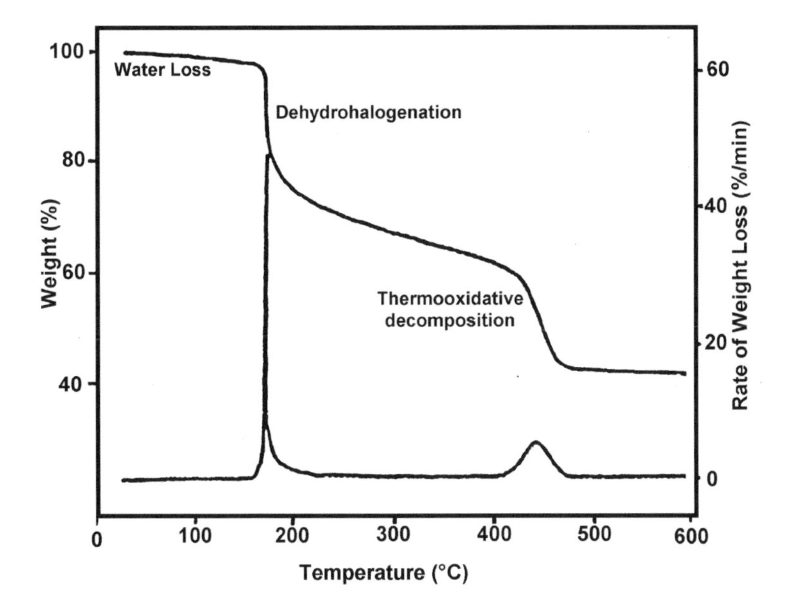

Figure 12. TG decomposition profile. (*Reprinted with permission.*)[22]

3.4 Inorganic Coatings

For inorganic coatings, thermoanalytical techniques can similarly be used to measure purity of raw materials including verification of the origin of some minerals, evaluation of thermal stability, estimation of the heats of reaction, determination of the nature of the reaction, establishment of processing temperatures, comparison of adhesion energies, and, by phase diagram determination, selection of a composition for the final coating. DTA is a more sensitive method for detection of many inorganic materials than for organic materials because inorganic phase changes and decompositions are generally faster and involve more energy. However, the technique has been a widely used tool for characterizing organic materials.

3.5 Miscellaneous Coatings

Dynamic mechanical analysis (DMA) and dynamic mechanical thermal analysis (DMTA) have been widely used to study polymeric materials ranging from organic coatings used in the beer and beverage can industry,[2][6] topcoats for automobiles,[27] coatings for glass-resin of optical fiber[28] to the structure-property relationship in polyester-urethane coatings.[29] The curing of thermoset coatings was also studied by Frey, et al.,[30] who monitored the cure of the coatings from the liquid paint to the fully cross-linked network by using tensile mode dynamic mechanical analysis. The time dependent isothermal cure studies yielded an insight into the cure behavior under realistic baking conditions. DMTA has been used to study the erosion of elastomeric protective coatings[31] and to measure differences between thin biodegradable polymer coatings for packaging applications.[32]

DMTA, DSC, and the TMA have been used to characterize two-stage acrylic latex coatings.[33] The DMTA, DSC, and TMA data together with particle size analysis techniques provide detailed information on copolymer compositions, latex particle morphologies, and coating morphologies.

Cypcar[34] used DEA to study acrylic and vinylidene fluoride/hexafluoropropylene (VF2/HFP) copolymer blends for high performance architectural coatings. In addition, DMTA and DSC were used to demonstrate that DEA was a useful technique for morphology characterization. The DTMA and DSC data suggested that the milky heterogeneous coating obtained (VF2/HFP) is due to the presence of a miscible fluoropolymer/acrylic phase, and the DEA data suggested that there was another morphological phase present for the VF2/HFP coating probably an acrylic-rich phase different from the miscible amorphous phase. The DEA results indicated that in the temperature range of the amorphous miscible phase there was only one relaxation in the clear homogeneous coating (VF2/HFP). It was suggested that this coating had reorganized into a completely miscible amorphous matrix. Although the results were not conclusive, the author reported that the study serves to illustrate the sensitivity of DEA measurements.

Schultz and co-workers[35] investigated the real-time kinetics of film formation of a floor polish polymer latex coating using DEA. They also used DSC and TMA. From the results of the study, it was reported that the drying DEA curves of latex film under study showed that at ambient temperature, the drying mechanism varied as a function of humidity. As the humidity increased, water evaporation slowed down and drying times increased. At a very high humidity, a rupture of the film was observed, which was attributed to internal stresses. They concluded that coalescence plays an important role in the film formation mechanism. The TMA and DSC results showed a long-drying time for the film as well as an increase in the T_g (>65°C). The second T_g was attributed to film phase separation. Evidence of a considerable amount of residual stress remaining in the film after two months was also reported. It was finally concluded that DEA can be used to measure the effect of variables on film formation.

Another useful thermoanalytical technique to characterize coatings and paints is TMA. For example, the thermomechanical effects of indentation of thin polymethyl (methamethacrylate) coatings attached to a steel substrate were investigated by Jayachandran.[36] He reported that it was informative to examine the stress field developed under the indenter to forecast potential sites and modes of failure during indentation as well as their dependence on loading conditions such as rate and friction. Moreover, TMA was used by Pindera, et al.,[37] to analyze functionally graded thermal barrier coatings.

Acrylic coatings have been used for asphalt-based roof applications. Thermoanalytical techniques can be used for characterizing this type of coating. A study carried out by Antrim, et al.,[38] reported that the acrylic coating prolonged the life of the asphaltic roofing material using low- and steep-slope roofing. Furthermore, the degradation mechanism of the asphalt is not only due to changes in the maltene/asphaltene ratio, but is more complex and of secondary dimension. It was mentioned that T_g is an important indicator of the tolerance for movement of a membrane or coating formulated with elastomeric coatings. However, it was not clear which thermoanalytical technique was used to measure T_g. It is, therefore, assumed that DSC, TMA, DMA, or DEA may have been used. Thermoanalytical techniques combined with the IR and SEM would have provided a greater amount of information on the degradation process of asphalt.

REFERENCES

1. Lambourne, R., *Paint and Surface Coatings: Theory and Practice,* Ellis Horwood Ltd. Publ., Chichester, UK (1987)
2. Myers, R. R., and Long, J. S., *Treatise on Coatings, Characterization of Coating: Physical Techniques*, Vol. 2, Ch. 7, Marcel Dekker, Inc., New York (1969)
3. Odlyha, M., Flint, C. D., and Simpson, C. F., The Application of Thermal Analysis (DSC) to Study of Paint Media, *Anal. Proc.*, 26:52–55 (1989)
4. Odlyha, M., Investigation of Binding Media of Paintings by Thermoanalytical and Spectroscopic Techniques, *Thermochimica Acta*, 269/270:705–727 (1995)
5. Wingard, C. D., Use of Several Thermal Analysis Techniques on a Hypalon Paint Coating for the Solid Rocket Booster (SRB) of the Space Shuttle, *Proc. 27th Conf. N. A. Thermal Analysis Soc., Savannah, GA,* p. 357–365 (Sept. 20–22, 1999)
6. Odlyha, M., Foster G., Hackney, S., and Townsend, J., Dynamic Mechanical Analysis for the Evaluation of the Deacidification Treatment of Painting Canvases, *J. Thermal Analysis*, 50(1–2):191–202 (1997)
7. Foster G., Odlyha M., and Hackney, S., Evaluation of the Effects of Environmental Conditions and Preventive Conservation Treatment on Painting Canvases, *Thermochimica Acta,* 294:81–89 (1997)
8. Odlyha, M., Craig, D. Q. M., and Hill, R. M., Dielectric Analysis of Relative Humidity Variations in Canvas Paintings, *J. Thermal Analysis,* 39:1181–1192 (1993)
9. Perera, D. Y., and Belgien L., Cure Characterization of Stoving Paints, *Materialpüfung,* 31(3):57–62 (1989)
10. Drouet-Fleurizelle, L., Gillereau, D., and Lacabanne, C., DSC and TSC Studies of Polyurethane Mono- and Bilayer Paints, *Thermochimica Acta*, 226:43–50, TCA 1284 (1993)
11. Clausse, D., Boursereau, F., and Pelisson, B., Contributions of Calorimetry and Thermal Analysis to the Characterization of Thermal Transitions of Thermal Hardening Paints, *Calorim. Anal. Therm.*, 27:28–33 (1996)
12. Sebastian, M., Thermoanalytical Control of Paint-Shop Wastes Prior to Disposal, *J. Thermal Analysis*, 38(9):2087–2093 (1992)
13. Bahra, M., Elliot, D., Reading, M., and Ryan, R., A Novel Approach to Thermal Analysis of Thin Films, *J. Thermal Analysis*, 38:543–555 (1992)
14. Pagella, C., Raffaghello, F., Def Averi, D. M., Differential Scanning Calorimetry of Intumescent Coatings, *Polymer Paint Colour J.*, 188(4402):16–18 (1998)

15. Trehan, R., and Kad, G. L., Thermal Analysis of Intumescent Coating System Based upon Cellulose and Ammonium Polyphosphate, *J. Polym. Mater.,* 11(4):289–293 (1994)
16. British Standards 476 Part 20 (1987)
17. Camino, G., Costa, L., and Trossarelly, L., *Polym. Deg. Stab.,* 12:203 (1985)
18. Jain, R. K., Lal, K., Bhatnagar, H. L., *J. Indian Chem. Soc.,* 57:620 (1980)
19. Taylor, A. P., and Sall, F. R., Thermal Analysis of Intumescent Coatings, *European Polymers Paint Colour J.,* 182(4301):122,124–125,130 (1992)
20. Meyer, E. F., Michalski, C. R., Bender, L. M., Burke, S. M., and Morris, E. E., MTDSC Applications in Coatings R&D, *Proc. 26th Conf. of the N. Am. Thermal Analysis Soc., Cleveland, OH,* (K. R. Williams, ed.), pp. 634–639 (Sept. 13–15, 1998)
21. Chang, E. P., Recrystallization of Silicone Release Coatings, *Polym. Bull.* 26:681–688 (1991)
22. Neag, M., Floyd, L., and Manzuk, S., The Application of Thermogravimetry to the Service Life Prediction of Coatings, Polymeric Materials: Science and Engineering, *Proc. ACS Div. Polym. Mater. Sci. Engin.,* 68:331–333 (1983)
23. Prime, R. B., Thermosets, *Thermal Characterization of Polymeric Materials,* pp. 478–480, Academic Press (1981)
24. Flynn, J. H., et al., *Polym. Lett.*, B4:323 (1966)
25. Toop, D. J., *IEEE Trans. Elec. Ins.*, EI-6–2 (1971)
26. Grentzer, T. H., Holsworth, R. M., and Provder, T., The Application of the Dynamic Mechanical Analyzer to Organic Coatings, *Polym. Preprints,* 22(1):254–255 (1981)
27. Hill, L. W., Korzeniowski, H. M., Ojunga-Andrew, M., and Wilson, R. C., Accelerated Clearcoat Weathering Studied by Dynamic Mechanical Analysis, *Progr. in Organic Coatings*, 24:147–173 (1994)
28. Hosoya, T., Nonaka, T., and Masude, S., Analysis of Transmission Characteristics of Dual Coated Fiber by Dynamic Mechanical Method, *Inst. of Elect. Eng., Conf. Publ., London, England,* pp. 522–525 (1964)
29. Scanlan, J. C., Webster, D. C., Crain, A. L., Correlation between Network Mechanical Properties and Physical Properties in Polyester-Urethane Coatings (R. R. Gould, ed.), *ACS Symp. Series,* No. 648:222–234 (1996)
30. Frey, T., Große-Brinkhaus, K. H., and Röckrath, V., Cure Monitoring of Thermoset Coatings, *Progr. in Organic Coatings,* 27:59–66 (1996)
31. Slikkerveer, P. J., van Dongen, M. H. A, and Touwslager, F. J., Erosion of Elastomeric Protective Coatings, *Wear* 236:189–198 (1999)

32. Rantanen, T., and Kimpimäki, T., Thermal and Mechanical Characterization of Biodegradable Polymer Coatings, *Annual Trans. of the Nordic Rheology Soc.*, 3:15–17 (1995)

33. Rearick, B., Swarup, S., and Kamarchik, P., Characterization of Two-Stage Latexes Using Dynamic Mechanical Thermal Analysis, *J. Coatings Technol.*, 68(862):25–31 (1996)

34. Cypcar, C. C., and Judovits, L., Dielectric Thermal Analysis of Acrylic and VF2/HFP Copolymer Blends for High Performance Coatings, *Polym. Preprints,* 39(2):861–862 (1998)

35. Schultz, J. W., and Chartoff, R. P., Dielectric and Thermal Analysis of the Film Formation of a Polymer Latex, *J. Coatings Technol.,* 68(861):97–106 (1996)

36. Jayachandran, R., Boyce, M. C., Montagut, E., and Argon, A. S., Thermomechanical Analysis of Identation Behavior of Thin PPMA Coatings, *J. Computer-Aided Mater. Design,* 2(1):23–48 (1995)

37. Pindera, M. J., Aboudi, J., and Arnold, S. M., Thermomechanical Analysis of Functionally Thermal Barrier Coatings with Different Microstructural Scales, *J. Am. Ceram. Soc.,* 81(6):1523–1536 (1998)

38. Antrim, R., Johnson, C., Kirn, W., Platek, W., and Sabo, K., Can Acrylic Coatings Save Your Next Roof? *RSI-Publ.,* pp. 30–34 (Oct. 1995)

Index

A

M

T

U

V

W

X

Y

Z